CAMBRIDGE LIBRARY COLLECTION

Books of enduring scholarly value

Life Sciences

Until the nineteenth century, the various subjects now known as the life sciences were regarded either as arcane studies which had little impact on ordinary daily life, or as a genteel hobby for the leisured classes. The increasing academic rigour and systematisation brought to the study of botany, zoology and other disciplines, and their adoption in university curricula, are reflected in the books reissued in this series.

British Fossil Brachiopoda

British palaeontologist Thomas Davidson (1817–85) was born in Edinburgh and began his studies at the city's university. Encouraged by German palaeontologist Leopold von Buch, he began to study brachiopod fossils at the age of twenty, and he quickly became the undisputed authority. He was elected fellow of the Geological Society of London in 1852, receiving the Wollaston medal in 1865. He became a Fellow of the Royal Society in 1857. Published between 1850 and 1886, this six-volume work became the definitive reference text on the subject. It includes more than two hundred hand-drawn plates and a comprehensive bibliography. This volume, the second of six, details the Permian and Carboniferous brachiopod species.

Cambridge University Press has long been a pioneer in the reissuing of out-of-print titles from its own backlist, producing digital reprints of books that are still sought after by scholars and students but could not be reprinted economically using traditional technology. The Cambridge Library Collection extends this activity to a wider range of books which are still of importance to researchers and professionals, either for the source material they contain, or as landmarks in the history of their academic discipline.

Drawing from the world-renowned collections in the Cambridge University Library, and guided by the advice of experts in each subject area, Cambridge University Press is using state-of-the-art scanning machines in its own Printing House to capture the content of each book selected for inclusion. The files are processed to give a consistently clear, crisp image, and the books finished to the high quality standard for which the Press is recognised around the world. The latest print-on-demand technology ensures that the books will remain available indefinitely, and that orders for single or multiple copies can quickly be supplied.

The Cambridge Library Collection will bring back to life books of enduring scholarly value (including out-of-copyright works originally issued by other publishers) across a wide range of disciplines in the humanities and social sciences and in science and technology.

British Fossil Brachiopoda

VOLUME 2: PERMIAN
AND CARBONIFEROUS SPECIES

THOMAS DAVIDSON
WITH INTRODUCTION BY
RICHARD OWEN &
W. B. CARPENTER

CAMBRIDGE UNIVERSITY PRESS

Cambridge, New York, Melbourne, Madrid, Cape Town,
Singapore, São Paolo, Delhi, Tokyo, Mexico City

Published in the United States of America by Cambridge University Press, New York

www.cambridge.org
Information on this title: www.cambridge.org/9781108038188

© in this compilation Cambridge University Press 2011

This edition first published 1858-63
This digitally printed version 2011

ISBN 978-1-108-03818-8 Paperback

BRITISH
FOSSIL BRACHIOPODA.

BY

THOMAS DAVIDSON, F.R.S., G.S.,

MEMBRE ETRANGER DE L'INSTITUT DES PROVINCES; OF THE GEOLOGICAL SOCIETY OF FRANCE; LINNEAN SOCIETY OF
NORMANDY; IMPERIAL MINERALOGICAL SOCIETY OF ST. PETERSBURGH; ROYAL SOCIETY OF LIÉGE;
ZOOLOGICAL SOCIETY OF VIENNA; GEOLOGICAL SOCIETY OF GLASGOW; ACADEMY
OF SCIENCES OF ST. LOUIS, AMERICA; PALÆONTOLOGICAL SOCIETY
OF BELGIUM, ETC.

VOL. II.
PERMIAN AND CARBONIFEROUS SPECIES.

LONDON:

PRINTED FOR THE PALÆONTOGRAPHICAL SOCIETY.

1858—1863.

A MONOGRAPH

OF THE

BRITISH FOSSIL BRACHIOPODA.

PART IV.

THE PERMIAN BRACHIOPODA.

BY

THOMAS DAVIDSON, F.R.S., G.S., &c.

LONDON:
PRINTED FOR THE PALÆONTOGRAPHICAL SOCIETY.
1858.

TO

SIR RODERICK IMPEY MURCHISON, K.C.B., G.C.St.S., D.C.L., LL.D.,

M.A., F.R.S., F.L.S., F.G.S., PRES. R.G.S., ETC.

DIRECTOR-GENERAL OF THE GEOLOGICAL SURVEY OF THE UNITED KINGDOM; A TRUSTEE OF THE BRITISH
AND HUNTERIAN MUSEUMS, OF THE BRITISH ASSOCIATION FOR THE ADVANCEMENT
OF SCIENCE, ETC., ETC.

My dear Sir Roderick,

In 1849, I undertook to prepare for the Palæontographical Society a series of Monographs to correspond with the seven great divisions in geology. I had then but a very imperfect idea of the magnitude and difficulty of the undertaking. It was at that time believed that the proposed work would occupy one volume, and that it might be completed in the course of four or five years; but I soon found that to attain even an approximate knowledge of the numerous species, and of the correct determination of their localities, it would require geological as well as palæontological researches to be made over nearly the whole extent of the British Islands. This portion of the undertaking would have proved an insurmountable difficulty but for the valuable assistance afforded me by the publication of your admirable 'Silurian System,' and its follower, 'Siluria.' Through the means of which, assisted by the works of other eminent geologists, I have been enabled to determine their proper positions in the respective formations of the numerous species with which I have become acquainted in the course of my researches. To whom, therefore, can I dedicate with greater propriety the second and third volumes of my work, containing the Silurian, Devonian, Carboniferous and Permian species, than to you who have so ably and completely extricated the Palæozoic rocks and fossils from the confusion in which they were involved, previous to the publication of your luminous and valuable works on those departments of geology.

I have the honour to remain,

My dear Sir Roderick,

With sincere respect and gratitude,

Yours most faithfully,

THOMAS DAVIDSON.

PRELIMINARY REMARKS.

THE first volume of the present work having been devoted almost entirely to the description and illustration of the Brachiopoda of the TERTIARY and SECONDARY, or Mesozoic series,[1] my second will be appropriated to those of the PERMIAN[2] and CARBONIFEROUS periods.

The Brachiopoda of the PERMIAN system of England are few in number, and have been more completely investigated than those of any of the other epochs: they are, with a single exception, described and illustrated in Professor King's valuable Monograph, issued in 1850 by the Palæontographical Society; this work being at the same time the largest and most complete that has hitherto appeared on British Permian fossils.

It may, therefore, be very naturally inquired why I should have taken upon myself to write on a subject apparently so completely investigated; my excuse must be based upon the necessity I found myself under to combine my series of Monographs by a few pages on the Permians, and especially so as the study of several publications prior and subsequent to 1850, as well as of a vast amount of new and very perfect material in the possession of Messrs. Howse and Kirkby, has made it desirable to propose a few small alterations to the works hitherto published, as well as to offer some additional details and illustrations, which will not, I trust, be considered entirely superfluous.

[1] For hitherto no species belonging to the class have been discovered in any of the TRIASSIC beds of Great Britain, which include the *variegated marls, Keuper* and *Bunter Sandstein.*

[2] It will not be necessary to enlarge upon the geology of the group, as this has been done already by different authors, and of which a full account will be found in one of the Society's volumes; but I will mention the subdivision of the beds, as latterly proposed by Mr. Howse in his valuable paper, published in the 'Annals of Nat. History' for January, 1857. In the descending order we find—

1. Upper yellow sandstone	} Upper	}
2. Conglobated or Botryoidal limestone		
3. Concretionary or cellular limestone	} Middle	} Magnesian
4. Shell limestone (Zechstein dolomit of the Germans		limestone.
5. Magnesian conglomerate	} Lower	}
6. Compact limestone		
7. Marl slate.		

The Brachiopoda are found in the *marl slate, compact limestone, conglomerate,* and *shell limestone* only.

The following pages will therefore require to be considered more as a supplement to the important labours of Professor King and Mr. Howse than that of a separate Monograph, as to those authors the chief credit is due of having worked out our English Permian species. My efforts have been especially directed to the minute illustration and study of every internal character which the perfect material at my disposal has enabled me to develop. I have also avoided reproducing long lists of synonyms, references, and certain other details which will be found in the works of the two gentlemen already named.[1]

In the preparation of the following pages, I have been most kindly assisted by several

[1] Professor King has published a long list of all the works relating to Permian fossils, from the year 1710 to 1850; to which I will now add a few others, so as to carry the catalogue down to the present time.

It will also be desirable to mention that both Professor King and Mr. Howse had for many years prior to 1848 been busily engaged collecting and studying the Permian fossils of the counties of Durham and Northumberland. In 1844 Professor King supplied M. De Verneuil with a manuscript list of the British species then known to him, which comprised the following Brachiopoda ('Bulletin de la Société Geologique de France,' vol. i, 2d series, p. 500, 3d of June, 1844): *Terebratula elongata, T. sufflata, T. pectinifera, T. Schlotheimi* (to this last M. De Verneuil added that Mr. King had proposed for it and *T. superstes* a new genus, named *Camarophoria*), *Spirifer undulatus, S. multiplicata, S. cristata, Productus horridus, Strophalosia Morrisiana, Stroph. spinifera,* and *Lingula mytiloides.* These names were also subsequently introduced into the first and second volumes of the 'Geology of Russia,' in 1845; and in 1846 Professor King's excellent memoir appeared, 'Remarks on certain Genera belonging to the Palliobranchia,' wherein, besides much important matter relating to the Brachiopoda in general, the genera *Camarophoria* and *Strophalosia* are for the first time explained. In 1847 Professor King prepared a Catalogue of the organic remains of the Permian rocks of Northumberland and Durham, which he presented to the Tyneside Naturalists' Club for publication, but which having been withdrawn by its author, another Catalogue was prepared by Mr. Howse at the request of the Club. Both were, however, printed during the month of August, 1848, and a delicate question arose as to the exact day of publication; but from evidence communicated by the publishers, it would appear that the one written by Mr. Howse was issued on the 17th, while that of Professor King appeared on the 19th of the same month.

They are both excellent and valuable productions, and prove the great knowledge possessed by their respective authors on the local Permian species, as well as their ability to write upon the subject; but in justice to Dr. Geinitz, I feel bound to observe (as has already been stated by Mr. Howse) that the 'Die Versteinerungen des deutschen Zechsteingebirges,' having appeared in April, 1848, does, as a matter of course, hold priority over both the catalogues of the above-named gentlemen for any new species it may contain.

In 1848 and 1854, a Russian work of considerable merit, but unfortunately little known, was published at Dorpat, 'Reise nach dem Nordosten des Europäischen Russlands durch die Tundren der Samojeden,' by Alexander Gustav Schrenk. In the first volume are mentioned several Permian fossils, which were well described and illustrated by Count Alex. von Keyserling, in pp. 81—114 of the second volume. *Productus hemisphæricum,* Kutorga, *Prod. Cancrini,* Vern. and Keyserling, *Strophalosia tholus,* Keys., *Spirifer Schrenkii,* Keys., *Tereb. Royssiana,* Keys., and *Terebratula concentrica?* Buch, var. *Permiensis,* and *Terebratula Geinitziana,* Vern., are the species of Brachiopoda discovered in 1837, by Dr. Schrenk, in that northern portion of the Russian empire.

We must also refer to Professor De Koninck's 'Nouvelle notice sur les Fossiles de Spitzberg,' published in the sixteenth volume of the Académie Royale de Belgique, in which the author has described and figured zealous friends, to whom it is a most pleasing duty to tender my grateful thanks. To Professor King, of Queen's College, Galway; to Mr. R. Howse, of South Shields; and to Mr. Kirkby, of Bishopwearwouth, I am equally and especially indebted for much valuable information, the liberal loan of the beautiful specimens of their collections, as well as for the indefatigable exertions they made in procuring the material required for the complete illustration of several of the species figured in the accompanying plates.

To Mr. G. Tate, of Alnwick; Mr. Hancock, of Newcastle; Mr. Binney, of Manchester; and Dr. Carpenter, of London, I am greatly obliged for much useful information. To Sir Roderick Murchison and Professor Huxley, for the use of Mr. Howse's original specimens, preserved in the Museum of the Geological Survey; and to Mr. S. P. Woodward, for some in the British Museum.

To Baron Schauroth, of Coburg, M. Eisel, jun., and to Professor Geinitz, I am indebted for valuable information, as well as the kind gift of a numerous series of German specimens, which have enabled me to compare our English forms with the equivalents found on the Continent, the shells from the Zechstein Dolomit of Pössnech, &c., being identical with those so abundantly distributed in the magnesian shell limestone of Humbleton and other of our British localities.

I must also express my warmest acknowledgments to Count Keyserling, of Raikull, near Reval, for the valuable information and most zealous endeavours he has made to procure for me several important Russian specimens, required for the perfect elucidation and identification of some of our English types; as well as to Dr. A. G. Schrenk and

---

Prod. horridus, P. Cancrini, P. Leplayi, P. Robertianus, Spirif. alatus, and *Sp. cristatus* from that distant region. Professor King had already noticed his first paper on the subject published in 1846.

In 1853, 'Ein Beitrag zur Fauna deutschen Zechsteingebirges,' by Dr. Baron Karl v. Schauroth, was published.

In 1854, 'Ein Beitrag zur Palæontologie des deutschen Zechsteingebirges,' by the same author, whose excellent publications on German Permian fossils has added very considerably to our knowledge on the subject.

In 1855, 'On the Permian Beds of the North-west of England,' by E. W. Binney; but no Brachiopoda have been discovered in the Permians of this part of England nor in any of the Irish similar beds.

In 1856, 'On the Occurrence of Permian Magnesian Limestone at Tullyconnel, near Artrea, in the County of Tyrone,' by Professor W. King.

In 1856, 'Ein neuer Beitrag zur Palæontologie des Zechsteingebirges,' by Baron v. Schauroth.

'Notes on Permian Fossil Palliobranchiata,' by Professor King.

In 1857, 'Notes on the Permian System of the Counties of Durham and Northumberland,' by R. Howse, Esq.

'Notes sur les genres *Athyris* (= *Spirigera*), *Camarophoria, Orthesina,* et *Strophalosia,*' par Mr. Thomas Davidson. ('Bulletin de la Société Linnéenne de Normandie,' vol. ii, pl. i and ii).

Besides these, several other memoirs on Permian rocks and fossils have been published by Dr. Geinitz, Mr. Coquand, and by the Geological Survey of Missouri; but which, not containing any matter in direct reference to the Brachiopoda, need not be at present more fully reverted to.

Professor Dr. C. Schmidt, of Dorpat, for the gift of some interesting specimens; also to General Von Helmersen, of St. Petersburgh; M. De Koninck, of Liege; and M. De Verneuil, of Paris, who have all facilitated and contributed to my researches among the species which will form the subject of the following pages.

The British localities, and stratifigraphical position here given, are those already published or contributed by Mr. Howse and Professor King.

TABLE OF THE BRACHIOPODA HITHERTO DISCOVERED IN THE BRITISH PERMIAN BEDS.

No.	Genus and Species.	Author.	Date.	PART IV. — Reference to my Plates and Figures.	Marl Slate.	Compact Limestone.	Magnesian Conglomerate.	Shell Limestone.	Botryoidal Limestone.	Upper Yellow Limestone.
1	Terebratula elongata	Schlotheim	1816	pl. i, f. 5—7, 12—14, 18—22		*	*	*		
	var. sufflata	Schlotheim	1816	pl. i, f. 8—11, 15—17						
2	Spirifera alata	Schlotheim	1813	pl. i, f. 23—36		*	*	*		
3	? clannyana	King	1848	pl. i, f. 47—49		*		*		
4	Spiriferina cristata	Schlotheim	1816	pl. i, f. 37—40, 45, 46; pl. ii, f. 43—45				*		
5	multiplicata	J. de C. Sow.	1829	pl. i, f. 41—44				*		
6	Athyris pectinifera	J. de C. Sow.	1840	pl. i, f. 50—56; pl. ii, f. 1—5			*	*		
7	Camarophoria Schlotheimi	Von Buch	1834	pl. ii, f. 16—27			*	*		
8	globulina	Phillips	1834	pl ii, f. 28—31			*	*		
9	Humbletonensis = multiplicata, King	Howse	1848	pl. ii, f. 9—18			*	*		
10	Streptorhynchus pelargonatus	Schlotheim	1816	pl. ii, f. 23—42				*		
11	Productus horridus	Sowerby	1822	pl. iv, f. 13—26		*	*	*		
12	latirostratus = umbonellatus, King	Howse	1848	pl. iv, f. 1—12				*		
13	Strophalosia Goldfusii	Münster	1839	pl. iii, f. 16—18		*		*		
	var. Lewisiana	De Koninck	1846	pl. iii, f. 19—22		*		*		
14	lamellosa	Geinitz	1848	pl. iii, f. 33		*		*		
	var. Morrisiana	King	1848	pl. iii, f. 24—31		*		*		
	var. Humbletonensis	King	1850	pl. iii, f. 34—41		*		*		
15	Crania Kirkbyii	Davidson	1857	See woodcut				*		
16	Discina Koninckii	Geinitz	1848	pl. iv, f. 27—29	*	*		*		
17	Lingula Credneri	Geinitz	1848	pl. iv, f. 30, 31	*	*				

MONOGRAPH

OF

BRITISH PERMIAN BRACHIOPODA.

Family—TEREBRATULIDÆ.

Genus—TEREBRATULA. (*Vide* General Introduction, Vol. I, p. 61, 1853.)

SEMINULA, M'Coy, 1844, 1855. EPITHYRIS, King, 1855.

Professors M'Coy and King are of opinion that the Palæozoic TEREBRATULÆ, *T. elongata, T. sacculus, T. hastata, T. vesicularis,* and *T. ficus,* &c., should be generically separated from TEREBRATULA proper, such as *T. vitrea, T. carnea, T. biplicata,* &c., on account of certain peculiarities to be hereafter described, and have respectively proposed *Seminula* and *Epithyris* as generic denominations for their reception.

The first mention of *Seminula* by M'Coy will be found at p. 158 of the 'Synopsis of the Carboniferous Fossils of Ireland,' 1844; but the characters therein are so vaguely expressed, that I did not consider it necessary to draw attention to the fact while writing my General Introduction. However, as Professor M'Coy has again introduced his genus,[1] with different characters and types from those made use of in 1844, it will be necessary to revert to the subject with some detail.

"Genus *Seminula*, M'Coy, 1844.—*Gen. Char.* Shell small, sub-pentagonal; smooth, or slightly plaited at the margin; beak of the dorsal valve small, with a minute perforation; no deltidium. The species of this genus are all small, nearly smooth shells; the margin frequently indented, but no distinct plaits on the surface; the outline is more or less pentagonal; the beak has a very minute foramen, for the passage of the muscle of attachment, but there is no deltidium separating the foramen from the hinge. This genus is peculiar to the Palæozoic rocks. Examples: *S. pentahedra,* Phillips, sp.; *S. pisum,* M'Coy; *S. rhomboidea,* Phillips."

It may, however, be observed, that none of the shells here enumerated present the

---

[1] 'British Palæozoic Fossils,' p. 408, 1855.

characters of any of the TEREBRATULIDÆ; for *T. pentahedra* appears to possess spirals, and is consequently an ATHYRIS; while *S. pisum*, M'Coy = *T. seminula*,[1] Phillips, and *S. rhomboidea*, Phillips, are RHYNCHONELLÆ.

In 1855, Professor M'Coy describes his genus thus: "*Gen. Char.* Ovate; a large oval perforation on the beak of the receiving valve (ventral or dental one), separated from the hinge-line by a portion of the valve, but apparently without deltidium; dental lamellæ strongly developed in beak of receiving valve, slightly diverging entering valve (dorsal, Owen), with a faint trace of mesial septum, and two cardinal teeth, from whence a small loop, with a very short recurved portion, arises; substance of the shell punctuated, usually without plaits. . . . . . . Lately, Professor King has given much excellent information on the genus, in his volume on the 'Permian Fossils of England,' under the name of *Epithyris* of Phillips, and pointed out the valuable and easily ascertained character of the strong dental lamellæ in the beak, bordering the foramen. From the observations in the middle of p. 54 of Phillips's 'Palæozoic Fossils,' it is obvious, however, that *Epithyris* was intended for the Oolitic *Terebratula*, congeneric with *T. maxillata*, having distinct deltidium, &c. The dental lamellæ leave slits in the beak of the casts, one on each side of the foramen, which are very characteristic of the genus, and in some states of exfoliation of the shell are likely to be confounded with the edges of the deltidium; indeed, the appearance thus produced in many specimens is so puzzling, that I prefer leaving the existence or non-existence of a deltidium an open question, the genus being well distinguished meanwhile from the more recent *Terebratulæ* by the rostrum being separate from the extension of the dental lamellæ, &c. This generic type seems confined to the Palæozoic rocks. Examples: *T. elongata*, Schloth.; *T. ficus*, M'Coy; *T. hastata*, Sow.; *T. juvenis*, Sow.; *T. sacculus*, Martin; *T. seminula*, Phillips;[2] *T. sufflata*, Schloth.; *T. virgoides*, M'Coy."

All these shells belong to the same generic type of *Terebratulidæ*; but the question to be determined is, whether they should be separated from Llhwyd's genus, and if so, whether Professor M'Coy's denomination should be the one selected. For the sake of comparison, I have tabulated the differences observable between *Terebratula*, Llhwyd, and those shells which are considered by M'Coy and King to constitute a separate genus.

---

[1] Professor M'Coy seems to have entirely misunderstood Professor Phillips's *T. seminula*, as will be found explained in my Monograph of Carboniferous species.

[2] The shell described as *T. seminula* (Phillips) by Professor M'Coy in 1855 is *T. vesicularis*, De Koninck, and not *T. seminula* of Phillips, nor the one made use of in the 'Synopsis.'

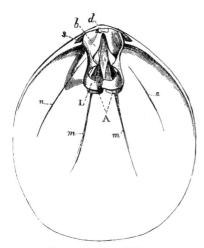

TEREBRATULA, Llhwyd.

Ter. vitrea (dorsal valve).

d. Cardinal process. *b.* Hinge-plate to loop. *s.* Sockets. *A.* Quadruple impression of the adductor. *m.* Inner groove. *n.* Outer groove.

Ventral valve.

1. No prominent rostral plates, only a simple thickening of the shell, at the dental projections, which leave no slits in the beak of internal casts.

2. In the interior two diverging grooves (*m*) extend from the extremity of the beak to a little more than half the length of the valve; there are also two other smaller lateral ones (*n*); so that in casts four diverging ridges may at times be perceived placed at nearly equal distances.

3. The muscular impressions are close together, and occupy a small space at a short distance from the extremity of the beak; they are more or less indented, according to the thickness of the shell, and consist of a small, central, oval scar left by the adductor, on either side of which are placed the larger cardinal ones, and outside of these again may be seen the pedicle muscular impressions, which vary in dimensions in different species according to the size of the pedicle and foramen.

Dorsal valve.

4. In the interior a small cardinal process or boss projects from under the extremity of the umbonal beak (*d*); the hinge-plate is divided, and consists of two rather wide and somewhat oblique shelly plates (*b*); from the prolonged extremities of these depart the

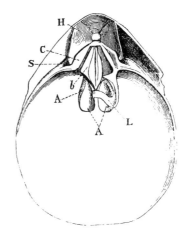

SEMINULA, M'Coy = *Epithyris*, King.

T. elongata, Schlotheim (interior of dorsal valve with part of the ventral one).

H. Rostral or dental plates of ventral valve. *S.* Sockets of dorsal valve. *C. b.* Hinge plates. *L.* Loops. *A.* Adductor impressions.

Ventral valve.

1. Well-defined dental or rostral plates, leaving slits in the beak of casts.

2. In the interior there exists a mesial longitudinal ridge, extending from the extremity of the beak to about two thirds of the length of the valve, and two shorter sub-parallel ones, situated at a small distance from the central one.

In casts these produce slight grooves, the central one being particularly evident.

3. The muscular impressions appear to be similar to those of *Terebratula* proper, and have left but slight impressions in the interior of the shell or on the casts of the specimens that have come under my notice.

Dorsal valve.

4. The cardinal process seems to be but slightly protruded; the hinge-plates varying somewhat in detail in different species, and even individuals. There exists first a testaceous ridge or plate (*C*), which forms at the same time the inner socket walls, and two

short longitudinal riband-shaped lamellæ, which are soon united by a transversal lamella, which is more or less bent upwards in the middle (L). This simply attached loop is confined to the posterior portion of the shell, and does not exceed much more than one third of the length of the valve. There are no sloping hinge-plates attached to the bottom of the valve, as in *T. elongata* and kindred forms, nor lozenge-shaped elevation. The quadruple impressions of the adductor muscle (A) being impressed on the bottom of the shell, exactly under the short simple loop, the diverging grooves (m) passing uninterruptedly through them.

other oblique and diverging ones (b) attached to those already described, and fixed to the bottom of the shell (Pl. I, fig. 19). From these proceed the longitudinal branches of a short simple loop (A), similar in character to that of *Terebratula* proper. A minute mesial ridge extends from under the cardinal process; these plates, with their central interspace and mesial ridge, form a lozenge-shaped elevation (Pl. I, fig. 18, also woodcut). On either side of this process, and extending further down, are seen the quadruple impressions left by the adductor (A), as dispayed in the accompanying cut.

From the above it will be perceived that the differences between *Terebratula* proper and *Epithyris*, as typified by Professor King in 1850, are chiefly confined to the presence of prominent dental or rostral plates in the one, and almost total absence in the other, as well as differences in certain details connected with the hinge-plate. On the other hand, the exterior characters are similar, the deltidium being more or less concealed in certain individuals than in others, from the greater or lesser approximation of the foramen to the umbonal beak.[1] This is also the case in many species of true *Terebratula*. In the interior the loop is the same, short, and simply attached, the longitudinal branches being united by a transversal band, more or less bent upwards in the middle.[2] The muscular impressions appear also similar, as well as the intimate shell-structure.

It will therefore be for palæontologists to determine whether the differences observable in the rostral cavity of the beak and hinge-plate of these few Palæozoic *Terebratula* should be considered of sufficient value to counterbalance the great resemblance they present with *Terebratula* proper in their more important dispositions and characters.

TEREBRATULA ELONGATA, *Schloth.*, sp. Plate I, figs. 5—22; and Plate II, fig. 2.

A difference in opinion has been expressed by some British and foreign palæontologists relative to the respective specific claims of *Ter. elongata* and *T. sufflata* of Schlotheim. These shells are so extremely variable in their shape, that Professor King could not help

---

[1] I am rather surprised that Professor M'Coy should have doubted the presence of a deltidium in *T. hastata*, *T. elongata*, and other similar shells, for it may be seen in many individuals, and more especially in young shells.

[2] Professor King states that "*Waldheimia* is most intimately related to *Epithyris*;" that "in *Waldheimia* the loop is elliptical, deeply recurved, and projecting about two thirds of the length of the shell; but that in *Epithyris* it is semi-elliptical, moderately recurved, and projecting about one third of the length of the shell." I do not perceive any important differences in its loop and that of *Terebratula*; moreover, Professor King admits that his illustration, pl. vi, fig. 45, was not quite correct.

observing, that Schlotheim considered some of their varieties as species,[1] and that he himself has no decided objection to that view, but feels utterly unable to separate one from another, as they merge so imperceptibly into each other. Professor King admits but two out of Schlotheim's several species, viz., *T. elongata* and *T. sufflata;* this view has also been reciprocated by Professor M'Coy.[2] On the other hand, Dr. Geinitz,[3] Baron Schauroth,[4] Mr. Howse,[5] and a few others, consider that there exists no valid grounds for even separating specifically *T. sufflata* from *T. elongata,* and either entirely amalgamate the two under the single denomination of the last-named shell, or consider *sufflata* in the light of a named variety.

Having had the opportunity of examining a very numerous series of both, I experienced the same difficulties in the attempt to separate *T. elongata* and *sufflata.* No doubt, if certain typical, or what might be considered typical shapes of both are selected, we might perceive certain peculiarities in each, and be tempted, perchance, to create more than one species; but these variations seem to exist only in certain individuals, while every intermediate form would be found in the same bed, and even quarry, to connect these different extremes. Such being the case, I have preferred to follow in the path of those authors who, while admitting but one species, have retained *sufflata* for the variety.

TEREBRATULA ELONGATA, *var.* GENUINA. Plate I, figs. 5—7, 12—14, and 18—22.
(King's Mon., pl. vi, figs. 30—45.)

TEREBRATULITES ELONGATUS et COMPLANATUS, *Schlotheim.* Akad. Münch., vol. vi, p. 27, pl. vii, figs. 7—14, 1816.

When full grown and well shaped, it is more or less elongated, widest near the middle, with almost equally deep, convex valves. The beak is more or less attenuated and incurved; the foramen rather small and circular, lying close to the umbone of the dorsal valve, so that the deltidium is but rarely exposed. The larger or ventral valve is moderately convex, presenting in profile a regularly arched curve from the extremity of the beak to the front, with a wide and gradually depressed or shallow sinus, commencing towards the middle of the valve, and extending to the front in almost all well-shaped examples; it produces in the frontal margin a convex and elevated curve, varying in degree according to age and individuals.

---

[1] 'Monograph of English Permian Fossils,' p. 148, 1850.
[2] 'British Palæozoic Fossils,' pp. 409, 412, 1855.
[3] 'Die Versteinerungen,' p. 11, pl. iv, fig. 27, April, 1848.
[4] 'Ein neuer Beitrag zur Palæontologie des deutschen Zechsteingebirges,' p. 213, 1856.
[5] 'Annals and Mag. of Nat. Hist.,' vol. xix, 2d series, p. 52, 1857.

The sinus is also sometimes narrow and more suddenly depressed (Pl. I, fig. 5 ; and pl. xlii *a* of King's Monograph), and hardly perceptible in certain middle-aged and young shells.

The dorsal valve is more or less regularly convex, with a mesial longitudinal elevation extending from the extremity of the umbonal beak to the front, from which the lateral portions of the valve slope more or less rapidly to the margins. External surface smooth ;[1] shell-structure minutely perforated.

The internal details having been already described under the genus, need not be repeated. The dimensions attained by this species are very variable.

The largest two British specimens I have seen measured—

Length 17½, width 15, depth 8 lines.

    „   19,   „   13 lines.

a. Var. SUFFLATA.   Plate I, figs. 8, 9, 10, 11, 15, 16, 17, and 21 ; Plate II, fig. 2. (King's Monog., pl. vii, figs. 1—9.)

TEREBRATULITES SUFFLATA, *Schlotheim.* Akad. Münch., vol. vi, pl. vii, figs. 10, 11, 1816.

This shell is smaller, relatively wider, and more inflated than the var. *elongata ;* ovate ; margins obtuse, and sometimes slightly emarginate in front. The *ventral valve* is either regularly convex (Pl. I, fig. 11), or presents a narrow mesial sinus, which is more or less excavated (Pl. I, figs. 16, 17 ; and Pl. II, fig. 2) in different specimens, so that the frontal line varies considerably in the convexity of its curve. The *dorsal valve* is more or less regularly inflated ; surface smooth. Interior exactly similar to that of *elongata.* In dimensions this variety does not appear to greatly exceed—

Length 7, width 6, depth 5 lines.

Professors King and M'Coy are of opinion that this shell should be specifically separated from *T. elongata,* as has been already observed. They state that it is smaller and more tumid, with a greater gibbosity, and with more obtuse angles ; the umbone more gibbous and prominent (?), a small definite lobe in the front margin, and corresponding long, narrow mesial sulcus in the ventral valve ; the posterior part of the shell not so tapering, and the sides of the beak more obtusely rounded and less angulated.

[1] In his 'Notes on Permian Palliobranchiata,' published in the 'Annals of Natural History' for March and April, 1856, Professor King observes that "specimens occasionally occurring at Glücksbrunn show *T. elongata* to have been a prettily coloured species ; in one example several dark bands interradiating with others of a lighter colour almost continuously from the umbone to the margin, and increasing in width in their forward progress ; in another, the dark bands reduced to dark lines are only developed near the margin."

It is not very difficult to find a number of examples presenting all these differences; but again it would be as easy to procure others which, from their intermediate character, leave one in the greatest doubt as to which of the two they should be referred; thus a vast number of *sufflata* present not a trace of sinus or sulcus in the ventral valve, both being regularly and equally inflated. Some also vary considerably in their depth, and with smaller proportions, resemble the two large adult examples of *T. elongata*, illustrated in our plate (Pl. I, figs. 5 and 18); and I can compare them to nothing better than to the differences we would perceive between a tall, thin, and a short, thick, stumpy man, the first representing *T. elongata*, the second *T. sufflata*.

Professor King is of opinion that *T. sufflata* closely resembles the Carboniferous *T. sacculus* of Martin; and, having kindly forwarded for my examination a specimen of both, they appeared to me undistinguishable; but having compared a numerous series of the Carboniferous and Permian varieties of *T. elongata*, *sufflata*, and *sacculus*, I could perceive in none of the Permian specimens the depression which commonly exists near the front of the dorsal valve of most individuals of the Carboniferous species; there remains but little doubt as to the intimate resemblance existing between certain examples of Martin's and Schlotheim's shell. *T. elongata* is certainly specifically distinct both from *T. hastata* and *sacculus*; and if we are to consider *T. sufflata* as a var. of *elongata*, then the individual similarity presented between some young examples of *sacculus* and *sufflata* cannot be considered of paramount importance, as the general facies of the species would be very different; but if, on the contrary, *T. sufflata* is to be considered as specifically different from *T. elongata*, then it would become a delicate question to determine whether *sufflata* is in reality more than a variety or race of Martin's *T. sacculus*.[1]

Loc., &c. Both *T. elongata* and its var. *sufflata* are common to the same beds and localities. It is also one of our commonest English Permian species in the shell limestone of Tunstall-hill, Humbleton, Ryhope-Fieldhouse farm, Hilton Castle, Clack's Heugh, Dalton-le-dale, &c. It has also been found in the magnesian conglomerate of Tynemouth. On the Continent it is common to several localities—Corbusen, Pössneck, &c., in Germany; at Nikefur, Orenbourg, Ilschalki, &c., in Russia.

[1] Professor King observes, while describing *Ter. sufflata*, that he has elsewhere stated that "this species appears to be identical with a shell found in the mountain limestone of Bolland, probably hitherto considered a var. of *T. sacculus*, a distinct, although closely allied species ('Monograph,' p. 150). M'Coy supposes that the shell here referred to is identical with *T. virgoides*, but this is not the case. The Bolland specimen, noticed under the last head as resembling *T. elongata*, has more affinity to M'Coy's species. I have been led to re-examine the shell found in the neighbourhood of Bolland, and I cannot but say that it agrees most remarkably with some specimens of the Permian species, particularly the testiferous one represented under fig. 7, pl. vii, of my 'Monograph.' On the other hand, there are specimens figured on the same plate closely approximating to true forms of *T. sacculus* in its mesial depression and emarginate front. The only difference I perceive between the Bolland shell alluded to and the Permian fossil quoted is, that on the former there are faint traces of longitudinal lines on the anterior half of the valves. I perceive nothing of the kind on any of the Permian forms, nor do I recognise any on normal specimens of *T. sacculus:*

*Family—*SPIRIFERIDÆ.

In the French edition of my General Introduction, published in vol. x of the 'Transactions of the Linnean Society of Normandy,' I have provisionally divided the family SPIRIFERIDÆ into three principal genera—SPIRIFERA, Sow., ATHYRIS, M'Coy = *Spirigera*, D'Orb., ATRYPA, Dalman; and into six sub-genera—*Cyrtia*, Dalman, *Spiriferina*, D'Orb., *Suessia*, E. Deslong., *Retzia*, King, *Merista*, Suess, and *Uncites*, Defr. But before a genus or sub-genus can occupy a definite and permanent position in science, it is necessary to be acquainted with all its characters, both internal and external, and to have appreciated these characters so as to be able to compare them to those of other genera in the same family.

Some short time since only have all the internal arrangements of the three above-named genera been completely ascertained, that of *Athyris* in particular having resisted for many years the most persevering researches.[1] The differences presented by these genera are most satisfactory; they represent three *well-defined types*, around which certain modifications of comparatively smaller value naturally converge. Of the six sub-genera, *Spiriferina* alone has been thoroughly investigated; but of the others much still requires to be learnt before the value of their respective characters or distinctions can be satisfactorily established. MM. Suess and Deslongchamps have already done much towards the elucidation of the interiors of *Merista* and *Suessia*, but we are not yet in a condition to furnish a completely restored illustration of all the parts of which their interior is composed. *Merista* was in all probability closely related to *Athyris*, and what little we know of *Suessia* would appear to denote that it possessed a very remarkable interior, since, with the external shape of a true Spirifer, it presents many dissimilarities in its internal organization. In *Suessia* the two branches which constitute the first spiral coils are united by a transversal, shelly band, from the centre of which proceeds another short lamella, which is directed towards the bottom of the valve. The species that compose this small group

there appears to be no difference between them in their histological perforations." ('Annals and Mag. of Nat. Hist.,' vol. xvii, 2d series, March, 1856.)

In speaking of *T. sacculus*, Professor M'Coy states ('British Palæozoic Fossils,' p. 411) that "several writers mention their inability to distinguish this species from some of the varieties of *Seminula elongata* and *S. sufflata* of the Permian rocks; but specimens perfectly identical in form and size may be readily distinguished by a small but distinctly marked upward wave in the front margin towards the anterior valve in the Permian fossil, while the margin of the Carboniferous species is nearly or quite horizontal. In *S. sufflata*, also, the mesial septum is much longer and more strongly marked in the receiving valve, extending to within one third of the length of the front margin."

[1] When I published my Introduction in the first volume of this work, the character of *Athyris* had not been fully ascertained.

possess likewise an unusually large hinge-plate, as well as two singularly shaped appendages, which, arising from the inner socket walls, follow an inward direction. No other member of the Spiriferidæ have presented those arrangements; and it is possible that when the interior shall be completely known, that they may be considered of more than sub-generic importance. As to *Cyrtia, Retzia,* and *Uncites,* much requires to be done before their characters or value as sub-genera can be completely determined.[1] In the British Permians, the genera *Spirifera, Athyris,* and the sub-genus *Spiriferina,* are hitherto alone represented.

Genus—Spirifera, *Sowerby.*

(See General Introduction, Vol. I, p. 79, 1853.)

Spirifera alata, *Schlotheim,* sp. Plate I, figs. 23—36; Pl. II, figs. 6, 7. (King's Monog., pl. ix, figs. 1—17.)

Terebratulites alatus, *Schlotheim.* Leonhards. Taschenbuch, vol. vii, p. 58, pl. ii, figs. 1, 2, 3, 1813.

— undulatus, *J. de C. Sowerby.* Mineral Conchology, vol. vi, p. 119, pl. 562, fig. 1, March, 1827.

— Cordieri, *Robert.* Atlas du Voyage de la Commission scientifique du Nord, pl. xix, fig. k, 1845.

S. alata varies considerably in shape, according to age and individual. When adult or full grown it is transversely fusiform, being twice, and even three times, as wide as long (Pl. I, figs. 23 and 27). Valves convex, deepest at a short distance from the umbone; hinge-line as long as the greatest width of the shell, the cardinal extremities being more or less attenuated in different individuals. The area is wide, with sub-parallel sides; fissure triangular, and in great measure covered by a convex pseudo-deltidium; a narrow rudimentary area may be seen likewise in the smaller valve; beak small and incurved. The mesial fold is simple, of variable width, and flattened along its upper surface; while in the ventral valve there exists a shallow sinus, interrupted by the presence of a rounded, slightly elevated mesial rib. The valves are likewise ornamented by a variable number of rounded, or but slightly angular, ribs; these are simple, or here and there augmented by an occasional intercalation. In number they vary from about eight to thirty on each valve, the larger number occurring on the most adult individuals. The

---

[1] I have published these few observations in the second volume of the 'Bulletin de la Soc. Linnéenne de Normandie,' 1857.

ribs are also at times of unequal width, even on the same example; and the entire surface of the shell is ornamented by close and regular scale-like, concentric, imbricated laminæ.

In the interior, the spiral cones fill the larger portion of the shell, as may be perceived by a glance at the illustration (Pl. I, fig. 27), drawn from a beautiful specimen in the collection of Mr. Kirkby. The principal lamellæ are here attached, as in all *Spiriferas*, to prolongations departing from the base of the inner socket walls.

The shell-structure has been stated by Professor King to be minutely punctuated, but neither Professor M'Coy nor myself have been able to recognise any trace of those tubular perforations;[1] nor does the interior itself present those peculiarities which accompany the perforated test of *Spiriferina*. The interior of the ventral valve does not show a trace of that elevated mesial septum which is always present in *Spiriferina cristata, Sp. octoplicata, Sp. Münsteri, rostrata, Tessoni*, and other forms composing that sub-genus. The dental or rostral plates in *S. alata* are also much smaller, and I might almost say rudimentary; the muscular impressions are likewise exactly similar to those peculiar to the genus *Spirifera*. In the ventral valve the adductor (A) forms a small, lengthened, oval impression, apparently divided by a minute mesial ridge or raised line, and on either side of the adductor are seen the larger scars left by the cardinal muscle (R): these are well displayed on the numerous internal casts found at Humbleton Hill, and of which Pl. II, figs. 6, 7, are illustrations. The ovarian spaces (o) are likewise clearly defined on most specimens; and some of the vascular markings have been described by Professor King.

In the dorsal valve, under the extremity of the umbone, there exists a small striated cardinal process or boss, but no hinge-plate, and a little lower down is seen the quadruple impression left by the adductor (Pl. I, figs. 31, 32, 33 A).

Professor King seems to have misunderstood the impressions visible on the internal casts of this valve, for he describes the smooth space above the central pair as the "posterior division of the valvular muscle," and our central pair as the "anterior division of the same muscle,"[2] thus placing one pair above the other, while in reality both are situated almost on a level (figs. 32 and 33). If the reader will kindly refer to the casts of the Carboniferous *Spirifera trigonalis* (Part V, Pl. V, figs. 26, 27) he will perceive how strikingly they agree with those of the Permian shell. It is, therefore, evident that the internal arrangements alone denote an imperforated species. Most authors are now of opinion that *Sp. undulata*, Sowerby, is only one of the numerous variations in shape of Schlotheim's *Sp. alata*, to which I must certainly add one or two of Professor King's illustrations of his *Sp. Permiana* (pl. ix, figs. 18, 19, and 20).

The study of a large number of specimens of *Sp. alata*, collected by Messrs. Howse

---

[1] Dr. Carpenter has examined the intimate shell-structure, and, having scaled off large flakes from a well-preserved specimen (a thing that could not be done to any perforated shell, as it does not split thus into laminæ), he did not find the least vestige of anything that could be called perforations, the shell-structure being in all respects analogous to the *Rhynchonella* type.

[2] 'Monograph,' page facing pl. ix, fig. 6.

and Kirkby, have proven to my entire satisfaction that the hinge-line was at times shorter than the greatest width, nor was the shell always so extremely transverse (Pl. I, figs. 24, 25, 30) as is commonly the case with full-grown individuals (Pl. I, figs. 23, 27); but I feel less certain regarding Professor King's figs. 21 and 23. The distinctive characters of *Sp. Permiana* have not, in my humble opinion, been sufficiently established to warrant the present adoption of that species.[1]

Sp. alata is distinguished from *Sp. laminosa* of the Carboniferous period, which it sometimes resembles by its more transverse shape.

The largest British specimen I have been able to examine measured 14 lines in length, 32 in breadth, and 11 in depth.

T. alata is not a very common species in England. It has been found at Tunstall and Humbleton Hill, Midderidge and Tynemouth Cliff. On the Continent it occurs at Pössneck, Röpsen, &c.; Bell-Sound, Spitzberg. (De Koninck.)

SPIRIFERA ? CLANNYANA, *King*. Plate I, figs. 47—49.

> MARTINIA CLANNYANA, *King*. Catalogue of the Organic Remains of the Permian Rocks of Northumberland and Durham, 19th August, 1848; and Monograph of English Permian Fossils, p. 134, pl. x, figs. 11—13, 1850.
> — WINCHIANA? *King*. Catalogue, p. 8; and Monograph, p. 135, pl. x, figs. 14—17.

This small shell is almost circular, and sometimes slightly emarginate in front; as wide, or a little wider, than long; valves unequally convex; hinge-line shorter than the

---

[1] Professor King believes his species well characterised, and kindly forwarded for my inspection the young example, fig. 23 of his plate; but Sowerby's drawing is not very correct, for the specimen is much more regularly semicircular, and possesses seven ribs, comprising the mesial one. It measures—length $1\frac{1}{2}$, width $2\frac{1}{2}$, depth $1\frac{1}{4}$ lines, and might perhaps be the fry of *Sp. alata?*

Professor King diagnoses his *Sp. Permiana*—"Margin semi-elliptical, twice as wide as long. Lateral surfaces with four or more sharpish, rather distant ribs; mesial furrow or ridge not much larger than the adjoining folds. Beak erect in casts, but gibbous in testiferous specimens; valves marked with regular lamellæ of growth, crossed with hair-like striæ, rarely exceeding half an inch in length; differs from *Trigonostrata undulata* in having a narrower mesial furrow or elevation, and only half the number of folds, which are broader and more angulated than those of the latter; the valves are not so tumid, and the lateral extremities are rounded instead of pointed." My figs. 35 and 36, Pl. I, approach most to Professor King's *Sp. Permiana*.

Spirifer Schrenkii, Keyserling, described and illustrated in Dr. A. G. Schrenk's excellent work, 'Reise nach dem Pordosten des Europöischen Russlands durch die Tundren der Samogeden,' vol. i, p. 88, 1848; and vol. ii, p. 106, pl. iii, figs. 20—30, 1854, appears to be closely related to *Spirifera alata*, but is well distinguished by the absence of the rib which exists in the sinus of the ventral valve of Schlotheim's species.

width of the shell; cardinal angles rounded, area triangular, fissure large and partly concealed by a pseudo-deltidium; beak rounded and elevated. The dorsal valve is but slightly convex, with a mesial depression or furrow, commencing at a short distance from the moderately inflated umbone, and extending to the front; it also possesses a small triangular area. The ventral valve is by far the deepest and most convex, with a mesial furrow originating at a short distance from the extremity of the beak, and extending to the front. The external surface of the shell is covered with numerous closely set and inclined hair-like spinules.[1] In dimensions it does not in general exceed—length 2, width 2, and depth 1½ lines.

This interesting little species appears to be so closely related to the Carboniferous *Sp. Urei* of Fleming,[2] as well as to the Devonian *Sp. unguiculata* of J. de C. Sowerby and Phillips,[3] that I am still uncertain whether it is in reality distinct, or simply a variety or race slightly modified by time? After a minute examination of some well-preserved examples of *Sp. Urei*, which I had obtained through the kindness of Dr. Fleming and of another friend in Scotland, and having ascertained that they had been likewise covered with spinules, I requested Professor King to compare my Carboniferous specimens with those of his Permian shell, and he has transmitted the following observations: "*Sp. Urei* and *Sp. Clannyana* are, I am decidedly of opinion, distinct species, though apparently allied to each other. *Urei* differs from *Clannyana* in being a wider shell; it has the umbone more incurved, the area of the small valve not so deep. The dorsal valve is more excavated, and, as it were, towards the postero-lateral angles. The spines decidedly less numerous, and the median sulcus more pronounced on both valves."

The double area, so like that of some *Orthis*, is well displayed in this little shell; but it

Sp. Clannyana,
seen from the beaks,
enlarged.
M. Area of ventral
valve. *D.* Deltidium.
N. Area of dorsal
valve.

cannot be considered a character of generic value, as such likewise occurs to a greater or lesser extent in many species of *Spirifera*, such as in *Sp. striata* and *Sp. alata*, but more especially so in *Sp. decora* of Phillips. Professor King and Mr. Howse are of opinion that *Sp. Clannyana* should be generically separated from *Spirifera* proper, and have placed it in M'Coy's *Martinia*, a genus I have hitherto declined adopting, from its not appearing to be founded upon any important or valid internal character. I have therefore deemed it preferable to allow the shell under description to remain (provisionally at least) under *Spiri-*

---

[1] These spines were first noticed by Baron Schauroth, in his 'Ein Beitrag zur des deutschen Zechsteingebirges,' fig. 16, 1853, and afterwards in his 'Ein neuer Beitrag zur Palæontologie,' 1856. Their existence is now also admitted by Professor King, in his Notes on Permian Fossils ('Annals and Mag. of Nat. Hist.,' March and April, 1856); and Mr. Howse informs me that he was in error when he stated in his paper in the 'Annals' for 1847, that the surface does not appear to be covered with spines.

[2] 'History of British Animals,' p. 376, 1828; and Ure's 'History of Rutherglen,' pl. xiv, fig. 12, 1793.

[3] 'Geol. Trans.,' 2d series, vol. v, pl. liv, fig. 8; and Phillips's 'Palæozoic Fossils of Cornwall,' &c., p. 69, pl. xxviii, fig. 119, 1841.

fera, from which it may be hereafter removed, should the study of its interior demonstrate the necessity. Mr. Morris has combined *Martinia Winchiana*, King, with *M. Clannyana*,[1] a view likewise adopted both by Baron Schauroth and Mr. Howse, and in which I should also feel inclined to concur. Professor King, however, still insists upon the separation, but the distinctions proffered do not appear to me to have been satisfactorily made out, and seem more individual than specific.

Loc. Sp. Clannyana has been collected rather abundantly at Ryhope-field House. One example was also found at Tunstall-hill by Mr. Kirkby, and another from Pallion may be seen in the Sunderland Museum. Professor King names Whitley as the locality whence he obtained his *Sp. Winchiana*. German examples have also been discovered at Pössneck by Baron Schauroth.

Sub-Genus—SPIRIFERINA, *D'Orbigny.*

(Introduction, Vol. I, p. 82, 1853.)

SPIRIFERINA CRISTATA, *Schlotheim*, sp. Plate I, figs. 37—40, 45, 46; Plate II, figs. 43—45. (King's Monog., pl. viii, figs. 9—14.)

TEREBRATULITES CRISTATA, *Schlotheim*. Beitr. z. Naturg. d. verst in Akademie der Wissenschaften zu München, tab. i, fig. 3, 1816.

This variable shell is more or less transversely semicircular, moderately convex or inflated; the hinge-line as wide or slightly shorter than the greatest width of the shell. The cardinal angles are more often rounded (Pl. I, fig. 40), rarely prolonged with acute terminations (Pl. II, fig. 43). Area large, triangular, flat, or slightly concave, and placed at almost right angles to the plane of the smaller valve, so that the beak is not often seen to protrude to any extent beyond or above its termination. The fissure is rather wide, triangular, and partially covered by a pseudo-deltidium. The number of ribs which ornament the shell varies considerably, both according to age and individual; the central one in the dorsal valve (which represents the mesial fold) is, in general, twice as wide as those which cover the lateral portions of the valve; its crest is angular throughout, or more or less flattened, especially towards the front. In profile it presents a regularly convex curve, but in some individuals is slightly bent upwards near the front. In the ventral valve the sinus is more often deep and angular, but is also sometimes flattened along its centre. From four to fourteen angular or rounded ribs, of greater or lesser width and depth, ornament each valve. When quite young, with dimensions not exceeding one to one and a half line in length, and about the same, or a trifle more, in width

---

[1] 'Catalogue of British Fossils,' 2d edition, 1854.

(Pl. II, fig. 45), from four to six ribs were only developed, and in this condition resemble the var. *Jonesiana* of King; but most examples of *Sp. cristata* possess from eight to ten ribs; while a remarkable and unusually large individual, obtained at Tunstall hill by Mr. Kirkby (Pl. II, fig. 43[1]), presented fourteen on either valve, and so closely did this specimen resemble some of Sowerby's typical examples of the Carboniferous *S. octoplicata*, that it is very probable if not entirely certain that *Sp. cristata* is at most but a variety or race, slightly modified by time and circumstance, of the Carboniferous species.[2] In the Permian period it was, however, in general a smaller shell, the number of ribs likewise frequently less numerous (Pl. I, fig. 38; and Pl. II, fig. 44). The external surface and ribs are intersected at close intervals by many concentric laminæ, or ridges of growth. Its shell-structure has been described by Professor King as closely perforated, the canals being large, and (according to Professor M'Coy) half their diameter apart.[3] In the interior of the ventral valve two short diverging dental or rostral shelly plates form the fissure-walls, and between these a sharp elevated mesial septum arises, and extends along the bottom of the valve from the extremity of the beak to less than half its length, the sharp elevated extremity dividing the upper portions of the two spiral coils. In the dorsal valve the spiral cones are attached to prolongations of the inner socket-walls, and occupy a large portion of the interior of the shell with their numerous convolutions,[4] as may be seen in the enlarged illustration (Pl. I, fig. 40), taken from a beautifully perfect individual from Humbleton hill, and forming part of Mr. Kirkby's valuable collection. A short and small mesial septum seems likewise to exist in the smaller valve. The dimensions taken from two British examples have produced—

Length 5, width 9½, depth 5 lines (the largest known).

„   4, „ 7,   „ 4 lines.

Loc. Sp. cristata does not appear to have been a very common fossil in England, and especially so with its shell preserved. Professors Quenstedt and King, Mr. Howse, and others have found it in the shell limestone of Humbleton and Tunstall hills, Hylton, North Farm, and Tynemouth Cliff. On the Continent it is mentioned by Schlotheim, Dr. Geinitz, and Baron Schauroth, from Glücksbrunn, Könitz, Pössneck, Altenstein, Schwaara, and Röpsen; and was found by Mr. E. Robert at Bell Sound, Spitzberg.[5]

---

[1] This specimen was first figured by Mr. Howse, in pl. iv, figs. 5 and 6, of the nineteenth volume of the 'Annals and Mag. of Nat. Hist.,' January, 1857.

[2] See *Spiriferina cristata*, var. *octoplicata*, in our Part V, p. 38.

[3] Professor King has described singular casts of the tubular perforations visible on the surface of the greater number of internal casts from Humbleton hill.

[4] This and other individuals which I have been able to examine, have presented from eleven to twelve convolutions in each of the spiral cones. The crural processes were not united by a shelly band, as was found to be the case in some examples of *Spiriferina rostrata*.

[5] Refer to De Koninck's 'Nouvelle notice sur les fossiles du Spitzberg,' vol. xvi of the Académie Royale de Belgique, No. 12 des Bulletins.

SPIRIFERINA MULTIPLICATA, *J. de C. Sowerby.* Plate I, figs. 41—44.

SPIRIFER MULTIPLICATUS, *J. de C. Sowerby.* Geol. Trans., 2d series, vol. iii, p. 119, 1829; but described and illustrated for the first time by Professor King, p. 129, pl. viii, figs. 15—18, of his Monograph of English Permian Fossils, 1850.

— — var. JONESIANA, *King.* Mon., p. 129, pl. viii, fig. 19.

Shell small, rarely exceeding 5 lines in length, by 6 in breadth and 4 in depth; is more or less transversely oval, and often very much rounded in outline; valves convex, and sometimes much inflated. The hinge-line is shorter than the width of the shell; beak more or less produced, elevated, and incurved; area triangular, slightly concave, and varying in dimensions (Pl. I, fig. 44[1]); fissure wide, and partially covered by a pseudo-deltidium. From the narrowness of the area, the sides of the beak are more or less visible, as well as its incurved extremity. The ribs are small, rounded, and rarely exceeding ten in number; mesial fold not much produced, and either rounded or flattened along its crest. The sinus in the ventral valve presents a moderate depth, and the entire surface of the shell is covered by numerous concentric laminæ or ridges of growth; shell-structure perforated; the canals are (stated by Mr. Howse to be) smaller than those in *Sp. cristata.* The internal details, according to Mr. Kirkby, differ a little from those already described in the preceding species; the whorls of the spiral are not so numerous, the spirals are not so obliquely placed, and the first branch of the coil is not so angulated.

I have felt considerable uncertainty as to whether the present shell should or not be considered specifically distinct from *Sp. cristata.* Some authors, while admitting the relationship existing between the two, are still of opinion that they should not be united, on account of the smaller dimensions of *Sp. multiplicata,* its rounded outline, more elevated and incurved beak, smaller and less angular ribs, and proportionately wider and more flattened mésial fold. That, on the contrary, *Sp. cristata* is more acutely triangular, the beak less elevated, the ribs more numerous, their sharpness and depth greater, and the intimate shell-structure somewhat different; but, after a minute examination of a very numerous series of both, I found that, although many examples of the shell under description were easily distinguishable from *S. cristata* by the means of the characters above specified, a great number left me in much uncertainty as to which of the two they in reality did belong, for many individuals possessed the same number of ribs, and the other distinctive characters became also much attenuated in their respective values. Professor King allows me to state that he now begins to think that *Sp. multiplicata* may perhaps be a variety of *cristata,* an opinion in which I entirely concur; but, as some uncertainty still

[1] This remarkable specimen, from Tunstall hill, forms part of the collection of the Geological Survey of London. It measures—length 6½, width 6, and depth 6 lines.

exists relative to the question of absolute identity, it will be preferable for the present to allow both Schlotheim's and Sowerby's shells to retain their distinctive appellations. The *Trigonostrata Jonesiana* of King appears to Messrs. Howse, Kirkby, and myself to be only a more rounded form or variety in shape of *Sp. multiplicata*, and was so considered by Professor King himself until 1850, when he removed it on account of its lesser width, more prominent umbone, and higher area; the ribs being also more evenly rounded and at a greater distance from each other, the median one on the smaller valve more evenly convex, and its corresponding furrow in the opposite valve more evenly concave; but these differences brought forward by that author seem to be those common to many young individuals of the Sowerby shell, and cannot, in my opinion, claim a specific distinctive denomination.

Sp. multiplicata is not very rare in the shell and compact limestone of Tunstall and Humbleton hills, and at Dalton-le-Dale.

Genus—ATHYRIS, *M'Coy*. = *Spirigera*, D'Orb.

SPIRIGERA. Introduction, vol. i, p. 87, pl. vi, figs. 65—70, 79, 1853.

Obs. With the desire to obviate and correct a zoological misnomer, I proposed, in 1853, to adopt the generic designation *Spirigera*, D'Orb., for such shells as *T. concentrica, pectinifera, de Roissyi*, &c., and to retain that of *Athyris*, M'Coy, for such as *T. tumida, Herculea, scalprum*, &c. But this arrangement not having met with the entire approval of several distinguished friends, I have willingly complied with their desire by re-establishing *Athyris* as typified by *A. concentrica, pectinifera*, &c., while for such shells as *T. tumida, Herculea*, &c., that of *Merista*, Suess, has been adopted.[1] The external and some of the internal characters are already well known; but it is not until lately that the arrangement, attachment, and connecting process of the two spiral cones has been discovered, and which will be found fully described under *A. pectinifera*.

In France and in Germany the term *Spirigera* is generally adopted and preferred to that of *Athyris*, and I cannot but say with good reason; but the law of priority obliges us to retain that of *Athyris* as a mere generic appellation, for it would not be just or fair to repudiate that of M'Coy on account of its erroneous interpretation or etymology, as long as so many other equally objectionable names are allowed currency in science.

[1] French edition of my General Introduction, in vol. x of the 'Transactions of the Linnean Society of Normandy.' Also in the German edition, published by M. Suess.

ATHYRIS PECTINIFERA, *J. de C. Sowerby.*   Plate I, figs. 50—56; Plate II, figs. 1—5.

> ATRYPA PECTINIFERA, *J. de C. Sowerby.*   Min. Con., vol. vii, p. 14, pl. 616, 1840.
> CLEIOTHYRIS PECTINIFERA, *King.*   Monograph of British Permian Fossils, p. 138,
> pl. x, figs. 1—10, 1850.
> SPIRIGERA PECTINIFERA, *Howse.*   Annals of Nat. Hist., vol. xix, 2d series, p. 51.
> ATHYRIS PECTINIFERA, *Dav.*   Bulletin de la Soc. Linnéenne de Normandie, vol. ii, p. 14,
> pl. i, figs. 1—7.

This shell assumes in general a more or less rounded shape, is transversely oval, rather wider than long, and in dimensions (with its fringe) rarely attaining 12 lines in length by 14 in breadth. The valves are almost equally deep and convex; the beak of the ventral valve is short, incurved, and perforated at its extremity by a small circular foramen. The external surface is regularly covered by numerous concentric scaly ridges of growth, from each of which radiate closely set fringes of elongated, somewhat flattened, spines (Pl. I, figs. 50, 52, 54; Pl. II, fig. 2). In the interior, the hinge is strongly articulated, the dental or rostral plates offering by their position much solidity to the beak of the ventral valve; between these we find located the extremity of the beak of the smaller valve, as well as the cardinal or hinge-plate, which is not very largely developed in the present species. This hinge-plate is likewise perforated close to its summit (under the umbone) by a minute circular aperture,[1] destined in all probability for the passage of the intestine? as we know that this organ occupies a place about similar in other genera (*Rhynchonella*, &c.) of which the animal has already been studied.

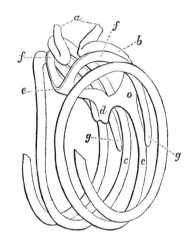

On each side of this small hinge-plate and of its aperture, exists a prolonged testaceous plate, situated at a higher level, and forming the ledge or rim of the inner socket-walls. These two prolongations become lengthened, and give birth (while serving as points of attachment) to the two lamellæ (*a*), which, by their convolutions on either side, constitute the spires, as well as the intermediate process by which they are united. These spiral lamellæ, disposed vertically to the plane of the valve, converge at first, then bending suddenly upon themselves, at a short distance from their origin, outwards and backwards (*b*), in the shape of a half circle, which, passing first close to the sockets (*b*), then follows in the direction of the bottom of the valve (*o, c*), to become

---

[1] The hinge, the rostral plates, and the position of the minute circular aperture in the cardinal or hinge-plate, have been described and figured by Professor King in pl. x, figs. 7 and 9, of his 'Monograph of British Permian Fossils.'

recurved again, and thus producing the first of the seven or eight convolutions of which each cone is composed. From the first two spiral lamellæ spring up perpendicularly other two secondary ones (*o*, *d*), which, by being bent, become united towards the middle of the shell (*d*) between the two spiral cones, and afterwards form but a single branch (from *d* to *e*). This last, after having attained the upper level of the spire (at *e*), becomes again bifurcated, and forms other two lamellæ (*f*) recurved in half circle, and terminating backwards by a free extremity (*g*) between the first and second coil of each spiral cone.

Obs. The discovery of the perfect example from which the above description has been taken is entirely due to the zeal of Mr. Howse, who, along with Mr. Kirkby, had kindly (and at my especial request) worked for several days in the quarry of Humbleton hill, in the hopes of obtaining a specimen wherein the connecting process of the two spirals could be satisfactorily viewed. It was not, therefore, until the 17th of September, 1856, that material sufficiently perfect could be found so as to admit of a clear definition of this singularly complicated system of lamellæ, so difficult to describe, but which will be readily understood by a glance at our several illustrations.[1]

In 1840 Mr. J. de C. Sowerby figured (in tab. 616 of the 'Mineral Conchology') an imperfect specimen (54 of our Pl. I), in which the perpendicular lamellæ (T) are seen to spring up from the first two spiral coils, as we have already described, and another (Pl. I, fig. 53) in which a small unexplained portion of the connecting process was exhibited (L). In this last example the attachment of the short bent lamellæ to the prolongations of the hinge-plate are likewise imperfectly displayed.

Mr. J. de C. Sowerby and Professor King have referred to the pectinated character of the spiral process, and to this I have likewise devoted some attention. The appearance is considered by Mr. Howse to be more deceptive than real, and entirely due to a foreign substance which encrusts the internal portions of the larger number of specimens. Although this is to a certain extent correct, and that some spirals have not presented a trace of pectinations, others were so distinctly marked, that after a careful examination I felt disposed to believe that the appearance was in a measure due to short spines (Pl. II, fig. 5), similar to those first described by myself as existing on and near the edge of the coils of *Spiriferina rostrata*, *S. Münsteri*, *Ter resupinata*, and other species, as will be seen by a reference to Part I of the present work.

The fringes which so closely encircle both valves are likewise of much interest; and it will be perceived (Pl. I, fig. 55) that the spinose extremities become at times flattened and expanded, so as to produce small plates of a limited extent.

M. De Verneuil and Professor King have alluded to the resemblance which appears to exist between the Carboniferous *Athyris Roissyi* and the Permian *A. pectinifera*; but, besides the more developed sinus peculiar to the Carboniferous shell, there would seem to

---

[1] A small example had also been observed in the collection of the British Museum by Mr. S. P. Woodward, but not nearly as perfect as the one discovered by Mr. Howse (Pl. II, fig. 2).

exist also a slight difference in the shape of the intermediate process which unites the spiral coils ; and until this point has been clearly ascertained, it will, I think, be preferable to consider both separate, although closely allied, species. There also appears to exist some difference in the external spines.[1]

Loc. *A. pectinifera* appears to be a rather uncommon shell in the limited number of localities where it has been hitherto discovered. In England it has been found from the shell limestone of Tunstall, Humbleton, and Hylton, and in the magnesian conglomerate of Tynemouth, and by Messrs. Howse, Kirkby, and Professor King. In the German 'Zeichstein' it is described from Milbite, Corbusen, and Gera. Also from Kirilof, Tioplova, and Bielebei, in Russia.

Family—RHYNCHONELLIDÆ.[2]

Genus—CAMAROPHORIA, *King*, 1846.[3]   Plate II.

(*Vide* General Introduction, Vol. I, p. 96, 1853.)

The external and internal characters of this excellent genus have been fully described by Professor King, and again detailed in the General Introduction to the present work, so that there remains but little further to be added to what has been already published.

The excellent material in the possession of Messrs. Howse and Kirkby has, however,

---

[1] M. De Verneuil observes that " *T. pectinifera* differs from *T. Roissyi* but by the absence of a sinus in either valve, and by the horizontality of its margin, comprised in a same plane. It is probable enough that it is but a variety, and that specimens will be found which will connect the two species together ; but it would deserve not the less to be considered as a remarkable variety, by its preponderance more at one epoch than at another ; for in Russia, as in England, *T. pectinifera* is peculiar to the Permian system, and if it exists in the Carboniferous system of Belgium, it is at least exceedingly rare."

Terebratula Royssiana, Keyserling, 1846 ('Reise i d. Samojedenland' of Dr. Schrenk, vol. ii, p. 109, pl. iv, figs. 31—33), belongs to the genus Athyris, and is specifically different from *A. Royssii*, Lev., as well as *A. pectinifera*.

[2] The occurrence of the genus *Rhynchonella* in the Permian period was noted by D'Orbigny at p. 167 of the first volume of his 'Prodrome de Paléontologie Stratigraphique' (1849), as typified by *Terebratula Geinitziana*, De Verneuil, a shell found by the authors of the 'Geology of Russia,' at Schedrova, near the mouth of the Vaga (Dvina), as well as by Count Keyserling in the uninhabited forests near the river Oukhta (province of Archangel), Russia.

From not being acquainted with this shell, and having overlooked its mention as a *Rhynchonella*, I queried the existence of the genus in the Permian period when publishing the Tables contained in the English, French, and German editions of my General Introduction. It must, however, be observed, that

---

[3] 'Annals and Mag. of Nat. History,' vol. xviii, August, 1846.

enabled me to offer a few additional illustrations, in which certain parts are perhaps more completely expressed. Thus, fig. 25 shows the hinge and different shelly processes in a very distinct manner, as well as do the detached valves, figs. 15 and 24; but the point on which I should desire to direct more particular attention is to those defined and often deeply indented impressions visible or either side of the septa (as seen in the casts, figs. 11, 12, 13, 14, and 23), and which I am disposed to believe are in a great measure due to the adductor (A) in the dorsal, and to the cardinal muscles (B) in the ventral valves. Professor King does not appear to coincide in this interpretation; but when I compare these markings with those observable in the interior of *Rhynchonella*, I am naturally led to conclude from their shape and position that they may be attributable to a similar origin.

Three species only have been discovered in British Permian localities, and they present the same internal details, which appear to be simply more or less developed, according to the dimensions and age of the individual. In the ventral valve the dental plates are

from D'Orbigny having placed *Camarophoria Schlotheimi*, along with *Rhynchonella Geinitziana*, in the same genus, it is evident the French author was neither acquainted with their internal character nor dissimilarities. More merit is therefore due to M. De Verneuil and to Count Keyserling for having placed the two above-named shells in separate genera, and particularly to Professor King, who first described their differences.

Rh. Geinitziana was likewise found in Germany by Dr. Geinitz and Baron Schauroth; and it was while describing some examples of this last from Röpsen, near Gera (in the 'Annals of Nat. Hist.' for March and April, 1856), Professor King announced for the first time that, in his opinion, the shell-structure was regularly perforated as in *Terebratula*. So anomalous a condition in a genus and family where all the species hitherto examined by Dr. Carpenter had proved to possess a fibrous *non-perforated* texture, made me very desirous that the subject should be further investigated before the statements made by Professor King should be admitted as an established fact. I therefore obtained, through the kind medium of Count Keyserling and General Helmersen, the loan of those Russian individuals collected by Count Keyserling near the river Oukhta, and which form part of the collection of the "Corps des Mines" of St. Petersburgh. M. De Verneuil likewise kindly communicated his original specimen, and to which were added others from German individuals, presented to me by Baron Schauroth and Professor King. These I submitted to Dr. Carpenter, who, on the 17th of February, 1857, after a careful and minute microscopical examination of *shreds*, removed in my presence from the exterior of the best preserved of the Russian shells, declared that he could perceive no perforations in their outer layer, but that in the German examples the inner surface was covered with minute pits, such as are seen on the outer surface of Poranbonites; but at that period Dr. Carpenter had not made any vertical sections through the entire thickness of the shell. But as Professor King had observed numerous dots on the Russian and German specimens which he considered to be due to perforations passing directly through the entire thickness of the shell, the specimens and preparations made were shown by Dr. Carpenter to Professors Quekett and Salter, who both entirely coincided in the opinion expressed on the 17th of February, viz., that the outer or first-formed layer (in perfectly preserved specimens) was not perforated; but also agreed that a considerable thickness of the inner ones were perforated, and which last corresponded with the portion of the shell examined by Professor King. Therefore, although it appears certain that a part of the thickness of the shell is traversed by passages, there exists an external layer quite free from such, so that the shell-structure cannot be considered exactly the same as in *Terebratula*.

conjoined at their dorsal margins, forming a trough-shaped process, affixed to a low medio-longitudinal plate (Pl. I, figs. 24, 25, 27). In the dorsal valve the space between the sockets is occupied by a small cardinal muscular process or boss, on either side of which are two long, slender, curved processes, and to which were no doubt attached the free cirriated spiral arms: from beneath the cardinal process a high vertical mesial septum extends to a little more than a third of the length of the valve, supporting along and close to its upper edge a spatula-shaped process, considerably dilated towards its free extremity, and projecting with a slight upward curve to nearly the centre of the shell (figs. 15, 26, 27).

The vascular markings (figs. 13, 14) have been described by Professor King, as well as in my Introduction, and are beautifully preserved on many examples of the larger species from our Permian beds.

CAMAROPHORIA SCHLOTHEIMI, *Von Buch*, sp. Plate II, figs. 16—27 ; and King's Mon., pl. vii, figs. 10—21 ; and pl. viii, fig. 8.

TEREBRATULITES LACUNOSUS (part), *Schlotheim.* In Dr. Leonhard's Min. Taschenbuch, 1813.
TEREBRATULA SCHLOTHEIMI, *Von Buch.* Ueber Terebrateln, p. 39, pl. ii, fig. 32, 1834.

This abundant and characteristic species varies less in shape as in the number of its ribs; is wider than long, and somewhat obscurely rhomboidal or sub-pentagonal, the greatest width being situated at about the anterior portion of the shell. The valves are unequally convex, and possess marginal expansions, which, when fully developed, produce a flattened frill, as wide as half the length and breadth of the shell. The dorsal valve is often gibbous, and presents in profile a convex arch, with its marginal expansions slightly bent upwards. The mesial fold differs in width and elevation, according to the number of ribs which ornament its surface, these varying from two to seven in different ages and individuals. On the lateral portion of the valves the ribs are also at times obscurely defined, and vary in number from two to six on either side. The ventral valve is most convex about the beak, this last being small, slightly produced, and incurved, so that the deltidial plates which margin the foramen are but rarely exposed, except in very young individuals: the beak ridges are also but slightly defined; the sinus is of greater or lesser depth, and ornamented by from one to five ribs, which first appear at about the middle of the shell, and extend thence to the front. Shell-structure not perforated.[1] The measurement, taken from a specimen with its marginal expansions fully developed, offered

[1] In page 117 of his Monograph of 'English Permian Fossils,' Professor King describes the shell-texture of *Camarophoria* as possessing extremely minute punctures, but Dr. Carpenter could find none, and pronounced the genus and present species to be "not perforated," at p. 35 of our General Introduction.

—length 10, width 14, depth 6 lines; the same without the expansions—length $6\frac{1}{2}$, width $8\frac{1}{2}$, depth $5\frac{1}{2}$ lines.

In 1816, Schlotheim published his first illustration of the shell under description, but with the mistaken idea that it belonged to the same genus and species as *Terebratulites lacunosus*, a Jurassic shell; and in 1834 Von Buch applied to the Permian shell the distinctive appellation now generally adopted. Professor M'Coy states that *C. multiplicata* of King should be merged into the species under description,[1] and with which Baron Schauroth would likewise combine *C. globulina* of Phillips,[2] but I am not prepared to subscribe to either opinion, because specific differences appear to have been satisfactorily made out both by Mr. Howse and Professor King. The first is a much larger shell, more transversely oval, and possesses a greater number of ribs; nor have I been able to discover the passages which are said to connect it with the Schlotheim species; and it appears also to be distinguished from *C. Schlotheimi* by its much narrower marginal expansions (Pl. II, figs. 15 and 22), a point first mentioned to me by Mr. Howse. *C. globulina* is likewise a well-characterised shell, more globular in shape than either of the others, with only two or three ribs on the mesial fold, and one in the sinus, and has not hitherto exhibited marginal expansions; but I am ready to admit that certain adult examples of Phillips's shell very closely resemble some young individuals of that variety of *C. Sclotheimi* which possess only two ribs on the mesial fold (fig. 19), but both are well distinguished in the adult condition. I do not, therefore, perceive the advantage to be derived from combining these three shells under a single denomination.

It has been supposed that *Ter. superstes*, De Verneuil, might likewise be united to the species under description, and certainly some of our British examples of *C. Schlotheimi*

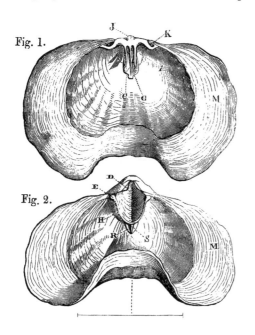

Fig. 1. *C. Schlotheimi*. Interior of the dorsal valve.

Fig. 2. *C. Schlotheimi*. Interior of the ventral valve.

A. Adductor muscular impressions? *C.* Curved processes, to which were affixed the fleshy spiral arms. *D.* Deltidium. *E.* Teeth. *H.* Conjoined dental plates, or trough-shaped process. *J.* Cardinal process. *K.* Sockets. *M.* Marginal expansions. *O.* Spatula-shaped process, or visceral support? *R.* Cardinal muscular scars? *S.* Septum.

---

[1] 'British Palæozoic Fossils,' p. 445.

[2] 'Ein neur Beitrag zur Palæontologie des deutschen Zechsteingebirges,' p. 218. Baron Schauroth admits three varieties—*a* var. *multiplicata*, *b* var. *genuina*, and *c* var. *globulina*.

appear to closely resemble De Verneuil's figures; but not having hitherto been able to examine the Russian type, 1 cannot comment upon its specific merits. M. De Verneuil observes, in the second volume of the 'Geology of Russia,' p. 101, that his species is distinguished by certain slight external characters, such as the more rounded shape and less prominent ribs on the lateral portions of the shell, the last being at times completely wanting, as well as the general contour, which is more triangular than pentagonal.

Loc.   *C. Schlotheimi* is one of the most common species in the Permian rocks of Humbleton and Tunstall hills, less abundant at Hylton Castle, Silksworth, Ryhope, Dalton-le-Dale, Tynemouth Castle-hill, &c. On the Continent it appears equally abundant at Röpsen, Milbitz, Poëssneck, Corbusen, Schmerbach, &c., in Germany. It is stated by M. De Verneuil to occur in the Carboniferous strata of Russia.

CAMAROPHORIA GLOBULINA, *Phillips.*   Plate II, figs. 28—31; and King's Mon., pl. vii, figs. 22—25.

TEREBRATULA GLOBULINA, *Phillips.*   Encyc. Met. Geology, vol. iv, pl. iii, fig. 3, 1834.

Shell small, globular, almost as wide as long, with convex inflated valves, ornamented by from six to nine short angular ribs on the dorsal, and five to eight on the ventral valve; of these two, more rarely three, occupy the mesial fold, and one or two the corresponding sinus, thus producing a bisinuated or trisinuated frontal wave. The ribs take birth at about half the length of the shell, and extend to the margin. Beak small, entire, rounded and incurved; foramen minute, with small deltidial plates. No marginal expansions?

Length 4, width 4, depth 3 lines.

C. globulina is common in the shell limestone of Tunstall and Humbleton hills, Hylton Castle, Dalton-le-Dale, and Ryhope-house Farm; as also, but more rarely, in the conglomerate of Tynemouth. It was named *T. corymbosa* by Mr. Howse in his 'Catalogue,' but afterwards referred to that of Phillips.

CAMAROPHORIA HUMBLETONENSIS, *Howse.*   Plate II, figs. 9—15; and King's Mon., pl. vii, figs. 26—32; and pl. viii, figs. 1—7.

TEREBRATULA HUMBLETONENSIS, *Howse.*   Catalogue of the Fossils of the Permian system of the Counties of Northumberland and Durham, T. N. T. C., vol. i, part iii, p. 252, 17th August, 1848; and Annals of Nat. Hist., vol. xix, 2d series, p. 50, pl. iv, figs. 3, 4, January, 1857.

CAMAROPHORIA MULTIPLICATA, *King.* Annals and Mag. of Nat. Hist., vol. xviii, 1st series, p. 28, July, 1846;[1] and a Catalogue of the Organic Remains of the Permian Rocks of Northumberland and Durham, p. 7, 19th August, 1848; a Monograph of the Permian Fossils of England, p. 121, pl. vii, figs. 26—32; and pl. viii, figs. 1—7, 1850.

Shell sub-trigonal or obovate, wider than long, with narrow marginal expansions; dorsal valve inflated and gibbous, presenting in profile a very convex curve; mesial fold wide, rounded, and somewhat flattened along the middle. The ventral valve is most convex at and near the beak, which is rounded and incurved; foramen small, circular. The sinus varies in depth and width, and is likewise flattened along its middle, producing a more or less elevated convex curve in front. About thirty simple or intercalated radiating ribs ornament the surface of each valve, of which from eight to ten occupy the surface of the mesial fold, and from seven to nine that of the sinus of the opposite valve. The dimensions taken from two examples have produced—

Length 13, width 17, depth 11 lines.

„ 11, „ 14, „ 7½ lines.

No testiferous specimens of this large *Camarophoria* have been discovered as yet, and all we know of the species is taken from beautifully preserved internal casts. It is stated to have been found abundantly only in a few localities, and but at times in the shell limestone of Humbleton and Dalton, once in the magnesian conglomerate of Tynemouth.

Mr. Howse has observed that, "after examining an extensive series (of the shell under description and *C. Schlotheimi*), we are obliged to conclude, from the following constant characters, that they are quite distinct. In the present species the form is more ovate, and the sinus much wider and less elevated in front. The ribs are always more numerous, and when they are not bifurcated they are more nearly parallel than in *C. Schlotheimi*. It is also of a much larger size."[2] In p. 121 of his 'Monograph,' Professor King enters upon other details to demonstrate that the *Camarophoria* in question cannot be considered as a large state of Von Buch's species, *C. Schlotheimi*, an opinion in which I entirely coincide. Baron Schauroth sent me several specimens from Ilmenau, under the name of *C. multiplicata;* but these do not agree sufficiently, either in size or detail, with our British shell so as to enable me to pass a decided opinion on the matter, and especially so as the specimens were not in a very perfect or satisfactory state of preservation. It is, therefore, still uncertain whether this large *Camarophoria* has been found in any of the Continental Permian localities.

[1] Unfortunately, in 1846, the term *multiplicata* was neither described nor accompanied with reference to locality or stratigraphical position; it therefore falls under the class of *Catalogue names,* which cannot claim priority should some subsequent author *describe* the same shell under a different name. I am therefore compelled to adopt that given by Mr. Howse on the 17th of August, 1848.

[2] 'Catalogue,' p. 252, 1848.

Family—STROPHOMENIDÆ.

Sub-genus—STREPTORHYNCHUS, *King.*[1]

In 1853 I was tempted to suppose Professor King's genus STREPTORHYNCHUS synonymous of ORTHESINA, *D'Orbigny*, but the perfect interiors, which I have recently studied, have led me to abandon that view, and to consider both to be characterised by certain well-defined peculiarities, which will be better understood by the comparison of the subjoined illustrations.

Fig. 1.   Fig. 3.   Fig. 2.

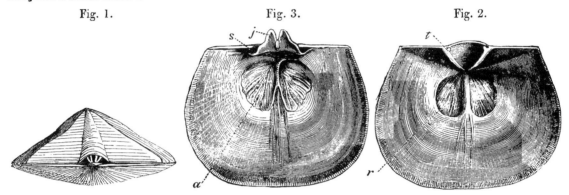

Streptorhynchus Devonicus, D'Orbigny, species.

Fig. 1. Both valves, seen from the beaks.   Fig. 2. Interior of the larger or ventral valve, as seen under the area.   Fig. 3. Interior of the smaller or dorsal valve.

Fig. 4.   Fig. 6.   Fig. 5.

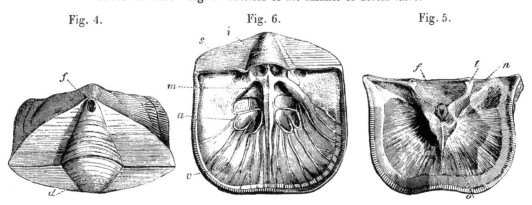

Orthesina anomala, Schlotheim, species.

Fig. 4. Both valves, seen from the beaks.   Fig. 5. Interior of larger or ventral valve, seen under the area.   Fig. 6. Interior of the smaller or dorsal valve.   *a.* Adductor scar.   *d.* A deltidium of the smaller valve.   *f.* Foramenal aperture, under side of deltidium.   *j.* Cardinal process.   *m.* Cavity on either side of the mesial septum.   *n.* Dental plates.   *o.* Horizontal plate, supported by a septum.   *r.* Cardinal muscular impressions.   *s.* Sockets.   *t.* Teeth.   *v.* Vascular impressions.

A 'Monograph of English Permian Fossils,' Palæontog. Soc., p. 109, 1850. Etym. στρεπτω, I bend or twist, and ρὑγχος, a beak.

The sub-genus *Streptorhynchus,* as typified by *Terebratulites pelargonatus* of Schlotheim (Pl. II, figs. 32—42), or by *Leptæna devonica,* D'Orb.,[1] may be characterised as an inequivalved shell, convex or concavo-convex, externally striated. The smaller valve is semicircular, the larger or ventral one possessing a prolonged and oftentimes bent or twisted beak; hinge-line rather shorter than the width of the shell. The area in the larger valve is triangular, with a fissure covered by a convex pseudo-deltidium. A small, narrow rudimentary area exists, likewise, in the smaller valve. No foramen is observable, but the cardinal process is at times seen partially extending under the deltidium, as in the woodcut (fig. 1, p. 29).

In the interior of the larger or ventral valve (Pl. II, figs. 40, 41; and woodcut, fig. 2) a strong hinge-tooth is situated on either side at the base of the fissure, supported by a dental ridge or plate; these diminish in size as they converge under the area towards the extremity of the beak (fig. 40; woodcut, fig. 2 *t*). At the bottom of the valve, under the beak, and extending a little beyond, are the impressions left by the cardinal and adductor muscles, which occupy about one third of the length of the valve; they form two elongated oval scars, more or less deeply excavated, and separated by a rather wide mesial ridge (fig. 40; and woodcut, fig. 2 R).

In the interior of the smaller or dorsal valve the cardinal process is largely developed (figs. 38, 39; and woodcut, 3), being composed of two testaceous projections, which are either slightly convex or concave on the side facing the interior, but grooved or bidentated towards the extremity of their outer surface (fig. 39); the socket-plates are large, and partly united to the lower portion of the cardinal process. Under these, on the bottom of the valve, are seen the quadruple impressions left by the adductor, which occupy more than a third of the length of the valve, and are arranged in pairs, divided by a short rounded mesial ridge.

---

[1] *Leptæna Devonica,* D'Orbigny, 'Prodrome,' vol. i, p. 90, 1849 = *Orthis crenistria,* var. *Devonica,* De Keyserling, from the Devonian beds of Ferques, which has long been confounded with *Orthis umbraculum,* V. Buch, from the Eifel, certainly belongs to the same section as that typified by *Streptorhynchus pelargonatus.* The woodcuts, figs. 2 and 3, have been made from interiors obtained at Ferques by Mr. Bouchard and myself, and are essentially similar in internal character to those of *S. pelargonata* (Pl. II, figs. 38—41).

The following differences have been pointed out by Mr. Bouchard as occurring between *O. umbraculum,* V. Buch, from the Eifel, and *O. Devonica,* from the Devonian beds of Ferques:

"The first is very constant in its shape, and in that of its area; is concavo-convex—that is to say, its smaller valve is convex, while the larger or ventral one is concave in all fully developed or adult individuals; on both valves the striæ are strongly marked and granulated; the area always regular and narrow.

"*Streptorhynchus Devonicus,* D'Orb., is, on the contrary, very inconstant in its external form: one does not meet with two similarly shaped individuals. It is always bi-convex—that is to say, both its valves are convex, their striæ smooth and not strongly marked. The area assumes every kind of shape, both in height and width; is very often irregularly twisted, being wider than long on one side than on the other; the beak being curved backwards, or inclined to one or to the other side, nothing appearing regular in the shell. Interiorly the details of both are similar."

Such are the general dispositions presented by the shells composing this sub-genus, and which denote its intermediate position between *Orthis* and *Strophomena*. Professor King lays great stress upon the peculiar twisted character of the beak, but it remains still to be ascertained whether this can be considered a common feature of the species which compose the group, or peculiar only to some specimens.

Now, if we compare the shell above described with those of D'Orbigny's *Orthesina*, of which woodcuts figs. 4, 5, and 6 are faithful representations,[1] the differences will be obvious. It will be seen that both valves possess large areas, with convex deltidiums, leaving no open or free space whatsoever between them at the hinge-line (woodcut, fig. 4), but in that of the larger valve, and towards its extremity, is seen a circular or oval aperture, which is in general cicatrized or closed in full-grown shells, but which was open up to a certain age, and evidently affording passages to pedoncular fibres of attachment. No similar aperture ever existed in *Streptorhynchus*, for no trace of such a foramen can be seen in any of the species and numerous specimens that have come under my observation, while it would appear to be a constant character in *Orthesina anomala*, *O. Verneuili*, *O. adscendens*, &c.; and if any fibres of attachment ever existed in Professor King's sub-genus, they must have passed through the small space left between the deltidiums and the cardinal process of the ventral and dorsal valves, a point not hitherto satisfactorily established.

In the interior of the larger valve of *Orthesina* there exists, as in *Streptorhynchus*, a strong hinge-tooth on either side at the base of the fissure; but instead of the small or rudimentary dental plate or ridge peculiar to the last-named sub-genus, in *Orthesina* these last are largely developed, and of a remarkable shape: they first project considerably into the interior, and converge towards the extremity of the beak (woodcut, fig. 5 N), and afterwards bend slightly outwards, and inwards, to form a third curved and connecting plate (o). This last is supported by a vertical mesial septum, which extends along the bottom of the valve to about half its length. On either side of this last are faintly-marked impressions, probably due to the cardinal muscle. Towards the extremity of the beak, on the under surface of the deltidium, is seen the foraminal aperture (*f*), surrounded by a raised margin; and the inner surface of the dental plates (N) probably afforded a surface for the attachment of the pedicle muscle, which must have been large, from the extent of the grooved surface. No such characters can be found in the larger valve of *Streptorhynchus*.

The smaller or dorsal valve of *Orthesina* has also its well-marked and peculiar features. Its large area and convex deltidium have already been described. At its base, on either side, are situated two sockets for the articulation of the teeth of the ventral valve: the cardinal process (J) is very remarkable; it is formed of a single projection, with two small

[1] I must express my warmest thanks to Count Alex. V. Keyserling for the kind exertions he made to procure me the means of studying the interior of *Orthesina*; and also to Professor Dr. Schmidt, of Dorpat, for the two beautiful interiors of *Orthesina anomala*, which he kindly gave me, and from which the woodcut representations, figs. 5 and 6, have been taken.

lateral depressions, which served no doubt as points of attachment to the cardinal muscles, and is, moreover, entirely covered by the deltidium, leaving a deep conical hole on either side. At a short distance under this is seen a large, strong, mesial ridge,[1] with two lateral ones; between these are situated, on the bottom of the valve, the quadruple impressions left by the adductor (A); two deep holes (*m*), similar to those situated on either side of the cardinal process, are likewise visible on either side, of the commencement of the mesial ridge last described.

These dispositions are also very different from those of *Streptorhynchus*, where the cardinal process is bilobed and completely exposed; nor are the other internal impressions similar in their details. Professor King was therefore justified in his views, although he was unacquainted at the time with several of the distinguishing internal characters of D'Orbigny's genus. Much more, however, remains to be discovered before the numerous species composing the family STROPHOMENIDÆ can be correctly classed, and we cannot be too cautious in our inferences, as the example just given will sufficiently demonstrate.

STREPTORHYNCHUS PELARGONATUS, *Schlotheim*, sp. Plate II, figs. 32—42; and King's Mon., pl. x, figs. 18—28.

TEREBRATULITES PELARGONATUS, *Schlotheim*. Akad. Münch., vol. vi, p. 28, pl. viii, figs. 21—24, 1816.

This little species, which in external shape bears much resemblance to several species of *Thecidium*, is more or less sub-conical or triangular. The smaller or dorsal valve is semicircular, and slightly indented in front, with a shallow mesial depression or furrow, commencing at a short distance from the extremity of the umbone; it possesses, likewise, a narrow rudimentary area. The hinge-line is shorter than the greatest width of the shell. The ventral valve is convex, semi-conical; the beak, in general largely developed, and at times exceeding the length of the smaller valve, is produced, elevated, bent backwards, and often irregularly twisted either to the one or other side. The area is large, triangular, with a fissure covered by a convex pseudo-deltidium. Externally the valves are ornamented by numerous small radiating raised striæ, which augment in number at variable distances from the beaks, both by intercalation and bifurcation, and are likewise intersected by numerous concentric lines of growth. The internal details have been alluded to under the generic characters.

The dimensions attained by this little shell vary considerably in different countries and

---

[1] Septa are rarely developed in the smaller valve of *Strophomenidæ*; but in one species from the Devonian beds of Ferques, *Orthis Deshaysii*, Bouchard, there exists a largely produced mesial plate, which, commencing under the cardinal process, extends almost to the frontal margin.

localities. Some British examples have measured 7 lines in length, 5½ in width, and 4 in depth, but is commonly a smaller shell; while a few foreign specimens have exceeded the dimensions above given.

The genus, or rather sub-genus, to which this interesting species should be referred has been the subject of considerable difference in opinion. Schlotheim placed it among his *Terebratulites*; J. de C. Sowerby made it his *Spirifer minutus*; Brongniart considered it to be a *Terebratula*; Von Buch named it *Orthis Laspii*; and Professor King created for its reception a new genus, *Streptorhynchus*. In 1853 I left it provisionally with *Orthesina*; and Professor M'Coy places the shell in *Leptæna* (1855). Thus, since 1816, this little species has been located into six different genera by different authors!

The fortunate discovery of several perfect detached valves by Messrs. Howse and Kirkby has enabled me to furnish some additional figures, which will serve to complete our knowledge of its internal characters.

Mr. Howse and Professor King state to have obtained the shell from the shell limestone of Tunstall and Humbleton hills, Dalton-le-Dale, Silksworth, Ryhope-field House, and in the Breccia of Tynemouth Cliff. In Germany Dr. Geinitz, Von Buch, Decken, and Professor Quenstedt mention it from the Lower Zechstein of Röpsen, Corbusen, and Schmerback, as well as from the Zechstein conglomerate of Könitz and Altenstein. Baron Schauroth describes the shell from Pössneck, and M. De Verneuil mentions having found it at Bielagorskaia.

Family—PRODUCTIDÆ.

Genus—PRODUCTUS, *Sowerby.*

(Introduction, Vol. I, p. 117, 1853.)

PRODUCTUS HORRIDUS, *J. Sowerby.* Plate IV, figs. 13—26. (King's Mon., pl. ix, figs. 29—31; and pl. xi, figs. 1—13.)

> PRODUCTUS HORRIDUS, *Sowerby.* Min. Con., vol. iv, p. 17, pl. 319, fig. 1, Jan., 1822
> = *Prod. clava*, Sow., Min. Con., pl. 560, figs. 2—6.

This well-known and characteristic fossil has been often described and illustrated.[1]

---

[1] De Koninck, 'Recherches sur les Animaux fossiles de la Belgique,' and 'Monographie du Genre Productus,' p. 158, pl. xv, fig. 1, &c., 1847. Geinitz, 'Versteinerungen,' p. 15, pl. vi, figs. 1—14, 1848. King, 'A Monograph of the Permian Fossils of England,' p. 87, 1850. M'Coy, 'British Palæozoic Fossils,'

It varies considerably in external shape, but may be said more or less sub-pentagonal or quadrangular marginally; wider than long in most individuals, but sometimes the reverse. The ventral valve is convex and gibbous, with prominent auricular expansions extending on either side of its somewhat attenuated but inflated and incurved beak: there exists, also, a mesial furrow, commencing at the extremity of the beak, and extending to the frontal margin. The smaller or dorsal valve is convex, and of moderate depth, with one or three undulations upon its surface, viz., a mesial and two lateral ones. The hinge-line is as long, or a little shorter, than the greatest width of the shell. There is no area on either valve,[1] nor articulating condyles. On the external surface of the ventral valve a variable number of long, rather large, and hollow spines are regularly scattered; on the dorsal valve they are confined to the vicinity of the cardinal edge, where they assume much regularity in both valves, from being arranged in one or two rows in the proximity of the hinge-line; they also project externally with an oblique angle, and become larger and longer as they approach the lateral margin.[2] The ventral valve is also obscurely ribbed in some specimens, and marked by numerous concentric lines of growth.

In the interior of the ventral valve (fig. 22) the muscular impressions occupy the larger portion of a pear-shaped space at the bottom of the valve, which is chiefly situated under the cavity of the beak. This space is longitudinally divided by a raised, slightly convex callosity, along the upper surface of which are impressed the ramified dendritic impressions of the adductor, longitudinally divided by a small mesial ridge (A of the internal cast, fig. 20, as well as on the beak of fig. 19). On either side there is a deep and strongly grooved or striated subquadrate impression, which is supposed to be due to the cardinal muscle (B of fig. 20[3]), the remaining surface of the interior of the

p. 465, 1855. Quenstedt, 'Handbuch der Petref.,' p. 490, tab. 39, figs. 26—30, 1851. Howse, 'Annals and Mag. of Nat. Hist.,' vol. xix, 2d series, p. 44, 1857, &c. To these works we must refer the reader for more copious details. This *Productus* has also received several names which will be found recorded in the synonyma published by Professors De Koninck and King, and which dates back as far as Hoppe, who, in 1745, was the first to describe the shell.

[1] M. De Koninck, who minutely and carefully describes every particular connected with the external form of this species, states, at p. 160 of his work, that most specimens are deprived of areas, but that with a certain number it exists in a rudimentary state; it appears to me to be but a thickening of the hinge-line.

[2] These spines are well described by Professor King, at p. 90 of his Monograph.

[3] Beautifully preserved internal casts of this *Productus* are not of rare occurrence in the shell limestone of Humbleton hill, and of which figs. 19 and 20 are careful illustrations. Fig. 19 represents the same view of a similar but much less perfect cast, figured by Sowerby under the name of *Productus clava*, in pl. 560, fig. 5, of the 'Mineral Conchology.' I considered it, therefore, desirable to offer faithful representations of these, as such had not been done by other authors, and the more especially so as upon them are seen, in relief, all those impressions which in the shell itself would appear in hollow, as in figs. 21 and 22, which are also drawn from admirably preserved detached valves, in the collections of Messrs. Howse and Kirkby. The internal details relating to this interesting fossil have been more or less completely described by M. De Koninck, Dr. Geinitz, Professor King, Mr. Howse, and others: but I trust that my series of illustrations may also be considered of some interest, as they have been drawn from the best

valve being covered by small pits and rugosities. In the interior of the smaller or dorsal valve (fig. 21) the cardinal process is produced beyond the level of the hinge-line, and varies somewhat in shape in different individuals (figs. 21 and 25, as seen from the interior of the valve; figs. 23 and 24, from the exterior). This boss is bifid at its extremity, with a slit or groove along the middle in each of the lobes facing the exterior of the valve. From the base of this process proceed three ridges or crests; one on each side runs almost parallel to the hinge-line, and the third forms a central mesial ridge or septum, which, becoming more acute and elevated at its extremity, extends to about two thirds of the length of the valve. On either side of this ridge are situated the muscular and reniform impressions; those left by the adductor are of a triangular shape, deeply grooved, and more or less elevated above the level of the valve.

In front of the muscular scars, and about the middle of the valve, are seen two elongated reniform impressions or callosities, bounded by small ridges, which, after dividing the adductors in a perpendicular manner near to the mesial septum, proceed outwardly in an almost horizontal direction, when turning abruptly, form an elongated half circle, which, becoming again horizontal, terminates close to the septum (fig. 21; and cast, fig. 19).[1]

Nearly half of the posterior internal surface of the valve unoccupied by the impressions above enumerated is minutely pitted, denoting, in all probability, ovarian spaces, while the remaining portion is covered with a multitude of short spinose tubercles, which become especially visible near the marginal portions of the shell. There is, also, a row of perforations, due to the cardinal spines, between the hinge-line and the upper pair of ridges which depart from the base of the cardinal process.

material procurable, and in them will be found certain details which have perhaps not hitherto been as carefully delineated. Professor King's description and figures of this shell in his Monograph are very excellent, but, with the exception of his fig. 10, chiefly relate to differences observable in the external shape; so that, by combining mine with his, every external and internal feature, I trust, will have been satisfactorily exemplified.

It must also be observed, that although the adductor impressions in the ventral valve are not so broad as the cardinal ones, they are much more elongated, and placed at variable levels in different examples; thus their lower extremity does not attain the level of the lower margin of the cardinal impressions, or extends beyond them in either direction.

[1] In his excellent description of the interior of the smaller valve of this *Productus*, Mr. Howse observes—" It is generally supposed that the reniform callosities are connected with the vascular system, but this supposition appears to be unsupported by a comparison of these processes with the corresponding parts of other Brachiopods. If we compare them, for example, with the same valve of *Argiope* or of *Thecidium*, genera which show points of resemblance to this in several particulars, we find that these processes have served for the attachment of the oral arms; and this view is supported by all that we know of the position of these arms in both recent and fossil genera,—at least these structures cannot be attributed to the vascular system, or they would undoubtedly be present in both valves, whereas they are confined to one. On some casts of this *Productus* a great number of small parallel grooves or furrows are seen running from the central ovarian region to the anterior margins of the valves; they are not very distinct, but

Productus horridus abounds in the shell limestone of Humbleton and Tunstall hills; at the North Point, near Shields; Westol, Hylton Castle, Dalton-le-Dale, Durham; and in the compact limestone of Midderidge, Garmundsway, as well as in the Breccia at Tynemouth Cliff. It is also mentioned from Derbyshire. On the Continent it is found in the lower Zechstein of Röpsen, near Gera; Ilmanau, Könitz, between Bucha and Gosswitz, Glücksbrunn; and on the meridial border of the Shüringer Wald, Pössneck; Corbusen, near Ronnebourg; in Silesia, at Lauban on the Gneis. It was also recognised by M. De Koninck among the fossils brought back by M. E. Robert from Bell Sound, at Spitzberg, &c.

PRODUCTUS LATIROSTRATUS, *Howse.*[1]   Plate IV, figs. 1—12.

> PRODUCTUS LATIROSTRATUS, *Howse.* Catalogue of the Fossils of the Permian System of the Counties of Northumberland and Durham, in the Transactions of the Tyneside Naturalists' Field Club, vol. i, part iii, p. 256, 17th of August, 1848; and Ann. and Mag. of Nat Hist., vol. xix, 2d series, p. 46, pl. iv, figs. 1, 2, 1857.

they may perhaps hereafter be found to have been connected with the vascular system. This idea is somewhat strengthened by the appearance of similar lines on some casts of *Spirifer*, which few persons would hesitate to pronounce as vascular sinuses." (' Annals and Mag. of Nat. Hist.' vol. xix, p. 46, 1837.)

This view of Mr. Howse's relating to the interpretation of the reniform impressions is new, and would require some further investigation and confirmation before final admission, as it does not entirely agree with the notion entertained by Mr. Woodward, myself, and others, as to the probable form and position of the fleshy arms, which appear to have produced peculiar markings visible in the interior of both valves of several species of *Productus;* and in *P. giganteus* in particular, would appear to have been small, and vertically coiled. The singular markings seen in the larger valve (*s*) are no doubt imprints of those processes; and similar appearances are also visible in valves of *Strophomena rhomboidalis* and *Davidsonia Verneuilii*, where no reniform impressions exist.

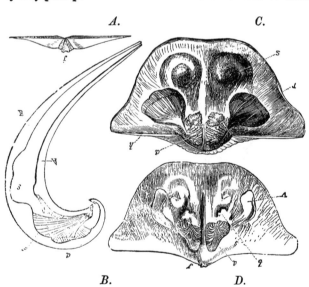

Productus giganteus (from Mr. Woodward's drawings).

A. Interior of dorsal valve.   *B.* Interior of ventral valve, with the umbo removed.   *C.* Ideal section of both valves.   *D.* Hinge-line of dorsal valve.   *J.* Cardinal process.   *a.* Adductor.   *r.* Cardinal muscles.   *s.* Hollows occupied by spiral arms.   *v.* Reniform impressions.   *b.* Brachial processes.   *h.* Hinge-area.

---

[1] This species seems to have been discovered independently by Mr. Howse and Professor King. Mr. Howse claims the adoption of his name on account of the two days' priority of publication.

PRODUCTUS UMBONILLATUS, *King*. A Catalogue of the Organic Remains of the Permian
Rocks of Northumberland and Durham, p. 8, 19th of
August, 1848; Monograph of English Permian Fossils,
p. 92, pl. xi, figs. 14—18, 1850; *Anlosteges umbonillatus*, King, Annals and Mag. of Nat. Hist., vol. xvii,
2d series, pl. xii, fig. 6, March and April, 1856.

More or less sub-quadrate and concavo-convex, without articulating condyles; the larger
valve is slightly convex, and rather flattened when young (fig. 1); more gibbous, and
sometimes geniculated, when adult (fig. 4), with a shallow sinus commencing at about the
middle of the valve, and extending to the frontal margin. The beak is not much inflated,
nor is it produced beyond the extremity of the cardinal edge; the auricular expansions
are hardly defined, the hinge-line being in general shorter than the greatest width of the
shell; the margin is also more or less thickened, so as to simulate a false area, with a
narrow uncovered fissure? The external surface of the valve is ornamented by a small
number of irregularly scattered spines, which attain at times a considerable length. The
dorsal or smaller valve is very slightly concave, with a small mesial wave.

In the interior of the ventral valve the muscular impressions occupy a pyriform space
at the bottom of the valve; those produced by the adductor are oval, narrow, and placed
on a mesial elevation; on either side of these are seen the longitudinally striated sub-
quadrate impressions left by the cardinal muscle. In the interior of the dorsal valve the
hinge-line margin is more or less flattened (figs. 9 and 10); the cardinal process, which
projects at almost right angles to the plane of the valve, is bifid, with a deep slit or groove
along the upper surface of each lobe. Under this a narrow longitudinal ridge extends to
nearly two thirds of the length of the valve, and on either side are seen the scars produced
by the adductor and reniform impression (figs. 5, 6). The remaining unoccupied surface
of the interior is minutely pitted on the posterior half, and covered with produced tubercles
on the anterior portion, and especially in the vicinity of the frontal margin.

The largest British examples I have seen measured 12 lines in length by 16 in breadth,
but the shell is in general of smaller dimensions.

This species is well distinguished from *Productus horridus* by its external and internal
details, as will be at once perceived from a comparison of the illustrative series of figures
given in my Pl. IV, and which were drawn from the best examples hitherto discovered by
Mr. Kirkby in the shell limestone of Tunstall hill, and by Mr. Howse in that of Dalton-le-
Dale, Durham.[1] Specimens in which the shell is preserved appear rare; but beautifully
perfect internal casts are rather more abundant, and upon them, in relief, may be studied

[1] Mr. Howse observes, that "it differs from its congener in several important particulars. The boss
or muscular fulcrum, the shape of the muscular impressions, the greater size of the oral arms" (Mr. Howse
alludes here to the reniform impressions), "the absence of cardinal spines on the upper valve, the flanging of
the hinge-margin of the upper valve, are so strongly characterised, that it cannot be mistaken for any other
species."

all the peculiarities of its internal organization (figs. 5 and 11). Although correctly described in 1848 and subsequently, it has nowhere been properly illustrated, for Professor King's figures were evidently taken from very imperfect material. The species was also discovered at Pössneck, in Germany, by Baron Schauroth, who published some illustrations of it in his first 'Beitrage.'[1]

Professor King is of opinion that the shell under description should be referred to General Helmersen's *Aulosteges*,[2] on account of the rudimentary area visible in some examples, but this appearance seems to be more deceptive than real in the greater number of specimens; nor have I any grounds to suppose that the narrow fissure observable in some specimens was covered by a deltidium, and the internal details, although slightly different, are essentially the same as those peculiar to *Productus;* and Professor King seems to be somewhat imbued with the same opinion, since he states, in his paper in the 'Annals' (1856), " I have represented the interior of the small valve of (?) *Aulosteges umbonillatus*, on which it will be seen these structures (the reniform impressions) are situated as in *Productus*."

Sub-Genus—STROPHALOSIA, *King.*

(Introduction, Vol. I, p. 115, 1853.)

The identification or fixing of the species to which our English *Strophalosias* should belong has given rise to considerable difference of opinion. I have therefore used every effort in my power, in the hopes of arriving at a satisfactory conclusion, by a prolonged examination not only of a vast number of specimens of our British and foreign equivalents, but also by a direct correspondence and discussion with those who have studied the subject with most attention.[3]

It must, however, be confessed that I have experienced considerable uncertainty and difficulty in arriving at a conclusion relative to some of the points upon which the difference of opinion principally prevails, and would therefore invite the reader to satisfy himself before adopting the suggestions here proposed. Some authors would divide the

[1] 'Ein Beitrag zur Fauna des Deutschen Zechsteingebirges,' 1853.

[2] Professor King considers the sub-genus *Aulosteges* to be represented by the following species : *A. Wangenheimi*, Vern. and Keyserling, sp., *P. umbonillatus*, King, and *Strophalosia tholus*, Keyserling (in Schrenk's 'Reise durch die Tundren der Samojeden,' vol. ii, p. 103, p. ii, figs. 18—21, 1854.

[3] The authors whom I have particularly consulted are Professor King, Mr. Howse, Mr. Kirkby, Count A. Von Keyserling, Professor L. de Koninck, Dr. Geinitz, and Baron Schauroth; but it has been impossible to arrive at a unanimous conclusion, probably from the different manner in which a species is viewed, some considering certain differences of specific value, while others view the same as accidental, or at most varietal.

English *Strophalosias* at present known into four or five species,[1] while others would feel disposed to admit but two or three. Assuming, therefore, that the *Strophalosias* in question belong but to two species, and that these are properly identified as the *Strophalosia Goldfussi*, Münster, and *S. lamellosa*, Geinitz, there exists, in my opinion, two or three varieties deserving a special notice, as well as a varietal designation. I have therefore described these separately, and given the different opinions entertained, in order that the reader may have a better opportunity of judging for himself as to the grounds upon which a larger number of species have been maintained, as the subject may still afford grounds for further discussion.

STROPHALOSIA GOLDFUSSI, *Münster.*   Plate III, figs. 1—10.

> SPONDYLUS GOLDFUSSI, *Münster.*   Beitrage, vol. i, p. 43, pl. iv, fig. 3 *a*, *b*, 1839.
> ORTHIS EXCAVATA, *Geinitz.*   Neues Jahrback for 1842, pl. x, p. 578, figs. 13, 14.
> PRODUCTUS GOLDFUSSI, *De Koninck.*   Monographe du genre Productus, p. 148, pl. xi,
> fig. 4 ; and pl. xv, fig. 4, 1846.
> ORTHOTRIX GOLDFUSSI, *Geinitz.*   Versteinerungen, pl. v, figs. 27, 28, April, 1848.
> STROPHALOSIA GOLDFUSSI, *King.*   Monograph, p. 96, pl. xii, figs. 1—12, and 23?
> 1850.
> —   PARVA ? *King.*   Mon., p. 102, pl. xii, fig. 33, 1850.
> —   GOLDFUSSI, *Howse.*   Annals and Mag. of Nat. Hist., vol. xix, 2d series,
> p. 47, 1857.

Very variable in its external shape, generally more or less sub-triangular (figs. 5, 7), subquadrate (fig. 6) or oval, but less commonly circular; its anterior angles are rounded,[2] and at times slightly indented in front. The hinge-line is always much shorter than the width of the shell; the beak of the larger valve tapering a little, and often twisted more to the one than to the other side; area narrow (figs. 1, 6, 14), with the extremity of the beak slightly protruding beyond its extremity, or large, triangular, and more or less bent backwards (figs. 3, 5, 8), varying both in height and breadth, with a narrow fissure covered by a

---

[1] Professor King writes me on the 29th of July, 1857—"My present view of Permian North of England *Strophalosias* is, that there are the following species : *S. excavata*, *S. Goldfussi*, *S. Humbletonensis*, *S. Morrisiana*, and (?) *S. parva*. I doubt much that *Orthotrix lamellosa* and *Cancrini* of Geinitz and Germany are the same as my *Morrisiana* and *Humbletonensis*. The *lamellosa* does not appear to occur in the North of England."

[2] Professor King alludes to the great variability presented by the different individuals composing the present species. At p. 97 of his Monograph he also mentions that the casts so abundant at Ryhope-field House differ from those found at Humbleton hill, in having the larger valve more rounded, with a very small indication in some specimens of a median furrow. This fact has also been confirmed by Messrs. Howse and Kirkby, who have assembled a vast number of specimens from both localities. The length of the hinge-line is often not more than one third of the breadth of the shell. There appears also to exist every passage from those examples which are longer or shorter than wide to those which are almost circular ; some are irregularly and scantily spined, while others have the spines closely and regularly placed.

convex pseudo-deltidium. There is also a small area in the dorsal valve. The ventral
or dental valve is moderately convex, with or without a shallow mesial furrow or sinus;
the dorsal or socket one being moderately concave, and following the curves of the opposite
valve, that is to say, with or without a slight mesial elevation corresponding with the sinus
in the ventral valve. Externally the whole surface (areas excepted) is covered with a
multitude of curved slender tubular spines, which attain a considerable length, and are
irregularly placed at not much more than their thickness one from the other (fig. 13).
These spines in the larger valve lie, from their origin, either close to the surface, or vertically
implanted, become suddenly bent downwards so as to become entangled one with the
other;[1] but those on the smaller valve seem in general to converge towards the centre of
the shell,[2] the valves being likewise marked by numerous incremental lines or wrinkles of
growth. When quite young the shell is much depressed, the smaller valve being almost
flat and circular, while the umbone is comparatively produced. *Strophalosia Goldfussi*
varies also in proportions, but does not often exceed 1 inch in length, and somewhat less
in width; but when covered with its spinulose appendages, presented larger proportions.

In the interior of the larger valve a tooth is situated on each side at the base of the
deltidium, which articulates by the means of sockets placed on both sides of a cardinal
process in the opposite valve (fig. 12); and at a small distance from the extremity of the
beak there is a short, raised, slightly curved platform, with a somewhat obtuse abrupt
termination towards the centre of the valve. On this are placed the two small oval scars
produced by the adductor, and separated by a minutely elevated line or ridge (fig. 14),
better seen in hollow or on internal casts (figs. 10 and 13 A). Immediately under, but
outside, there are two longitudinal, slightly indented, and sub-quadrate semicircular
impressions, produced by the cardinal muscle (fig. 10[3]), and which seem to have been

[1] M. De Koninck observes that sometimes the spines, while interlacing each other, surround foreign
objects which may lie within their reach, and that those situated on the beak appear to have possessed
that faculty in particular. He possesses specimens of which the spines envelop the fragment of a large
spine of *Productus horridus*, and Dr. Geinitz has figured another similar example.

Mr. Howse informs me that the typical German *Goldfussi* from the Zechstein differ somewhat from
those we are accustomed to call *Goldfussi* in England. They are shapeless at first until well cleaned, like
Münster's figure, and somewhat more circular, and densely covered with very curved spines. They occur
in little marly nodules, out of which they are washed with a tooth-brush into very fine specimens.

[2] At least such is seen to be the case in many German specimens, and has been so described by
Professor De Koninck in his 'Monographie du genre Productus.' A beautiful illustration by M. Suess
will be found in pl. v, fig. 10 *b*, of the German edition of my General Introduction.

[3] The remarkable internal cast, of which I offer an enlarged illustration in Pl. III, figs. 9, 10,
was discovered by myself at Humbleton quarry, during a visit to that locality in company of Messrs. Howse
and Kirkby, and it is worthy of remark that, although those gentlemen and Professor King have collected
some hundred of specimens in the same and in other localities, as well as by Baron Schauroth in the
Zechstein Dolomit of Pössneck, in Germany, in none were these scars so well or so distinctly displayed as
in the specimen here represented, and which I had already figured and described in my few notes on
Permian genera, published in vol. ii of the 'Bulletin de la Société Linneenne de Normandie,' Feb., 1857.

overlooked or unobserved by those who have described the interior of this species, as they are not represented in any of their illustrations. These impressions are also much less deeply marked in *Strophalosia* than in *Productus*, but prove that they were in general character essentially similar to those of that genus, varying only to a small extent in shape and depth.[1] The position of the adductor in this valve has been noted by several writers, and is represented in pl. xii, fig. 3, of Professor King's 'Monograph;' but the author describes them " as the cardinal and valvular muscular scars confluent," which appears to be a mistaken view from the evidence afforded by my specimen, in which both are distinctly separated, as in *Productus*.

In the interior of the smaller or dorsal valve the cardinal process is of moderate dimensions, but considerably developed and produced in certain individuals (figs. 12, 16, 17), from the base of each of which proceed three strong ridges, viz., a median longitudinal one, which extends to about half the length of the valve, and two short lateral ones, which constitute the inner socket-walls. On either side of the median ridge or septum there are two small longitudinally divided pear-shaped impressions, produced by the adductor, the one close to the central ridge being the largest (figs. 8 and 12 A). The reniform impressions are large, their prominent outer edge issuing from between the adductor scars, gradually arch forwards and outwards on each side, then turning backwards for about half their length, and finally run inwards horizontally to meet each other near the extremity of the median septum or ridge. The interior surface of this valve is either uniformly convex, or presents a small medial depression or sinus towards the front (fig. 12), while in a few young individuals there is a raised margin, somewhat similar to that seen in different forms of *Thecidium*, and of some other Brachiopoda (fig. 17[2]). The internal surface is also minutely pitted all over, with the exception of those parts occupied by the muscular scars.

Strophalosia Goldfussi is stated by Professor King and Mr. Howse to have been found in the compact and shell limestone of Humbleton and Tunstall hills, Clack's Heugh, Ryhope-field House, Dalton-le-Dale, Garmundsway, Castle-Eden Dene, and in the Breccia of Tynemouth.

On the Continent it seems equally common. Count Münster's original example was obtained from the lower Zechstein of Röpsen, near Gera. Dr. Geinitz mentions Milbitz,

[1] The muscles were, perhaps, stronger and more deeply implanted in *Productus*, to supply the deficiency of the articulated hinge which *Strophalosia* possesses. The cardinal process appears to be also rather larger in *Productus* than in the last-named genus. Mr. Howse observes, that " the boss or cardinal muscular fulcrum (in *Productus*) when *in situ*, fills nearly the whole of the umbonal cavity of the lower valve, and presses against its inner surface; it may thus assist in keeping the valves in position."

[2] I have been able to study this curious peculiarity in four or five examples from Tunstall hill, in the collection of Mr. Howse. They consist of beautifully preserved detached valves of young individuals, not exceeding three and a half lines in length; but in all the larger individuals which have come under my observation, the raised margin has either entirely disappeared, or exists only in a very obscure and rudimentary condition.

Corbusen, &c.   Baron Schauroth has found it abundantly in the Zechstein dolomit of Pössneck, where it appears associated with the same species found at Humbleton and in some other of our magnesian shell limestone localities.

In 1839 Münster figured, under the name of *Spondylus Goldfussi*, a specimen evidently belonging to the shell under description.   It is longer than wide, concavo-convex, and covered with spines; but as much of the beak and area was obscured by matrix, the German author was not quite certain as to the exact place his shell should occupy; so that, while placing it in *Spondylus*, he at the same time took care to add, that had it been free he should have taken it for a *Productus*.   The shell under description is the one very generally known on the Continent to represent the species of the German Count, which either did or did not possess a mesial furrow in the larger valve, so that this last-named feature could not be considered as a character of any specific importance.

In 1842, Dr. Geinitz described and figured a small shell under the name of *Orthis excavata*.[1]   It was longer than wide, with a straight and large triangular area, narrow fissure, pointed beak, and plano-convex spinulose valves.   This shell has appeared to me, as well as to Professor De Koninck and Mr. Howse, to be only a young condition of *Strophalosia Goldfussi,* and in our opinion should therefore be placed among the synonyms of that species; but Professor King dissents from this view, and seems disposed to consider it not only a distinct species, but also the same as that subsequently named *Prod. Lewisiana* by Professor De Koninck, 1846.   At a later period, in 1848, it is true Dr. Geinitz published figures of his so-termed *excavata*, differing somewhat from the original type, and which might be either taken for *S. Goldfussi* or its variety *S. Lewisiana*, and from this unfortunate coincidence much of the confusion and uncertainty has arisen that prevails at present among these Strophalosias.

I therefore consider it necessary to restrict my inferences upon this matter to the

---

[1] In the 'Neues Jahrbuch' for 1842, p. 578, figs. 12 and 13 *a*, *b*, Dr. Geinitz has omitted to represent the tubercular prominences left by the broken spines, so that his figures might give the idea of a smooth species; but in the accompanying description the author does not omit to allude to their existence.   Professor King kindly forwarded for my examination three or four examples closely agreeing with Geinitz's figure, and which he had obtained when in Germany; but after a minute examination of these (evidently young shells) with fig. 3 of my Pl. III, which certainly belong to Münster's species, I could perceive no differences of sufficient importance to warrant my supposing them specifically distinct.   In a letter written by Dr. Geinitz to Professor King, dated the 11th of March, 1857, the German author states that *Stroph. Goldfussi* "*is without sinus,*" which observation cannot be considered strictly correct, for although it may not be perceptible on certain individuals, it exists to a greater or lesser degree on the larger number.   Dr. Geinitz also mentions that his *Orthis excavata* of 1842 is the same as the one described under the name of *Lewisiana* in 1846, and by Professor King under that of *Stroph. excavata* in 1850; but I fear I must object to that interpretation.   Dr. Geinitz further observes that *Stroph. excavata* possesses "*a sinus more or less only in the upper Zechstein,*" but his figures of *O. excavata* show none; and in the character published in 1842, he distinctly states that the larger valve is regularly convex, and that the smaller one is similarly concave, and there is no mention of any sinus.

original representations published in 1839 and 1842, and finding these to belong to the same species, it appears to me desirable, in order to obviate further confusion, to erase the term *excavata* altogether; and if the shell described as *St. excavata* by King (figs. 19—22 of my plate) is to be either viewed in the light of a separate species or variety of *Goldfussi*, the term *Lewisiana*, De Koninck, is the one which should be adopted. So little is known of the shell termed *Strophalosia parva*, King,[1] that I should not consider myself justified in admitting it as a distinct or well made out species, and have therefore followed Professor M'Coy and Mr. Howse in leaving it, provisionally, among the synonyms of *St. Goldfussi*, which it appears to resemble; for the specimen in the Museum of Practical Geology (fig. 18) proves that it was not always " irregularly circular marginally," as stated by Professor King, since the individual before us is sub-trigonal, as is the case with the larger number of *St. Goldfussi*. It has also been ascertained that the spines in Münster's species very often attain the comparative length displayed in King's illustration, as well as in the one I have represented.

Var. LEWISIANA, *De Koninck.*   Plate III, figs. 19—22.

> PRODUCTUS LEWISIANUS, *De Koninck.* Monographie du genre Productus, p. 150, pl. xv, fig. 5, 1846.
> ORTHOTRIX EXCAVATUS, part, *Geinitz.* Versteinerunger, pl. v, fig. 38; pl. vi, figs. 20, 21, 1848.
> PRODUCTUS SPINIFERUS, *King.* Catalogue of the Organic Remains of the Permian Rocks of Durham, &c., p. 8, 1848.
> STROPHALOSIA EXCAVATA, *King.* Monograph, pl. xii, figs. 13—17, 1850.

Almost circular or slightly transversely oval marginally, and about as wide as long, regularly concavo-convex : hinge-line shorter than the width of the shell, with a narrow area in both valves, that of the larger one possessing a small fissure, covered by a pseudo-deltidium. The ventral valve is generally uniformly convex, but in some examples presents a narrow, very slightly marked, median depression on the anterior portion of the valve. The beak is regularly rounded, incurved, and varying in its proportions : the smaller valve is concave, following the curves of the opposite one. Externally the entire surface of the valves is covered with a vast number of long, slender, tubular spines, which are situated sometimes very uniformly in quincunx, but in the larger number of individuals this extreme regularity does not appear to exist. The internal details are exactly similar to those of the typical *Goldfussi*. In dimensions this variety does not in general exceed 10 lines in length, by about the same in width.

Professor De Koninck, Dr. Geinitz, and Professor King consider the shell under

[1] King's 'Monograph,' pl. xii, fig. 33, 1850.

description to be specifically distinct from *St. Goldfussi*; and Baron Schauroth informs me that he is also of opinion that, although *Strophalosia Lewisiana* and *St. Goldfussi* are closely related, the first can always be distinguished by its more circular shape, as well as by its lower and smaller area. Messrs. Howse and Kirkby, on the contrary, state that, after the careful study of this form, they cannot find one fixed character by which to distinguish the shell we are now describing; that the two extreme forms I represent (Pl. III, figs. 3 and 19—22) graduate by almost imperceptible degrees into each other, having in the regular form a narrow area and incurved beak, and in the other a larger area and a depressed beak. I have therefore considered the shell under description as a variety of *Goldfussi*, with the denomination of *Lewisiana*, so that it may be thus retained, or specifically separated, should such a course be considered desirable.[1]

In a paper published in the 'Annals of Natural History' for March and April, 1856, Professor King has proposed to consider the shell (fig. 23 of our Pl. III) as a variety of *Strophalosia excavata*, to which he has applied the varietal term *Whitleyensis*, but from such imperfect and insufficient material as the cast of a single dorsal valve, we can decide with but little certainty; it may or not belong to *Stroph. Goldfussi* or its var. *Lewisiana*.

The variety *Lewisiana* has been found in the same beds and localities already recorded for *Stroph. Goldfussi*, both in this country as well as on the Continent, but is not so abundant as the typical form of Münster's species.

STROPHALOSIA LAMELLOSA, *Geinitz*. Plate III, figs. 24—41.

> ORTHOTRIX LAMELLOSUS, *Geinitz*. Versteinerungen, p. 14, pl. v, figs. 16—26, April, 1848.
> STREPHALOSIA MORRISIANA, *King*. Catalogue of the Organic Remains of the Permian Rocks of Northumberland and Durham, p. 9, 19th of August, 1848; Monograph of English Permian Fossils, p. 99, pl. xii, figs. 18—25, and 27—32; Annals and Mag. of Nat. Hist., vol. xvii, 2d series, March and April, 1846.
> LEPTÆNA CANCRINI, *M'Coy*. British Palæozoic Fossils, p. 457, 1855 (not *Productus Cancrini*, Verneuil and Keyserling).

[1] It has been argued that the regular form of a shell should be taken as the type of the species, and not the irregular growth, and I quite coincide in the opinion; but when we view a species under its general aspect, and find that the great majority of individuals of which it is composed are not perfectly regular, the beak being more often twisted a little to the one side or to the other, and that the shell is more often sub-trigonal, and rarely circular as in *Stroph. Goldfussi*, with or without a slight mesial elevation in the smaller valve, and corresponding depression or sinus in the opposite one, we feel bound to select our typical shape from the normal condition, or the one that gives the best representation of the general character of the species, and not only from one particularly favoured individual who may have been perfectly symmetrical in all its parts.

STROPHALOSIA MORRISIANA, *Schauroth.* Ein neuer Beitrag zur Palæontologie des
deutschen Zechsteingebirges, p. 221, 1856.
CANCRINI, *Howse.* Annals and Mag. of Nat. Hist., vol. xix, 2d series,
p. 49, 1857 (not *Productus Cancrini,* Verneuil and
Keyserling).

Shell nearly circular, or slightly oval marginally when full grown : concavo-convex,
and in general about as wide as long, the greatest breadth being situated towards the
middle or anterior portion of the shell. The ventral or larger valve varies much in degree
of convexity. It is evenly convex and without sinus in well-shaped individuals, while in
others the greatest gibbosity is situated towards the beak. The margin of the valve is
likewise sometimes either slightly bent upwards, or suddenly downwards, while in other
specimens the convexity is very moderate. The beak is regularly shaped, evenly rounded
and incurved in the larger number of full-grown specimens from certain localities (fig. 35),
showing no indication whatever of having adhered by any portion of its surface (var.
Humbletonensis, King) ; but other examples less favorably placed or developed in the
same and in other localities, show that the shell adhered or was adpressed to a greater or
lesser extent (fig. 25) ; and this is particularly the case with distorted shells (figs. 28—30),
where the beak is truncated, and more or less irregular in shape and growth (*Stroph.
lamellosa,* type, and var. *Morrisiana,* King). The hinge-line is much shorter than the
width of the shell ; the area in the larger valve is narrow, and sometimes hardly visible in
those shells which possess an evenly rounded and incurved beak. It is also in such cases
linear and scarcely perceptible in the opposite valve, but becomes larger and better
developed in both valves in those less-perfectly shaped individuals, in which the beak
adhered to some foreign body, or did not extend beyond the cardinal edge (figs. 25 and 31).
The external surface of the large valve is ornamented at variable intervals by long,
adpressed, tubular, creeping spines, directed outwardly and downwardly, and often
adpressed for a distance of eight or nine lines before rising from the surface of the valve
(figs. 34 and 36), so that, when perfect, some of the spines must have exceeded an inch
in length ; but those on the lateral portions of the beak and cardinal edge stand
erect from their origin, or are slightly directed backwards. The surface of the valve
is also more or less distinctly covered by a multitude of minute, radiating, raised striæ,
which increase in number by intercalation and bifurcation (figs. 40, 41), and are sometimes
so closely packed, that towards the middle of a specimen measuring one inch in diameter,
from eighteen to twenty may at times be counted in the space occupied by a couple of
lines (fig. 40) : they are also more or less irregular in their respective widths, and at times,
by the uniting of two or three, form the base of a spine (fig. 41).

The dorsal valve is, in general, moderately or very slightly concave, and almost flat,
in very young shells, with a slight convexity at the umbone. The external surface also
presents numerous radiating striæ, with small indented impressions at various intervals
(fig. 35). In the generality of specimens no spines existed on this valve, but that some

few short ones occasionally occurred seems demonstrated from their presence in a certain number of specimens recently discovered at Ryhope by Mr. Kirkby (fig. 42). Numerous lines, laminæ, or wrinkles of growth, occasionally occur on the surface of both valves, and particularly so in specimens from certain localities in the lower Zechstein, where the shell assumed dwarfish dimensions, and lived under different conditions to those that existed during the deposition of the upper shell limestone.[1]

Interiorly the valves articulate by means of teeth and sockets; and as the muscular and reniform impressions agree in arrangement and character with those of *St. Goldfussi*, already described, they need not be again reproduced (figs. 37 and 39).

In dimensions this species does not often appear to exceed 15 or 16 lines in length, by about the same in breadth, the greatest depth between the valves rarely exceeding 6 lines.

Considerable difference of opinion has been expressed regarding the shells here classed under *Orthotrix lamellosus*, of Geinitz, a species founded on a dwarf race, of which *Strophalosia Morrisiana*, var. *Humbletonensis*, may be considered the full grown and most favoured form. Geinitz's shell (of which I have seen several typical examples, found near Gera by Mr. Howse, as well as others from Moderwitz, &c., forwarded by Baron Schauroth) never appears to have greatly exceeded 7 or 8 lines in length and breadth; and they also very often exhibit indications of having adhered by portions of their beak, but otherwise possess all the essential or important characters of the more favoured varieties, for their larger valve is sparingly covered with long adpressed spines, and more or less distinctly marked radiating striæ. The smaller valve also is generally spineless, but striated as in the more perfect shells; but in this dwarf race or condition the surfaces of both valves seems more closely marked with larger and more prominent incremental lines, or laminose projections, than in the var. *Humbletonensis*, King. While describing his *Strophalosia Morrisiana*, in p. 100 of the 'Monograph of English Permian Fossils,' Professor King places *Orthotrix lamellosus* with a point of doubt among the synonyms of his species, and adds—"In making the *Orthotrix lamellosus* of Geinitz synonymous with this species, notwithstanding the former is stated to be without spines on the small valve, I have been influenced by certain of my specimens displaying lamella on this valve, somewhat similar to those represented in the 'Versteinerungen,' at figs. 15 a, 16, 17 a, 21, pl. v. The fossils identified by Dr. Geinitz (*vide* pl. vi, figs. 16—18) with the *Stroph.*

[1] It may be remembered that Professor King, in p. 101 of his 'Monograph,' has alluded to a singular peculiarity presented by some examples found at Humbleton hill, of appearing possessed of three valves, a circumstance he feels at a loss to account for satisfactorily. Mr. Howse (in his paper published in the 'Annals' for 1857) remarks, "there is also a tendency in this species to form a new internal surface behind the old upper valve, for the purpose of contracting the interior of the shell. It is not an additional third valve as King has supposed, for it is essentially connected with the upper valve, and must have been formed by the upper lobe of the mantle." I have seen several examples presenting this peculiarity, but it is only of rare occurrence, and is therefore not a general character of the species.

Cancrini, I readily recognize as belonging to *S. Morrisiana.*" Again, in his paper in the
' Annals' for March and April, 1856, Professor King observes—" I suspect all the specimens described and figured by Geinitz as *Orthotrix lamellosus* and *Prod. Cancrini* belong
to var. *Humbletonensis.*" From the above it is evident that Professor King suspected
Stroph. Morrisiana and *lamellosus* to be synonymous, but from not being cognizant
with Geinitz's priority, had placed the German author's name among the synonyms of
his own. Baron Schauroth, in his late publication, considers *S. lamellosa* a variety of
Morrisiana; and Mr. Howse is now strongly convinced that the two are synonymous.

Strophalosia Morrisiana has been several times confounded with *Productus Cancrini*,
a shell not only specifically but generically distinct. In 1855, Professor M'Coy felt
confident as to King's species being a synonym of *Cancrini*, which opinion was also reproduced in January, 1847, by Mr. Howse; and Professor De Koninck was strongly biassed
towards a similar conclusion. Professor King, on the contrary, has always maintained
that both were distinct although closely related species.

In order to arrive at a positive conclusion, I obtained from Dr. Schrenk, through the
mediation of Count A. von Keyserling, a perfect example of the Russian *P. Cancrini*,
brought from Usty Joshuga, near Archangel (see woodcut in p. 48), and for which kindness
I feel much indebted to my two Russian friends. On its arrival in England, I lost no time
in submitting the specimen to the attentive examination of Professor King, Mr. Howse, and
Mr. Kirkby, and they all agreed with me that it was a true *Productus*, while Geinitz's
lamellosus and King's *Morrisiana* belonged to the genus *Strophalosia*, these shells differing
in every particular except in that of the striæ, a character that appears to have been the
source of all the mistaken identifications.

Productus Cancrini may be at once distinguished from all the varieties of *Strophalosia
lamellosa* by its total absence of area and denticulated hinge;[1] for although the first is
small and scarcely perceptible, on account of the incurvation of the beak, in the larger
number of *Strophalosia lamellosa*, var. *Humbletonensis*, it is clearly visible; the hinge-line,
which in the last is comparatively much shorter, is also distinctly articulated by the means
of teeth and sockets; it differs also in general character, the spines on the back of the
larger valve in *P. Cancrini* do not appear to have been so long, adpressed, or creeping, as
in *St. lamellosa* and its varieties, and the numerous erect cardinal spines are altogether

---

[1] In order to facilitate comparison, I have here described the Russian example of *Productus Cancrini*,
kindly forwarded by Dr. Schrenk and Count Keyserling:

> *Productus Cancrini*, Murch. De Verneuil and De Keyserling, Russsia and the Oural
> Mountains, vol. ii, p. 273, pl. xvi, fig. 8 *a, b;* pl. xviii, fig. 7,
> 1845.
> — — De Koninck. Monographie du genre Productus, p. 105, pl. xi,
> fig. 3, 1847.

Shell concavo-convex, as wide as long, slightly transverse or elongated, and rarely exceeding one inch peculiar. Its relationship to more than one Carboniferous species has also been noticed both by M. De Verneuil, De Koninck, and others.

There exists, perhaps, one or two varieties of Geinitz's *Stroph. lamellosa*, which it might be desirable to distinguish by varietal appellations;[1] but it would be incorrect to

in length, by about the same in width. The larger or ventral valve is very convex and swollen out posteriorly, without any mesial depression or furrow; the lateral slopes are rapid, and almost perpendicular to the back of the valve. The beak is of moderate dimensions, rounded, and projecting but very slightly beyond the cardinal edge. The hinge-line is straight, and a little shorter than the greatest width of the shell, without area or hinge-teeth. The ear-shaped expansions are small, and terminated by a peculiar angular plait. The outer surface of the larger

(The radiating striæ have not been sufficiently expressed in this cut.)

valve is covered by a great number of minute, irregular-sized striæ, which are simple or dichotomous, one, two, or more forming the elongated base of a spine; these last are irregularly scattered over all the surface, at variable distances; those on the back are more sparingly distributed than on the ears, and project somewhat outwardly and downwardly; but those on the lateral slopes are perpendicular to the striæ. The spines on the cardinal region and ears are quite erect, long, and so numerous that they completely conceal the surface of the shell. The length attained by the spines on different portions of the valve is very variable; some are of great length, hollow, and ornamented by delicate rings. The smaller or dorsal valve is concave, following the curves of the opposite one; no area nor hinge-sockets are visible; the surface is covered by a multitude of fine, slightly raised, radiating striæ, and numerous wrinkles or lines of growth. This *Productus* was well described by M. De Verneuil and Count Keyserling, as also by M. De Koninck, from the Permian beds of Russia. The first two authors discovered it at Kicherma, on the river Wel; and Ouchta, at Chidrova, Arramas Ilschalki, Kidash Nikefur, Grebeni, &c. Dr. Schrenk found it at Ustj-joshuga, near Pinega, in the Government of Archangel. Baron Schauroth obtained the same shell at Milbitz, in Germany; and Professor De Koninck describes the species from Bell Sound, Spitzberg; but it has not hitherto been discovered in our British Permian strata.

[1] In a very interesting paper published in the 'Annals and Mag. of Nat. Hist.' for March and April, 1856, Professor King describes his two varieties as follows:

"*Strophalosia Morrisiana*, taking the Tunstall hill specimen as its type, may be described as follows: *General form* flatly concavo-convex, transversely elliptical. *Large valve* slightly convex, evenly rounded, often with one or more contracted longitudinal furrows; wrinkles on the sides, and furnished with a number of long, rather distant, somewhat irregularly arranged creeping or adpressed spines directed forwards; both inner and outer surface marked with nearly obsolete striæ, radiating from the umbone; also with numerous well-defined incremental lines. *Umbone* slightly affecting the even roundness of the valve; decidedly impressed or truncated, and scarcely converging down to the cardinal edge. *Area* a little more in length than half the width of the valve; rather low, but well defined, being in the form of a very obtuse triangle, the sides of which are about equal to $\frac{3}{5}$ths of the length of the base; faintly lineated transversely, and furnished with a narrow deltidium. *Small valve* slightly concave, here and there exhibiting a few nearly obsolete, slightly elongated, indented impressions, a little raised at their anterior end, which causes them to appear as if produced by a blunt-pointed instrument; both inner and outer surface marked with fine radiating striæ, a little more strongly marked than those on the large valve. *Nucleus* raised a little above the general surface of the valve; area little more than rudimentary. All the specimens I collected of this species are a little under an inch in width, and about $\frac{3}{4}$ths of an inch in length. The smaller valve, I am strongly inclined to think, cannot be considered spiriferous; probably the nearly obsolete indented

suppose that the typical shape of *lamellosa* does not occur in England, as examples identical with some represented in Geinitz's Versteinerungen have been found at Dalton.[1]

Loc. *Strophalosia lamellosa* or *Morrisiana* has been found by Professor King, Messrs. Howse, Kirkby, and others, in the shell and compact limestone of Humbleton and Tunstall hills, Ryhope-field House, Weston, Dalton-le-Dale, Clack's Heugh, &c. A very remarkable specimen has also been found by a gentleman in the lowest beds of limestone at Midderidge. The species likewise occurs in several German localities near Gera, at Pössneck, &c.

Family—CRANIADÆ.

Genus—CRANIA.

(*Vide* Introduction, Vol. I, p. 122.)

CRANIA KIRKBYI, *Davidson.*

The upper valve only of this shell has been discovered. It is sub-quadrate, with rounded angles, and is sometimes a little indented at its anterior margin: is slightly conical, with a sub-central vertex, and presents a small longitudinal depression on its anterior portion. Externally the entire surface is closely covered by a multitude of minute, short, hollow, spinulose tubercles, which produce a granulated aspect. The interior was not completely exposed in the specimens; but the

impressions may be modified bases of abortive spines. I am not aware that this species has been found anywhere except at Tunstall hill.   *   *   *   *   *   *

"Var. *Humbletonensis.* *General form* rather strongly convex, as wide as long, sometimes longer than wide, rarely the converse. Large valve rather strongly convex, and evenly rounded; occasionally with one or more longitudinal wrinkles on the sides; furnished with numerous long, somewhat irregularly arranged spines, creeping, and directed forward on the back, erect, and bent backwards on the sides and adjacent to the hinge; both inner and (?) outer surface marked with numerous fine striæ, radiating from the umbone, also with well-defined incremental lines; umbone somewhat tumid, occasionally a little impressed, and incurving over or below the cardinal edge; area small, scarcely perceptible through the incurvation of the umbone; teeth well developed. *Small valve* slightly concave, marked with numerous elongated, indented impressions, which are somewhat regularly arranged, and deepest at their anterior end; both inner and outer surface marked with distinct radiating striæ; those on the sides near the hinge dichotomous, and arcuated or curving posteriorly; they pass uninterruptedly over the indented impressions, and are crossed by rather strong incremental lines. *Nucleus* raised a little above the general surface of the valve."

[1] Dr. Geinitz has also declared (in a letter to Professor King, dated Dresden, the 11th of March, 1857) that his *St. lamellosa* is distinct from King's *Morrisiana*, but he has also confounded this last with *Productus Cancrini*, and is evidently not well acquainted with the numerous forms and variations presented by his *lamellosa* in England.

two posterior muscular impressions were perfectly displayed, while the anterior or central ones were much concealed by the matrix. The dimensions of the largest example were —length 4, breadth $5\frac{1}{2}$ lines.

About fifteen examples of this interesting species were discovered by Mr. Kirkby in the shell limestone of Tunstall hill. In external form it bears much resemblance to many species of *Cranium*; but the peculiar spinulose character, combined with its sub-quadrate shape, distinguishes it from those species to which it bears the nearest affinity. This is also the first time that a *Cranium* has been recorded from the Permian rocks, a discovery entirely due to the active researches of Mr. James Kirkby, to whom I take the greatest pleasure in dedicating the species.

Family—DISCINIDÆ.

(*Vide* General Introduction, Vol. I, p. 126.)

DISCINI KONINCKII, *Geinitz.*   Plate IV, figs. 27—29.

> ORBICULA KONINCKII, *Geinitz.* Grundriss. d. Verst., p. 495; and Versteinerungen,
> pl. iv, figs. 25, 26, April, 1848.
> DISCINA SPELUNCARIA, *King.* Mon., pl. vi, figs. 28, 29.

Circular, or oval marginally; the upper valve conical, and of moderate elevation; the vertex sub-marginal, situated at a variable distance from the posterior margin. The lower valve is almost flat, with a fissure extending from the centre to nearly the edge of the shell. The valves are strongly marked by numerous concentric lines of growth. In dimensions it rarely exceeds 3 or 4 lines in length, by 3 or $3\frac{1}{2}$ in width.

Discina Koninckii appears to be one of the rarest of our British Permian fossils, as I have never seen more than five or six examples. It has been collected at Thrislington Gap, in the marl slate; at Garmundsway, in the overlying beds of compact limestone; in the shell limestone of Tunstall hill, by Professor King; and in the last-named locality, and at Humbleton hill, by both Messrs. Howse and Kirkby. On the Continent it appears also rare; and I am indebted to Baron Schauroth for examples from Ilmenau, in Germany.

The only authors who have hitherto described and figured the species are Dr. Geinitz and Professor King;[1] but I entirely coincide with Mr. Howse[2] and M. De Koninck,

---

[1] In the 'English Permian Fossils,' p. 85, Professor King makes use of the name *Discina speluncaria* in preference to that of *Koninckii*.

[2] 'Annals and Mag. of Nat Hist.,' vol. xix, 2d series, p. 44, 1857.

while adopting the term *Koninckii* in preference to the apocryphal one of *O. speluncaria*, simply mentioned in the German edition of Sir H. de la Beche's 'Manual' (1832).[1]

Family—LINGULIDÆ.

(*Vide* Introduction, Vol. I, p. 133; and Appendix p. 8.)

LINGULA CREDNERI, *Geinitz.* Plate IV, figs. 30, 31. (King's Monog., pl. vi, figs. 25—27.)

> LINGULA CREDNERI, *Geinitz.* Versteinerungen des Zechsteingebirges, p. 11, pl. iv, figs. 23—29, April, 1848.

A small oval shell, rarely exceeding 5 lines in length, by 3½ in width. The valves are very slightly convex, thin, and marked by numerous raised lines of growth.

This species has been well described by Dr. Geinitz, Professor King, and Mr. Howse. It is found in the marl slate of Ferry hill, Thrislington Gap, and Trickley. Mr. Howse possesses a bivalve example, in which the attenuated beak of the ventral valve is clearly exhibited. It may still remain a question to be determined hereafter, whether this form did not also occur in the Carboniferous period, for it is very difficult to separate several species of *Lingula*, which closely resemble each other. *L. Credneri* is not a rare fossil, either at Ilmenau or Corbusen, in Germany.

---

[1] It is a *Catalogue name*, which Goldfuss communicated to Von Decken for the German translation, and all that is said consists of "*Orbicula speluncaria*, Schlotheim, Glücksbrunn." But the name is nowhere to be found in any of Schlotheim's numerous Memoirs, and it is probable that Goldfuss took the denomination from Schlotheim's collection, and nothing warrants the assertion made that it is the same as that dedicated by Geinitz to M. De Koninck. It was mentioned by M. De Verneuil as "un corps tres douteux," in the 'Bulletin de la Société Geologique de France,' vol. i, 2d ser., p. 504, 1844, and might be a little *Patella*, or anything else than a *Discina*, for all we know; and therefore we feel in justice bound to adopt the name given by Geinitz, as the species was first described and illustrated by him. *Catalogue names* are so injurious to the progress of science, that their introduction cannot be too strongly deprecated, nor can they ever claim priority over any subsequent description or illustration of the same object by another author.

PLATE I.

(Permian Species.)

Fig.

1. Terebratula punctata, *Sowerby.* ⎱ These specimens were figured in this place by mistake; they
2—4. Rhynchonella tetraedra, *Sow.* ⎰ belong to the Liassic period.

5. Terebratula elongata, *Schlotheim.* A large specimen. Humbleton hill. Collection of Mr. Kirkby.

6, 7. — — Two average-sized specimens. Tunstall hill.

8—11. — — var. *sufflata.* Different ages. Tunstall hill.

12—14. — — Young specimen. Tunstall hill.

15—17. — — var. *sufflata.* Casts. Humbleton. The letter N indicates the slit in the beak, due to the rostral plates.

18. — — A remarkable internal cast from Humbleton, formerly in the collection of Mr. Howse, now in that of the Museum of Practical Geology. The letter *o* refers to the curious impression explained in fig. 20.

19. — — A very fine interior, showing the rostral plates of the dorsal valve, and loop in the ventral one: enlarged from a specimen in the collection of Mr. Kirkby.

20. — — A beautiful fragment, considerably enlarged, from Humbleton. In the collection of Mr. Howse.

21, 22. — — Two interiors of the dorsal valve, to show small dissimilarities in the shape of the loop.

23. Spirifera alata, *Schlotheim.* Internal cast of a large example. Humbleton hill.

24. — — A trigonal specimen from Humbleton hill. Collection of Mr. Kirkby.

25. — — A testiferous specimen. Tunstall hill. Collection of Mr. Kirkby.

26, 27. — — Two specimens from Humbleton, showing the spirals. Collection of Mr. Kirkby.

28—31. — — Different specimens and ages from Humbleton and Tunstall hills. In the collection of Mr. Kirkby.

32. — — Portion of the internal cast of the dorsal valve (enlarged), to show the position of the adductor impressions, A A; also of the cardinal process. Collection of Mr. Howse.

33. — — A gutta-percha impression from specimen, fig. 32 (less enlarged), showing the interior of the valve itself.

35, 36. — — var. *Permiana,* King? Humbleton hill.

37. Spiriferina cristata, *Schlotheim.* Tunstall hill. Collection of Mr. Kirkby.

38, 39. — — Internal casts. Humbleton hill. N, slits left by the dental plates.

40. — — Interior (magnified), showing the spirals *in situ,* from a perfect specimen. Humbleton. Collection of Mr. Kirkby.

41. — multiplicata, *Sowerby.* (Enlarged.) Tunstall hill.

42. — — Interior of the ventral valve (enlarged), to show the position of the central septum and dental plates. Collection of Mr. Kirkby.

43. — — From Tunstall hill.

44. — — A remarkable example, with unusually developed area, from Tunstall hill. Museum of Practical Geology.

45, 46. — cristata? Young shells. Tunstall hill.

47—49. Spirifera Clannyana, *King.* (Enlarged.) Ryhope-field House.

PERMIAN.

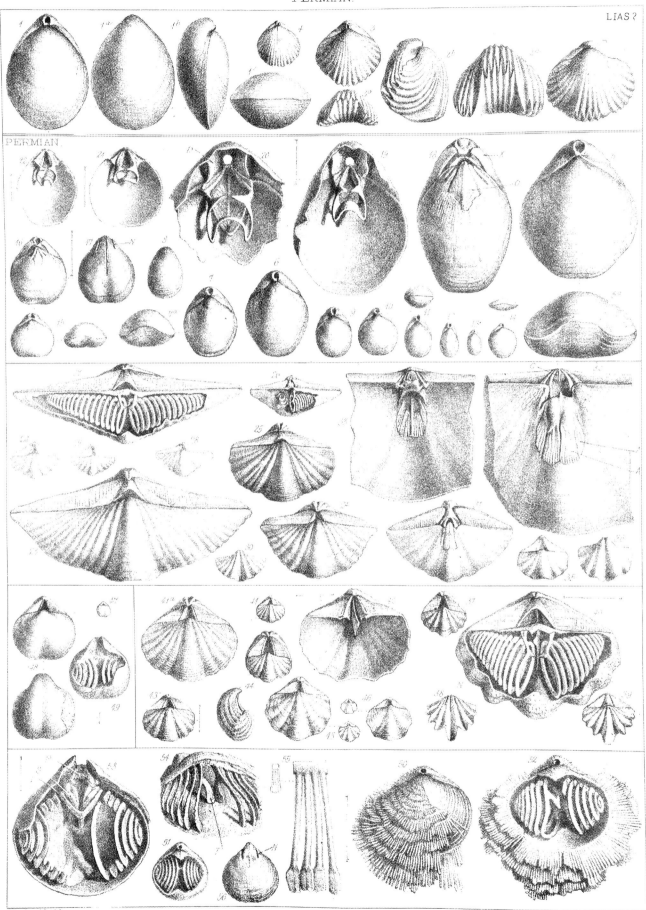

PLATE II.

(Permian Species.)

Fig.

1. Athyris pectinifera, *J. de C. Sowerby*. Interior. Humbleton.

2. — — This beautiful (magnified) illustration of the interior is entirely drawn from nature, but not from a single specimen. The hinge-plate and its small circular aperture, as well as the dental or rostral plates, have been taken from specimens in the possession of Professor King and Mr. Kirkby. The spirals are from a perfect example discovered by Mr. Howse at Humbleton hill. The spinose expansions are copied from two or three specimens in the cabinet of Mr. Kirkby. This illustration, while correct in its different parts, conveys a clear idea of the complicated internal arrangement.

3. — — The central portion of the spirals, considerably magnified, to show the connecting processes more clearly than in fig. 2.

4. — — A portion of the spiral and connecting process, seen as if the beaks had been removed.

5. — — A fragment of the spiral lamella, showing the spines on its outer edge.

6, 7. Spirifera alata, *Schlotheim*. Internal cast of the ventral valve, to show the adductor (A) and cardinal (R) muscular impressions, as well as the ovarian spaces (o), from specimens in the collections of Mr. Howse and Mr. Kirkby.

8. Terebratula elongata, *Schloth.*, var. *sufflata*. Internal cast from a typical example. Humbleton hill.

9. Camarophoria Humbletonensis, *Howse* = *multiplicata*, King. Internal cast, showing portions of the narrow marginal expansions peculiar to this species. Humbleton. Collection of Mr. Howse.

10. — — Front of a very large individual.

11. — — Internal cast, seen from the beaks, in which the impressions referred to the adductor (A) and cardinal muscles (R)? are exposed. Humbleton hill. Collection of Mr. Howse.

12. — — Internal cast of the ventral valve, showing the cardinal (?) muscular impressions, and the raised band (T) observable in many individuals, but which remains still unexplained. Humbleton. Collection of Mr. Howse.

13, 14. — — Internal casts, from Humbleton, exhibiting the vascular impressions. Collection of Mr. Howse.

15. — — Interior of the dorsal valve, enlarged.

16—18. — Schlotheimi, *V. Buch*. Different ages. Tunstall hill.

19. — — A var. with only two ribs on the mesial fold, and thus hardly distinguishable, except in dimensions, from *C. globulina*. Tunstall.

20. — — A trigonal-shaped specimen, with hardly any indication of ribs, and thus approaching to *T. superstes*, De Verneuil. Tunstall.

21. — — A specimen, showing the large dimensions of the marginal expansions. Tunstall. Collection of Mr. Kirkby.

22. — — Cast of the ventral valve, exhibiting the marginal expansions. Humbleton hill. Collection of Mr. Kirkby.

PLATE II (*continued*).

Fig.

23. CAMAROPHORIA SCHLOTHEIMI, *V. Buch*. Internal cast, seen from the beak, to show the muscular impressions, as in fig. 11. Humbleton hill.

24. — — Interior of the ventral valve, enlarged, minus the marginal expansions. Tunstall hill. Collection of Mr. Howse.

25. — — Fragment, considerably enlarged, to show the hinge and other internal processes, from a very perfect example in the collection of Mr. Howse. Tunstall hill.

26. — — Another fragment, seen more in front.

27. — — Profile section, to show the position of the different parts. Tunstall hill.

28. — GLOBULINA, *Phillips*. Natural size. Tunstall hill. Collection of Mr. Kirkby.

29, 30. — — Two other examples (enlarged), with two and three ribs on the mesial fold. Tunstall hill.

31. — — A fragment of the interior, enlarged, in the collection of Mr. Kirkby. Humbleton hill.

32—36. STREPTORHYNCHUS PELARGONATUS, *Schloth*. Different specimens and variations in shape, from Tunstall and Humbleton hills, in the collections of Messrs. Howse and Kirkby.

37. — — A specimen from Dalton, in the collection of Mr. Howse (enlarged).

38. — — Interior of the dorsal valve, enlarged. Tunstall hill. Collection of Mr. Howse.

39. — — Dorsal valve, seen from the umbone, to show the bifid cardinal process. Tunstall. Collection of Mr. Kirkby.

40. — — Interior of the ventral valve. Tunstall. Collection of Mr. Howse.

41. — — Interior of the ventral valve, seen under the area, to show the rudimentary dental or rostral plates.

42. — — Internal cast of the ventral valve, showing the position and shape of the muscular scars; from Tunstall hill. Collection of Mr. Kirkby.

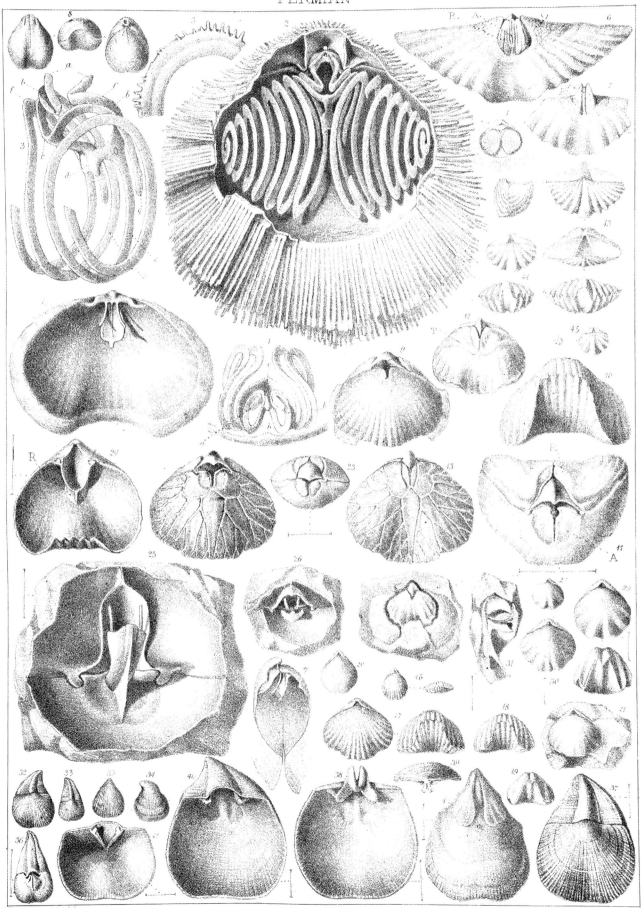

PLATE III.

(Permian Species.)

Fig.

1—4. Strophalosia Goldfussi, *Münster*. Young shells, from Tunstall hill, in the collection of Mr. Howse.

5, 6. — — Internal casts, from Ryhope-field House.

7. — — Internal cast of the ventral valve, natural size. Humbleton; in the collection of Queen's College, Galway. This specimen exhibits the ordinary condition in which the internal casts are preserved; the adductor impressions are indicated, but no trace of the cardinal muscular ones can be perceived.

8. — — External cast of the dorsal valve, from the same specimen as fig. 7, but enlarged. This specimen was figured by Professor King in his 'Monograph,' pl. xi, fig. 9; but my drawing shows more clearly the manner in which the quadruple impression of the adductor is disposed, and how the reniform impressions divide each pair.

9, 10. — — A remarkable internal cast, from Humbleton, in my collection. Fig. 9 shows the interior of the dorsal valve; fig. 10 that of the ventral one, in which the adductor (a) and cardinal muscular impressions (k) are beautifully defined.

11, 12. — — Internal cast of the dorsal valve. Fig. 12 is a gutta-percha impression, taken from the same, showing the articulation of the valves, &c.

13. — — A very much enlarged representation of the ventral valve, with its long slender spines, a small portion of the shell being removed to show, on the internal cast, the adductor and cardinal muscular impressions; from specimens in the collection of Mr. Kirkby. Humbleton hill.

14, 15. — — Interior of the ventral valve. Humbleton hill. Collection of Mr. Howse.

16, 17. — — A dorsal valve, from Tunstall hill; in the collection of Mr. Kirkby. Fig. 17 shows the remarkable internal margin peculiar to some young individuals.

18. — — S. *parva*, King. From an internal cast, natural size. It adheres to a cast of *Productus horridus*, from Humbleton. In the Museum of Practical Geology; formerly in that of Mr. Howse.

19—22. Strophalosia Goldfussi? var. *Lewisiana*, De Koninck. Slightly enlarged; from Humbleton and Ryhope-field House. Fig. 19 is a cast of the interior of the dorsal valve; 20—22 represent a testiferous specimen, of which the spines are broken.

23. — — var. *Whitleyensis*, King. This is the original specimen from Whitley, now in the collection of Queen's College, Galway, and the one from which Professor King published a gutta-percha impression in pl. xii, fig. 26, of his Monograph.

24—31. Strophalosia lamellosa, *Geinitz*, var. *Morrisiana*, King. From Tunstall hill. Collections of Professor King and Mr. Kirkby.

32—33. — — From Dalton-le-Dale. Collection of Mr. Howse.

PLATE III (*continued*).

Fig.

34. STROPHALOSIA LAMELLOSA, var. *Humbletonensis*, King.   Natural size, from specimens in the collection of Mr. Kirkby.   Humbleton hill.

35. — — var. *Humbletonensis*, King.   Is an enlarged representation of the dorsal valve and beak of the ventral one, from specimens in the collections of Messrs. Howse and Kirkby.   In this specimen the area is visible, but it is sometimes concealed from the incurvation of the beak.

36. — — Ventral valve, to show the great length of the adpressed spines. Humbleton.   Collection of Mr. Kirkby.

37. — — Internal cast of the interior of the dorsal valve. Humbleton. Collection of Mr. Kirkby.

38. — — Impression of the dorsal valve (enlarged), from Ryhope-field House, proving that, although the valve in question is not spiniferous in the great majority of individuals, they did exist in some few examples.   Collection of Mr. Kirkby.

39. — — A gutta-percha impression of the interior of the dorsal valve, to show the articulation of the valves, muscular and reniform impressions, from a specimen in Queen's College, Galway.   Humbleton.   The internal cast from which the impression here represented was taken is the same figured by Professor King in pl. xii, fig. 30, of his Monograph; but the Professor's illustration was not quite correctly drawn.

40. — — A fragment of the dorsal valve, greatly magnified, to show how the striæ, which cover the surface, augment by intercalation and bifurcation the distance between M and N, representing 2 lines.

41. — — Enlarged portion of the ventral valve, to show how the long adpressed spines sometimes originated from two or three striæ, forming the base of a single spine.

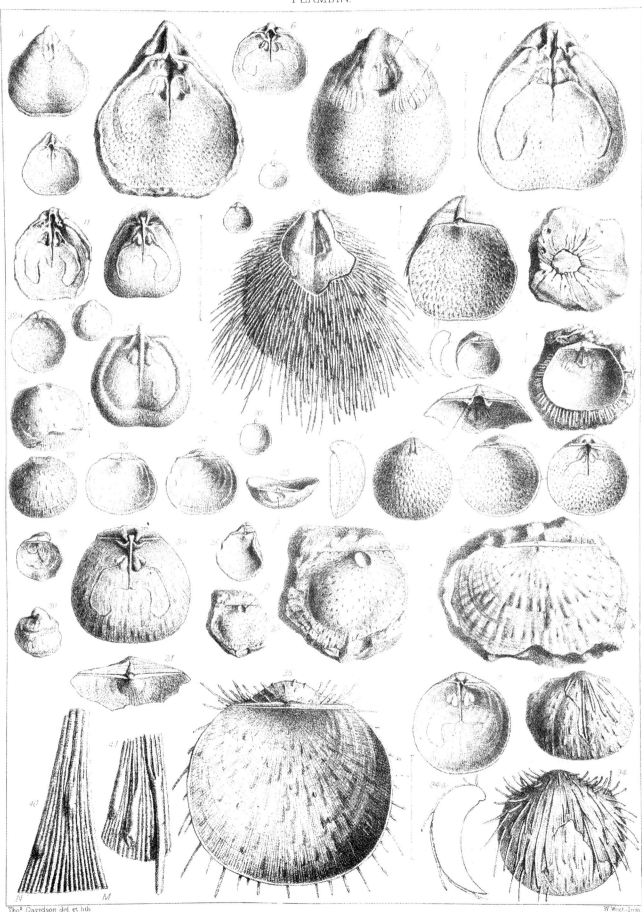

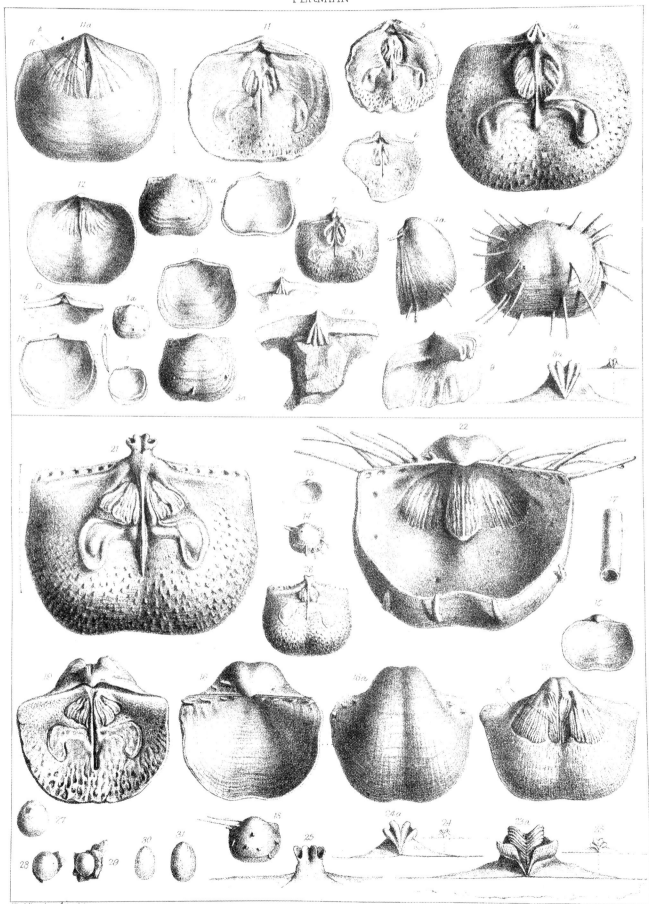

A MONOGRAPH

OF THE

BRITISH FOSSIL BRACHIOPODA.

PART V.

THE CARBONIFEROUS BRACHIOPODA.

BY

THOMAS DAVIDSON, F R.S., G.S., &c.

LONDON:

PRINTED FOR THE PALÆONTOGRAPHICAL SOCIETY.

1858—1863.

PRELIMINARY REMARKS.

In the descending order we now arrive at that extensive and important series of rocks to which the term CARBONIFEROUS SYSTEM has been so emphatically applied; a period unequalled for the extraordinary luxuriance of a flora, which in its present fossil condition has proved a source of so much wealth and prosperity to this and other nations: but it is not only the vegetation of its land that has proved so remarkable, the inhabitants of its waters were perhaps almost equally so; and in every respect deserving of the most careful and complete investigation. The system occupies a considerable area in England, Scotland, and Ireland, as may be perceived by a glance at any geological map.[1] It is

[1] We extract the following passage from Mr. Marcou's excellent Memoir on the American Carboniferous Deposits ('Bulletin de la Soc. Geol. de France,' vol. xii, 2d series, p. 844):

" The Carboniferous period is composed of a series of rocks, the importance of which, whether it be considered in the scientific, industrial, or commercial point of view, is neither equalled nor even attained by any of the other sedimentary deposits. In a scientific point of view the Carboniferous group presents the most extended geognostic horizon, of which the characters are constant over all the surface of our terrestrial globe. In Europe, in Asia, in Africa, in the two Americas, as well as in Australia, we meet with the same rocks, and often also with the same fossils; and one is at a loss to know which should be most admired, this consistency in the lithological characters of its strata, or the presence of the same fossils buried in contemporaneous beds, and often situated at the antipodes. From the glacial zone of Spitzberg, the Bear, and Melville Islands, to Australia, Tasmania, and New Zealand, the Carboniferous strata form islands, mountains, table-lands, plains, and even half continents, where the identity and unity of the lithological and palæontological characters exhibit the surest marks of recognition and the most certain horizon that can be found in geological investigations. . . .

" The Carboniferous deposits of America may be divided into an upper and lower deposit:

" *a. Upper Carboniferous,* or *Coal measures,* above the mountain limestone, is composed of a series of beds of sandstone and argillaceous schists, containing beds of coal, and which constitutes the Coal formation properly so termed. In America it contains no Brachiopoda.

" *b.* The *Lower Carboniferous,* of which the general character is, so to say, universal; since we observe that it presents well-stratified beds of a hard grayish limestone, replete with numerous marine fossils, similarly found in Europe, America, and Australia. In America it contains *Athyris Roissyi, planosulcata; Sp. striata, lineata;* the *Orthis crenistria, Michelini; Prod. semireticulatus, Cora, Flemingii, costatus, scabriculus, pixidiformis, pustulosus,* and others."

1

made up of a vast accumulation of beds of limestone, shales, indurated clays, grits, coal and sandstone, alternating to a large extent ; and which may in different districts or countries be more or less successfully divided into two, three, or even four principal groups, viz. :

1. *The Coal measures* (the uppermost bed of which lies under the *Lower New Red Sandstone* of the Permian series) are composed of a vast accumulation of beds of clay, shale, coal, and sandstone, divided more or less by layers of ironstone and chert. These beds intercalate with each other, and at times also with—

2. *The Millstone grit,* which underlies the Coal measures properly so termed. It is chiefly composed of a coarse-grained, gritty sandstone, not very dissimilar in character from the one found above, but it also alternates with many beds of limestone, shale, and even coal, and presents some of the features of—

3. *The Carboniferous or Mountain Limestone.* These large bands of hard limestone principally prevail in the Yoredale and Scar Limestones of Yorkshire, but they are likewise interstratified with beds of grits, &c. ; the series being terminated by lower Carboniferous shales and sandstones, which at times assume a red colour.

But the basement line of the Carboniferous system is still a subject of some divergence in opinion, as certain geologists would comprise within its limits all or portions of the *Old Red Sandstone,* while others consider the strata last named to form part of the Devonian system.[1]

[1] In a most interesting memoir, entitled 'Researches among the Palæozoic Rocks of Ireland,' published in the seventh volume of the 'Journal of the Geol. Soc. of Dublin' (1856), Mr. Kelly is a strong advocate for the admission of what he terms the *Old Red* Sandstone into the *Carboniferous system,* and even seems doubtful as to the propriety of admitting the *Devonian system* in total. It must, however, be here remembered that Mr. Kelly founds his views chiefly on appearances peculiar to Ireland ; he also refers to statements made by Dr. M'Culloch (in his 'Treatise on Geology,' 1831), and to the table published by Professor Phillips, at p. 11 of the 'Geology of Yorkshire,' in which the learned Professor divides the Carboniferous system into three principal parts—the *Coal formation,* the *Carboniferous Limestone,* and the *Old Red Sandstone,*—adding at the same time that "the Carboniferous system does undoubtedly permit itself to be considered in three series, characterised by the prevalence of coal, limestone, and red sandstone." Mr. Kelly then states that the Carboniferous formations of Ireland are likewise divisible into a similar triple system, which he seeks to develop in his valuable memoir, to which we must refer the reader for more ample details. It may, however, be observed that bands of red sandstone no doubt occur and form part of the Carboniferous group, but it remains to be a question for further consideration whether the *Old Red Sandstone* of the Devonian period should or not, in part or in total, be considered as a portion of the great Carboniferous system? Sir R. Murchison kindly informs me that there exists in Ireland a series of many thousand feet of shales and grits, &c., above the highest *Upper Silurian,* with all the characteristic fossils, and which represents precisely *in time* the mass of the *Devonian rocks ;* and that, from his last year's direct survey of Ireland, his belief is that the Old Red system, Devonian, in Ireland, has undergone a great separation into two parts, which has not occurred in England or Scotland. Professor De Koninck seems also inclined to admit a portion of the

It has been observed by several geologists that "the triple arrangement becomes considerably modified as we proceed towards the north of England and Scotland; for in Derbyshire and Yorkshire the true coal-bearing strata do not cease with the Millstone grit, but are intercalated with it; and in Yorkshire the Limestone (Yoredale rocks) contains several coal seams, flagstones, ironstones, &c.; and still further northwards, in Northumberland and the south of Scotland, the lower Limestone becomes frequently divided by intervening beds of grit, shale, and coal."[1]

To ascertain the true stratigraphical limits of the group is therefore a matter of great importance not only to the geologist but likewise to the palæontologist, because the number of species varies considerably in the different portions of the system; and to offer at present an example, without anticipating the conclusions we may arrive at ourselves after having completed the review of the different species, we will borrow a few lines from p. 244 of the 2d vol. of the 'Geology of Yorkshire,' in order to show what were the numerical proportions arrived at in 1836 by its distinguished author:

> " Coal formation . . 3 species of Brachiopoda.
> Millstone grit . . 4 ,, ,,
> Yoredale rocks . . 29 ,, ,,
> Lower Scar Limestone . 96 ,, ,,

"It appears to be in the upper part of the *Lower Scar Limestone* that the greatest number of fossils of all kinds occur; they grow continually less and less plentiful as we ascend in the series of the *Yoredale rocks, Millstone grit*, and *lower part of the Coal measures*. In the upper part of the Coal series *all* the species vanish."

It was, therefore, during the deposition of the lower divisions of the Carboniferous groups, *i. e.*, of the *Mountain Limestone* and its accompanying shales and sandstones, that the Brachiopoda flourished; while in the upper divisions of the period are entombed the remains of that stupendous vegetation which we all so much admire.

The Brachiopoda of the Carboniferous period are therefore both numerous and widely distributed, and have been partially the subject of several important works published at various intervals both at home and abroad. Still of British fossils they seem to be those perhaps the less generally or completely understood, and especially so if we are to infer from the great extent of misnaming prevalent in almost every public and private collection, not merely in this country but also on the Continent. This unfortunate state of things seems in great measure attributable to the vague and unsatisfactory manner in which certain species have been described and illustrated, as if a few words of description, taken from some obscure fragment at times even imperfectly represented, were sufficient to

---

Old Red Sandstone into the Carboniferous period; but this is not the place for either discussing or enlarging upon such matters, and we must content ourselves by referring the reader for further information to the numerous and excellent geological works and memoirs by Conybeare and Phillips, Buckland, Sir R. Murchison, De la Beche, Prestwich, De Verneuil, De Koninck, and others.

[1] Tennant, 'A Stratigraphical List of British Fossils,' p. 90, 1847.

enable any but the possessor to understand the shape and character of the object thus treated. It, therefore, required a considerable amount of labour and research before I felt myself at all in a competent condition to endeavour to undertake the publication of the *numerous* and *variable* forms which will compose the present monograph.

Having traced on separate sheets the original figures of all those species *said* to have been found in Great Britain, and classed these under their respective genera, according to general affinities and resemblances; my next effort was to procure the loan of the *original specimens* themselves still extant in the United Kingdom, and which were at once communicated in the most kind and liberal manner by their respective possessors. By this means I have been enabled to class, with a greater or less degree of success, the thousands of specimens forwarded from various quarters around the *named* and *original types* of their respective species.

I have stated this in order that the reader may feel assured that, however faulty and imperfect the present work may appear, no effort has been neglected to ensure as far as possible a correct identification of the species by a direct comparison with the original figured types. In a few cases it has not, however, been possible to procure the originals, such as some of those described by Martin, Portlock, and M'Coy— the specimens being no longer to be found; but in those comparatively exceptional cases wherein doubt might prevail, I have invariably reproduced not only the author's descriptions but likewise their figures, as in the instance of *Spirifera transiens, Sp. mesogonia, Sp. subconica,* &c., but without, however, warranting their specific claims.

After a careful investigation of the various works, I found that considerably more than two hundred species of Carboniferous Brachiopoda were stated to have been found in the United Kingdom, but, as will be hereafter demonstrated, many of these are not British, while others are mere synonyms, or in a great measure attributable to incorrect identifications with Devonian shells, which have been supposed by Professor M'Coy and a few others to be common to both systems. I am very far, however, from wishing to deny that certain forms did continue to live during both the Devonian and Carboniferous periods as well as during the Carboniferous and Permian epochs; but it would be, I believe, a mistaken notion to suppose that they occurred in that numerical abundance which we should be led to believe from the names introduced into Professor M'Coy's otherwise important work on 'Irish Carboniferous Fossils.' Many of these *said* to be Devonian shells were identified from undeterminable or obscure fragments or specimens, at times distorted by pressure and cleavage ; and as a large number of them are still extant in Dr. Griffith's collection, Mr. Salter and myself have (through the kindness of their possessor) been enabled to minutely examine and compare a certain number, which invariably turned out to be true *Carboniferous* and not Devonian types. Thus, for instance, M'Coy's so termed Sp. simplex belonged to *Sp. cuspidata,* Sp. speciosa to *Sp. laminosa,* Athyris concentrica to *A. Roissyi,* &c. These and other incorrect determinations have tended to add considerable confusion to our notions as to the repartition of species in the two respective systems.[1]

It is well known that the careful stratigraphical and palæontological investigations made in the vast Silurian system by Sir R. Murchison, Mr. Barrande, M. De Verneuil, and others, as well as those conducted throughout the Jurassic, Cretaceous, and Tertiary periods by different competent naturalists, have nowhere exhibited that wholesale mixture or general longevity in time or existence of the numerous species which we should be led to believe did take place in the two above-named epochs.[2] Our researches have hitherto unmistakingly led us to repudiate such an assumption, and to urge us on the contrary to believe that the law which has regulated the vertical distribution of animal forms did not differ in those extended palæozoic systems any further than it did in those of newer or of more ancient date. M. De Koninck states "that he has succeeded in tracing, in the Carboniferous formations of England and Scotland, two great different faunas; the one corresponding to the Carboniferous fauna of Visé and Bleiberg, the other to the fauna of the Tournay coal basin. These two faunas, although contemporaneous, are said to be nowhere found coexistent."[3] But it will be preferable to reserve what we may have to add on this subject until the completion of the descriptions of the species which will compose the present monograph.

So variable do we find the individuals of the same species to be, especially when our examinations are not restricted to a small number of examples (and this more so in certain forms than in others), that we are sadly at a loss in many cases to know how to define and where to confine the limits of variety, and even how to appreciate the value of the characters which are to be brought forward in the discrimination of two different, although closely allied forms. There generally, however, exists a certain *facies* or peculiarity in each combination of individuals that leads the experienced palæontologist to separate, with more or less success, forms which could hardly be identified if unaccom-

[1] These erroneous identifications with Devonian species were published in 1844, and I believe that Professor M'Coy himself repudiates at present the larger number so inscribed; but I have felt myself compelled to draw attention to the point in question, as it is of great importance in our geological and palæontological inductions. The same must be said relative to Mr. Kelly's excellent and most valuable memoir and synoptical table, 'On the Localities of Fossils of the Carboniferous Limestone of Ireland,' published in the 'Journal of the Geological Society of Dublin,' March 14th, 1855, but in which Professor M'Coy's mistaken identifications are reproduced. This most useful work, to which we shall have so often occasion to refer, was published by its author in a great measure to fill up a sad omission in the 'Synopsis,' in which the localities of almost every species had been purposely omitted.

[2] It is true that, in Professor Phillips's work, 'Figures and Descriptions of the Palæozoic Fossils of Cornwall, Devon, and West Somerset,' a very large number of true Carboniferous forms are described as occurring in Devonshire *Devonian strata*, such as at Barton, near Torquay, &c.; but perhaps some of the beds taken as Devonian may be in reality Carboniferous, a point which will require hereafter to be investigated with all possible attention.

[3] 'Proceedings of the Geol. Inst. of Vienna,' 1856; 'Quarterly Journal of the Geol. Soc.,' vol. xiii; and 'Bulletin de l'Académie Royale de Belgique,' vol. xxiii, No. 9.

panied by copious illustrations. The difficulty of discriminating becomes also even more sensible when a genus is largely represented in any single or contiguous group of strata, as are, for example, the *Spirifers* and *Productuses*, which in no other epoch were more varied or half so abundant as in the Carboniferous period; they also here attain their maximum of development both in number of species and proportions, as may be exemplified by the well-known full-grown individuals of *Sp. striata* and *Pr. giganteus*. The comparative length of the hinge line, as well as number of ribs, is so variable, not only according to age but from other natural and extraneous causes, that it becomes at times most perplexing to know where to find words sufficiently precise or explicit, so as to convey a clear idea of those *minute* differences which exist, and which suffice in many cases to warrant the specific or varietal separation of two seemingly allied forms.

It has been my strenuous endeavour, as far as the space at my command would admit, to figure not merely one marked individual, but likewise others less characteristic in their features; that the general observer may feel less embarrassed how to deal with the more numerous intermediate variations in shape, so prevalent among the species of this class of Mollusca.

Among the numerous works and papers which have been consulted during the preparation of the present monograph, I feel bound to call attention to a few British ones in particular, not merely from their importance, but also to satisfy the reader as to the source whence a large portion of our material was derived.

In the valuable 'History of Rutherglen and East Kilbride,'[1] published by David Ure, in 1793, we find some very passable figures of from eleven to twelve species of Carboniferous Brachiopoda, which illustrate in a satisfactory manner a not inconsiderable number of those forms so abundantly distributed in the parishes of Kilbride and Carluke, Scotland; but the author unfortunately does not apply to them any distinctive specific appellations: he classes his specimens into *Anomitæ læves, Striatæ*, and *Echinatæ*, the last comprising those species with spines, such as *Productus*. In the 'Mineral Conchology' we find described and illustrated a few more Scottish species, while others are briefly described, but not illustrated, by Dr. Fleming, in his excellent book on 'British Animals,' 1828. In these works, and that of Sowerby, will be found the principal records and illustrations of the Carboniferous Brachiopoda observed in Scotland; but the researches I have undertaken, with the assistance of several kind friends, will enable me to considerably augment the list of the species from that portion of the British Empire, although it has hitherto proved much less prolific than both England and Ireland.

Some years after Ure's publication, W. Martin's work, 'Petrifacta Derbiensia' (1809), appeared, in which about nineteen species of Brachiopoda were described and carefully

---

[1] It is now a scarce volume, and not generally known, but is well deserving of an honorable mention, being the first work in which Scotch fossils were correctly described and delineated. Specimens identical with and from the localities of David Ure were both lent and presented to me by several Scottish friends, and will be found illustrated in the present work.

illustrated, and of which only two, viz., *Anomites subconica* and *acutus*, have not as yet been properly understood.[1] In the seven volumes of the 'Mineral Conchology,' published at different intervals by James Sowerby and his son, J. de C. Sowerby, will be found good descriptions and illustrations of many excellent species;[2] and in 1836 the second volume of the 'Geology of Yorkshire' was published, in which its distinguished author briefly describes ninety-six forms of Carboniferous Brachiopoda, and of which sixty-three are said to be new, the remaining number being made up from those already described by Martin and Sowerby. All the new species and a few of the old ones are there illustrated, but in general by a single figure only, which was not, however, always sufficient for the clear understanding of the species. Professor Phillips's material was chiefly derived from his own personal exertions, to which was added the Gilbertsonian collection now in the British Museum; and owing to the kind assistance of the author, I have been able to define and completely illustrate certain of his less clearly figured species, such as *Sp. radialis, Sp. septosa, Sp. humerosa*, &c., which could not have been understood without a personal inspection of the originals, fortunately still extant in the author's cabinet. In 1843 also appeared a 'Report on the Geology of the Coast of Londonderry and part of Tyrone and Fermanagh,' by Colonel Portlock, in which a few new British species were described and illustrated.

While Professor De Koninck was publishing his celebrated work on 'Belgian Carboniferous Fossils,'[3] Professor M'Coy was likewise hard at work on the Irish species,

---

[1] After the death of Martin a certain number of his original types came into the possession of Sowerby, in whose collection they may be still distinguished.

[2] I beg to refer the reader for information relative to the exact dates of each species published in the 'Mineral Conchology,' to M. E. Renevier's useful list printed in the 'Bulletin de la Société Vaudoise des Sciences Naturelles,' May 2d, 1855. M. Renevier calls attention to the fact that the portion of the work published from June, 1812, to November, 1822, was due to James Sowerby, while the remainder, dating from January, 1823, to January, 1845, is the work of his son, Mr. J. de C. Sowerby, a distinction which should always be attended to.

[3] It is not my present intention to enumerate all the foreign works and papers that have treated directly or indirectly of Carboniferous Brachiopoda, as they will be referred to at their proper places under the respective species; but I cannot pass in silence two most important works published by Professor L. de Koninck, viz., 'Description des Animaux fossiles qui se trouvent dans le Terrain Caronifère de la Belgique,' 1843, and 'Mongraphie des Genres Productus et Chonetes,' 1847. In these works many British species have been clearly described and illustrated, and, indeed, much more so than in most of our English books. It is true that in certain cases the learned Belgian author did misunderstand a small number of our British types, but when I consider how imperfectly the forms in question were both described and illustrated, without the possibility of a direct reference to the original specimens, I cannot feel surprised that a few mistakes should have occurred, which were unavoidable under the circumstances. Of these some have already been corrected by M. De Koninck himself, in the supplement to his great work, issued in 1851; while a few others will be put right, I trust, in the present monograph.

I may likewise refer the reader to the memoirs of Baron Von Buch; to the 2d vol. of 'The Geology of Russia,' by M. De Verneuil and Count Keyserling, 1845; as well as to the important memoir by Von P. V. Semenow, entitled 'Ueber die Fossilien des Schlesischen Kohlenkalkes,' 1854, &c.

which formed part of Dr. Griffith's valuable collection. His work, 'A Synopsis of the characters of the Carboniferous Limestone Fossils of Ireland,' appeared in 1844, just one year subsequent to the foreign one above referred to. Professor M'Coy describes about two hundred and twenty-eight species of Brachiopoda, stated to belong to the Carboniferous deposits of Ireland! Of these, fifty-eight or fifty-nine are said to be new, and are illustrated in the accompanying plates; but, as already stated, this list has been considerably increased by the unfortunate introduction of a large number of Devonian shells, based upon incorrect identifications. There occurs also many synonyms and other species founded on undeterminable fragments or doubtful malformations, so that when cleared from those intruders the catalogue will be found to be considerably reduced. It is, however, worthy of notice that, although its author has introduced the names of Phillips's, Sowerby's, and some foreign authors' Devonian species, those described and figured by himself are all true Carboniferous specimens, although in several cases not in reality new. The loan of the original examples made use of by Professor M'Coy is a service for which I cannot sufficiently thank their liberal possessor, as it has enabled me to determine a great number of obscure and uncertain forms, which must have continued so had I not possessed the originals to work with. Professor M'Coy has, however, made large amends for the shortcomings of his early book, by the publication of his able and most valuable work, 'On the British Palæozoic Fossils in the Geological Museum of the University of Cambridge.' It contains, among other things, the careful and elaborate description of from ninety-six to ninety-seven species of Brachiopoda of the Carboniferous period, thus adding considerable information to what was already known; but unfortunately, as has been so commonly the case with most authors, he has illustrated but those fifteen which he considered new.

I must now conclude this short notice of the most remarkable British works that have treated of Carboniferous Brachiopoda, by alluding to the valuable 'Catalogue of British Fossils,' prepared with infinite care and patience by Mr. Morris. In his last edition (1854), the author has devoted a certain number of pages to the enumeration and classification of the Carboniferous species into families and genera.

It must, however, have been very generally felt by all who have had occasion to refer to the works above mentioned, how great is the want of correct and more copious illustrations; and it is doubly so in those cases where shells hitherto considered as foreign to the period are introduced for the first time. Under *such* circumstances, a figure is, in my opinion, of *as much importance* as if the shell were actually new, nor can the reader be satisfied as to the correctness of the identification, if he be not provided with some ocular demonstration in the shape of a figure; for it is a well-known fact, that no description, however elaborate in its details, will in natural history compensate for the want of an illustration; or in other words, that with a good figure a short description will often convey to the mind a much clearer idea of the object intended, than the most lengthened one, unprovided with that necessary appendage. I have generally omitted, among my references, to allude to authors who have furnished simple lists or catalogues of species, because, having no means of ascertaining that these identifications are correct, it would be merely burdening the list of references to no useful purpose.

It now remains for me to express my warmest thanks to those numerous friends who have so zealously assisted me, and who have contributed so largely to the success of the present work by the loan of specimens, and by the advice they have afforded.

To Professor Owen, Mr. Waterhouse, Mr. Woodward, Dr. Gray, and Dr. Baird, for the use of the specimens in the British Museum, wherein so many of Professor Phillips's original figured types are preserved. To Sir R. Murchison and Professor Huxley for the liberal assistance and communication of specimens contained in the Museum of Practical Geology, or Geological Survey; to Mr. Salter and Mr. Baily, of the same institution, for much valued help and useful information. To the Council of the Geological Society, for the kind manner in which they placed the valuable collection of Carboniferous species, forming part of the Society's Museum, at my disposal, for the use of the present work. To Mr. Bowerbank, for all the trouble he has taken relating to this monograph. To Professor Tennant, Mr. Rupert Jones, and Mr. Morris, of London, I am indebted for the loan of many specimens, but in particular to Mr. J. de C. Sowerby, who, in the most liberal manner, gave me the unlimited loan and use of the original specimens described by his father and himself in the 'Mineral Conchology.'

To Professor Sedgwick, for the obliging manner in which he has allowed me free access and use of the specimens in the Cambridge University Museum, and in which are assembled a large proportion of Professor M'Coy's named and original types. Also to Professor M'Coy, Mr. Carter, and L. Barrett, who at various times have kindly assisted in my researches while at Cambridge.

To Professor Phillips, of Oxford, for the loan of his private collection, containing many of the original examples, published in the 'Geology of Yorkshire,' as well as for the kind advice and interest he has taken in the success of the present monograph. To Mr. Etheridge, and the directors of the Bristol Institution Museum for the use of their specimens, as well as to Mr. Charlesworth, and the directors of the York Museum.

To the Earl of Ducie, Dr. Wright, of Cheltenham, Messrs. Walton and Moore, of Bath; Mr. Parker, Mr. Binney, and Mr. Ormerod, of Manchester; Mr. Reed, of York, Mr. Howse, of South Shields, Mr. Tate, of Alnwick, Mr. Muschen, of Birmingham, and the Rev. J. G. Cumming, of Lichfield (formerly of the Isle of Man), for the loan and gift of many specimens, as well as for much valued information connected with the species from their respective districts, or collections. I must, however, here tender my especial and warmest thanks to my valued friend, Mr. Edward Wood, of Richmond (Yorkshire), for the indefatigable and zealous exertions he has displayed in assembling material and information of various kinds; to his kind and liberal assistance this monograph is much indebted, as he has enabled its author to illustrate and fully describe many important species hitherto involved in much obscurity.

To my Scottish friends and countrymen, it is now my pleasing duty to acknowledge the liberal assistance I have received from them, and among whom I must particularly mention Professor Fleming, who lent all the important Scottish Carboniferous Brachiopoda preserved in his collection, and among which are the originals of those referred to in his work on British animals, as well as several of those figured by Sowerby in the 'Mineral Conchology.' This collection contains the most numerous series of Scottish species I have as yet been able to consult. My thanks are likewise due to the late Hugh Miller, to Professor G. Wilson, Mrs. Rogers, Mr. A. Bryson, and Mr. Rose, of Edinburgh, for the loan and gift of many specimens, as well as for much useful information; to Professor Nicol, of Aberdeen, Mr. Fraser, of Glasgow, Mr. J. Young, and Mr. A. Cowan; and to another friend, who, although unnamed, I feel it a most pleasing duty to express my warmest thanks, for the liberal assistance he has afforded by the loan and gift of specimens derived from Lanarkshire, one of the most important and interesting of our Scottish Carboniferous districts.

To my Irish friends, I am likewise indebted for considerable and most liberal assistance. First, to Dr. Griffith, for the kind manner in which he has lent the Carboniferous Brachiopoda contained in his valuable collection; to Mr. Jukes, for the use of many specimens out of the collection of the Geological Survey of Ireland. To Mr. Carte, for the kind communication of those in the Royal Dublin Museum; to the Rev. Professor Haughton, of Trinity College, Dublin, to Mr. R. Nelson, and to Professor King, of Queen's College, Galway; to Mr. Kelly, of Dublin, I must tender my warmest thanks for his zealous and indefatigable exertions, not only in procuring me a vast number of important specimens, but also for the valuable information he has at all times conveyed on the distribution and localities from which he had himself collected a large number of the types published by Professor M'Coy in the 'Synopsis.'

Among my foreign friends, I might name many who have expressed the most lively interest in the success of the present undertaking, but I must confine myself to those who have communicated specimens and information in connection with Carboniferous species. And among these, it is a most pleasing duty for me to express my grateful thanks to Professor L. de Koninck, of Liege, whose excellent works and extensive knowledge of Carboniferous fossils have proved of so much value in the present investigation. To Count A. V. Keyserling, M. De Verneuil, M. Bouchard, and to all the kind and disinterested friends above named, I again tender my most grateful acknowledgments and thanks.

THOMAS DAVIDSON.

LONDON; JUNE, 1857.

MONOGRAPH

OF

BRITISH CARBONIFEROUS BRACHIOPODA.

Family—TEREBRATULIDÆ.

Genus—TEREBRATULA, *Llhwyd.*

(General Introduction, Vol. I, p. 62 ; and Part IV, article Terebratula.)

THE species belonging to this genus in the Carboniferous period are but few in number, and, as far as we at present know, are characterised by the short, simple loop peculiar to TEREBRATULA *proper.* The ventral valve, it is true, possesses dental or rostral plates, more developed than in the recent or Jurassic type, but this peculiarity does not appear to constitute a character of sufficient importance for the creation of a separate genus; I must therefore decline adopting the genus *Seminula*, M'Coy, for reasons already assigned.

TEREBRATULA·HASTATA, *Sowerby*. Plate I, figs. 1—12.

> TEREBRATULA HASTATA, *J. de C. Sowerby*. Min. Con., tab. 446, figs. 1, 2, 3, Jan., 1824.
> — — *Fleming*. Brit. Animals, p. 371, 1828.
> — — *Phillips*. Geol. of Yorkshire, vol. ii, p. 221, pl. xii, fig. 1, 1836.
> — SACCULUS (part), *De Koninck*. Animaux Fossiles de la Belgique, p. 293, pl. xx, figs. 3 $^{a, b}$, 1843.
> ATRYPA HASTATA, *M'Coy*. Synopsis of the Carb. Foss. of Ireland, p. 153, 1844.
> — VIRGOIDES, *M'Coy*. Ibid., p. 158, pl. xxii, fig. 21.
> SEMINULA HASTATA, *M'Coy*. British Palæozoic Fossils, p. 409, 1855.
> — VIRGOIDES, *M'Coy*. Ibid., p. 413, Pl. 3 D, fig. 23.

Spec. Char.   Shell of an elongated, oval, or obscurely pentagonal shape, more or less truncated in front, tapering at the beaks, and widest towards its middle or frontal region; valves almost equally convex, rarely gibbous, with a wide mesial depression or concavity towards the front in the larger number of full grown individuals; beak moderately produced, and but slightly incurved; foramen rather large, oval, and in adult shells approximating the umbone of the smaller valve, so as to conceal much of the deltidium which partly surrounds the aperture; the lateral portions of the beak are somewhat flattened, the ridges being indistinctly defined; the lateral margin of the ventral valve indents the opposite one in the proximity of the beak, and is straight or slightly raised in front; external surface smooth, marked only by a few concentric lines of growth; shell structure minutely perforated.   In the interior of the ventral valve there exists two short, diverging, dental or rostral shelly plates, while in the interior of the dorsal one a short, simple loop is observable, occupying about one third or less of the length of the valve.   Dimensions variable.

Length 24, width 18½, depth 13 lines.

„   22½, „ 14,   „   10 lines.

Obs.   M. De Verneuil and several other authors are of opinion that *T. hastata* should be considered but as a variety of *T. sacculus* of Martin.   This view was also advocated, in 1843, by Professor De Koninck, but since abandoned from having observed differences which appeared to him incompatible with the idea that both should be confounded under a single denomination.   In his opinion, as well as in that of Professor M'Coy, there exists two adult types, the one being infinitely larger than the other, " which is proved by the thickening of the margin at or under one inch in length (*T. sacculus*), the deeper notch in front, and the mesial sulcus existing at a much smaller distance from the beak than in *T. hastata ;* and, above all, the species are distinguished by the great difference of angle at which the valves meet at the margin, the front and lateral margin of *T. hastata* being sharp and wedge-like, but those of *Sp. sacculus* being extremely blunt and obtuse."   Such are the characters by which Professor M'Coy distinguishes the two shells; but I confess that at times, and before a large series of both, one feels tempted to look upon these differences as of little value, and to consider the one as a variety of the other.   A point worthy of notice is, however, that in no example of the true *T. sacculus* have we observed any trace of those stripes due to colour which are so beautifully preserved in many specimens of Sowerby's shell, such as in those we have illustrated from Derbyshire (figs. 6, 8, 9, 16).   It is the opinion of Professor De Koninck and of other authors, that the shell under description may have been intended for " *Anomia attenuata*," Martin,[1] which was simply characterised by a few vague Latin words; but as

---

[1] Martin, while alluding to species unfigured in his work, states that "five belong to the Perforati, Fam. *ff*, but only one there is any danger of mistaking for *Sacculus*, it may be distinguished by the following characters : Conch. *Anomites attenuatus.*—C. anomites longitudinaliter ovatus lævis compressus, margine acuto integerrimo."

no one would be able to identify a species from that alone, authors have justly agreed to retain Sowerby's name for the shell under description. *T. virgoides*, M'Coy, does not appear either to Mr. Salter or to myself to present any distinctive characters of sufficient value to warrant its separation from Sowerby's species, as proposed by the Irish author, because it is not uncommon to find that the depression described in *T. hastata* occurs only in one of the valves, while in other cases it is hardly perceptible in either; the front margin is also very variable, being quite straight, and even slightly convex, in some individuals, while in others it presents many degrees of concavity, so much so that some individuals in this respect resemble *T. cornuta*, Sowerby. For similar reasons we are doubtful as to the propriety of admitting *T. ficus*, M'Coy, established from a single shell in the Cambridge Museum; but as we have since been able to examine some other examples, more elongated it is true, but otherwise approaching M'Coy's type, and differing by their extreme convexity and non-indented front from the generality of full-grown examples of *T. hastata*, that it may perhaps appear desirable to provisionally retain for these exceptional forms the varietal denomination of *ficus*, given by Professor M'Coy to his species.

T. hastata was ornamented by stripes, in all probability of a red colour, similar to those we find in several recent forms, such as *T. rubella*, *T. pulchella*, &c. They are also very similar to those seen on some specimens of *Terebratula biplicata*, from the Upper Green-sand of Cambridge, as may be perceived by referring to Vol. I, Part II, pl. vi, fig. 6. *T. hastata* has been stated to have been found also in the Permian rocks, but I must confess that I have never observed any specimens that would satisfactorily prove the assertion; nor am I disposed to admit that it and *T. elongata* did belong to the same species, as it has been more than once hinted.

Loc. *T. hastata* abounds in many English and Irish localities, but seems to be a much rarer shell in Scotland. I have it from Derbyshire, and in particular from Park Hill, Longnor, and a beautiful series, with their colour-markings, may be seen in the Museum of Practical Geology. It is common at Lowick, Kendal, Settle, Bolland, the Isle of Man, &c. In Ireland it abounds at Millecent, Lisnapaste, Little Ireland, &c. In Scotland at Nellfield, Lanarkshire. It is not a rare fossil on the Continent.

Var. *ficus*, M'Coy. Plate I, figs. 13—16.

> Seminula ficus, *M'Coy*. Annals and Mag. of Nat. Hist., vol. x, 2d series; and British Palæozoic Fossils, p. 409, pl. iii D, fig. 22, 1855.

The variety *ficus* is longer than wide, ovate, with convex and gibbous valves, presenting in old individuals a tendency to become obscurely triundate towards the front. The margin in front forms a convex outward curve, which in the ventral valve is likewise slightly raised, and indents to a lesser or greater degree that of the opposite one. The beak is rather more incurved than in *T. hastata* proper. Dimensions from two examples have produced—

Length 17½, width 15, depth 10 lines (type).

    „   22½,   „   16,   „   12 lines.

Loc. Derbyshire, the Isle of Man, &c.

TEREBRATULA SACCULUS, *Martin*, Sp.   Plate I, figs. 23, 24, 27, 29, 30.

> CONCHYLIOLITHUS ANOMITES (SACCULUS), *Martin.* Petref. Derbesiana, tab. xlvi, figs. 1
>   and 2, 1809.
> TEREBRATULA SACCULUS, *J. de C. Sow.* Min. Con., tab. 446, fig. 1, Jan. 1824.
>   —   —   *Fleming.* Brit. Animals, p. 371, 1828.
>   —   —   *Phillips.* Geol. York., vol. ii, p. 222; pl. xii, fig. 2, 1836.
>   —   HASTATA (part), *De Koninck.* Animaux Fossiles de la Belgique, p. 294;
>   pl. xx, fig. 3 *c, d, e, f, g, h, l*; not *a, b*, 1843.
>   —   SACCULUS, *M'Coy.* Carb. Foss. of Ireland, p. 156, 1844; and British
>   Palæozoic Fossils, p. 411, 1855.

Spec. Char. Obovate, or somewhat obscurely pentagonal, notched and emarginated in front; surface smooth, marked only by a few lines of growth; valves nearly equally deep, and more or less inflated; dorsal valve regularly convex, or with a slight depression near the front. The ventral valve presents a rather deep and concave mesial furrow, commencing at about half the length of the valve, and extending to the front. The margin of the ventral valve is straight in front, or indents by a convex curve the corresponding portion of the opposite one; the beak and foramen are of moderate dimensions, incurved with obscurely marked ridges; shell-structure punctuated. Dimensions variable.

Length 14, width 11, depth 7 lines

    „   8,   „   6,   „   4 lines (Martin's type).

Obs. Martin states that "the form of the shell is purse-like, its margin blunt, hollowed out opposite the beak by an obtuse indentation, which is sometimes continued along the back of the beaked valve, in the form of a slight hollow furrow or wave."[1] The last-named character is that which generally distinguishes it best from *T. hastata* and *T. vesicularis;* but, although this peculiar sinus is well and deeply marked in many individuals, it is at times but obscurely so in others, and which occurrence has, no doubt, tempted some authors to unite both Sowerby's and Martin's shell under a single denomi-

---

[1] In p. 14, of the Systematic Arrangement of the Petrifactions of Martin's species, described in the Petrifacta Derbiensis,' the author again alludes to his *Anomia sacculus* as follows:

> *Sacculus.* 24. Conch. Anomites subscrotiformis lævis, margine obtuso: sinu exsculpto.
>   Tab. xlvi, figs. 1, 2.
> *a.* v. Sinu à margine ad valvulæ perforatæ dorsum ducto.
> *b.* v. Sinu subobsoleto. Tab. xlvi, figs. 1, 2.

nation. The frontal margin of the ventral valve, in the greater number of well-shaped and adult individuals of *T. sacculus*, presents a convex curve, indenting to a lesser or greater degree that of the opposite valve, but without producing in it any sensible mesial elevation. Many examples, wherein the sinus presents a gradual concave curve, bear much resemblance to some young conditions of the Permian *T. elongata* and its var. *sufflata*, so much so that it would be difficult to verbally describe the minute differences which seem to distinguish both species, but which, when full grown, are sufficiently evident.[1] From *T. vesicularis*, De Koninck, *T. sacculus* is very often still more difficultly distinguished; and although both shells are pronounced distinct by Professors De Koninck and M'Coy, I have not been able to arrive at so decided an opinion, for although the frontal wave in most examples of Martin's shell is formed by a single curve, there appears to exist a very gradual passage leading to the triundate wave and frontal plication observable in well-characterised examples of *T. vesicularis*.

As the above-named authors seem to differ with me in this particular, I have provisionally described both under a separate denomination. Some naturalists have proposed to consider *T. pentaedra*, Phillips, and *T. didyma*, of Dalman, as synonyms of Martin's *T. sacculus*, but this mistaken notion has already been objected to by M. De Verneuil, at p. 65 of vol. ii of the 'Geol. of Russia." I am likewise compelled to observe, that M. De Koninck is mistaken while identifying *Anomiæ lævis*, in Ure's 'Hist. of Rutherglen,' p. 313, pl. 16, fig. 9 (1793), with the species under description. It belongs to *Athyris ambigua*, Sowerby, a form that abounds at Lawrieston, as well as in the neighbourhood of Carluke, whence Ure's examples were derived.

Loc. *T. sacculus* is common in the mountain limestone. In England it is found at Eyem, Middleton, Moneyash, Cronkstone, Matlock, and other Derbyshire localities; it occurs also at Bolland, Otterburn, Kendal, Malham-moor, &c. In Scotland it is found at Westlothian. Mr. Kelly mentions Ardagh, Millecent, and Little Ireland as Irish localities. M. De Koninck has obtained it at Visé and Chokier in Belgium; but, according to M. De Verneuil and Count Keyserling, it would be rare in Russia, and has hitherto been found but at Cosatchi-datchi, to the east of Miask, in the Oural, and at the mines of Gerichof, Gouvernement of Tomsk, in Siberia.

TEREBRATULA VESICULARIS, *De Koninck.* Plate I, figs. 25, 26, 28, 31, 32; Plate II, figs. 1—8.

> TEREBRATULA VESICULARIS, *De Koninck.* Animaux Fossiles de la Belgique (Sup.), p. 666, pl. lvi, fig. 10, 1851.
> SEMINULA SEMINULA, *M'Coy* (not of Phillips). British Palæozoic Fossils in the Camb. Mus., p. 412, 1855.

Spec. Char. Shell small, ovato-pentagonal, longer than wide, its greatest breadth

---

[1] Refer to what we have stated on this subject in Part IV, under *T. elongata*.

near the middle. In some adult, and in all young shells, the valves are regularly convex, and moderately inflated, but after a certain age a sinus with two lateral ridges is developed, while a mesial depression, with two lateral ridges and a smaller central elevation or rib, exists in the dorsal one, so that this valve towards the front becomes triundate, forming a W-shaped frontal line, of which the central point is either higher or lower than the lateral ones. The ventral valve is deeper and more inflated than the opposite one, the beak rounded and incurved, ridges obscurely defined; foramen small, oval or circular, and partly surrounded, and separated from the hinge by a small deltidium. Surface smooth, or marked by either few or numerous lines and ridges of growth; shell-structure minutely perforated. In the interior of the ventral valve there exists two moderately developed dental or rostral shelly plates, while in the dorsal one a short, simply attached loop extends to about one third of the length of the valve. Dimensions variable. Three individuals have measured—

Length, 7, width 6½, depth 5 lines.
   ,,   5,   ,,   5,   ,,   3½ lines.
   ,,   4½ ,,   4,   ,,   3 lines.

Obs. This shell is extremely variable, both in shape and character, so much so that, to my eyes, certain examples are undistinguishable from others of Martin's *T. sacculus*, and to which M. De Koninck admits it to be nearly related, but distinguishable in well-grown examples by the triplicated aspect of its smaller valve, as well as by its W-shaped frontal margin. This last appearance is, however, exceedingly variable, for in many specimens the triundate wave is imperceptible, or existing simply in a rudimentary condition. Professor M'Coy seems to have been unfortunate in his researches and appreciations regarding the present form, as appears evident from a glance at p. 412 of the 'British Palæozoic Fossils.' Therein the author refers the shell in question to *Terebratula seminula* of Phillips, which is not only specifically different, but belongs likewise to another genus; for, on the admission of Professor Phillips, as well as in the opinion of Professors De Koninck and Morris, *T. seminula*, Phil., is a true *Rhynchonella*, while the shell so described by Professor M'Coy, in the work above mentioned, is a *Terebratula*. In the 'Synopsis of the Carboniferous Fossils of Ireland,' p. 158, the same author changes Phillips's name *T. seminula* into *Ter. pisum*, but which last is subsequently repudiated in the Cambridge work, the author again returning to that of Phillips. Having, through the kindness of Dr. Griffith, been able to examine the original example upon which the Irish professor had founded his views, it has appeared to me evident that *T. pisum*, M'Coy, is a *Rhynchonella*, which either belongs to Phillips's species, or is closely related to that form, for it is similarly plicated, and does not present the appearance of any example of the shell under description, which Professor M'Coy allows to be the same as *T. vesicularis*, De Koninck, and to which the five specimens in the Cambridge Museum so labelled certainly belong.

Loc. In England, it is common in the Craven district, as well as in the Yoredale rocks of Wensleydale, Yorkshire; at Pilsbury Castle, Longnor, Derbyshire, &c. In Scotland Dr. Fleming has the shell from Westlothian. I am not acquainted with any Irish specimens. In Belgium it is stated by M. De Koninck to occur in the lower Carboniferous limestone of Visé, where it is common.

TEREBRATULA GILLINGENSIS, *Dav.* Plate I, figs. 18—20; Plate II, fig. 1.

Spec. Char. Obovate, depressed, smooth, slightly indented in front, widest towards the middle, and frontal region. The dorsal valve is convex at the umbone, but thence to within a short distance of the frontal margin presents a straight or even inward curve. The ventral valve exhibits a flatness in the vicinity of the front, so that this portion of the margin is produced, and forms a convex curve, indenting that of the opposite valve. Beak rounded and incurved, projecting beyond the umbone of the dorsal valve; beak ridges moderately defined; foramen circular, generally approximating the hinge-line, and partially margined by a small deltidium; shell-structure minutely perforated.

Length 9, width 7½, depth 4 lines.
,, 7, ,, 6, ,, 3 lines.

Obs. Numerous examples of this small Terebratula have been collected by Mr. E. Wood, in the Carboniferous beds of Yorkshire. It has been supposed by some palæontologists to be a variety or young state of *Ter. hastata;* while by others it would be referred to the Devonian *Atrypa juvenis* of J. de C. Sowerby. We may also notice that, in the work on the 'Palæozoic Fossils of Devon and West Somerset,' Professor Phillips observes (p. 90), while describing the last-named shell, that, "Mr. Sowerby's figure represents a young specimen; mine is, perhaps, that of a full-grown individual. It is a well-characterised species, though I have specimens supposed to be varieties of *T. hastata,* from the mountain limestone of Yorkshire, whose only distinction is that the widest part of the shell is nearer the front, while in Devonshire forms it is nearer the beaks." Sowerby's figure of *T. juvenis*[1] resembles the young example (Pl. I, fig. 20) of our Carboniferous deposits, but from shells of that age it would not be safe to arrive at a conclusion respecting their specific identity, and especially when we find that there exists so much dissimilarity between the adult condition of *T. juvenis* (as figured by Professor Phillips) and that of our Carboniferous fossil. In the first it is the margin of the dorsal valve that is depressed and convex, while the reverse is the case with the species under description; it exhibits also that dissimilarity in the relative widest part, as was noticed by Professor Phillips. In the work on the 'British

[1] In the 'Trans. of the Geol. Soc.,' 2d series, vol. v, tab. xxxv, Mr. Sowerby describes his *Atrypa juvenis*—"Broad ovate, slightly convex, smooth, curved, longer than wide; front somewhat pointed; valves nearly equal; the lower curved upwards with a minute beak. A small, rather flat species, distinguished by its narrow front, and being curved."

Palæozoic Fossils' (p. 410), Professor M'Coy assimilates both Sowerby's and Phillips's representations of *T. juvenis* with the Carboniferous shell, but I have not been so fortunate as to meet with any agreeing with the adult condition described by Professor Phillips; and the Carboniferous specimens so named in the Cambridge Museum (Plate I, fig. 17) do not certainly represent the Devonian species. This species seems to be easily distinguished from well-authenticated young of *T. hastata* by its more depressed appearance, as well as by the shape and curve of its frontal margin. We have named it after the locality where it abounds, and it will be as well also to notice that all the examples had a reddish tinge, which may perhaps be due to remains of colour.

Loc. It abounds at Gilling, in Yorkshire. Dr. Fleming has the species from West-lothian, Scotland (Plate III, fig. 1).

TEREBRATULA (?) SUBTILITA, *J. Hall.* Plate I, figs. 21, 22.

> TEREBRATULA SUBTILITA, *J. Hall.* In Howard Stransbury's work, 'Explanation of the
> Valley of the Great Salt Lake of Utaty,' p. 409, pl. ii,
> figs. 1 *a, b,* and 2 *a, b, c,* Philadelphia, 1852.

Spec. Char. Ovate, longer than wide, and somewhat tapering at the beaks; valves almost equally deep or convex, but most inflated at and about the umbone of the dorsal one; the mesial fold forms a moderately elevated curve, whence the lateral portions of the valves rapidly decline. In the ventral valve, a shallow sinus commences at about half the length of the valve, and extends to the front, which there presents an elevated convex curve, indenting to a lesser or greater degree that of the opposite valve. Beak moderate in dimensions, and but little incurved; the ridges are obscurely defined; foramen circular, and in general contiguous to the umbone of the opposite valve. External surface smooth, marked only by a few lines of growth; interior unknown.

Length 12, width 10, depth 7 lines.
 ,,   8, ,, 7, ,, 5 lines.

Obs. Of this species I have hitherto been able to examine but two British individuals, obtained by Professor Phillips, in a yellow Carboniferous grit at Mayen Wais, and which appear to be identical with *T. subtilita* of Hall, an opinion first expressed by Professor De Koninck, from the inspection of my figures, before I possessed the means of direct comparison. *T. subtilita* appears to abound in the Carboniferous limestone near the village of Pecos, in the Rocky Mountains of New Mexico, where it is associated with *Prod. semireticulatus* and *Spirifera striata.* It also occurs in the same beds at Sierra Madre and Sierra de Mogoyon, and M. Marcou possesses specimens from the mouth of the Rio San Pedro, in the Rio Gila (Sonora). Professor Hall mentions it from the Missouri.

I am, however, very uncertain whether this shell belongs to the genus *Terebratula,* as

I have not had the opportunity of studying either its interior or intimate shell-structure; this last appearing to be so altered, that I am unable positively to decide whether it was perforated or otherwise. Indeed, some of the American examples in my possession lead me to suppose the structure to be fibrous, and that the species may belong to the genus *Athyris* or *Merista*. It is therefore placed provisionally only into the genus *Terebratula*.

Family—SPIRIFERIDÆ.

Genus—SPIRIFERA, *Sowerby*.

(General Introduction, Vol. I, p. 79.)

SPIRIFERA STRIATA, *Martin*. Plate II, figs. 12—21; Plate III, figs. 2 to 6.

> ANOMITES STRIATUS, *Martin*. Pet. Derb., tab. xxiii, 1809.
> TEREBRATULA STRIATA, *Sow*. Lin. Trans., xii, part ii, p. 515, pl. xxviii, figs. 1 and 2, 1815.
> — SPIRIFERA, *Val.* in *Lamarck*. An. sans Vert., vol. vi, No. 59, 1819; and *Dav.*, Annals and Mag. of Nat. Hist., vol. v, 2d series, p. 449, 1850.
> SPIRIFER STRIATUS, *Sow*. Min. Con., tab. 270, May, 1820.
> — ATTENUATUS, *J. de C. Sow*. Min. Con., vol. v, p. 151, tab. 493, figs. 3, 4, 5, May, 1825.
> — STRIATUS and ATTENUATUS, *Fleming*. Brit. Animals, p. 375, 1828.
> — — *Davreux*. Const. Géogn. de la Province de Liege, p. 273, pl. vii, fig. 2, 1831.
> — — and ATTENUATA, *Phillips*. Geol. of Yorkshire, vol. ii, pp. 217, 218, pl. ix, fig. 13, 1836.
> — — — *Von Buch*. Ueber Delthyris, p. 47, 1837 and 1840.
> — — — *De Koninck*. Animaux foss. de la Belgique, p. 256, pl. xv *bis*, fig. 4, 1843.
> — STRIATA, *M'Coy*. Synopsis of the Carb. Foss. of Ireland, p. 135, 1844.
> — PRINCEPS, *M'Coy*. Ibid., p. 133, pl. xxi, fig. 1.
> — ATTENUATA, *M'Coy*. Ibid., p. 129.
> — CLATHARATA, *M'Coy*. Ibid., p. 130, pl. xix, fig. 9.
> — STRIATA, *De Vern*. Russia and Oural, vol. ii, p. 167, pl. vi, fig. 4, 1845.
> — — *M'Coy*. British Palæozoic Fossils, p. 222, 1855.

Spec. Char. A very large and variably shaped shell, transversely semicircular, or sub-rhomboidal; valves almost equally convex. In the dorsal valve the mesial fold is of moderate elevation, while the sinus in the opposite one is both variable in its width and depth. The hinge-line is either a little shorter, or as long as the greatest width of the shell, the cardinal angles being more or less rounded in adult individuals. The area is of moderate width, with sub-parallel sides; fissure triangular, and partially covered by a pseudo-deltidium. The external surface of the shell is ornamented by a variable number of radiating ribs, which augment in number, to a greater or lesser extent, from intercalations at unequal distances from the beaks; so that from seventy to ninety may be counted round the margin of each valve in adult individuals. The ribs on the fold and sinus are likewise more flattened than on the lateral portions of the shell. The surface is closely and finely reticulated. In the interior of the dorsal valve, under the extremity of the incurved umbonal beak, there exists a small cardinal process or muscular fulcrum, and on either side are situated the dental sockets. The spiral cones which fill the larger portion of the shell are attached to the extremities of the inner socket-walls. The lamellæ, after having converged and given birth to the crural processes, diverge, and form the first of the twenty or twenty-two convolutions of which each spiral is composed. Four impressions left by the adductor muscle are visible in this valve. In the interior of the ventral valve a strong hinge-tooth is situated on either side at the base of the fissure, and is supported by a vertical shelly plate of much strength, but not advancing to any great length into the interior of the valve. Between these a large portion of the free space at the bottom of the shell is occupied by the adductor and cardinal muscular impressions, which are divided by a blunt, central, longitudinal ridge. The dimensions of one of the largest examples are—

Length 4½ inches, width 6 inches 1 line, depth 3 inches 1 line.

Obs. In the opinion of M. De Koninck, De Verneuil, M'Coy (1855), as well as of other authors, *Sp. attenuata*, Sow., must be considered as a synonym of *Sp. striata* of Martin; and in this view I am the more disposed to concur, from having been able to assemble and study upwards of four hundred individuals of this shell, from the dimensions of a few lines to those of the largest example above recorded. With so numerous a series before me, all minor differences, so remarkable when particular or exceptional forms are placed in comparison, soon vanished, and I can see in them nothing further than those dissimilarities so common to individuals of every species composing the animal kingdom; for it is a well-known fact that no two examples are ever found to be so exactly similar as to induce one to suppose that they were cast in the same mould, and it has also been well ascertained that under certain conditions individuals of most species may become adult with much smaller dimensions in one locality than in another. The number of ribs or external sculpture varies also to a considerable extent in different specimens. This is no doubt the reason why at Millecent (Kildare) we may meet with upwards of a thousand middle-sized specimens for one large individual, while in other localities the proportions are on an average much larger. The name *Sp. princeps* was given by Professor M'Coy to a full-grown example of Martin's *striata*, and must therefore be added to the synonyms of that form. M. De Verneuil is also of opinion that *Sp. condor* (D'Orb),[1] represents an individual of *Sp. attenuata;* and I possess specimens from the same locality as those described by the French author, which entirely agree with the species under description. Professor M'Coy is also of opinion that what he described in 1844 as *Sp. clatharata,* must be added to the list of synonyms.

The long hinge-line of *Sp. striata* distinguishes it from *Sp. duplicicostata* of Phillips, but to which it is, however, very closely related by several peculiarities, for many examples of *Sp. striata* present not only similarly bifurcated and intercalated ribs, but these are also at times arranged in clusters, as is generally the case in Phillips's species. The elevation and proportions of the mesial fold varies also considerably, and in some young and middle-aged shells from Millicent it is badly defined; these last being also much more elongated and spindle-shaped than is commonly the case; while others are rhomboidal, with or without prolonged cardinal angles. The ribs on the mesial fold present at times (although much more rarely than in *Sp. bisulcata*) a tendency to divide into three groups separated by deeper sulci; but, as above observed, all these minor differences so remarkable, in single individuals, are, specifically, of little importance.

Sp. striata is the largest Spirifer known, and must always be considered as the type of the genus. There are many varieties, but I shall simply allude to the one found at Dovedale, Derbyshire (Plate II, figs. 15, 16), in which the ribs become so narrow and numerous, that one hundred may be counted round the margin of each valve in certain individuals, giving the shell much the appearance of the Silurian *Sp. cyrtæna.*

Loc. This species is found at Castleton, Dovedale, and other Derbyshire localities; at Bolland, and different parts of Yorkshire. In Ireland Mr. Kelly mentions Ratheline, Millecent, Little Ireland, Lisnapaste, Cornacarrow, Ardclogh, &c. In Scotland it appears extremely rare, for I have not been able to find any well-authenticated example in the different collections I have been able to examine. It seems to be a rare fossil at Visé, in Belgium, but more common at Ratingen. M. De Verneuil found it at Cosatchi-datchi (east of Miask), on the other side of the Oural Mountains; also at Pos-en-Cavales (Spain). It is one of the most characteristic fossils of the lower Carboniferous limestone, being found in various American localities, such as Yarbichambi, at 4000 métres of elevation, north of La Paz (Bolivia), as well as in the Quebaya Island, in the Sea of Titicaca. M. Marcou obtained it abundantly in the Rocky Mountains, as well as from New Scotland to Vancouver's Island.

---

[1] 'Voyage dans l'Amérique Méridionale,' vol. iv, pl. xlvi, figs. 11—15.

SPIRIFERA MOSQUENSIS, *Fischer de Waldheim*. Plate IV, figs. 13, 14.

> CHORISTITES MOSQUENSIS, *Fischer*. Programme sur les Choristite, p. 8, No. 1, 1837;
> and Oryctogr. du Gouv. de Moscow, p. 140, pl. xxii,
> fig. 3; pl. xxiv, figs. 1—4, 1837.
> — SOWERBYI, *Fischer*. Ibid., pl. xxiv, figs. 5, 6, 7, pl. xxv, fig. 6. (Not
> Spirif. id. *Defrance*.)
> — KLEINII, *Fischer*. Ibid., pl. xxiv, figs. 8, 9.
> DELTHYRIS INCISA, *Goldf.* German Translation of the Manual of De la Beche, p. 526,
> 1832.
> SPIRIFER CHORISTITES, *Von Buch*. Ueber Delthyris, p. 45, 1837; and Mém. de la
> Soc. Géol. de France, vol. iv, p. 191, pl. ix, fig. 16.
> — PRISCUS, *Eichwald*. Urwelt Russl., Heft i, p. 97, pl. iv, fig. 12, 1840.
> — CHORISTITES, *M'Coy*. Synopsis, p. 130, 1844.
> — SOWERBYI, *De Koninck*. Desc. des Animaux fossiles de la Belgique, p. 252,
> pl. xvi, fig. 1, 1843.
> — MOSQUENSIS, *De Verneuil* and *Keyserling*. Russia and Oural, vol. ii, p. 161,
> pl. v, fig. 2.

Spec. Char. Semi-oval, or of a lengthened semicircular shape, transverse when young, more or less elongated in the adult condition; valves convex, and somewhat inflated with a rather wide but not greatly elevated fold in the dorsal, and shallow sinus in the ventral valve. The hinge-line is generally as long as the greatest width of the shell; the area has sub-parallel sides, with a triangular fissure partially covered by a pseudo-deltidium. The beaks are incurved and approximate. Externally each valve is ornamented by from fifty-five to sixty-five narrow, simple, or intercalated radiating ribs, of which from twelve to sixteen occupy the sinus and mesial fold. The ribs in both valves are likewise intersected by numerous concentric lines or ridges of growth, which become the more approximate as they approach the margin. In the interior of the ventral valve a strong tooth is situated on eitherside at the base of the fissure, supported by two large, vertical, shelly plates, which extend from the extremity of the beak to the bottom of the valve, first forming the fissure walls, then converging to diverge again, and extending to a distance of two thirds of the length of the valve. Dimensions variable.

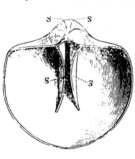

Spirifera mosquensis.
Interior of the ventral valve.
S, septa.

Length 22½, width 23, depth 14½ lines.

Obs. This species has been described to considerable length by both MM. De Koninck and De Verneuil, the last-named author having drawn particular attention to those large dental or rostral plates observable in the ventral valve, which distinguish it from *Sp. striata* of Martin. These shelly plates or septa have been illustrated in Pl. VI of our General Introduction, as well as in pl. v, fig. 2ᵇ, of the 'Geology of Russia.'

The very few British examples of this species which I have been able to examine did not present any interior, but the external characters are so exactly similar to some Russian examples from the Donetz at Moscow and Viterga, that I have felt no doubt as to their identity. *Sp. mosquensis* is also very variable in its proportions, and (as observed by one of the above-named authors) shows a marked tendency to elongation, especially in the adult condition. Thus, some Russian specimens measured 27 lines in length by 22 in breadth; and a Belgian example, figured by M. De Koninck, attained still larger dimensions (38 lines in length by 39 in breadth).

Sp. mosquensis is distinguished from *Sp. striata* by its much more elongated appear-ance; from *Sp. bisulcata* it is separable by its more numerous and smaller ribs, and approaches most to *Sp. humerosa* of Phillips, but from which it is likewise distinguished by a smaller and less inflated beak than that peculiar to Phillips's species.

Both MM. De Koninck and De Verneuil have remarked that the shell under descrip-tion must not be confounded with that named *Sp. Sowerbyi* by Defrance, in the 'Dictionnaire des Sciences Naturelles,' vol. 50, p. 295, pl. lxxvi, fig. 2, which belongs to a specimen of *A. reticulata* of Linnæus.

Loc. In England it is found near Bristol. A specimen is also labelled Derbyshire in the Cambridge Museum. In Ireland it has been collected from Little Ireland; but I have not hitherto observed any example from Scotland. In Belgium it is stated by M. De Koninck to occur abundantly near Tournay, d'Ath, de Soignies, de Felay, &c. Fischer de Waldheim obtained it from a white Carboniferous limestone at Grigorievo, Podolsk, Miatchkovo, and other localities in the neighbourhood of Moscow. M. De Verneuil found it to be one of the most characteristic shells of the middle beds of the Carboniferous system throughout Russia and Oural, being there associated with *Productus gigas* and other shells.

SPIRIFERA HUMEROSA, *Phillips.* Plate IV, figs. 15, 16.

SPIRIFERA HUMEROSA, *Phillips.* Geol. of Yorkshire, vol. ii, p. 218, pl. xi, fig. 8.

Spec. Char. A ponderous shell, variable in shape, sometimes transverse, but more often elongated; valves convex; beak large and considerably inflated, much incurved, with a wide, shallow sinus, extending from the extremity of the beak to the front. Hinge-line shorter than the greatest width of the shell. Area narrow. Dorsal valve convex, with a produced, rather angular mesial fold; frontal wave strongly marked, the margin of the sinus indenting to a considerable extent the corresponding portion of the dorsal valve. Surface ornamented by numerous small ribs, which augment rapidly from numerous intercalations; the ribs are more or less flattened, especially on the mesial fold, and are at the same time intersected by numerous concentric lines or ridges of growth. Dimensions very variable.

Length 2 inches 7 lines, width 2 inches 2½ lines, depth 1 inch 6½ lines.
  ,, 2 inches,   ,, 2 inches 6 lines, ,, 1 inch 6 lines.

Obs. Professor Phillips describes this species with "a lower (ventral) valve swollen near the beaks, and *produced* in a mesial furrow receiving the angular ridge of the upper valve; radiating ribs small, duplicate. *Loc.* Greenhow Hill." The figure published in the 'Geology of Yorkshire' is very unsatisfactory; but, having received the loan of the original specimen, I have drawn it correctly in Pl. IV, fig. 15, so that no doubt may exist regarding its identification with the perfect example, fig. 16, in the possession of Mr. Wood, of Richmond. Professor Phillips's specimen is very incomplete and fragmentary, but still exhibits the characters of the species; and what gives his figure such an extraordinary beak is, that the valves were widely separated while the shell was filling with matrix, so that, in addition to the real beak, a larger portion of the ventral valve is seen than would have been the case had the valves been closed. In fig. 15ᵃ I have given a profile view of the specimen, to satisfy the reader of the correctness of this assertion.

I have been able to examine six individuals of this remarkable shell, but they did not all possess the elongated shape figured in our plate. Some were much shorter and transverse, as will be seen from the measurements already given, and selected from two extreme examples. Its ponderous shape, shorter hinge-line, and smaller ribs distinguish it from Martin's *Sp. striata.*

Loc. Greenhow Hill, and Wensleydale, Yorkshire.

SPIRIFERA DUPLICICOSTA, *Phillips.* Plate III, figs. 7—10; Plate IV, figs. 3, 5—11.

> SPIRIFERA DUPLICICOSTA, *Phillips.* Geol. of Yorkshire, vol. ii, p. 218, pl. x, fig. 1, 1836.
> — FASCIGER, *Keyserling.* Wissenschaft Beobach. Petchora Land., tab. viii, fig. 3.
> — FASCICULATA, *M'Coy.* British Palæozoic Fossils, tab. iii D, fig. 25, 1855.

Spec. Char. Transversely sub-rhomboidal when adult, longer than wide, or almost circular when quite young; valves moderately convex, with a more or less produced mesial fold in the dorsal, and a corresponding sinus in the ventral one. The hinge-line is shorter than the width of the shell, the area of moderate breadth, beak incurved. Valves ornamented by numerous radiating ribs, which rapidly augment at various distances from the beaks by intercalation as well as bifurcation. Two examples have afforded the following measurements:

Length 16, width 20, depth 11 lines.

    „   16½, „   17½, „   10½ lines.

Obs. Professor Phillips describes his species with "mesial fold angular; radiating ribs numerous, duplicate towards the margin." No form is, however, more variable than the one under description, both as to shape, dimensions, and character of plication; and although almost circular when young (Pl. IV, figs. 7, 9, 10, 11), it is in general more or less transversely oval, or sub-rhomboidal, in the adult condition (Pl. III, figs. 9—11 ; Pl. IV, fig. 6), the last having been considered by Professor Phillips to represent the typical form of his species. The ribs differ also much, being either only here and there duplicose, or, as is the case with many examples, having three or four ribs clustered together, and thus producing a very remarkable appearance (Pl. IV, fig. 6). The term *duplicicosta* is not, however, very appropriate, since many species of *Spirifera* present that peculiarity, which is also observable to a lesser degree in *Sp. striata*, to which Phillips's form is very closely related. Professor M'Coy has likewise proposed to unite to the present species the one named *Sp. crassa* by M. De Koninck = *planicosta*, M'Coy ; and although the learned Irish author may perhaps be correct in his identification, I do not as yet feel myself prepared to admit the point as an established fact. *Sp. crassa* seems to constitute a form intermediate in character between *Sp. duplicicosta* and *Sp. bisulcata* of Sowerby, but I have not been able to examine a sufficient number of species to decide the point to my own satisfaction. Professor M'Coy's illustrations of *Sp. faciculata* (Pl. IV, fig. 11) so closely resemble many well-authenticated examples of the shell under description, that I have, with the sanction of both Professors Phillips and De Koninck, added it to the synonyms of *Sp. duplicicosta*. *Sp. fasciger*, Keyserling, evidently belongs to the same species, and has been so admitted by Professor M'Coy.

Loc. Common in many Carboniferous localities, such as in the Great Scar Limestone of Park Hill, Longnor, Derbyshire ; at Bolland and the Craven district ; at Lowick, in Northumberland ; at Kendal ; and Poolvash, Isle of Man. In Scotland, Mr. J. Young and Dr. Fleming possess it from Corieburn (Campsie), in Lanarkshire, as well as from Westlothian. In Ireland, it is mentioned by Mr. Kelly as occurring at Lisnapaste, Malahide, and Mullaghfin.

SPIRIFERA CRASSA, *De Koninck.* Plate VI, figs. 20—22 ; Pl. VII, figs. 1, 2, 3.

SPIRIFER CRASSUS, *De Koninck.* Animaux fossiles de la Belgique, p. 262, pl. xv *bis*, fig. 5, 1843.
BRACHYTHYRIS PLANICOSTA, *M'Coy.* Synopsis of the Carb. Foss. of Ireland, p. 146, pl. xxi, fig. 5, 1844.
SPIRIFERA CRASSA, *D'Orbigny.* Prodrome, vol. i, p. 149.
— DUPLICICOSTA, *M'Coy.* British Palæozoic Fossils, p. 415, 1855.

Spec. Char. Transversely oval; valves almost equally convex, and somewhat inflated ; hinge-line shorter than the greatest width of the shell ; area triangular, of moderate width, with cardinal angles rounded off. Beak small, incurved, not much produced. The mesial fold is but slightly raised above the level of the lateral portions of the valve ; sinus wide, not very deep. The valves are ornamented with from forty-five to seventy narrow, rounded, unequal, bifurcated, or intercalated ribs, of which from ten to twelve occupy the surface of the mesial fold and sinus. The following are the measurements from two examples :

Length 28, width 36, depth 23 lines.

„ 22, „ 30, „ 13 lines.

Obs. I have been able to examine very few examples of the present form, but these varied considerably both in degree of convexity as well as depth. The mesial fold is never much elevated, and is at times scarcely defined, as in Pl. VII, figs. 1 and 3 ; so that some examples show hardly any frontal wave. The ribs also vary much in their respective widths, even on the same individual, so much so that some specimens seem hardly distinguishable from *Sp. bisulcata*, while others approach most to *Sp. duplicicosta ;* it seems however, to be much more regularly oval than is commonly the case with Phillips's species, in which the mesial fold is also generally more elevated, giving to the shell a different aspect. From *Sp. bisulcata* it seems distinguished by the shortness of its hinge-line.

Loc. Dr. Griffith's examples are from the lower limestone of Mullaghfin and Milverton, in Ireland. A fine specimen in the Cambridge Museum is labelled Derbyshire. It is not very rare at Visé, in Belgium, whence Professor De Koninck's types were derived.

SPIRIFERA PLANATA, *Phillips.* Plate VII, figs. 25—36.

> SPIRIFERA PLANATA, *Phillips*, Geol. of Yorkshire, vol. ii, p. 219, pl. x, fig. 3, 1836.
> BRACHYTHYRIS PLANATA, *M'Coy.* Synopsis of Carb. Foss. of Ireland, p. 146, 1844.
> SPIRIFER ROTUNDATUS ? *Semenow* (non *Martin*). Ueber die Fossilien des Schlesischen Kohlenkalkes, 1854.
> — RECURVATUS, *M'Coy* (non *De Koninck*). British Palæozoic Fossils, p. 421, 1855.

Spec. Char. Almost circular or ovato-subtrigonal, often as wide as long; hinge-line shorter than the width of the shell; area triangular, of rather small dimensions. Dorsal valve moderately convex, uniformly so in young and even aged individuals, so that there hardly ever exists a regularly defined mesial fold, its position being in general indicated by two deeper sulci. The ventral valve is much deeper and more convex, with a narrow longitudinal sinus, producing but a slight frontal wave; beak prominent and incurved. The surface of the valves are ornamented by numerous delicate radiating ribs, four or five occupying the place of the fold, while from thirteen to fifteen exist on either side, several of which being due to intercalation at various distances from the beaks.

Length and width rarely exceeding 9 or 10 lines, depth from 5 to 6 lines.

Obs. This pretty little shell is a well-characterised and readily recognisable species, but from not having been sufficiently described in the 'Geology of Yorkshire,' has given rise to some false identifications, not only abroad but likewise at home; for in the

Cambridge Museum we find the species, both labelled and described by Professor M'Coy as *Sp. recurvatus* of Professor De Koninck. For a long time I considered the Belgian form last named to be synonymous with the British species, but am positively assured by Professor De Koninck that they are distinct; and that, although both are found in Belgium, they do not occur in the same beds. The characteristic shape of the adult shell is represented by figs. 26, 27, 31, and 35; fig. 26 being drawn from one of Professor Phillips's typical examples in the British Museum; fig. 31 is from another in his own collection. The specimens selected for illustration exhibit the different aspects and character of striation presented by this species, and it will be observed that in the young state (fig. 36), and often even up to an advanced age (fig. 33), the regular convexity of the valves is undisturbed by either sinus or fold, both valves being covered by numerous slightly produced striæ. In other examples (figs. 26 and 35) a distinct, narrow, and obscurely defined flattened fold is visible; a sinus in the opposite valve being also perceptible, which is, however, deeper in some exceptional examples (fig. 28). These variations have their influence on the frontal wave, which is at times hardly perceptible; while at others, as in figs. 31 and 38, it is clearly defined. The number of ribs is also very variable, being smaller or larger according to individuals, and increased by intercalation. I have counted from thirty to thirty-eight round the margin of the smaller valve, four or five occupying the place of the fold where such is regularly defined.

In the 'Synopsis of Carb. Fossils of Ireland,' Professor M'Coy states—"It appears to me that this species (*Brachythyris planata*) is the shell intended by Martin as his *Anomites rotundatus*, and not *Spirifer rotundatus* of Sow." This is, however, a mistake, as the young of *rotundatus* is certainly that of the species of which Sowerby published the adult at a later period, and does not possess the character of *planata* of Phillips. Professor M'Coy gives to his Irish specimens the dimensions of 1 inch in length and width, which surpasses that of any of the numerous English examples I have been able to examine. M. Semenow has also fallen into the same mistake.

Loc. Common at Bolland, whence Professor Phillips obtained his types. Those in the Cambridge Museum are labelled Derbyshire. It occurs also at Settle, in Yorkshire, as well as in the Isle of Man (Dr. Cumming). Mr. Kelly quotes Bundoran, Ardclogh, and Little Ireland, as the Irish localities. I am not acquainted with any Scotch example. M. De Koninck has it from Vaulsort, near Dinant, in Belgium.

Spirifera triangularis, *Martin*, sp. Plate V, figs. 16—24.

Conchyliolithus anomites triangularis, *Martin*. Pet. Derb., pl. xxxvi, fig. 2, 1809.
Spirifer triangularis, *Sowerby*. Min. Con., tab. 562, figs. 5, 6, May, 1827.
— — *Von Buch*. Mém. Soc. Géol. de France, vol. iv, p. 182, pl. viii, fig. 5, 1840.

SPIRIFER TRIANGULARIS, *De Koninck.* Animaux fossiles de la Belgique, p. 234, 1843.

—   ORNITHORHYNCHA, *M'Coy.* Synopsis of the Carb. Foss. of Ireland, p. 133, pl. xxi, fig. 2, 1844; and British Palæozoic Fossils in the Camb. Museum, p. 418, pl. iii D, fig. 27, 1855.

—   TRIANGULARIS, *Semenow.* Ueber die Fossilien des Schlesischen Kohlenkalkes, 1854.

Spec. Char. Triangular, twice as wide as long, with a straight, elongated hinge-line, and slightly concave, nearly parallel-sided area, towards the attenuated extremities of which the lateral margins of each valve converge, forming acute angles with the hinge. The fissure is triangular, and partly covered by a pseudo-deltidium. The dorsal valve is less convex than the opposite one, with an elevated mesial fold, which commonly assumes the character of a single produced and acutely angular cuneiform ridge or rib, at times considerably prolonged beyond the frontal level of the lateral portions of the valve. On either side of this central ridge from six to ten smaller ribs ornament the lateral portions of the valve. The beak of the ventral valve is narrow, produced, and incurved. A shallow mesial sinus commences at the extremity of the beak, and extends to the front; but at a short distance from its origin a mesial or central rib originates, which becomes wider and more elevated and produced as it approaches the front, and corresponds with the central ridge of the dorsal valve. Seven to eleven smaller ribs exist also on the lateral portions of the valve, on either side of the sinus. The dimensions taken from a perfect individual have produced—

Length 10½, width 21½, depth 6½ lines.

Obs. This elegant shell was first described and figured by Martin, from a specimen which, at a subsequent period, became the property of the late J. Sowerby, and of which he gave a somewhat restored illustration in the 'Mineral Conchology.' Through the kindness of Mr. J. de C. Sowerby, I am myself enabled to portray this individual (figs. 16, 17), so that there can exist no doubt as to the identity of the original type. I have said so much to be able to prove that Professor M'Coy was decidedly mistaken when, in 1844 and 1855, he considered his so-termed *Sp. ornithorhyncha* to be distinct from Martin's species;[1] for, after a careful inspection of the last-named author's type (fig. 23), it will be easily seen that the only visible difference between it and Martin's specimen consists in the more or less developed condition of the mesial ridge in either valve. Martin's specimen clearly shows the ridge that originates in the sinus of the larger valve, so characteristic a feature of his species. *Sp. triangularis* appears to be a rare fossil, both in Great Britain and on the Continent, the only perfect example I have been able to examine being the one in the possession of Mr. Reed, of York (figs. 18—21). It here also assumes much of the cruciform aspect described by Professor M'Coy, and is intermediate

---

[1] Professor De Koninck allows me to state that he entirely coincides in the view here expressed.

in character between Martin's specimen and that figured by Professor M'Coy. The number of ribs also varies a little in different individuals.

This shell was well described and figured in 1843 by Professor De Koninck, in his excellent work on Belgian Carboniferous fossils, and through that gentleman's kindness I have been able to examine the very few individuals he could obtain from the quarries of Visé. M. De Koninck states that "this shell, whose denomination sufficiently denotes the shape it usually assumes, is, above all, characterised by the singular conformation of its sinus, to which no attention has been paid by the authors who have described it. This sinus is smooth, and presents nothing particular in young individuals; but when they have acquired half their growth, a small ridge appears, which becomes wider with age, and is never absent in the numerous varieties under which this species presents itself, and of which we have figured the most dissimilar shapes," &c.

The surface of the valves is ornamented, in addition to the ribs, by numerous concentric lines of growth, which become the more approximate as they approach the margin of the valves. I have, through the kindness of Dr. Griffith and Professor Sedgwick, been enabled to examine the original examples on which Professor M'Coy founded his *Sp. ornithorhyncha*, and to compare them with Martin's specimen of *T. triangularis*. Professor Phillips's figure, in the 'Geol. of Yorkshire,' vol. ii, pl. ix, fig. 12, is larger than any example I have seen, and cannot be considered a characteristic representation of the species.

Loc. This shell was obtained at Buxton by Martin. It occurs also in Derbyshire; at Settle, in Yorkshire; at Bolland, Kirkby Lonsdale, &c. Professor M'Coy's specimens were from Millecent (Clare), Ireland. I have not seen any Scotch specimens.

SPIRIFERA TRIGONALIS, *Martin.* Plate V, figs. 25—34; 35—37?

CONCHYLIOLITES ANOMITES TRIGONALIS, *Martin.* Pet. Derb., tabl. xxxvi, fig. 1, 1809.
SPIRIFER TRIGONALIS, *Sowerby.* Min. Con., tab. 265, fig. 1 (not 2 and 3), 1820.
SPIRIFERA TRIGONALIS, var. a, *M'Coy.* British Palæozoic Fossils, p. 423, 1855.

Spec. Char. Transversely trigonal; hinge-line almost as long, or a little longer, than the width of the shell, the lateral angles being either rounded off or acute, and slightly prolonged. Area sub-parallel, of moderate width, and divided by a triangular fissure, covered in part by a pseudo-deltidium. Beak rounded, moderately produced, and incurved; valves almost equally convex; the mesial fold in the dorsal valve is elevated, angular, and extended beyond the level of the lateral portions of the valve; it is, in general, divided by three principal ribs, of which the central one is at the same time the largest and most extended. In the ventral valve the sinus is deep, and likewise divided by three longitudinal ribs, the central one being (as in the dorsal valve) the most developed.

In addition to these, the surface of each valve is ornamented by from twenty to twenty-two simple ribs. The dimensions, taken from two examples, measured—

Length 11, width 13½, depth 8 lines.

„ 15, „ 18, „ 10 lines.

Obs. Martin states that "the general form or outline of this shell is trigonal or three-cornered, with the angles rounded off (our fig. 25); its surface longitudinally furrowed and sinuated; the furrows rounded, their number varying from twenty to thirty; the sinus continued, rounded, and extending the breadth of three or four furrows, hence striated like the other parts of the surface," &c. The species was not sufficiently illustrated, which has led subsequent authors to combine with it other allied forms, so as to obscure its distinctive characters. Thus, for example, in tab. 265 of the 'Mineral Conchology,' fig. 1 alone would belong to Martin's shell, while figs. 2, 3, and 4 are referable to Sowerby's own *Sp. bisulcata*, a fact I was able to confirm from the close examination of type examples of each species.[1] I am quite ready to admit *Sp. bisulcata* to be a nearly allied form, but am not yet entirely convinced that Professor M'Coy is right while considering Sowerby's shell as a simple variety of that of Martin's. Both may be distinguished by differences in general shape, the mesial fold in *Sp. trigonalis* being much more elevated, angular, prolonged, and possessing fewer ribs than in *Sp. bisulcata*, where the fold is much flatter, or more regularly rounded. Its hinge-line is also more extended, and the contour of the shell more semicircular and less trigonal than in Martin's shell. Nor can I agree with Professor M'Coy when he places *Sp. rhomboidea*, Phillips, among the varieties of the species under description.

Some young and exceptional examples of *Sp. triangularis* approach much to *Sp. trigonalis*, but in full-grown individuals the differences are well defined. The illustration attributed to *Sp. trigonalis* by Professor De Koninck, in his work on the 'Animaux fossiles de la Belgique,' pl. xvii, fig. 1, does not convey an adequate idea of Martin's species, to which in all probability it does not belong. I agree with Professor M'Coy that Eichwald's *Sp. incrassata*, as figured in the 'Geol. of Russia,' pl. vi, fig. 3, may be referred to *Sp. trigonalis*, but De Verneuil's *Sp. Strangwaysi*, fig. 1 of the same plate, is not equally certain.

Loc. This species is very abundant in the lower Carboniferous limestone of many British localities. Martin's type was derived from Derbyshire. Mr. Tate has it from Denwick and several other localities in Northumberland. It is found at Buxton. At Bakewell, in Derbyshire, several beautiful examples were obtained by Mr. Binney, from a

---

[1] Professor M'Coy seems to have hesitated to admit as *Sp. trigonalis* the above-named figures published by Sowerby, for we find stated, at p. 424 of the work on the 'British Palæozoic Fossils'—"I have not quoted Sowerby's figures ('Min. Con.,' tab. 265), as he gives no definition to the mesial ridge, and makes the ribs so much broader than any of the great number of specimens I have examined, that it is scarcely recognisable as a portrait of the ordinary forms." Professor M'Coy's description of Martin's shell is both detailed and correct.

cherty Carboniferous limestone, in which the spirals were beautifully preserved, while the internal casts exhibit the muscular and ovarian impressions (figs. 26, 27). It is also common, in this shape, in a yellow sandstone near Richmond, in Yorkshire, as well as Kendal. It abounds in the dark Carboniferous limestone of Lowick, and at Dent, in Yorkshire. It is not rare in Scotland, it is found at Courland, near Dalkeith; at Dryden; and Braidwood, near Carluke; and is also mentioned from several Irish localities.

SPIRIFERA BISULCATA, *Sowerby.* Plate IV, fig. 1? Plate V, fig. 1; Plate VI, figs. 1—19; Plate VII, fig. 4.

SPIRIFER BISULCATUS, *J. de C. Sowerby.* Min. Con., tab. 492, figs. 1 and 2, 1825.
— TRIGONALIS, *Sowerby* (not *Martin*). Min. Con., tab. 265, figs. 2 and 3, 1820.
— BISULCATUS, *Davreux.* Const. Geog. de la Province de Liege, p. 272, pl. vii, fig. 3, 1831.
— — *De Koninck,* Animaux fossiles de la Belgique, pl. xiv, fig. 4, 1843.
Phillips. Geol. of Yorkshire, vol. ii, pl. ix, fig. 14, 1836.
— SEMICIRCULARIS, *Phillips.* Ibid., pl. ix, figs. 15, 16.
— CALCARATA, *M'Coy* (not *Sow.*) Synopsis of the Carb. Foss. of Ireland, p. 130, pl. xxi, fig. 3.
— TRIGONALIS, var. BISULCATA and SEMICIRCULARIS, *M'Coy.* British Palæozoic Fossils, p. 424, 1855.

Spec. Char. Semicircular or sub-rhomboidal, commonly wider than long, with valves almost equally convex. The hinge-line is, in general, longer than the greatest width of the shell, the cardinal extremities being rounded, or forming angles of variable projection. Area moderately wide, divided by a triangular fissure, which is closed in part by a pseudo-deltidium. Beaks incurved, and at times considerably approximate. The sinus presents a moderate depth; the mesial fold, regularly rounded, is not much elevated, nor does it project to any extent beyond the level of the lateral margins. Each valve is ornamented by from thirty to forty obtusely rounded ribs; these are simple, and rarely bifurcated, but increase occasionally by intercalations at various distances from the beaks. The ribs on the mesial fold are arranged into three groups, separated by sulci of greater depth. Dimensions very variable—

Length 15½, width 18, depth 13 lines (Sowerby's type).
,, 24, ,, 41, ,, 20 lines (a very large individual).

Obs. This common Spirifer is so variable in its general shape, as well as in the number and width of its ribs, that it is not always easily separated from certain exceptional individuals of two or three approximate species. It is, however, distinguished from *Sp. trigonalis*, with which it has been often confounded, by the characters of its mesial fold, which does not project much beyond the level of the lateral margins, and is also regularly rounded, and not acutely elevated as in Martin's shell.

Professor M'Coy considers the form under description to be a simple variety of *Sp. trigonalis*, and states that "the most distinctive character of this variety (*bisulcata*), besides its gibbosity, is the abrupt rising of the mesial sinus at the margin, nearly at right angles to the plane of lateral edges, shortening the length of the middle of the ridge, notching the front margin, and giving a nearly semicircular curve to the profile of the receiving valve," &c.

It is distinguished from *Sp. mosquensis* and *striata* by the simplicity of its ribs, and from the Devonian *Sp. aperturata*, to which it has been assimilated, by the greater depth of the sinus, as well as the elevation of the fold of Schlotheim's species. Sowerby's type, which I have illustrated from the original specimen (Pl. VI, figs. 6 to 9), is very gibbous, and much less transverse than is the case with a vast number of individuals of the species, which also present wing-shaped expansions of greater or lesser extension, as may be seen from the series of examples which have been selected for illustration. It seems rather surprising that Sowerby did not perceive that the specimens, tab. 565, figs. 2 and 3 (13 and 14 of our plate), did belong to his species, and not to *Sp. trigonalis* to which he refers them, a mistake subsequently copied by various authors.

Sp. semicircularis, Phillips, does not seem to possess any characters distinguishable from *Sp. bisulcata*, and I have therefore followed Professor De Koninck in adding the name to the synonyms of the species under description.

The most remarkable peculiarity in *Sp. bisulcata* is the tendency of the ribs on the mesial fold to divide into three distinct groups, separated by sulci of greater depth; and although this may be also observed at times in other Spirifers, such as on some young examples of *Sp. striata*, still it is nowhere better exemplified than in the shell under description. *Sp. bisulcata* appears to have attained much larger dimensions than any example of *Sp. trigonalis* I have hitherto examined.

The shell figured as *Sp. calcarata* by Professor M'Coy, in the 'Synopsis,' pl. xxi, fig. 3 (Pl. VII, fig. 4, of our plates), does not belong to the species so named by Sowerby: it is probably a malformation of *Sp. bisulcata*. The fine specimen, Pl. IV, fig. 1, seems likewise referable to one of the forms of Sowerby's variable species, and *Sp. transiens*, M'Coy, may also, perhaps, be an allied form; but as the original specimen is no longer to be found in Dr. Griffith's collection, and knowing of no other individual, there may exist some doubts as to the correctness of this last supposition.

Loc. Sowerby's original specimens are said to be from Dublin and Derbyshire; it occurs abundantly in several localities of the last-named county, such as at Dovedale; is also common in the Wensleydale and Yorkshire districts, at Lowick, Northumberland, in the Isle of Man, &c.

In Scotland it is found at Gare, Barrhead, Craigenglen, and Dallmellington, Dumfries-shire, &c. In Ireland it is said to occur near Dublin, at Ballintrillic, and Millecent. M. De Koninck mentions Visé, Chokier, and Ratingen among the foreign localities.

SPIRIFERA TRANSIENS, *M'Coy*.[1] Plate IV, fig. 2.

> SPIRIFERA TRANSIENS, *M'Coy*. Synopsis of the Carboniferous Fossils of Ireland, p. 135, pl. xix, fig. 14, 1844.

"*Spec. Char.* Triangular or rhomboidal, including the beak, twice as wide as long, gibbous; mesial fold very large, prominent, rounded, undivided, producing a very deep sinus in the front margin; sides radiated, with about ten or twelve large, thick, rounded ribs, equal or irregularly duplicate; mesial fold with about six or seven ribs, equal in size to those of the sides; cardinal angles acute; cardinal area low, triangular.

"This species is nearly allied to *Sp. grandæva* of the Devonian rocks, but is distinguished by its very large, undefined mesial fold, and more tumid sides. It differs from *Sp. bisulcata* and *Sp. attenuata* in its very large, undefined mesial fold, and the smaller number and greater size of its radiatory ribs, and most importantly by the cardinal area, as in *Cyrtia*. Length one inch seven lines, width two inches six lines. Young specimens are not so wide in proportion to the length ('Synopsis,' p. 135)."

Mr. Kelly states that it is abundant at Clonalvy, Ireland.

SPIRIFERA GRANDICOSTATA, *M'Coy*. Plate V, figs. 38, 39; Plate VII, figs. 7—16.

> SPIRIFERA GRANDICOSTATA, *M'Coy*. Annals and Mag. of Nat. Hist., 2d series, vol. x, 1853; and British Palæozoic Fossils in the Cambridge Museum, p. 417, pl. iii D, fig. 29, 1855.

Spec. Char. Transversely sub-rhomboidal or obtusely triangular; hinge-line as long or longer than the greatest width of the shell; the lateral margins are regularly curved until they reach the extremities of the hinge-line, or are abruptly attenuated towards the cardinal angles, so as to produce acutely prolonged extremities; valves moderately convex. The area, with sub-parallel sides, is divided by a rather wide triangular fissure. Beak small, incurved, and but slightly produced. The four or five ribs which ornament the mesial fold are, in general, smaller and less defined than those which cover the lateral portions of the shell; the last, varying from ten to twenty-four on each valve, are very large, and either

[1] From not having been able to procure any example of this species, I am compelled to reproduce the description taken from the 'Synopsis.'

simple, or here and there bifurcated, or trifurcated near the margin.   The sinus is of moderate depth.   The measurements taken from two examples have produced—

Length 20, width 34, depth 14 lines.

„   12,   „   22   „   11 lines.

Obs. Few species seem to be more variable in shape and character than the one I am now describing, and it was only after considerable hesitation—and the examination of a vast number of specimens—that I could make up my mind to consider the extreme forms delineated in Pl. VII, figs. 7—16, as belonging to a single species, viz., *Sp. grandicostata*, and of which I have also reproduced the original illustration (Pl. V, figs. 38, 39).   In the single representation given by the author of ('British Palæozoic Fossils') the lateral margins are abruptly attenuated, as in some of the figures in my Pl. VII, but this is not the constant peculiarity of the larger number of individuals, in which the cardinal extremities are not extended to the same extent as in the representation of the original type.   It is further observed by Professor M'Coy that "this shell is allied to *Sp. trigonalis* of Martin, but differs from it by its abruptly narrowed and attenuated sides, and by its few very large angular ribs occupying the body of the shell, and the abrupt diminution in size of the five or six outer ridges on each side.   A very young specimen, nine lines wide, has the three ridges in the mesial hollow distinctly marked, but nearly as large as the lateral ones, of which there are three or four great ones on each side, but scarcely a trace of any additional ones on the flattened cardinal angles, which are strongly striated parallel to the angles." I may here observe that, although I have observed the apparent diminution in width of the five or six outer ribs alluded to, they are far from being so in the greater number of specimens, in which the ribs gradually diminish in width and dimensions from either side of the mesial fold, as in most other species of Spirifer.   The proportion and dimensions of the ribs on the mesial fold is also very dissimilar in different individuals, but in no case have I observed them as large as upon the lateral portions of the valves.   Professor De Koninck, to whom I have forwarded a proof of my plates, thinks that perhaps the shells in question might belong to *Spirifer Kelhavii* (V. Buch) found at Bear Island, and published in the 'Memoirs of the Academy of Sciences of Berlin,' in 1846 ; but the shell there delineated is so very much more elongated than any British example I have seen, that I should not feel myself authorised to attribute to it our English shells without having been able to examine some specimens of the Prussian author's species.

Loc.   Professor M'Coy states that his specimens (now in the Cambridge Museum) are from Derbyshire ; it abounds at Park Hill, Longnor, whence a beautiful series may be seen in the Museum of the Geological Survey, as well as the British Museum ; and I have received the loan of several fine examples obtained at Bolland and the Isle of Man, from Messrs. E. Wood, Parker, and Muschen, and the Rev. Dr. Cumming.   Professor M'Coy states the shell to be common in the Irish limestone at Ardagh.   It has often been erroneously labelled *Sp. triangularis* in various collections.

SPIRIFERA CONVOLUTA, *Phillips*. Plate V, figs. 2—15.

> SPIRIFERA CONVOLUTA, *Phillips*. Geol. of Yorkshire, vol. ii, p. 217, pl. ix, fig. 7, 1836.
> — RHOMBOIDEA, *Phillips* (?). Ibid., figs. 8, 9.
> — CONVOLUTA, *De Koninck*. Animaux foss. de la Belgique, p. 247, pl. xvii, fig. 2, 1843.
> — — *M'Coy*. Synopsis of Carb. Foss. of Ireland, p. 130, 1844.

Spec. Char. Fusiform, three or four times as wide as long, with a straight hinge-line, to which the lateral margins of each valve rapidly converge, producing acute angles at the extremities. Area rather narrow, and parallel-sided, with a triangular fissure partly covered by a pseudo-deltidium. Beak small, and but slightly produced beyond the cardinal edge. Ventral valve a little more convex than the dorsal one, with a sinus of variable depth and width, and which corresponds in the small valve with a mesial fold at times considerably elevated above the general convexity of the valve, and to which is due the strong frontal wave observable in some examples. Each valve is ornamented by from thirty to forty simple or intercalated ribs of unequal width, three or four of which occupying the mesial fold and sinus. Full-grown examples measured—

Length 11, width 45, depth 9 lines.
„   14, „ 36, „ 15 lines.

Obs. This species is easily distinguished from other Carboniferous Spirifers by its great width compared to its length, which gives to the shell much the appearance of a weaver's spindle. The ribs in adult individuals are often much contorted and irregular in their width, as may be seen in specimens figs. 9 and 12. Professor Phillips represents a specimen in which one of the wings was imperfect. This I have replaced in the more correct illustration of the same example, fig. 9; here, as in fig. 12, the mesial fold has but little elevation above the general convexity of the valve; but in another of Professor Phillips's type specimens in the British Museum, as well as in one (figs. 14 and 15) belonging to the Royal Dublin Society, the mesial fold is considerably elevated. Professor De Koninck has published a very good description and illustration of this species, and mentions that in the young specimens the shell is far from presenting so extreme a transverse shape as it afterwards assumed. The lines of growth on the specimens from which figs. 9 and 12 were drawn denote that, while the length was $3\frac{1}{2}$ lines, the width did not exceed 9, and others seem to have been still less transverse. Professor De Koninck is also of opinion that *S. rhomboidea* (figs. 2—8) and *S. fusiformis*, Phillips, cannot be considered distinct from *S. convoluta*, and proposes to place both names among the synonyms. Having been able to study numerous examples of *S. rhomboidea*, I feel disposed to agree so far with my learned friend, while refusing to the last-named shell a separate specific value. The name may, however, perhaps be retained as a varietal denomination for *S. convoluta*, in which the ribs on the mesial fold are less defined than on the lateral portions of the valve; it is also rather more rhomboidal, and in some examples the sinus is less excavated towards the front than is the case with typical examples of the species under description.[1]   M. Semenow adopts Professor De Koninck's views on this subject; but in order that each person may be better able to form his own opinion, I may here mention that figs. 2 to 8 represent what Professor Phillips considers his *Sp. rhomboidea*, while 9 to 15 are types of *Sp. convoluta*.   There would, however, exist some risk in adopting Professor De Koninck's supposition relative to *Sp. fusiformis*, from the fact that the original type specimen of this species, now in the British Museum, is smooth, and not ribbed, as Professor Phillips's illustration would seem to imply.   Lastly, M. De Koninck further observes that "it is very probable that Martin may have created his *Anomites acutus* (figs. 5 and 6 of our Pl. VII) on a similar specimen (*Sp. rhomboidea*)," but I do not consider we have sufficient grounds for this supposition.   Martin's figure has been so differently interpreted, and there appears to exist so much doubt as to the species to which it really belongs, that I have considered it preferable to completely erase the name from the nomenclature, than to continue to discuss a matter on which no certain conclusions can be arrived at.

Loc.   Bolland and Kildare are given by Phillips as the localities whence his specimens were derived.   Mr. Parker has it from Clitheroe Quarry, Lancashire, and it is also found at Visé, in Belgium.   The var. *rhomboidea* seems to be more abundant.

SRIRIFERA LAMINOSA, *M'Coy*.   Plate VII, figs. 17—22.

> SPIRIFER HYSTERICUS, *De Koninck*.   Animaux fossiles de la Belgique, p. 236, pl. xv, fig. 3, 1843 (not of *Schlotheim*).
> CYRTIA LAMINOSA, *M'Coy*.   Synopsis of the Carb. Fossils of Ireland, p. 137, pl. xxi, fig. 4, 1844.
> —   SPECIOSA, *M'Coy*.   Synopsis, p. 134 (not of *Schlotheim*).
> SPIRIFER TRICORNIS, *De Koninck*.   Animaux fossiles de la Belgique (Supplement), p. 657, 1851.
> —   LAMINOSA, *M'Coy*.   British Palæozoic Fossils, p. 426, 1855.

Spec. Char.   Transversely sub-rhomboidal; valves unequally convex, the ventral one

---

[1] Professor Phillips's description in the ' Geol. of Yorkshire,' vol. ii, p. 217, is as follows:

"*Sp. rhomboidea*, pl. ix, figs. 8, 9, Bolland, Ireland.

" Width fully double the length, extremities sub-cylindrical, cardinal area very wide, mesial fold defined; surface radiated with obtuse smooth sulci.   The great proportionate width of the cardinal area is a strong character, yet it very much resembles both *Sp. convoluta*, Ph., and *Sp. attenuata*, Sow., 'M. C.,' t. 562, but the ribs of that species are bolder and the mesial fold is different."

Professor Phillips also describes as follows his—

"*Sp. convoluta*, Phil., pl. ix, fig. 7, Bolland.

" Width four times the length; cardinal area concave, surface obtusely and unequally radiated."

by far the deepest. The lateral portions of the shell are regularly curved, forming, with the extremities of the hinge-line, acute, but not prolonged cardinal extremities; area large, triangular, more or less elevated, and divided by a fissure of moderate width. Beak small, not much produced above or beyond the level of the area. The mesial fold in the dorsal valve is broad, and more or less elevated, without ribs, and corresponding with a deep and rather wide longitudinal sinus in the ventral one. Each valve is ornamented by about twenty or twenty-two narrow radiating ribs, intersected by closely disposed, sharp, concentric, undulating laminæ. The measurements from two examples have produced—

Length 12, width 21, depth 10 lines.

    „   8, „ 11, „ $6\frac{1}{2}$ lines.

Obs. This beautiful shell was correctly described and illustrated by Professor De Koninck in 1843, but unfortunately under the mistaken denomination of *Sp. hystericus*, but which the same author replaced by that of *S. tricornis* in 1851. It had, however, been named *Sp. laminosa*, in 1844, by Professor M'Coy, a denomination we are bound to maintain; notwithstanding that the term is in itself hardly appropriate, from the fact that several other Carboniferous, Devonian, Silurian, and even Permian species, are similarly ornamented by scale-like imbricated laminæ. The illustrations of this shell, published in the 'Synopsis,' are so incomplete and unsatisfactory, that we can easily understand the doubts entertained by some of those who have not, like ourselves, had the opportunity of studying the Irish specimens; and I feel at a loss to divine the reason which could have tempted Professor M'Coy to place this and several other species into the sub-genus *Cyrtia*, with which it and they possess so little affinity. *Sp. laminosa* has some points of resemblance to *Spiriferina octoplicata;* but it is readily distinguished by the greater number of its ribs.

In our British localities the shell under description is more often found with its valves disunited; the mesial fold not extending, in general, beyond the level of the lateral portions of the valve; but in some exceptional examples preserved in the Cambridge Museum, this portion of the shell "is produced in front into a long tongue-shaped flattened lobe" (figs. 21, 22), as was observed by Professor M'Coy in his work on 'British Palæozoic Fossils.' The specimens described in the 'Synopsis' under the name of *Sp. speciosa* undoubtedly belong to *Sp. laminosa*, an opinion in which Mr. Salter fully concurs.

Loc. It is stated by Professor M'Coy not to be uncommon in Derbyshire; Mr. Tate has the shell from Denwick, Northumberland; Mr. Howse from Redesdale. In Ireland it is said to occur at Stridagh Point, Malahide, Hook, Abbey Bay, Ballintrillic, and Ballyshanns, from most of which localities I have been able to examine examples, through the kindness of Dr. Griffith. It is not rare at Tournay; and has been found, although less commonly, at Visé and in other Belgian localities, by Professor De Koninck.

SPIRIFERINA CRISTATA, *var.* OCTOPLICATA, *J. de C. Sowerby.*   Plate VII, figs. 37—47.

SPIRIFER OCTOPLICATUS, *Sowerby.*   Min. Con., p. 120, pl. 562, tabs. 2, 3, 4, May, 1827.

—   CRISTATUS, *V. Buch.*   Ueber Delthyris, p. 39, 1837.

—   —   *M'Coy.*   Synopsis of the Carb. Foss. of Ireland, p. 133; and British Palæozoic Fossils, p. 418, 1855.

Spec. Char.   Transversely sub-rhomboidal, valves about equally convex, and at times rather gibbous; hinge-line as long as the greatest width of the shell.   Cardinal angles acute or slightly rounded; area concave, triangular, and of variable width, fissure partly covered by a pseudo-deltidium; beak small and incurved.   The mesial fold of the dorsal valve is more often composed of a single rib which is much larger than those situated on the lateral portions of the shell; its crest being in general rounded from the umbone to about half its length, when it gradually becomes more and more flattened as it approaches the frontal margin (fig. 38), but at times it remains angular during its entire length, with a tendency to the formation of a rudimentary plait on either of its slopes, so that in these rarer cases the fold assumes towards the front an obscurely triplicated appearance (fig. 37).   The sinus in the ventral valve is deep, acute, and generally simple, but also more rarely interrupted by a rudimentary rib, which becomes visible in the proximity of the front.

The valves are ornamented by from eight to twelve angular ribs, which are, as well as the sinus and fold, intersected by closely disposed, concentric, scale-like laminæ.   The surface of the shell is also closely beset by numerous small granular (spinose) asperities; the shell-structure being likewise perforated by minute tubuli or perforations.

In the interior of the ventral valve there exists a sharp elevated mesial septum, which rises from the bottom of the valve, and partly divides the spiral cones.   Dimensions very variable.   Three examples, of which the first two are Sowerby's original types, have afforded the following measurements :

Length 9, width 13, depth 8 lines.

    „   6,   „   11,   „   6 lines.

    „   5,   „   8,   „   5 lines.

Obs.   The shell under description seems to have been mistaken and misunderstood by the various authors who have commented upon its characters and affinities.   It is very variable in shape, and but rarely possesses the definite number of ribs which its name would imply.   The area is also very variable in its dimensions as well as in the angle it forms with the level of the smaller valve, as may be seen from a glance at figures 41 and 43.   Its affinities with the Permian *Sp. cristata* of Schlotheim ('Akad. Münch.,' vol. vi, p. 28, pl. i, fig. 3, 1816) did not escape the notice of Baron Von Buch in 1837, nor that of Professor King in 1849, for we find stated at p. 128 of his work on 'British Permian Fossils,' "*Trigonotreta cristata* closely resembles one or more so-called species found in the Carboniferous and other formations, particularly *S. octoplicata* of J. Sowerby. Having examined in Mr. J. de C. Sowerby's collection the originals (from Derbyshire) of the figures in the 'Mineral Conchology,' the only difference I could perceive is that they are wider than any examples which have occurred to me of the present species. The specimens bearing the name of *Sp. insculpta*, in the Gilbertson Collection of the British Museum, appear to be undistinguishable from *Trigonostrata cristata*. The Jurassic fossil, which Zieten has identified with *T. octoplicata*, is another closely analogous species."

In 1843, Professor De Koninck had placed Sowerby's *Sp. octoplicata* ('M. C.,' tab. 562, figs. 2 and 3 only) with a mark of doubt among the synonyms of *Sp. cristata*,[1] while fig. 4 of the same author's plate was considered by him to be referable to the *Sp. crispus*; but the examination of Sowerby's three examples will convince any observer that they belong to a single species. The Belgian author further remarks that according to M. De Buch, *Sp. cristata* would possess but four ribs on each side of the sinus, while our Carboniferous one possesses a much larger number; this being the only difference he was able to discover, and which did not appear to him of sufficient importance to warrant a separation of the two shells. However, in the Supplement to the great work on the 'Carboniferous Fossils of Belgium,' M. De Koninck retracts his former identification, adding that the Carboniferous shell was distinct and distinguishable by its larger number of ribs.

In 1855 we find Professor M'Coy refuting the inferences published by the author of the Permian monograph; he states, "Professor King mentions that the only differences he could perceive between this species (*Sp. cristata*) and *Sp. octoplicata* is the greater width of the latter; but I observe that specimens of the latter have a proportionately much lower cardinal area, longer hinge-line, and are of nearly double the average size, and have five or six lateral plaits on each side of the mesial sinus, which are comparatively so small, that at the margin three of them would be required to equal the mesial ridge in width, while in the present species the mesial ridge is little wider than the adjoining lateral ones. The Carboniferous *Sp. insculpta*, Phillips, seems also to him to be undistinguishable from the Permian *Sp. cristata*; but besides the greater height of the cardinal area, I have never seen more than one or two lateral ribs on each side of the median one, and the punctuations seem slightly closer." From the statements above recorded it will be perceived how very dissimilar are the opinions entertained respecting the affinities of the form under description.

Having been able to assemble, through the kindness of many zealous friends, a vast number of both the Permian and Carboniferous shells above mentioned, comprising the original specimens illustrated by the author of the 'Mineral Conchology,' I have been able

[1] 'Animaux fossiles de la Belgique,' p. 211.

to institute between them a minute and searching comparison, which enables me to substantiate to a large extent the opinion expressed by Baron Von Buch and Professor King as to the close resemblance existing between the Carboniferous and Permian shells, and to declare that, in my humble opinion, *Sp. octoplicata* cannot claim to be considered more than a variety of *Sp. cristata*.[1]

The discovery lately made by Mr. Kirkby, at Tunstall Hill, of a magnificent and unusually large example of Schlotheim's species (first illustrated by Mr. Howse in the 'Annals of Nat. Hist.,'[2] and subsequently more fully so, during the same year, in Pl. II, fig. 43, of our Permian illustrations), places this matter beyond even the range of doubt; for, with the dimensions of 5 lines in length, 9½ in width, and 5 in depth, a mesial fold, of which the crest is depressed towards the front, and fourteen ribs on each valve, it presents all the characters and appearances of several of Sowerby's type examples, from which it is distinguished only by the light-yellow colour peculiar to our Permian fossils. I feel, therefore, disposed to maintain for the Carboniferous shell the varietal designation of *octoplicata*, to distinguish it from Schlotheim's type, which is, in general, of smaller dimensions, with a minor number of plaits. The variety *octoplicata* has likewise shown a tendency to triplication in the mesial fold, which I have not hitherto observed on any of the numerous individuals of the Permian shell that have passed under my notice. From *Sp. laminosa* the shell under description is easily distinguished by its less numerous, comparatively larger, and more angular ribs; from *Sp. insculpta* (a closely allied form) by the greater disproportion of the mesial plait relative to the lateral ones, which are, at the same time, more numerous and smaller in Sowerby's shell. I am still uncertain whether *Sp. minima* (Sowerby) and *Sp. partita* (Portlock) should be considered as specifically distinct, or forming part of the variety of *Sp. cristata*. The material in my possession has not been sufficient to allow me to determine that point satisfactorily.

Loc. Sowerby mentions his specimens as having been derived from Derbyshire. Gare, Lanarkshire; Dr. Fleming has examples from Westlothian, Scotland; and a fine Irish series from Hook Point and the shores of Lough Hill, county of Sligo, may be studied in the Museum of the Geological Society. Professor M'Coy mentions the shell from Flintshire, and Mr. Kelly quotes Bundoran, Carrowmably, and Cregg, Ireland. It has also been found in several foreign localities.

SPIRIFERINA MINIMA, *Sowerby*. Plate VII, figs. 56—59.

SPIRIFER MINIMUS, *Sowerby*. Min. Con., p. 105, tab. 377, fig. 1, Nov., 1822.

Spec. Char. Rhomboidal, a little wider than long; hinge-line rather shorter than the

---

[1] As *Sp. octoplicata* was first introduced into existence, and therefore the oldest form, it is in reality the type of which *Sp. cristata* would be the variety, but the law of priority as to names obliges us to retain Schlotheim's in preference.

[2] Vol. xix, 2d series, pl. iv, figs. 5 and 6, 1857.

greatest width of the shell, cardinal angles rounded; valves moderately convex; area triangular, rather elevated; fissure large; beak small and incurved. Each valve is ornamented by from seventeen to twenty-three radiating ribs, of which three smaller ones occupy the mesial fold of the dorsal valve, which is but moderately elevated, and longitudinally flattened along its middle. In the ventral valve the sinus is rather deep, and exhibits also two or three smaller ribs. The frontal wave is strongly marked; the ribs are intersected in both valves by concentric laminæ of growth. The dimensions of Sowerby's two original examples have given—

Length 7, width 8, depth 6 lines.

„ 6, „ 7, „ 4½ lines.

Obs. The only specimens I have been able to examine are those figured in the 'Mineral Conchology,' and from such scanty material I would hardly consider myself warranted to offer any decided opinion as to their specific claims. After comparing the specimens with the originals of *S. octoplicata,* the differences consisted in the last-named shell being larger and more transverse, with fewer and bolder ribs; there exists also a dissimilarity in the character of the mesial fold, which is, in *S. minima,* much flatter and more regularly divided into three flattened ribs, than in any examples of true *octoplicata* that have passed under my notice. It is, therefore, probable that after a more extended examination of a larger number of specimens these slight differences may be considered valueless, and that it may be found desirable to add it also to the varieties of *Sp. cristata* of Schlotheim.

Professor M'Coy is, however, mistaken when he describes the shell under description as possessing an angular mesial fold or ridge, this last being most distinctly flattened in the two original examples preserved in the collection of Mr. J. de C. Sowerby, and which I have drawn afresh under figs. 56 and 59 of my plate. For similar reasons I must object to the identification of the shell in question, as proposed by Mr. Morris,[1] with *Anomites acutus* of Martin, which seems to me much more probably the young of some other species.

Loc. Sowerby's specimens were obtained, along with others of *Athyris ambiguus,* in decomposed limestone near Bakewell; the shell being silicefied.

SPIRIFERINA (?) PARTITA, *Portlock.*[2] Plate VII, figs. 60, 61.

SPIRIFERA PARTITA, *Portlock.* Report on the Geology of the County of Londonderry, Tyrone, and Fermanagh, p. 567, pl. xxxviii, fig. 3, 1843.

"*Spec. Char.* A small shell, 2″ long and 35″ wide; dorsal valve (*our ventral one*) with

---

[1] 'A Catalogue of British Fossils,' p. 150, 1854.

[2] Not having been able to procure a specimen of the above-named shell, and as the description and illustrations are not sufficiently detailed to allow of my forming a positive opinion as to its specific claims and affinities, I have preferred to simply reproduce Colonel Portlock's description and figures.

a deep sinus extending to the beak, and in the casts marked by a central linear furrow; the ventral valve (*our dorsal one*) with a rounded central rise marked in the centre by a linear furrow approximating it to *Sp. pingens*. The sulci between the central plait and the lateral costæ strongly marked; the number of the lateral costæ variable, being sometimes three and at other times six on each side; the general form rounded, appoaching to that of *Sp. speciosa*."

Loc. Kildress, Ireland.

SPIRIFERINA (?) INSCULPTA, *Phillips.* Plate VII, figs. 48—55.

> SPIRIFERA INSCULPTA, *Phillips.* Geol. of Yorkshire, vol. ii, p. 216, pl. ix, figs. 2 and 3, 1836.
> — CRISPUS and HETEROCLYTUS, *De Koninck* (not *Linnæus* nor *Defrance*). Animaux fossiles de la Belgique, pp. 257 and 259, pl. xv, figs. 7, 8, and pl. xv *bis*, fig. 1, 1843.
> — QUINQUELOBA, *M'Coy.* Synopsis of the Carb. Fossils of Ireland, p. 134, pl. xxii, fig. 7, 1844.
> — KONINCKIANA, *D'Orbigny.* Prodrome, vol. i, p. 119.
> — INSCULPTA, *Semenow.* Ueber die Fossilien des Schlesischen Kohlenkalkes, 1854.
> — — *De Koninck.* Animaux fossiles de la Belgique (Supplement), p. 658, 1851.

Spec. Char. More or less semicircular, about one third wider than long; hinge-line straight, and as wide as the greatest width of the shell. Area large, triangular, and but slightly curved; fissure wide; beak very small, and not much produced above the level of the area. Valves almost equally convex. The ventral one is ornamented by five (rarely seven) large, bold, angular ribs, the central one exceeding the others somewhat in proportions, and corresponding with the deep angular sinus of the opposite valve. Six angular ribs exist on the ventral valve; these, as well as those on the dorsal one, being intersected by close, concentric laminæ of growth. Shell-structure minutely perforated. Dimensions taken from two individuals have produced—

Length 8, width 11, depth 6 lines.

„ 6½, „ 11, „ 7 lines.

Obs. Phillips's short description and illustrations convey a clear idea of the adult or full-grown condition of the shell under description. Its few bold ribs distinguish it from *Sp. cristata* and its var. *octoplicata*, as well as from *Sp. minima* and *Sp. laminosa*. It has been taken for the *Anomia crispa* of Linnæus, as well as for the *Terebratulitis cristata* of Schlotheim, but seems specifically distinguished from both by the small difference that exists in the proportions of its central rib or plait relative to the lateral ones, which in the above-named species and varieties are much more disproportionate. *Sp. quinqueloba* belongs undoubtedly to Phillips's *Sp. insculpta*, of which I was able to convince myself by the comparison of the original specimens of both, Professor M'Coy's very imperfect specimen having been kindly lent for the purpose by Dr. Griffith, of Dublin.

Loc. Professor Phillips obtained his examples from Bolland. Mr. Parker has it from the Clitheroe quarries, in Lancashire. The Geological Survey possesses a fine series, of different ages, from Yorkshire, and Longnor, Derbyshire. In Ireland it is found at Ardagh (Drumcondra). M. De Koninck mentions the shell from Visé and Tournay, in Belgium.

SPIRIFERA REEDII, *Dav.* Plate V, figs. 40—47.

Spec. Char. Longitudinally oval, valves almost equally convex; hinge-line shorter than the greatest breadth of the shell; lateral margins and cardinal angles rounded; area small, triangular, elevated; beak produced and incurved. Each valve is ornamented by from nineteen to twenty-three small, radiating ribs, of which the central one in the dorsal valve is rather wider, and a little more produced, than the lateral ones: to this corresponds a shallow, small, longitudinal sinus in the opposite valve. Dimensions taken from two examples have produced—

Length 9½, width 7, depth 6½ lines.

„ 5, „ 4, „ 3 lines.

Obs. I have been able to examine but three individuals of this little species, one adult (fig. 40), in the British Museum, and two younger shells (figs 43 and 47), in the possession of Mr. Reed, of York. They are distinguished from *Sp. sexradialis*, Phillips, by their larger number of ribs, smaller fold and sinus, and lesser width near the hinge-line.

Loc. Settle, Yorkshire.

SPIRIFERA DECEMCOSTATA, *M'Coy.*[1] Plate VII, fig. 23.

SPIRIFERA DECEMCOSTATA, *M'Coy.* Synopsis of the Carb. Fossils of Ireland, p. 131, pl. xxii, fig. 9, 1844.

"*Spec. Char.* Semicircular, gibbous, smooth, twice as wide as long; front rounded,

---

[1] The *single* and *imperfect* valve upon which this species was founded, was kindly lent to me by Dr. Griffith; but from such incomplete material it would be more than hazardous to decide as to its specific value. Under these circumstances I have preferred to simply reproduce the author's description, that the reader may have before him all that has been published on the subject. M. De Koninck informs me that he has found similar decorticated valves near Dinant, in Belgium, and seems inclined to refer *Sp. decemcostata* to *S. laminosa*, M'Coy; but the evidence does not permit my confirming the statement.

sides suddenly attenuate, cuspidate; mesial lobe large, round, projecting, with five large rounded ribs on each side, leaving a broad space at the cardinal angles smooth, or only striated transversely. This species is so very distinct from any other *Spirifera* with which I am acquainted, that it would be unnecessary to point out any peculiar difference. The whole surface is smooth except the cuspidate sides, which are marked with strong lines of growth."

Loc. Lower limestone, Millecent (Clare), Ireland.

SPIRIFERA CUSPIDATA, *Martin*, sp.[1] Plate VIII, figs. 19—24; Plate IX, figs. 1 and 2.

> ANOMITES CUSPIDATUS, *Martin*. Trans. Linnean Soc., vol. iv, p. 44, pl. iii, figs. 1—4, 5, 6, 1796.
> CONCHYLIOLITHUS ANOMITES CUSPIDATUS, *Martin*. Petref. Derb., tab. xlvi, fig. 34, and tab. xlvii, fig. 5, 1809.
> SPIRIFER CUSPIDATUS, *Sowerby*. Min. Con., tab. 120, figs. 1, 2, 3, Feb., 1816.
> CYRTIA SIMPLEX, *M'Coy*. Synopsis Carb. Foss. of Ireland, 1844 (not of *Phillips*).
> — CUSPIDATA, *M'Coy*. British Palæozoic Fossils, p. 466, 1855.

Spec. Char. Transverse and pyramidal; hinge-line straight, in general rather shorter than the greatest width of the shell. The cardinal angles, formed by the junction of the lateral portions of the valves with the extremities of the hinge-line, are rounded off. *Dorsal* valve semicircular, moderately convex; mesial fold large, smooth, regularly convex, and elevated, with a slight longitudinal depression or groove (especially visible in casts) extending from the extremity of the umbone to about half the length of the fold. Each valve is ornamented by numerous small, simple ribs, varying from thirty to forty-four in number, according to age and individual, these being at the same time intersected by numerous concentric lines of growth. The ventral valve is pyramidal, much deeper than the opposite one, and possesses a large, wide, concave sinus, which extends from the extremity of the frontal margin, producing in front a strongly marked wave. Area very large, triangular, and acute, situated at a right angle to the general level of the dorsal valve; fissure large, but comparatively narrow in proportion to its length.

In the interior of the ventral valve a strong tooth exists on either side, at each extremity of the fissure. These are supported by dental plates, which diverge from the extremity of the beak, forming the fissure walls, and occupying about one third of the length of the bottom of the valve. The larger portion of the interior of the shell is

[1] This well-known Carboniferous fossil has been very often referred to and figured by various authors. By Parkinson (1811), Bronn (1824), Krüger (1825), Defrance (1827), Fleming (1828), Holl (1830), Keferst (1834), Deshayes *apud* Lamarck (1836), Phillips (1836), V. Buch (1837), Conrad (1838), &c., a full reference to which will be found in page 243 of Professor De Koninck's works on the Belgian Carboniferous Fossils.

occupied by the spiral cones, which were fixed in the usual manner to the projections of the inner socket walls of the dorsal valve. Two examples measured—

Length 2 inches, width 4 inches, depth 3 inches.

Length 2 inches 9 lines, width 2 inches 10 lines, depth 2 inches 5 lines.

Obs. The external characters of *Sp. cuspidata* were correctly described and illustrated by Martin as early as 1796, and but little has been added concerning its internal arrangements. Having, through the kindness of Mr. Kelly, obtained a certain number of Irish specimens, I fractured a few in order to ascertain more correctly the true position and dimensions of the dental or rostral shelly plates, which have already been described, and which will be found illustrated in the accompanying cut. It is quite evident that the shell under description belongs to the genus *Spirifer* proper, and not to the subgenus *Cyrtia*, as supposed by Professor M'Coy, and in which opinion I had coincided before having become fully aware of its complete internal arrangements. No specimen of *Sp. cuspidata* I have hitherto been able to examine has exhibited the deltidium in its entire condition, but which, in all probability, was not perforated by a circular foramen, as is seen in true types of the subgenus *Cyrtia*, such as in *C. trapezoidalis* and *C. Murchisoniana*. Nor do I perceive upon what grounds Professor M'Coy asserts that the fissure displays a deep-seated pseudo-deltidium.

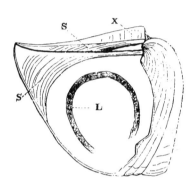

Spirifera cuspidata.

X. Area. *S.* Dental or rostral plates. *L.* Coil of the spirals.

The position of the dental plates is defined on the exterior of the shell by two diverging lines (Pl. VIII, fig. 22 *s*) departing from the extremity of the beak, and are more clearly visible in those examples (by far the more numerous) wherein the shell is but imperfectly preserved. It was to compressed shells from the lower limestone of Black Rock (Cork), on which the ribs were almost obliterated, that Professor M'Coy identified the Devonian *Sp. simplex* of Phillips, a mistake which Mr. Salter and myself were able to correct from the inspection of the original examples so named, and kindly communicated by Dr. Griffith. Other species have also been inaccurately referred to Martin's shell; these will be found enumerated in Professor De Koninck's work on Belgian fossils.

Some varieties of *Sp. cuspidata* (fig. 20) have been referred to the Devonian *Sp. macroptera* of Hall, in which the area is narrower and less developed than is usually the case in typical examples of Martin's shell. These varieties occur plentifully in limestone and millstone grit, near Kendal, but at the same time present every intermediate link connecting those individuals with a narrow area to the typical form of the species under description. This fact was pointed out by Professor M'Coy, at p. 426 of his work on 'British Palæozoic Fossils,' but I do not perceive that there exists any proportional difference in the width of the fissure as described by the Professor, and many of the examples are evidently malformations, the beak being twisted more to the one than to the other side, as well as otherwise misshaped. The regular convexity and smoothness of the mesial fold and sinus distinguish the species under description from *Sp. distans* of Sowerby, as will be shortly explained.

Loc. This species occurs plentifully in various localities, such as in the lower Carboniferous limestone of Castleton (Derbyshire), at Twiston (Lancashire), at Preston, Bolland, Settle, near Bristol, &c. Mr. Kelly mentions the shell from Lisnapaste, Millecent, Ballyduff, Malahide, Cork, &c., in Ireland. I have not yet seen any Scotch example. In Belgium it is found at Tournay, and in other Continental localities.

SPIRIFERA DISTANS, *Sowerby*.   Plate VIII, figs. 1—17.

> SPIRIFER DISTANS, *J. de C. Sowerby*. Tab. 494, fig. 3, May, 1825.
> —   —   *Phillips*. Geol. Yorkshire, vol. ii, p. 217, 1836.
> CYRTIA DISTANS, *M'Coy*. Synopsis of the Carb. Fossils of Ireland, p. 136, 1844.

Spec. Char. Very variable in shape and proportions, more often imperfectly rhomboidal and transverse, with unequal convex valves. Hinge-line as long as the greatest width of the shell. The dorsal valve is more or less convex, but not in general gibbous; semicircular, and somewhat indented in front. The lateral margins form a convex curve to within a short distance of the cardinal angles, where they are often prolonged with acute terminations. The fold is wide, and but little elevated, its upper surface being much flattened, with a longitudinal sulcus extending along its middle, either quite to the front, or soon becoming converted into a rounded central rib. The lateral portions of the shell, on either side of the fold, are ornamented by from fourteen to fifteen small, single, or bifurcated ribs. The *ventral valve* is much deeper than the opposite one; the area is triangular and variable in its proportions, as well as in its development; it either presents a large flat surface at right angles to the level of the dorsal valve, or is more or less concave and elevated, showing or not a portion of the beak above and beyond its angular termination. The fissure is narrow, and in great part covered by a convex pseudo-deltidium. The sinus is of moderate depth, and, in general, exhibits along its centre a well-defined rib, which is larger than those which ornament both slopes of the sinus and lateral portions of the valve. In the interior of the ventral valve there exists two largely developed, diverging, dental or rostral shelly plates, which extend from the beak along the bottom of the shell to rather less than half its length. The dimensions are exceedingly variable, on account of the great difference in depth of the ventral valve, as may be perceived from the following measurements:

Length 13, width 18½, depth 13 lines (Sowerby's type).
  „   14, „   21, „   13 lines.
  „   9, „   22, „   14 lines.

Obs. Sowerby's description and illustration does not convey a complete idea of all the peculiarities of this interesting species. His figure represents only one of its numerous variations, as will be better understood by a glance at our plate than by any lengthened description. In it we have been able to illustrate those connecting shapes which unite and explain the great differences observable between such specimens as figs. 1, 15, and 16. *Sp. distans* is easily distinguishable from *Sp. cuspidata*, to which at times it bears some resemblance by the shape of its fold and sinus. In Martin's shell the fold is regularly convex, the sinus concave, without any ribs, while in Sowerby's species the fold is hardly produced above the level of the lateral portions of the shell, with a central longitudinal rib. The sinus is likewise more or less obscurely ribbed, with a mesial one of larger dimensions. The hinge-line is also much longer in proportion than in *Sp. cuspidata*, the fissure being likewise more narrow.

Loc. Sowerby's specimen is stated to have been obtained near Dublin. Mr. Kelly mentions Bundoran, Malahide, and Millecent as Irish localities. Professor Phillips found it at Bolland. I am not acquainted with any Scottish examples.

SPIRIFERA BICARINATA, *M'Coy*.[1]   Plate VIII, fig. 18.

> SPIRIFERA BICARINATA, *M'Coy*. Synopsis of the Carboniferous Fossils of Ireland, p. 129, pl. xxii, fig. 10.

"*Spec. Char.* Rhomboidal, width more than twice the length, very gibbous; sides cylindrical; mesial fold wide, smooth, concave on both valves, bounded on both valves by two large, rounded, entire ribs on each side; cardinal area with parallel sides, very wide and hollow. This curious species is easily recognised by the mesial fold being concave, and bounded by large keels *on both* valves.

"Length 8 lines, width 1 inch 6 lines."

Loc. According to Mr. Kelly, Millecent (Ireland).

---

[1] As the original example is no longer to be found in Dr. Griffith's collection, it would be hardly safe to express an opinion on a species of which the author was able to illustrate but a very unsatisfactory and insufficient fragment; I therefore simply reproduce the original description and illustration.

SPIRIFERA MESOGONIA, *M'Coy*.[1] Plate VII, fig. 24.

CYSTIA MESOGONIA, *M'Coy*. Synopsis of the Carb. Fossils of Ireland, p. 137, pl. xxii, fig. 13, 1844.
SPIRIFER MESOGONIUS, *De Koninck*. Animaux Fossiles de la Belgique (supplement), p. 660, pl. lvi, fig. 4, 1851.

" *Spec. Char.* Semicircular, gibbous: surface smooth; cardinal area wide, concave, triangular; cardinal angles acute; mesial ridge very prominent, angular, both it and the angular mesial furrow without ribs; eight or nine strong rounded equal ribs on each side of the mesial fold.

"This species is allied to *Spirifer distans*, Sow., and *Sp. ostiolata*, Phillips, but is distinguished from both by the very prominent, angular mesial fold; *Sp. distans* has also the mesial hollow ribbed, while it is smooth in the present species.

" Length 1 inch, width 1 inch 5 lines."

Obs. Professor M'Coy's figure represents a shell having about twenty-two ribs on each valve and a smooth mesial fold, while Professor De Koninck's illustration of what he considers to belong to the same species from Belgium would only possess about fourteen, these being likewise much wider proportionately to the mesial fold than those of Professor M'Coy's example. I cannot, therefore, positively affirm that both did belong to the same species, although such in reality may have been the case.

Loc. Mr. Kelly mentions Millecent and Hook as the Irish localities. M. De Koninck found it at Chokier, in Belgium, where it is stated to be very rare.

SPIRIFERA SUBCONICA, *Martin*. Plate IX, fig. 3.

CONCHYLIOLITHUS SUBCONICUS, *Martin*. Petrefacta Derbiensia, tab. xlvii, figs. 6—8, 1809.

Spec. Char. Wider than long; dorsal valve semicircular; ventral, sub-pyramidal; hinge-line straight, as long as the greatest width of the shell; area large, triangular; fissure of moderate width. The smaller valve possesses an angular mesial fold, with corresponding sinus in the opposite one. Both valves are ornamented by about sixteen angular ribs.

Length 14, width 18, depth 11 lines.

[1] Not having had the opportunity of examining any example of this species, and as the original is no longer to be found in Dr. Griffith's collection, all that can be done is to reproduce the author's description and figure.

Obs. Although a description of Martin's *Anomites subconicus* is here appended, and which is taken from the original figure published in the 'Petrificata Derbiensia,' I am far from satisfied that the specific characters or claims of the species have been clearly established, and the more so since no other example could be discovered in our English collections. In 1843, Professor L. de Koninck described and represented *Sp. septosa*, Phillips, under the erroneous denomination of *Sp. subconicus*, Martin, from being at the time unacquainted with the British types of either species; but a few years later my friend had discovered his mistake, and he is now of opinion that Martin's figure may, perhaps, have been drawn from some exceptional specimen or variety of *Cuspidatus*. This may possibly prove a correct interpretation, although I am not at present acquainted with any example of the last-named species with quite so small a number of ribs, but we are all aware how much the specimens of the same species may vary in this particular. Martin observes that the difference between his two species is in the number of furrows or ribs, which in *Subconicus* are few and acute, the central rib or fold being angular instead of a rounded wave in the margin, and the beak of the conic valve is straight or not recurved, as in *Cuspidatus*; but had the author of the 'Petrificata Derbiensia' possessed a larger number of his own *Cuspidatus* he would soon have perceived the little real importance of the above distinctions.

Martin states that his fossil was found in the carboniferous limestone of Middleton, where the shell is said to be *very rare*.

SPIRIFERA TRIRADIALIS, *Phillips.* Plate IX, figs. 4—12.

> SPIRIFERA TRIRADIALIS, *Phillips.* Geol. of York., vol. ii, p. 219, pl. 10, fig. 7, 1836.
> — TRISULCOSA, *Phillips.* Ibid., pl. 10, fig. 6.
> — SEXRADIALIS, *Phillips.* Ibid., fig. 7.
> — TRISULCOSA = TRIRADIALIS, *De Koninck.* Descrip. des Animaux Fossiles de la
> Belgique, p. 266, pl. xvii, fig. 7, *a, b, c,* 1843.
> — SEXRADIALIS, *M'Coy.* British Palæozoic Fossils, p. 421, 1855.

Spec. Char. Circular or longitudinally oval; dorsal valve slightly convex, mesial fold smooth and sharply defined with two, four, or six rounded ribs on either valve; but when two or four only the remaining unoccupied lateral space is smooth. Ventral valve much deeper than the opposite one, with a sinus extending from the extremity of the beak to the front; beak moderately produced; hinge line not longer than about half the width of the shell. Cardinal angles rounded; area small, triangular, with its margins not always sharply defined, so that the lateral portions of the beak are visible on either side; fissure large and partly covered by a pseudo-deltidium. Numerous concentric lines of growth cover the surface of either valve. The dimensions are very variable, those taken from four specimens have produced—

7

Length 10, width 10½, depth 7 lines.

" 9, " 8, " 6½ "

" 6, " 4½, " 4½ "

" 5½, " 5, " 4 "

Obs. This spirifer appears to have rarely exceeded the dimensions given above; and although the number of its ribs is variable, it appears to be in general a well-characterised and easily recognisable species. After an attentive comparison of the original types upon which Professor Phillips had founded his three species, as well as of numerous other specimens of the same in the British Museum, &c., Professor L. de Koninck and myself have arrived at the conclusion that *Sp. trisulcosa,* figs. 7, 10, 11, and 12, *Sp. triradialis,* figs. 8 and 9, and *Sp. sexradialis,* figs. 4, 5, and 6, are only different states, conditions, or varieties of a single species. It has appeared to us evident that in *Sp. trisulcosa* and in *Sp. triradialis* the smooth or unoccupied space on either valve is occasioned by the last ribs, having been either obliterated or not produced by the animal, for in many individuals they are clearly indicated, although not completely developed.

In a few exceptional specimens, as many as eight ribs have been distinctly counted on either valve, while in a few still more rare instances, such as in the type specimen of *Sp. trisulcosa,* fig. 12, none of the lateral ribs are clearly defined. It would, therefore, appear desirable as well as rational to retain but a single name for the different conditions presented by this variable shell, for otherwise we would be obliged to make as many as four species, based upon the number of ribs each specimen might possess. I have, therefore, retained the term *Sp. triradialis* in preference to the other denominations, simply because it represents the intermediate and most common condition in which the species is found; but, as a general rule, it is always objectionable to give names to species derived from the number of ribs some individual example may possess.

Loc. The different varieties of this species have been found by Professor Phillips, the late Mr. Gilbertson, and others, in the carboniferous limestone of Bolland. It has also been found in Derbyshire. I am not acquainted with any Irish specimen, but it has been obtained from the Carboniferous shales of Lanarkshire in Scotland. It is a rare fossil at Visé, in Belgium, where it has been discovered by Professor de Koninck.

SPIRIFERA PINGUIS, *Sowerby.* Plate X, figs. 1—12.

SPIRIFERA PINGUIS, *Sow.* Min. Con., vol. iii, p. 125, tab. 271, 1820.

— ROTUNDATA, *J. de C. Sow.* Min. Con., vol. v, p. 89, tab. 461, fig. 1, 1824 (not *Anomites rotundatus* of Martin).

— — *Sow.* Davreux, Const. Géogn. de la Province de Liege, p. 272, pl. 7, fig. 8, A, B, 1831.

— — *Sow.* Phillips, Geol. of Yorkshire, vol. ii, p. 218, pl. ix, fig. 17, 1836.

SPIRIFERA PINGUIS, *Sow.* Ibid., figs, 18, 19.

—   —   *Von Buch.* Ueber Delthyris, p. 38, 1837, and Mém. Soc. Géol. de France, vol. iv, p. 184, pl. 8, fig. 7.

—   ROTUNDATUS, *Sow.* De Koninck, Animaux Fossiles de la Belgique, p. 263, pl. xiv, fig. 2, and pl. xvii, fig. 4, 1843.

—   PINGUIS, *Sow.* Ibid, p. 661, pl. lvi, fig. 5, 1851.

—   —   *Morris.* Catalogue, p. 153, 1854.

—   —   *M'Coy.* British Palæozoic Fossils, p. 420, 1855.

—   SUBROTUNDATA, *M'Coy.* Ibid., p. 423, 1855.

Spec. Char. Very variable in shape, dimensions, relative proportions, and degree of convexity. When full grown (under favorable circumstances) the shell is transversely oblong or oval. Sometimes as wide as long, very rarely longer than wide; hinge line rather shorter than the greatest width of the shell; cardinal angles rounded. Dorsal valve not so convex or deep as the opposite one, mesial fold wide, moderately raised, almost smooth, and divided along its middle by a shallow longitudinal depression or groove. Each valve is ornamented by from sixteen to thirty rounded or flattened ribs, which sometimes vary in their relative widths, a small rib being situated side to side by a larger one; they are also regular and simple in the larger number of individuals, but in some a few bifurcate or become more numerous by occasional intercalations. A narrow hinge area is likewise observable in this valve.

Dental or ventral valve at times very gibbous, beak of moderate dimensions, much incurved; area narrow, with a triangular fissure partially covered by a pseudo-deltidium. A sinus of greater or lesser depth extends from the extremity of the beak to the front, and is in general ornamented by two or more narrow longitudinal ribs of but small elevation. The spiral appendages are large, and occupy the greater part of the interior of the shell.

The relative proportions vary considerably, as will be perceived from the measurements taken from four adult individuals:

Length 2 inches 7 lines, width 3 inches 8 lines, depth 2 inches.

„   4   „   4   „   2   „   6   „   1   „   9 lines.

„   1   „   4   „   1   „   6   „   1   „

„   1   „   6   „   1   „   5½   „   1   „   4 lines.

Obs. How to deal with the innumerable variations in shape and proportions presented by this species seems to have been no easy matter. Sowerby and Phillips, while describing *Sp. pinguis* and *Sp. rotundata*, Sowerby, as distinct species, did not fail to observe that both were nearly allied forms, and that *Sp. pinguis* may be the young state or a simple variety of *Sp. rotundata*, Sowerby. In his work on the Belgian fossils, Professor de Koninck describes both as separate species, but now relinquishes that idea and is of opinion that they should be combined under a single denomination. The illustrations I have selected for Plate X demonstrate in the most incontestable manner not only the affinity existing between these two shapes, but their complete specific identity, an opinion founded upon the minute inspection and comparison of two hundred or more individuals. In his work on 'British Palæozoic Fossils,' Professor M'Coy advocates the separation, and states that in *Sp. rotundata*, Sowerby (*Sub-rotundata*, M'Coy), "the depression of the sides or sharpness of the margins form the acute angles at which the valves meet each other, less gibbosity and greater proportional width distinguish specimens of all sizes from the little *Sp. pinguis* of Sowerby, with which in other characters it is identical." Therefore, according to Professor M'Coy, both *are identical in all their characters*, excepting those of size, gibbosity, and the angles at which the valves meet each other. Any one, however, who has studied a number of specimens will soon feel convinced that dimensions, gibbosity, and angles at which the valves meet cannot be considered of specific importance in this present case, since these characters vary in almost every individual, and that every link may be traced connecting the most extreme variations presented by either forms. In Plate X, figs. 1—7 represent Sowerby's *Sp. pinguis*, figs. 8—12 his *Sp. rotundatus;* but the first is indubitably either the young of that author's *rotundatus*, or aged specimens stinted in their regular development, and which is due, no doubt, to local or accidental circumstances, the shell having acquired greater thickness or depth in proportion to its length and width. Therefore, since *all the other and more important characters are identical*, we cannot do better than to follow Mr. Morris,[1] while retaining but one name (*Sp. pinguis*) for both of Sowerby's species.[2] It should likewise be observed that on many full-grown specimens of *S. rotundata* can be seen interruptions in development which agree exactly with the form known as *Sp. pinguis*, but which, on the shell resuming its growth, assumed the shape and proportions of *Sp. rotundata*, Sow. Sometimes also during the earlier portion of their development the ribs were all regular and simple, when, after a sudden interruption, a certain number suddenly bifurcated, as will be perceived in fig. 8. In some examples the ribs are regular and simple on one half of the valve, while on the other several became bifurcated or irregular in their respective widths.

In all well-grown specimens that have come under my observation, the length of the hinge line has been rather shorter than the greatest width of the shell; still, in a few dwarfed individuals, such as in fig. 5, the hinge line is as long as the greatest width of the shell, and the cardinal angles are not rounded; this and other exceptional appearances cannot be taken as the normal condition of the shell, and require to be viewed in the light of those malformations to which all species in the animal or vegetable creation are more or less subjected.[3]

---

[1] 'Catalogue,' p. 153.

[2] The laws of priority oblige us to make use of the term *pinguis*, that of *rotundata* having been applied to another species.

[3] Geometrical measurements have been resorted to in their descriptions by several naturalists, such as Professor M'Coy, Mr. Kocklin Schlumberger, and others, but the apical or other angles in a Brachiopod are so exceedingly variable in different specimens of a same species that they become of no value whatso-

At all ages *Sp. pinguis* presents a distinct, well-defined, and almost smooth mesial fold, longitudinally divided by a shallow depression or furrow, but no such character is exemplified in Martin's figure or description of his *anomites rotundatus*, which no doubt belongs to another species. In his work on the 'Carboniferous Fossils of Belgium,' Professor de Koninck has included *Sp. ovalis* and *integricosta* of Phillips among the synonyms of *Sp. rotundata*, Sow. ;[1] but that distinguished author now admits the last two to be distinct and specifically separate from Sowerby's species. All three must, however, be considered as closely allied species, forming a well-defined group among the carboniferous species of the genus to which they belong.

Sp. pinguis is one of the most abundant fossils at Millecent, Little Island, Malahide, and other Irish mountain limestone localities. In England it occurs at Bolland, Castleton, and in the Isle of Man. I am not acquainted with any Scottish example. In Belgium it was found at Visé and Tournay, by Professor de Koninck, and occurs also in several other continental localities. [2]

SPIRIFERA OVALIS, *Phillips*. Plate IX, figs. 20—26.

SPIRIFERA EXARATA, *Fleming* (?) British Animals, p. 376, 1828.
— OVALIS, *Phillips*. Geol. of Yorkshire, vol. ii, p. 219, pl. x, fig. 5, 1836.
— ROTUNDATA, *Sow.* De Koninck, An. Foss. de la Belgique, p. 263, pl. xv, fig. 4, 1843.
BRACHYTHYRIS OVALIS, *M'Coy*. Synopsis of Carb. Foss. of Ireland, p. 145, 1844.
— HEMISPHERICA, *M'Coy*. Ibid, p. 145, pl. xix, fig. 10, 1844.
SPIRIFERA OVALIS, *Phil.*, = HÆMISPHERICA, *M'Coy*. British Palæozoic Fossils, p. 419, pl. 3 D, fig. 28, 1855.

Spec. Char. Marginally elongated or transversely oval; hinge line less than half the width of the shell, with rounded cardinal angles. Dorsal valve moderately convex and much less deep than the opposite one; mesial fold broad, smooth, and well defined at all ages; depressed or obtusely rounded. From eighteen to twenty simple flattened or rounded ribs ornament the surface of each valve. Ventral valve deep and gibbous; beak

ever as characters. As already stated, the measurements here given are those only of certain individual specimens remarkable for their dimensions, or to show how variable different examples are in their compative proportions.

[1] In his memoir 'Ueber die Fossilien Schlesischen Kohlenkalkes,' 1854, Von Semenow gives as synonyms of *Sp. rotundata*, Martin, *Sp. rotundata*, Sow., *S. ovalis*, *S. integricosta*, Phillips, *Sp. linguifera*, M'Coy, *Sp. hemisphericum*, M'Coy, *Sp. ostiolata*, V. Buch, and *Sp. exarata*, Fleming, but leaves out *Sp. pinguis*, Sow., which he considers to be a distinct species. This synonym is defective in several particulars.

[2] It is hardly necessary to observe that Baron Von Buch was decidedly in error while stating, at p. 184 of the French translation of his memoir on *Delthyrisis*, that *Sp. pinguis* had been found in the rocks of Dudley Castle and Wenlock Edge.

tapering, moderately produced and incurved; area triangular, wider than high; fissure large, partially covered by a pseudo-deltidium. The sinus is rather shallow; commencing at the extremity of the beak; it extends to the front, and is ornamented by one or two longitudinal ribs on each of its sides. There is also a small hinge area in the dorsal valve. Measurements taken from three specimens have produced—

Length 21, width 20, depth 13 lines.
„   20, „ 24, „   14 „
„   7,   „ 6,   „   4   „

Obs. There is no doubt that the shell under description is closely allied t *Sp. pinguis*, but it may be distinguished by the shortness of its hinge line and area, the last being much more triangular and higher in proportion to its width than what is found in any of the numerous examples of *Sp. pinguis* that have come under my observation. The area in *Sp. ovalis* is also at times so small and narrow that the fissure occupies more than half of its entire surface. The dorsal valve is likewise generally not so deep or convex as in *Sp. pinguis*, so that the inequality in convexity of the valves becomes very perceptible; the mesial fold is also more uniformly convex, with rarely any trace of that mesial groove or depression so prevalent in all specimens and ages of the species last mentioned; but it must also be remembered, that although the fold is in general evenly smooth, some exceptional specimens possess a tendency to obscure or undefined plication[1] at all periods of growth. In *Sp. ovalis* the mesial fold is sharply defined (figs. 20—26), while in *Sp. integricosta* it is always distinctly ribbed, so much so, that in many young shells the position of the fold can hardly be distinguished from the lateral plications of the valves (figs. 13—19). All three, *Sp. pinguis*, *Sp. ovalis*, *Sp integricosta*, present the same peculiarity of being sometimes longer than wide, and at other times the reverse. And it was from a transverse variety of *Sp. ovalis* that, in 1844, Professor M'Coy founded his *Sp. hemisphærica*, but which name the author abandoned in 1855. It appears to me also very probable that *Sp. exarata*, Fleming, belongs to the same type as *S. ovalis;* but, as Dr. Fleming's shell was never figured, and that his description " Perforated valve with broad, smooth, flattened ribs divided by shallow narrow furrows; beak gibbous, incurved, hinge very short," might apply equally well to several other species, I should question the propriety of adopting that name in preference to the well-known one by Professor Phillips, and especially so as Dr. Fleming has further observed that although he has frequently found the perforated valve it was always mutilated and without the other valve, with which he was not acquainted, as will be perceived from the representation of the original example (fig. 24) kindly communicated by the author. Under any circumstance, the extreme tenuity of the area excludes the possibility of its having belonged to *Sp. rotundata*, Sow., with which it has been erroneously identified.

Sp. ovalis is not a very common fossil in the carboniferous limestone. In England it

---

[1] This is also the case with some examples of *Sp. pinguis*.

has been collected at Bolland, in the Craven district, at Malham Moor, Lowick, in the Isle of Man, and in several parts of Derbyshire. In Scotland it occurs at Corieburn (Campsie) at Westlothian and Bleith (Ayrshire). In Ireland, Mr. Kelly mentions Ballinacourty, Armagh, and Ballyduff. On the Continent, it has been found at Visé, in Belgium, by M. De Koninck, and at Keokuk, Iowa (America) by Mr. Worthen.

SPIRIFERA INTEGRICOSTA, *Phillips*.   Plate IX, figs. 13—19.

> CONCHYLIOLITHUS ANOMITES ROTUNDATUS (*Martin*)? Petrificata Derbiensia, tab. 48, figs. 11, 12, 1809.
> SPIRIFERA INTEGRICOSTA, *Phillips*. Geol. of York., vol. ii, p. 219, pl. x, fig. 2, 1836.
> — ROTUNDATA (*Martin*), var. PLANATA, *De Koninck*. Animaux Fossiles de la Belgique, pl. xvii, fig. 4 (not *Sp. planata*, Phillips).
> — PAUCICOSTATA, *M'Coy*? (British Palæozoic Fossils, p. 420, pl. 3, D, fig. 26, 1855.

Spec. Char. Transversely or longitudinally oval, almost circular when young ; hinge line shorter than the greatest width of the shell. The dorsal or dental valve is not quite so deep as the opposite one, and ornamented by from twenty-one to twenty-five simple or bifurcated rounded ribs, of which the three larger or central ones compose the mesial fold, which is but slightly elevated above the regular convexity of the valve, except in the vicinity of the front. Ventral valve convex, beak of moderate dimensions, proportions, and incurvation. The surface is ornamented by from twenty to twenty-four rounded ribs ; the mesial sinus extending from the extremity of the beak to the front, and varying both in depth and width in different examples. Area wider than high, divided by a triangular fissure partially covered by a pseudo-deltidium. Measurements taken from three examples have produced—

Length 15, width 14½, depth 11 lines.

  ,, 14,   ,, 18,   ,, 11 ,,

  ,, 9,   ,, 9,   ,, 5 ,,

Obs. As I have already had occasion to remark, the principal, and indeed only important difference between *Sp. integricosta* and *S. ovalis* consists in the mesial fold of the first being divided by three or five longitudinal ribs, while in *Sp. ovalis* the same portion of the shell is sharply defined, convex, and generally smooth, the contrast being especially apparent in young shells. It has also appeared to me probable that *Anomites rotundatus* of Martin,[1] was founded on a young specimen of the species under descrip-

---

[1] Martin describes his species as follows: "Suborbiculatus longitudinaliter sulcatus, margine sinu obsoleto. S. P. A fossil shell. The original an *Anomia*. Perforate valves convex, hinge straight, patulose short ; foramen triangular. The general form of the shell somewhat orbicular and in a slight degree compressed, as the convexity of the valves does not equal that which is found in other Anomitæ of the same

tion;[1] but the figure is not sufficiently precise to warrant a decided assertion, and for which reason I have retained the name subsequently introduced by Professor Phillips. In his work on the 'British Palæozoic Fossils,' p. 423, Professor M'Coy justly observes that *Sp. rotundata*, of Martin, is quite distinct from *Sp. rotundata*, Sow ; but is decidedly in error while considering *Sp. planata*, of Phillips, a synonym of Martin's species, for a comparison, however slight, of the last-named shell (Pl. VII, figs. 28—36) with that of Martin, of which I have reproduced the original illustration (Pl. IX, fig. 27), will, I believe, convince any one of the greater probability of Martin's shell being that of a young specimen of *Sp. integricosta*, because, besides a difference in its general shape, the ribs of *Sp. planata* are always proportionately smaller and more numerous than in Martin's figure. *Spirifera paucicostata*, M'Coy,[2] of which I have reproduced the original drawings (Pl. IV, fig. 12), is, perhaps a variety of *Sp. integricosta*, but not of *Sp. pinguis*, as was erroneously printed in the description facing Pl. IV of the present monograph. The ribs on the mesial fold, as well as the general shape of the shell, are more those of Phillips's than of Sowerby's species. *Sp. integricosta* is not a very common species in the carboniferous limestone; it occurs at Bolland, in the Craven district, Northumberland, &c. It was also collected in the Isle of Man by the Rev. Mr. Cumming. In Ireland Mr. Kelly mentions Bundoran, Millecent, and Little Island. In Scotland it has been found at Gare, in Lanarkshire.

SPIRIFERA FUSIFORMIS, *Phillips*. Pl. XIII, fig. 15.

> SPIRIFERA FUSIFORMIS, *Phillips*. Geol. of Yorkshire, vol. ii, p. 217, pl. ix, figs. 10, 11, 1836.

Spec. Char. Fusiform, about three times as wide as long, the beak and umbone

tribe. Valves longitudinally furrowed; margin obtusely crenate, with a scarcely distinguishable sinus. The beak of the larger valve incurved. A small and not a very common species. Limestone, Middleton.

[1] Professor De Koninck informs me that he is now of the same opinion.

[2] Professor M'Coy describes his species as follows :

"*Spec. Char.* Globose, or very broad-ovate; hinge line slightly shorter than the width of the shell; cardinal angles slightly obtuse, sides and front moderately rounded, very obtuse from the meeting of the valves at a large angle, front abruptly raised into a wide semi-elliptical sinus. Entering valve evenly convex; sides tumid, with six or seven strong, rounded, obtuse simple ribs on each side; mesial ridge broad, prominent, very strongly defined from the beak, having three ridges about the size of the lateral ones, each of which dichotomoses close to the margin. Receiving valve very gibbous, semicircularly arched from the beak to middle of front margin. Mesial sinus deep, strongly defined from the beak, having at first three, subsequently six, small obscurely marked ribs; beak very large, incurved; cardinal area moderately wide." Length 7½, width 9, depth 6 lines.

"This species is most nearly allied to *S. trigonalis*, from which it is distinguished by the more spheroidal form, the obtuse rounding of the sides, and the very small number of its lateral ribs. The distinctly ribbed mesial ridge (fold) separates it from *S. pinguis*, as well as its more depressed form, and the fewer and more prominent radiations. Not very uncommon in the carboniferous limestone of Derbyshire." ('British Palæozoic Fossils,' p. 420.)

almost upon a level. The dorsal valve not quite so deep or convex as the opposite one; mesial fold defined, but not much elevated above the level of the valve. In the ventral valve the sinus extends from the extremity of the small incurved beak to the frontal margin; hinge line and area as long as the greatest width of the shell, to which the lateral margins of each valve rapidly converge, producing acute terminations. The area is of moderate width, and divided by a small triangular fissure. External surface finely striated (?).

Length 4½, width 14, depth 4 lines.

Obs. I am acquainted with but a single imperfect individual of this shell, the original type forming part of Gilbertson's collection now in the British Museum. Professor Phillips states the surface to be finely radiated; but the Museum specimen, which is deprived of almost all its shell, is nearly smooth, showing indications of radiating striæ, but at the umbone and in the vicinity of the cardinal edge; therefore, from such imperfect material, it is hardly safe to conjecture as to the condition of its external sculpture. It is, however, probable that in the perfect shell the surface may have been such as was described by Professor Phillips, and this opinion appears to be strengthened from the fact that an American mountain limestone Spirifer, collected at Clifton, Illinois, by Mr. Worthen, closely resembles in shape the shell under description, and the external surface of which is finely striated. *Sp. fusiformis* cannot be confounded with *Sp. convoluta,* Phillips, or *Sp. subconica,* Martin, on account of the difference in the character of its surface, which could not have been strongly ribbed.

Loc. Bolland.

SPIRIFERA RHOMBOIDALIS, *M'Coy.* Pl. XII, figs. 6, 7.

MARTINIA RHOMBOIDALIS, *M'Coy.* Synopsis of the Carboniferous Fossils of Ireland, p. 141, pl. xxii, fig. 11, 1844.

Spec. Char. Rhomboidal, gibbous, slightly wider than long; hinge line much shorter than the width of the shell; cardinal angles rounded; beak of the ventral valve small, much incurved, with a deep wide, angular sinus extending from the extremity of the beak to the front; area small, fissure partially covered by a pseudo-deltidium. In the dorsal valve the mesial fold is prominent and almost angular from rapid slope of its lateral portions; surface of both valves ornamented with numerous small radiating obscurely-rounded ribs.

Length 9, width 10, depth 7 lines.

Obs. This shell does not appear to have often exceeded the dimensions above given, and is easily distinguished by its rhomboidal shape, almost angular elevated fold, deep linguiform sinus, and small undefined rounded ribs. Professor M'Coy's original example was derived from the Carboniferous limestone of Cork, and I possess another from

8

Millecent, in Ireland.   No English or Scottish specimens appear to have been hitherto discovered.

SPIRIFERA URII, *Fleming*.   Pl. XII, figs. 13, 14.

> SPIRIFERA URII, *Fleming*. British Animals, p. 376, 1828, reference. *David Ure*, The Natural
> History, &c., of Rutherglen and Kilbride, p. 313, fig. 12, 1793.
> — UNGUILICUS, *Phillips*. Palæozoic Fossils, tab. xxviii, 119, according to Morris,
> Catalogue of British Fossils, p. 154.

Spec. Char.   Suborbicular, rather wider than long ; hinge line shorter than the greatest breadth of the shell; cardinal angles rounded.   Dorsal or socket valve semicircular, slightly indented in front, with a narrow hinge area; nearly flat or but slightly convex, most so at the umbone, with a shallow mesial furrow commencing at a short distance from the umbone and extending to the front.   Ventral valve much more convex and deep than the opposite one, with a lengthened incurved beak, and longitudinal furrow commencing at the extremity of the beak and extending to the front.   The area is triangular and of moderate length and width, the fissure being partly closed by a pseudo-deltidium.   The external surface is smooth in the generality of specimens ; but, when perfect, was covered with small spinules.   Dimensions variable; the largest British specimen I have seen measured—

Length 4, width $4\frac{1}{2}$, depth 2 lines.

Obs.   This interesting little shell was noticed and figured for the first time by David Ure, but not named or described, an omission which was filled up by Dr. Fleming thirty-five years later, in his excellent work on 'British Animals.'   Ure's illustration would convey the idea that the beak was not incurved ; but in all the numerous examples I have been able to examine the shell possessed a gibbous lengthened incurved beak, as described by Dr. Fleming; but Ure did not fail to observe and represent an area on either valve. *Sp. Urii* does not appear to have ever attained proportions much exceeding those here given, and was in general a much smaller shell, for, out of many hundred examples collected by a zealous friend at Carluke, the largest did not exceed 4 by $4\frac{1}{2}$ lines in length and width.   *Sp. Urii* closely resembles the Permian *Sp. Clannyana*, King, as I have already stated at p. 16 of my 'Monograph of British Permian Species;' and *Sp. unguiculus,* mentioned to occur in the Upper Devonian of Petherwin, Barnstaple, Pilton, and Brushford, is either the same or a closely allied species or variety.

Sp. Urii was stated by David Ure to be plentifully found in a lime quarry on the east bank of the Aven, a little below Strathaven.   It abounds near Carluke, and has been found at Corieburn (Campsie) by Mr. J. Young.   Dr. Fleming has it also from Westlothian, in Scotland.   In England it does not appear to be so common a fossil.   I have seen a specimen from Bolland, and some other examples were discovered at South Petherton by the late Mr. D. Sharpe, and form part of his collection.

On the Continent it has been discovered at Tournay, in Belgium, by Professor L. de Koninck, and where the shell attains rather larger dimensions than is common to our British individuals.

Although the external surface of all the specimens of *Sp. Urii* appeared smooth, by the help of the lens I was able to discern in some examples the broken base of numerous spinules, which must have covered its surface in the perfect condition.

SPIRIFERA CARLUKENSIS, *Davidson.* Pl. XIII, fig. 14.

Spec. Char. Shell minute, nearly circular, and smooth; valves almost equally deep. Dorsal valve regularly convex, most so at the umbone. Ventral valve convex, with a narrow mesial depression or furrow commencing at a short distance from the extremity of the beak and extending to the front, where it indents the margin of the opposite valve. Beak small, pointed, and but slightly incurved; hinge line much shorter than the greatest width of the shell, with its cardinal angles rounded; area small, triangular, with a comparatively large fissure.

Length 2, width 2¼, depth 1¼ lines.

Obs. This little shell, which I believe to be new, was discovered in the Carboniferous beds of Hill Head, Carluke parish, Scotland, by the same friend to whom I am indebted for so much information relative to the Lanarkshire species. It is easily distinguished from *Sp. Urii*, with which it is associated by the almost equal convexity of its valves, and by the absence of a mesial groove on the dorsal valve. The beak is likewise much smaller, more acute, and less incurved. *Sp. Carlukensis* does not appear to have been a very common shell in its locality, where a hundred or more of *Sp. Urii* may be collected for one of the species under description, nor am I acquainted with the shell from any other British or foreign locality.

SPIRIFERA GLABRA, *Martin.* Pl. XI, figs. 1—9; Pl. XII, figs. 1—5, 11, 12.

CONCHYLIOLITHUS ANOMITES GLABER, *Martin.* Petrif. Derb., pl. xlviii, figs. 9, 10, 1809.
SPIRIFER GLABER, *Sowerby.* Min. Con., vol. iii, p. 123, pl. cclxix, fig. 1, May, 1820.
— OBTUSUS, *Sowerby.* Ibid., p. 124, pl. cclxix, fig. 2.
— OBLATUS, *Sowerby.* Ibid., p. 123, pl. cclxviii.
— GLABER, *Davreux.* Const. Geogn. de la Province de Liege, p. 272, pl. vii, fig. 1, 1831.
TRIGONOTRETA OBLATA, *Bronn.* Leth. Geogn., i, p. 81, pl. ii, fig. 16, 1836.
SPIRIFERA GLABRA, *Phillips.* Geol. of Yorksh., vol. ii, p. 219, pl. x, figs. 10—12, 1836.
— LINGUIFERA, *Phillips.* Ibid., fig. 4.
— SYMMETRICA, *Phillips.* Ibid., fig. 13.

Spirifera decora ? Ibid., fig. 9.
Spirifer lævigatus, *V. Buch.* Mémoirs de la Soc. Geol. de France, vol. iv, p. 198, pl. x,
    fig. 25, 1840.
— glaber, *De Koninck.* Animaux fossiles de la Belgique, p. 267, pl. xviii, fig. 1.
Martinia glabra, *M'Coy.* Synopsis of Carb. Foss. of Ireland, p. 139, 1844.
— obtusus and oblatus, *M'Coy.* Ibid.
Spirifera glabra, *M'Coy.* British Palæozoic Fossils, p. 428, 1855.

Spec. Char. Very variable in shape and proportions; transversely oval, rarely as long
or longer than wide. Valves almost equally convex, with a mesial elevation or fold in the
dorsal, and a sinus in the ventral valve. Hinge line much shorter than the greatest width
of the shell; cardinal angles rounded; beaks rather approximate, that of the larger or
ventral valve prominent, incurved, and of moderate dimensions. A hinge area in the
dorsal valve, that of the ventral one triangular and of moderate dimensions, with its
lateral margins more or less sharply defined; fissure partially covered by a pseudo-
deltidium. The mesial fold in the dorsal valve is either slightly and evenly convex, rising
gradually from the lateral portions of the valve, or abruptly elevated, with a longitudinal
depression along its middle, which is also at times reproduced in the sinus of the ventral
one. The spiral appendages are large, and occupy the greater portion of the interior of
the shell.[1] Surface of valves in general smooth, but sometimes a few obscure rounded
ribs may be observed on their lateral portions. Dimensions taken from five examples
have produced—

Length 32, width 43, depth 26 lines.
  ,, 31, ,, 36, ,, 21 ,,
  ,, 24, ,, 34, ,, 13 ,,
  ,, 24, ,, 25, ,, 19 ,,
  ,, 14, ,, 13, ,, 8 ,,

Obs. Martin's illustration of *Sp. glabra* is one of the many modifications assumed by
this very variable species, and I feel disposed to agree with Professor De Koninck, while
considering *Sp. oblatus* and *S. obtusus* (Sow.), *Sp. symmetrica*, *Sp. linguifera*, and *Sp. decora*, of Phillips, as simple varieties or variations in shape of Martin's shell. In the
'Synopsis of Carboniferous Fossils,' Professor M'Coy retained all the above-named shells
as distinct species, but in his more recent work on 'British Palæozoic Fossils,' Sowerby's
two shells are reduced to the rank of varieties of *glabra*, and there can exist no

[1] I have already had occasion to remark, at p. 81 of my General Introduction, that in p. 139 of his
'Synopsis,' Professor M'Coy has described and represented the spiral appendages of *Spirifera (Martinia)*
glabra so small as only to occupy the rostral half of the shell, but this has been proved incorrect, for all
the specimens obtained in which the spirals were preserved, have shown them to be as large as in any
other species of the genus. Fig. 9 of my Plate XI is a representation drawn from the original example
figured by Sowerby in tab. 268 of the 'Min. Con.,' thirteen years prior to the publication of M'Coy's
'Synopsis.'

possible doubt as to the close affinity or specific identity between *Sp. oblatus, Sp. obtusus,* and Martin's shell; those of Phillips, therefore, will alone require some further consideration.

The original specimen on which *Sp. symmetrica* was founded forms part of the Gilbertsonian collection in the British Museum, and has appeared to Professor De Koninck, as well as to myself, to be a variation of *S. glabra,* in which the mesial fold is but feebly elevated above the regular convexity of the valve, with also a slight longitudinal depression or groove along its middle (Pl. XI, fig. 6). Nothing seems to be more variable than the development of the mesial fold, for it is entirely absent in some young individuals, while in others of a similar age it becomes sharply defined.

Professor M'Coy has strongly urged the maintaining of *Sp. symmetrica* as a distinct species, probably from not having had sufficient opportunity of studying the original type, for otherwise he would have seen that it has not those distinctive features he so emphatically announces. "This beautiful and very distinct species varies very little in its characters; . . . . it is remarkable for the nearly regular rhomboic outline of the receiving valve; for the broad, often minutely notched, sinus in the front margin, producing scarcely any distinct mesial ridge; and for the strong, filiform subregular, distant radiating lines from the beak to the margin of the interior, often appearing on the external surface. Two specimens differ remarkably (one from Lowick and one from Derbyshire), by the hinge line being only $\frac{45}{100}$ as compared to the width; and one of them, by the length nearly equalling the width, and the shortness of the beak of the receiving valve, so strongly approximates to *S. decora,* that I suspect additional experience may unite these species. Both differ from all the varieties of *Spirifera (Martinia) glabra* by the strong threadlike, subregular, internal ridging from the beak to the margin." To this I would observe, that the length of hinge line and area in many indubitable examples of *S. glabra* is not more than a third of the breadth of the shell, and, in such cases, entirely agreeing with what we find in typical shapes of *Sp. decora.* The rhomboic outline, also, is not constant in the last-named shell, as a series of specimens from the Island of Man has completely confirmed; and it is rare to meet with specimens so strongly marked as the one represented (Pl. XII, fig. 12), for every passage or intermediate link will be found connecting it with the more common shapes of *glabra.* The double longitudinal groove observable along the middle of the sinus and in the mesial fold is likewise to be seen in many specimens of Martin's shell (Pl. XI, fig. 1), as well as the filiform, subregular, distant radiating lines mentioned by Professor M'Coy. I therefore agree with the last-named author, while considering *Sp. decora* intimately connected with *Sp. symmetrica,* but must also go a step further, by uniting the last-named shell to Martin's species.

It is worthy of remark, that although the surface of *Sp. glabra* is in general entirely smooth, in some exceptional cases there is a tendency to the formation of rounded ribs on the lateral portions of the valves (Pl. XII, fig. 3), and to this variation must be referred Professor Phillips's *Sp. linguifera,* of which fig. 4 is a representation, drawn from the original Gilbertsonian example in the British Museum, and from which it will be perceived that up to a certain age the shell was entirely smooth, but that after an interruption in its regular growth, some slightly marked ribs were suddenly produced (figs. 3—5). It will therefore be necessary to look upon these and similar specimens as exceptional shapes, as we would do for figs. 1 and 2 of the same plate.

Professor De Koninck has recently informed me that he feels disposed to separate those more flattened specimens with smaller beak and finer shell-texture (?) from *Sp. glabra* proper (Pl. XI, figs. 3, 4), by the name of *Sp. glaberrimus;* that in the last-named shell the mesial fold is at all ages uniformly and evenly convex, while in Martin's *S. glabra* it is divided by a longitudinal depression or furrow; but as I have not been able to convince myself that these characters have any real permanency, I must leave to other palæontologists to decide whether we can adopt the learned Professor's suggestion.

Sp. glabra has been mentioned as occurring in the Devonian rocks of several localities, and I possess small specimens from Barton, in Devonshire, which appear undistinguishable from some from the Carboniferous limestone.

Loc. *Sp. glabra* is one of the most abundant of Carboniferous limestone fossils. Martin's specimens were obtained at Chelmerton, Tideswell, and in several other localities in the gray limestone of Derbyshire. Sowerby mentions Scaliber, near Settle, in Yorkshire, and Axton Quarry, south-west of Llanasa, in Flintshire. It is abundant at Bolland, and in lower dark Carboniferous limestone of the Isle of Man, in that of Lowick, Northumberland, at Kendal, and in numerous other English localities. In Scotland it occurs at Harestanes and Hill Head, near Carluke, as well as at Beith, Ayrshire. In Ireland, Mr. Kelly furnishes us with the following localities: Malahide, Little Island, Carrownanalt, Clonea, Mullaghboy, Mullaghfin, Cornacarrow, Millecent, &c. On the Continent it is also a very common Carboniferous limestone fossil at Visé, Tournay, &c., in Belgium; at Ratingen; at Sablé, in France; and it has also been collected in Russia, America, &c.

Spirifera lineata, *Martin.* Pl. XIII, figs. 1—13.

CONCHILIOLITHUS ANOMITES LINEATUS, *Martin.* Petrif. Derby., tab. xxxvi, fig. 3, 1809.
TEREBRATULA LINEATA, *Sow.* Min. Con., vol. iv, p. 39, tab. cccxliii, figs. 1, 2, March, 1822 (not tab. ccccxciii, fig. 1).
— IMBRICATA, *Sow.* Ibid., pl. cccxxxiv, fig. 3.
SPIRIFERA MARTINI, *Fleming.* British Animals, p. 376, 1828.
— LINEATA, *Phillips.* Geol. of Yorksh., p. 219, pl. x, fig. 17, 1836.
— ELLIPTICA, *Phillips.* Ibid., fig. 16.
— IMBRICATA, *Phillips.* Ibid., fig. 20.
— MESOLOBA, *Phillips.* Ibid., fig. 14.
— LINEATUS, *Von Buch.* Mémoirs de la Soc. Geol. de France, vol. iv, p. 199, pl. x, fig. 26, 1840.
— — *De Koninck.* Animaux fossiles de la Belgique, p. 270, pl. vi, fig. 5, and pl. xvii, fig. 8, 1843.

RETICULARIA RETICULATA, *M'Coy*. Synopsis of the Carb. Limestone Fossils of Ireland, pl. xix, fig. 15, 1844.
—   IMBRICATA, *M'Coy*. Ibid., p. 143.
—   LINEATA, *M'Coy*. Ibid.
MARTINIA STRINGOCEPHALOIDES, *M'Coy*. ? Ibid., p. 141, pl. xxii, fig. 8.
SPIRIFERA ELLIPTICA, *M'Coy*. British Palæozoic Fossils, p. 427, 1855.
—   IMBRICATA, *M'Coy*. Ibid., p. 429.
—   LINEATA, *M'Coy*. Ibid.

Of this species two principal varieties may be distinguished.

1st. Var. *a*, LINEATA, Martin = *imbricata*, Sow. = *Martini*, Fleming = *reticulata*, M'Coy. Pl. XIII, figs. 1—13.

Spec. Char. Transversely oval, or sub-orbicular; hinge line much shorter than the width of the shell, cardinal angles rounded; beaks more or less approximate and considerably incurved. Ventral valve gently and evenly convex, rarely exhibiting any mesial elevation. Dorsal valve rather deeper than the opposite one, uniformly convex or presenting a shallow longitudinal depression, apparent only in the proximity of the front, or extending to the extremity of the beak. Area small, with lateral margins obscurely defined, fissure triangular, partially covered by a pseudo-deltidium. Surface of both valves marked by numerous and regularly imbricated lines, the radiating striæ being either so close that they can hardly be distinguished or varying in their degree of proximity, but rarely in any place more than a line apart. The concentric lines differ likewise in a similar manner, but are in general more widely separated, in some examples being barely distinguishable, while at other times they form strong, broad, flattened, or slightly rounded ridges. The spiral appendages occupy the greater part of the interior, and do not differ in detail from those of other species composing the genus. Two examples have measured—

Length 12, width 16,   (Martin's type.)
„   20,   „   23, depth 13 lines.

2d. Var. *β*, ELLIPTICA, Phillips. Figs. 1—3.

Transversely elliptical, always wider than long, with an obtuse, slightly elevated mesial fold in the dorsal, and a defined longitudinal sinus in the ventral one; hinge line rather exceeding half the width of the shell; area triangular; fissure large, and partially closed by a pseudo-deltidium; beak rather small and incurved, but not contiguous to that of the opposite valve. Surface ornamented by small radiating striæ intersected by numerous small concentric ridges. A large specimen has measured—

Length 25, width 36, depth 18 lines.

Obs. A difference in opinion has been expressed as to the specific claims of the shells here combined under the single denomination of *Sp. lineata*, Martin; and I am aware that several authors still maintain *Sp. lineata*, *Sp. imbricata*, and *Sp. elliptica* to be specifically distinct; but, after a long and attentive examination of a very numerous series of all those shells, at different stages of growth and from various localities, I have discovered so many intermediate shapes that it has appeared to me impossible to arrive at any other conclusion but that they are all variations of a single species. This view was similarly expressed in 1843 by Professor L. de Koninck, but contested in 1855 by Professor M'Coy, who, having enumerated the specific characters of *Sp. elliptica*, observes—" I agree with Mr. Phillips in considering this quite a distinct species from *Sp. lineata* with which M. De Koninck has united it. At all ages and sizes it is more transverse, more depressed, the beaks are further apart, and, above all, the species is distinguished by the sinus in the front margin and the strong mesial hollow extending to the apex of the beak. It is only when the shell is removed that the comparatively strong radiating striæ figured by Phillips are seen. The reticulations exactly resemble that of *Sp. lineata*."

While speaking of *Sp. imbricata*, the same author further observes—" This species is very easily distinguished from *Sp. lineata*, with which some Continental authors unite it, by the great width and coarseness of the concentric lamellar ridges, and the much fewer, broad, obtuse, longitudinal fimbriations in a given space. It is also less wide, has generally some trace of mesial hollow, and has an unusually coarse fibrous tissue under the lens." And again, under *Sp. lineata*, he states—" This species, from the peculiar structure of the surface, and the slight divergence of the dental lamellæ with the strong mesial septum, was originally combined in my 'Synopsis' (of Carb. Foss.) with *Sp. imbricata*, *S. reticulata*, *S. microgemma*, &c., into a little group called *reticularia*. There is a fine submedian impressed line, apparently a fracture, visible in many specimens from the beak to the front margin."

From the statements here made it will be perceived that the most important distinctions Professor M'Coy can suggest between *Sp. elliptica* and the other two shells are that Phillips's species possesses "a sinus in the front margin and a strong mesial hollow extending to the apex of the beak;" but these distinctions lose much of their importance from the fact that in some specimens of true *Sp. lineata* and *imbricata* there also exists a sinus and mesial hollow extending to the apex of the beak. The beaks are likewise not always approximate in the last two shells, although more commonly so than in *Sp. elliptica*. All three are often extremely transverse, and, as admitted by Professor M'Coy, the reticulations of *Sp. elliptica* and *Sp. lineata* are exactly similar.

It will also be found that in different examples the radiating striæ and concentric lines or ridges vary in their degree of strength and proximity; in some (figs. 8 and 10) the decussating lines are so fine and so close that the shell appears almost smooth, while at other times the concentric ridges are much stronger and more separate than the radiating ones, which vary both in number and proximity, and thus constituting the only appreciable difference between the typical examples of *Sp. lineata*, figs. 4, 5, 6, 9, and *Sp. imbricata*, figs. 11 and 12, of authors.

Sp. mesoloba, Phillips, has been placed by Mr. Morris among the synonyms of *Sp. glabra* ('Catalogue,' p. 152); but the inspection of the original Gilbertsonian specimen in the British Museum has convinced both Professor de Koninck and myself that the shell in question will require to be considered as a variety of Martin's *Sp. lineata*, in which a small mesial elevation is somewhat unusually developed, but exaggerated in the original representation, of which fig. 10 of my Pl. XI is a reproduction.

I will only further remark that since the majority of authors seem inclined to maintain both *Sp. lineata* and *S. elliptica*, I have described them separately, but as varieties of a single species, and must leave for future observers to determine whether or not this is to be taken as a correct interpretation.

The specific claims of *Martinia stringocephaloides* are very uncertain, notwithstanding the lengthened description given by Professor M'Coy, at p. 141 of his 'Synopsis.' In Pl. XII, fig. 15, will be seen a reproduction of the author's original illustration, the specimen from Lisnapaste being no longer to be found in Mr. Griffith's collection. Fig. 16 is another example from Old Leighlin, in the Royal Dublin Society's Museum, and stated to have been so labelled by the author himself. In Professor Phillips's opinion, as well as in my own, these two shells bear so much external resemblance to some *S. lineata* that, until better evidence to the contrary arises, I have considered that it will be preferable to leave it with the last-named species, as it is always desirable to burden the nomenclature as little as possible. Professor M'Coy gives us the following specific character: " Suborbicular, gibbous; dorsal valve (our ventral one) produced in a lengthened acute beak; cardinal area narrow, acute, angular; no mesial fold; surface marked with regular concentric lines." The only objections that could be adduced against our interpretation is, that no radiating lines are mentioned, and that the beak of the larger valve is more produced or lengthened than is usually the case in Martin's shell; but this character is far from being constant, as seen by fig. 16. The radiating lines in certain examples of *Sp. lineata* are likewise so faintly marked and so minute that they may have become obliterated in the only two specimens that appear to have been found. Professor M'Coy states, moreover, that " this remarkable shell seems to conduct to the *Pentameræ* by means of *Stringocephalus*;" but I cannot perceive the gradation, for the shell in question possesses the spirals of a *Spirifer*, as represented by the author in his woodcuts (figs. 24 and 25) while *Pentamerus* and *Stringocephalus* have an entirely different internal arrangement; he would have been nearer the mark had he mentioned that by its external shape *S. stringocephaloides* conducted us to *Athyris*. The concentric lines or ridges in Professor M'Coy's species are also exactly similar to those observable in *S. lineata*.

Loc. Sp. lineata and its variety, *S. imbricata*, are among the most abundant of carboniferous limestone fossils. Martin found it at Castleton, Hope, Dovedale, and in other localities of the main carboniferous limestone of Derbyshire, in the Black Rock, Clifton, Bolland, and Settle, Kirkby Lonsdale, Crooklands, and the lower carboniferous limestone of Lowick, Northumberland, Malham Moor, about Berwick-on-Tweed, also in Pembrokeshire, &c. In Scotland, Dr. Fleming has collected the shell at Dreghorn and Ayr. Mr. H. Miller found it at Dryden, near Edinburgh, and Courland, near Dalkeith. It occurs also at Carluke and Balquarhage, near Campsie. In Ireland it is also very abundant. Mr. Kelly furnishes us with the following localities: Lisnapaste, Ardagh, Little Island, Millecent, Larganmow, Tornaroan, Armagh, and Bannaghagole.

The var. *elliptica* occurs in many localities along with the more common varieties of *lineata*, at Bolland and in the lower carboniferous limestone of Kendal, Westmoreland, at Millecent, in Ireland, &c.

Sp. lineata and its variety, *elliptica*, are also very common shells in many foreign carboniferous localities. Professor de Koninck names Visé, Males, near D'Ath, at Lives and Chockier, in Belgium; at Ratingen, &c. In the Geology of Russia, M. de Verneuil and Count Keyserling mention Podolie, Sterlitamach, Sarana, Simsk, Becheva, &c. In America it has been found in several localities, such as Keokuh, Iowa.

Sub-Genus—CYRTIA, *Dalman*, 1827, and CYRTINA, *Dav.*, 1858.

In the eighty-third page of my general introduction doubts are expressed as to the value of Dalman's CYRTIA, and his diagnosis is there stated to be unsatisfactory and equally applicable to several species of *Spirifer*.[1] In fact, the genus appears to have been created simply to receive those few species of Spirifer which possess a circular foramen in the deltidium of the larger valve, for the author did not furnish us with any information regarding the internal arrangements of his two named types, *C. exporrecta* and = *C. trapezoidalis*. Subsequently to 1827 several other species were added (by different authors) to the genus *Cyrtia*, and among these are some whose shell structure has been stated to be punctate, while that of Dalman's type is unpunctate, as in *Spirifer* proper, and although it has always appeared to me probable that a difference in shell structure would be accompanied by some important interior modification, it was not until very lately that I was enabled to discover some of the characters of the following species: 1, *C. exporrecta*; 2, *C. trapezoidalis*; 3, *C. Murchisoniana*; 4, *C. cuspidata*; 5, *C. heteroclyta*; 6, *C. Demarlii* and *C. septosa*.[2] The results of my examination will show that in the first four, which belong to Dalman's genus, the internal characters are similar, but different from the last three, which cannot be properly retained under the same generic denomination,

[1] As we progress with our investigations, and as obscure points are gradually made clear, it is sometimes necessary to correct or to modify conclusions which may have resulted from the study of imperfect or insufficient material.

[2] As I have had no opportunity of studying the interior of the other species classed with *Cyrtia*, it will not be necessary to mention their names in the present instance.

for it is evident that considerable dissimilarities in the arrangements of the plates of the ventral valve must have carried along with them some important difference in the soft portions of the animal, and I therefore propose at least provisionally to distinguish the little groups of spiriform shells of which *C. heteroclyta, C. Demarlii,* and *C. septosa* are examples under the generic or sub-generic appellation of CYRTINA,[1] and to leave that of CYRTIA to those shells which agree with Dalman's *C. exporrecta, C. trapezoidalis,* and *C. Murchisoniana,* &c.; but it is necessary to observe that the last-named genus is in itself of such little value that it will remain a question for further discussion whether it should be retained or added to the synonyms of Sowerby's *Spirifer.*

In CYRTIA a short hinge tooth is situated on either side at the base of the fissure, supported by vertical shelly plates which diverge and extend from the extremity of the beak forming the fissure walls and occupying about one-third of the length of the bottom of the valve, as may be perceived by a reference to the woodcut in page 45.[2] There exists in *Cyrtia* no median plate or septum, the arch-shaped deltidium which covers the entire fissure is generally, but not always, perforated by a circular foramen.[3] In the smaller valve the spiral appendages and their mode of attachment is exactly similar to what we find in *Spirifer,* and with which the plates in the ventral valve also very closely agree.

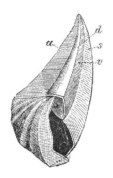

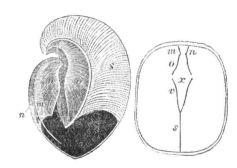

Pentamerus Knightii.

Cyrtina heteroclyta, slightly enlarged.   Fig. 2. Longitudinal section.   Fig. 3. Transversal section.

s. Septum.   *v.* Dental plates.   *x.* V-shaped chamber.   *m* and *n.* Diverging plates of the dorsal valve (in *Pentamerus*), to which are attached the curved plates *o.*   *a.* Area.   *d.* Deltidium.

---

[1] The diminutive of *Cyrtia* is *Cyrtidium,* but I have preferred bad Greek to a long name.

[2] The position and extent of these dental or rostral plates may sometimes be observed on the exterior of the shell, and especially on slightly worn specimens, by two diverging lines departing from the extremity of the beak (Pl. VII, fig. 22, s). In the upper Llandovery rock of May Hill are found many internal casts of *C. trapezoidalis,* which exhibit in a very beautiful manner the slits produced by the diverging plates; but these may be at any time exposed by grinding a portion of the beak of the Wenlock or Gothland specimens.

[3] The foramen is sometimes tubular, and especially so in certain Chinese examples of *C. Murchisoniana.*

Therefore *Cyrtia*, of Dalman, presents no other feature by which it can be separated from *Spirifer* proper, than that of its deltidium and foramen, which are characters of hardly sufficient importance to warrant the creation of a separate genus.

In our British carboniferous rocks, the only two forms known to me that could be referred to Dalman's *Cyrtia* are the *S. cuspidata* and *S. distans.*

In Cyrtina, the diverging plates already described do not exist, but we find in the interior of the ventral valve (of *C. heteroclyta* and *C. septosa*), two contiguous vertical septa (Pl. XIV, figs. 6, 7, 8), which coalesce into one median plate, which extends from the extremity of the beak to within a short distance of the frontal margin and then diverges to form the dental plates, in a very similar manner to what we perceive in *Pentamerus.*

The fissure is covered by an arch-shaped deltidium; but, in *C. Demarlii*, Mr. Bouchard has remarked that the median septum is continued as far as the under surface of the deltidium, and the dental plates are fixed to the sides, instead of the upper edge as in *C. heteroclyta* and *C. septosa.* The arrangements in the smaller or more important valve are still unknown, notwithstanding the many efforts I have made to pry into their interior; and it is certain that no vestige of spiral coils have hitherto been noticed by any author. Therefore, although we possess no proof that these three species of *Cyrtina* were possessed of spirals, and consequently true Spiriferidæ, it will be necessary to pause before admitting the shells in question into the genus *Pentamerus.*

Cyrtina septosa, *Phillips.* Pl. XIV, figs. 1—10, and Pl. XV, figs. 1, 2.

Spirifera septosa, *Phillips.* Geol. of Yorkshire, vol. ii, p. 216, pl. xi, fig. 7, 1836.
— subconicus, *De Koninck.* Animaux Fossiles de la Belgique, p. 255, pl. xii, *bis* fig. 5, *a, b, c*, 1843 (not *Anomites subconicus*, Martin, 1809).

Spec. Char. Very transverse, somewhat lozen-shaped; hinge-line as long as the greatest width of the shell; ventral valve moderately convex, subpyramidal, with a narrow sinus extending from the extremity of the beak to the frontal margin. Area very large, triangular, slightly curved upwards, and at an obtuse angle to the plane of the dorsal valve, the beak not protruding beyond or above the angular extremity of the area; fissure large, deltidium (?). Dorsal valve semicircular, convex, divided by a narrow, slightly elevated mesial fold, surface of both valves ornamented by from forty to seventy small angular ribs, which increase in number from bifurcation as well as intercalation as they proceed from the beaks to the frontal margin; six or seven of these form the mesial fold, and about a similar number that of the sinus, the radiating ribs being likewise intersected by numerous concentric lines or laminæ of growth.

In the interior of the ventral valve two large contiguous vertical plates or septa coalesce into one median plate, extending and augmenting in height from the extremity of the beak to nearly the margin of the shell, and separate to form the dental plates. Length and breadth very variable. One specimen measured—

Length 24, breadth 41, depth 15 lines; but some examples have attained rather larger proportions, and are not so very transverse.

Obs. This beautiful and very interesting species has sometimes been confounded with *Anomites subconicus* (Martin), but with which it bears no direct resemblance. Thus the last-named shell is described in the 'Petrificata Derbiensia' as differing from *Sp. cuspidata* "in the furrows, which are few in number and acute; in having a central angular fold instead of a rounded wave in the margin;" and it may be said to differ from *Sp. septosa* in the same particulars, for in Phillips's species the ribs are very numerous, small, " and divided into two, three, or four lesser ones towards the margin." There exists no acute mesial fold produced by a single prominent rib, that portion of the shell presenting a small, slightly convex, mesial elevation, composed of three principal ribs, but which from bifurcation become six or seven as they proceed towards the margin (Pl. XIV, figs. 1 and 2), and it would even appear that in certain individuals the fold is itself hardly distinguishable from the lateral portions of the valve. The sinus is also shallow; and, when perfectly shaped, margined on either side by a larger rib (Pl. XIV, fig. 4, and Pl. XV, fig. 2), between which may be seen a central and two smaller intercalated ones ; in other specimens the sinus is composed of as many as seven (Pl. XV, fig. 1). The number of ribs is likewise very variable in different examples; thus from forty to seventy may be counted round the margin of each valve of *C. septosa*, while seventeen only can be seen in the representations given by Martin of *A. subconicus ;* they are likewise very variable in their respective lengths and widths, which must be chiefly attributed to the intercalation of one or two smaller ribs next or between those first produced (Pl. XV, fig. 2).

The most important characters presented by this remarkable shell are, however, to be found in its interior arrangements, and these did not escape the notice of its first describer, who, after having briefly alluded to external appearances, pointedly remarks that "the septa in the lower (our ventral) valve divide it into three parts, as in *Pentamerus*, to which, *by this insufficient character*, it would be referred, but that many Spirifera exhibit less distinctly the same phenomenon."

Finding *S. septosa* to be a very rare species, but little understood or even known, the exterior of the ventral valve alone having been represented in the 'Geology of Yorkshire,' and those by Professor de Koninck being described under the mistaken denomination of *Sp. subconicus*, Martin, I did my best to assemble the few imperfect examples that were to be found in our English collections,[1] and to which my Belgian friend also kindly added

---

[1] No completely perfect adult specimens could be procured, the shell being generally found in separate valves.

the two or three fragments in his possession. This material has enabled me to satisfactorily develop and represent the internal details of the ventral valve (Pl. XIV, figs. 6, 7, 8); but all my efforts have hitherto proved ineffectual in making out those of the dorsal one.

Loc. *Cyrtina septosa* was discovered by Professor Phillips and by Mr. Salmond in the carboniferous limestone of Riddle Head, Burtonfell, Cumberland, and in the Museum of Practical Geology will be found a fine and instructive series of young and middle-aged specimens from the carboniferous limestone of Park Hill, Longnor, Derbyshire, as well as an internal cast (Pl. XIV, fig. 10), discovered in red dolometic limestone, at Ashby de la Zouch. The species has also been collected from the lower scar limestone of Settle, Yorkshire, by Mr. Burrow.

On the Continent, Professor De Koninck procured a few imperfect individuals in the carboniferous limestone of Visé, in Belgium, where the shell is, as in England, among the rarest species.

I am not acquainted with any Scottish examples, nor have any been hitherto discovered in Ireland, if *C. dorsata*, M'Coy, be not a variety of the shell under description.

CYRTINA DORSATA, *M'Coy*, &c.   Pl. XV, figs. 3, 4.

> CYRTIA DORSATA, *M'Coy*. Synopsis of the Carb. Limest. Fossils of Ireland, p. 136, pl. xxii, fig. 14, 1844.

Spec. Char. Subrhomboidal, nearly twice as wide as long; dorsal and ventral valves evenly convex; beak of the dorsal valve (our ventral) large, straight; cardinal angles very large, triangular, slightly concave; mesial fold indistinct or none; surface coarsely and regularly striated longitudinally.

"Length 2 inches, 4 lines; width 2 inches, 10 lines; height of cardinal angle 1 inch, 3 lines."—*M'Coy.*

Obs. As the original example figured by Professor M'Coy could no longer be found, and that the only other specimen, possessed by Mr. Griffiths, was in a very fragmentary condition, I could not determine to my entire satisfaction, the relationship of this form to *C. septosa*, but of which it may after all prove but a variety. The ribs seem, however, to be more numerous, simple, and of smaller proportions. The sinus and mesial folds are obsolete, and the general shape less transverse than in the specimens of Phillips's species that have come under my observation. I have therefore provisionally described *C. dorsata* under a separate head, where it had better remain until the discovery of more ample material will have confirmed or invalidated its specific claims.

Loc. The only two specimens hitherto recorded were obtained from the carboniferous limestone of Cork, in Ireland.

CYRTINA? CARBONARIUS, *M'Coy*, sp.   Pl. XV, figs. 5—14.

> PENTAMERUS CARBONARIUS, *M'Coy.* Annals and Mag. of Nat. Hist., vol. x, 2d series, and
> British Palæozoic Fossils, p. 442, pl. iii D, figs. 12—18,
> 1855.

Spec. Char.   Very variable in shape, globose, or imperfectly oval, generally longer than wide, hinge line rather shorter than the greatest width of the shell.   Dorsal valve semi-circular, moderately convex or gibbous, with its greatest depth about the middle; mesial fold narrow, of small elevation, regularly curved or with a longitudinal depression passing along its centre, and four or five smaller ribs originating near and extending to the margin. The lateral portions of the valve are furrowed by numerous angular ribs, which continue simple during their entire length, or become more or less subdivided from bifurcation or by intercalations originating at various distances from the margin.   The ventral valve is moderately convex or gibbous, beak large, slightly or greatly incurved, with its extremity straight, or twisted more to the one or to the other side.   Area large, at times higher than wide, generally concave, fissure triangular; deltidium (?).   The sinus is shallow and extends from the extremity of the beak to the front, being margined by larger ribs, while smaller ones occupy the intermediate space, and the lateral portions of the valve are similarly ornamented to those in the dorsal one.

In the interior of the larger valve a hinge tooth is placed on each side at the base of the fissure, the dental plates converge, and after forming the fissure walls, become conjoined so as to produce a single median plate or septum, which extends along the bottom of the valve to within a short distance of the frontal margin.   The internal details of the dorsal valve remain still to be determined.

Dimensions and proportions very variable.

Length 15, width 16, depth 13 lines (Professor M'Coy's type).

    ,, $17\frac{1}{2}$, ,, 17, ,, 14 ,,

    ,,   16,   ,, 13, ,, 10 ,,

    ,,   15,   ,, 13, ,,   7 ,,

Obs.   This remarkable species has been minutely described by Professor M'Coy at p. 442 of his 'British Palæozoic Fossils,' under the generic and specific denomination of *Pentamerus carbonarius*, the author observing, moreover, that " some of the specimens so nearly resemble *Spirifers*, that it was not until he had made sections in various directions of several specimens, demonstrating the invariable presence of two narrow longitudinal sub-parallel septa in the smaller valve, and the wide, extremely long, mesial septum in the ventral one, with its internal divaricating portions flanking the triangular opening in the cardinal area, perfectly agreeing with *Pentamerus*, as well as the absence of spiral appendages, that he was convinced of its true genus."   I must however observe that, without wishing to deny the possibility of the correctness of Professor M'Coy's conclusions, that the study of several specimens of this interesting species has not enabled me to arrive at so decided an opinion, and I am therefore still uncertain as to the propriety of classing the species in question with *Pentamerus*.

Professor M'Coy's illustrations of the interior (figs. 13, 14 of my plate) are, as he has himself admitted, evidently imperfect, for the dental plates are not represented extending so far as the inner upper extremity of the large mesial septum, which my preparations (figs. 11, 12) so completely exhibit. The appearance also of the internal cast of the dorsal valve of a specimen preserved in the Museum of Practical Geology (fig. 6), would lead me to infer that in the smaller valve there existed but a single median plate,[1] instead of two sub-parallel septa, as described and imperfectly represented in the work above quoted (fig. 14), and we have no certain evidence that the internal details of this valve were exactly similar to those of *Pentamerus*; on the contrary, there exists between the shell under description and *C. septosa* so much resemblance, both in external appearances as well as in the interior details of the ventral valve, that I have deemed it preferable (or at least so provisionally) to locate both Phillips's and M'Coy's species into my sub-genus *Cyrtina*, and from which the last-named author's shell may be hereafter removed should the discovery of the internal details of the dorsal valve determine the necessity.

It is possible that these *Spirifera*-shaped shells were not provided with spiral appendages, and that they formed a kind of passage between *Spirifer* and *Pentamerus*; but this must for the present remain an unsettled question.

A large area, similar to that seen in *C. carbonarius*, is not a character of *Pentamerus*, but it is necessary to remember that rudimentary areas occur in both *Pentamerus lens* and *P. liratus*, and nothing can be more variable than the extent and dimensions of the internal plates in different species of the genus. It is, therefore, to be hoped that ere long the discovery of some suitable specimens of *C. carbonarius* will enable us to determine the characters of the smaller valve, which are so important in the determination, not only of the genus, but also of its position in the classification of the group.

The figures I have selected for illustration will convey a good idea of the extreme variability in shapes presented by this species. In some examples the beak of the larger valve is so incurved as to come into contact with that of the smaller valve (fig. 5), while it assumes every degree of incurvature from this extreme condition to that in which the area is almost flat (figs. 9, 10). In relative width, breadth, as well as in degree of convexity, great differences are perceptible, as may be inferred from the measurements taken from four examples above noted. The ribs also vary considerably both in width and in the number of bifurcations, trichotomisings, or intercalations they may assume, from twenty to thirty being counted round the margin of each valve in different specimens; they are also at times much distorted, and, as mentioned by Professor M'Coy, " their surface is rather rugged and very coarsely granulo-punctate or minutely pustular under the lens;" but this

---

[1] A single median slit is observable in the cast (fig. 6). Had two sub-parallel septa existed, two slits would have been visible on the cast.

character is but rarely exhibited in the generality of specimens, on account of their external surface being considerably worn, so that the appearance can be detected only on those portions of the shell which are in their perfect condition.

Loc. Professor M'Coy's specimens were derived from the impure lower Carboniferous limestone of Kendal, Westmorland, where the species does not appear very rare.

Having now completed the description and illustration of all the species of *Spirifera*, *Spiriferina*, and *Cyrtina*, that are known to me as positively occurring in the carboniferous strata of Great Britain, it will be seen that not more than about thirty-one or two have been retained out of one hundred and seventeen that had been described or mentioned by various authors. A searching investigation has led me to infer that about eighty-six of the published names are made up of synonyms, of species not *positively* known to occur in Great Britain, or of Carboniferous shells erroneously identified with Devonian species, as well as of specimens belonging to other genera.[1]

[1] The following is a list of 117 so-termed British Carboniferous Spirifera ; the *species* adopted in this work are printed in roman type, the synonyms, &c., in italic letter, and a point of interrogation is placed before the more doubtful species:

? Spirifera acuta, Martin.
— *attenuata*, Sow.
— *aperturata*, Schloth. Not carboniferous.
— *arachnoida*, Phil. Strophomena.
— bisulcata, Sow.
? — bicarinata, M'Coy.
— *Bouchardii*, Murch. Not carb.
— carbonarius, M'Coy.
— *Crispa*, Linnæus. Not carb.
— *costata*, Sow. Not carb.
— *connivens*, Phil. Orthis.
— *clatharata*, M'Coy.
— *calcarata*, G. Sow. Not carb.
— convoluta, Phil.
? — crassa, De Koninck.
— cuspidata, Martin.
— *choristites*, V. Buch.
— carlukiensis, Dav.
— *crenistria*, Phil. Strophomena.
? — decemcostata, M'Coy.
— *decora*, Phil.
— distans, Sow.
? — dorsata, M'Coy.
— *duplicicosta*, Phil.
— *disjuncta*, Sow. Not carb.
— *elongata*, Phil.
? — exarata, Fleming.

Spirifera *expansa*, Phil. Athyris.
— *elliptica*, Phil.
— *extensa*, Sow. Not carb.
— *fasciculata*, M'Coy.
? — fusiformis, Phil.
— *furcata*, M'Coy.
— *ficiger*, Keyserling.
— *filiaria*, Phil. Orthis.
— glabra, Martin.
— *globularis*, Phil. Athyris.
— grandicostata, M'Coy.
— *glabristria*, Phil. Athyris.
— *gigantea*, Sow. Not carb.
— *grandæva*, Phil. Not carb.
— *heteroclytus*, Def. Not carb.
— *histericus*, Schloth. Not carb.
— humerosa, Phil.
— *hemisphærica*, M'Coy.
— *incisa*, Goldf.
— *imbricata*, Sow.
— insculpta, Phil.
— integricosta, Phil.
— *inornata*, Sow. Not carb.
— *Kleinii*, Fischer.
— *Koninckiana*, D'Orb.
— lineata, Martin.
— *linguifera*, Phil.

I must also state that notwithstanding the lengthened examination I have made of the Carboniferous *Spiriferæ*, there are a few even among those here retained that will still require some further investigation in order to ascertain whether or not they may be varieties of some of those already described, and among these I will mention *Sp. crassa*, De Kon., *Sp. grandicostata*, M'Coy, *Sp. minima*, Sow., *Sp. mesogonia*, M'Coy, and *Sp. fusiformis*, of Phillips. To these we must likewise add a few other doubtful forms, such as *Sp. transiens*, M'Coy, *Sp. partita*, Portlock, *Sp. decemcostata*, M'Coy, *Sp. bicarinata*, M'Coy, *Sp. subconicus*, Martin, *Sp. dorsata*, M'Coy, *Sp. rotundatus*, Martin, *Sp. exarata*, Fleming, *Sp. elongata*, Phil., and *Sp. similis*, M'Coy, whose relations or affinities could not be ascertained or established in a satisfactory manner from the absolute want of sufficient material, for not a single specimen or even fragment of some could be procured, and the descriptions and illustrations of the authors were not such as to lead to any satisfactory results. It is therefore probable, nay certain, that the larger number (if not all)

Spirifera laminosa, M'Coy.
— *mesoloba*, Phil.
— *microgemma*, Phil. Not carb.
— *megaloba*, Phil. Not carb.
? — mesogonia, M'Coy.
? — minima, Sow.
— *Martini*, Fleming.
— *macroptera*, Hall.
— *mesomala*, Phil. Not carb.
— mosquensis, Fischer.
— *nuda*, Sow. Not carb.
— octoplicata, Sow.
— *oblata*, Sow.
— *obtusa*, Sow.
— ovalis, Phil.
— *ornithorhyncha*, M'Coy.
— *ostiolata*, Steing. Not carb.
— *plebeia*, Phil. Not carb.
— *protensa*, Phil. Not carb.
? — partita, Portlock.
— pinguis, Sow.
— planata, Phil.
— *planicosta*, M'Coy.
— *paucicostata*, M'Coy.
— *princeps*, M'Coy.
— *pulchella*, Sow. Not carb.
— *phalœna*, Phil. Athyris.
— *prisca*, Eichw. Not carb.
— *papilionacea*, Phil. Chonetes.
— *recurvata*, De Kon. Not British.
— *reticulata*, M'Coy.
— *rotundata*, Sow.

? Spirifera *rotundata*, Martin.
— *rhomboidea*, Phil.
— *rudis*, Phil. Not carb.
— *Roemeri*, De Kon. Said to occur in Ireland. (?)
— Reedii, Dav.
— *radialis*, Phil. Strophomena.
— *resupinata*, Martin. Orthis.
— *semireticulata*, Phil.
— septosa, Phil.
— *sexradialis*, Phil.
— *Sowerbii*, Fischer.
— *similis*, Phil.
— striata, Martin.
— *striatella*, M'Coy. Not a Spirifer.
— *stringocephaloides*, M'Coy.
? — subconica, Martin.
— *subrotundata*, M'Coy.
— *symmetrica*, Phil.
— *senilis*, Phil. Streptorhynchus?
— *speciosa*, Schloth. Not carb.
— *semicircularis*, Phil.
— *simplex*, Phil. Not Dev.
— *spirifera*, Lam.
— *squamosa*, Phil. Athyris.
— *transiens*, M'Coy.
— triangularis, Martin.
— *trisulcosa*, Phil.
— trigonalis, Martin.
— *tricornis*, De Kon.
— *triradialis*, Phil.
— Urii, Fleming.

the above-named doubtful species will be found to be synonyms of some of the thirty already described, and that even three or four of this number may require, upon more extended examination, to be reduced to the rank of varieties.

The species of British *Spirifera*, *Spiriferina*, and *Cyrtina* at present known might perhaps be arranged in the following order.

1. SPIRIFERA STRIATA, *Martin*, sp. Pet. Derb., tab. xxiii, 1809; Dav. Brit. Foss. Brach., part v, pl. ii, figs. 12—21; pl. iii, figs. 2—6 = *Spirifera*, Lamarck = *attenuata,* J. de C. Sow. = *princeps*, M'Coy = *Clatharata*, M'Coy.

2. — MOSQUENSIS, *Fischer de Waldheim*. Programme sur les Choristite, p. 8, No. 1, 1837; and Dav., pl. iv, figs. 13, 14, and pl. xiii, fig. 16 = *Sowerbyi* and *Kleinii*, Fischer = *incisa*, Goldfuss = *Choristites*, V. Buch. = *priscus*, Eichwald.

3. — HUMEROSA, *Phillips*. Geol. York., vol. ii, pl. xi, fig. 8, 1836; and Dav., pl. iv, figs. 15, 16.

4. — DUPLICICOSTATA, *Phillips*. Geol. York., pl. x, fig. 1, 1836; and Dav., pl. iii, figs. 7—10; pl. iv, figs. 3, 5—11 = *faciger*, Keyserling = *fasciculata*, M'Coy.

? 5. — CRASSA, *De Koninck*. An. Foss. de la Belgique, pl. xv *bis*, fig. 5, 1843; and Dav., pl. vi, figs. 20—22; pl. vii, figs. 1—3 = *planicosta*, M'Coy.
N.B. By some this shell is placed among the synonyms of *Sp. duplicicostata*, by others among those of *Sp. bisulcata*.

6. — PLANATA, *Phillips*. Geol. of York., pl. x, fig. 3, 1836; and Dav., pl. vii, figs. 25—36.

? 7. — FUSIFORMIS, *Phillips*, pl. ix, figs, 10, 11; Dav., pl. xiii, fig. 15.
Of this only a single imperfect specimen has been discovered, so that the characters and value of the species cannot be considered finally established.

8. — TRIANGULARIS, *Martin*, sp. Pet. Derb., pl. xxxvi, fig. 2, 1809; and Dav., pl. v, figs. 16—24 = *ornithorhyncha*, M'Coy.

9. — TRIGONALIS, *Martin*, sp. Pet. Derb., tab. xxxvi, fig. 1, 1809; and Dav., pl. v, figs. 25—34.

10. — BISULCATA, *Sow.* M. C., tab. ccccxciv, figs. 1, 2, 1825; and Dav., pl. v, fig. 1; pl. vi, figs. 1—19; and pl. vii, fig. 4 = *semicircularis*, Phillips = *calcarata*, M'Coy = *transiens*, M'Coy?

11. — CONVOLUTA, *Phillips*. Geol. York., pl. ix, fig. 7, 1836; and Dav., pl. v, figs. 2—15 = *rhomboidea*, Phillips.

12. — GRANDICOSTATA, *M'Coy*. Brit. Pal. Fossils, pl. iii, D, fig. 29, 1855; and Dav., pl. v, figs. 38, 39, and pl. vii, figs. 7—16.

N.B. All these species are closely allied, and it may remain a question for further consideration whether they should be *all* specifically separated. It is possible that *Sp. convoluta* may after all be nothing more than a very transverse and exceptional condition of *Sp. bisulcata*.

13. — LAMINOSA, *M'Coy*. Synopsis, pl. xxi, fig. 4, 1844; and Dav., pl. vii, figs. 17—22 = *tricornis*, De Koninck.

14. Spirifera cuspidata, *Martin*, sp. Trans. Lin. Soc., pl. iii, figs. 1—6, 1796; and Dav., pl. viii, figs. 19—24; pl. ix, figs. 1, 2 = *simplex*, M'Coy, not of Phillips.

15. — distans, *Sow.* M. C., tab. 494, fig. 3, 1825; and Dav., pl. viii, figs. 1—17.

16. — mesogonia, *M'Coy.* Synopsis, pl. xxii, fig. 13, 1844; and Dav., pl. vii, fig. 24.

    This is said by Professor De Koninck to be a very good species, but I have never had an opportunity of studying the shell.

Doubtful Species.

? — *bicarinata*, *M'Coy.* Synopsis, pl. xxii, fig. 10; and Dav., pl. viii, fig. 18.

? — *decemcostata*, *M'Coy.* Ibid., fig. 9; and Dav., pl. vii, fig. 23.

? — *subconica*, *Martin.* Pet. Derb., tab. xlvii, figs. 6—8, 1809; Dav., pl. ix, fig. 3.

17. — Reedii, *Dav.*, pl. v, figs. 40—47.

18. — triradialis, *Phillips.* Geol. York., pl. x, fig. 7; and Dav., pl. ix, figs. 4—12 = *trisulcosa* and *sexradialis*, Phillips.

19. — pinguis, *Sow.* M. C., tab. cclxxi, 1820; and Dav., pl. x, figs. 1—12 = *rotundata*, Sow. (not Martin) = *subrotundata*, M'Coy.

20. — ovalis, *Phillips.* Geol. York., pl. x, fig. 5; Dav., pl. ix, figs. 20—26? = *exarata*, Fleming = *hemisphærica*, M'Coy.

21. — integricosta, *Phillips.* Geol. York., pl. x, fig. 2; Dav., pl. ix, figs. 13—19? = *rotundatus*, Martin? = *paucicostata*, M'Coy.

    N.B. The species forming this little group are all very nearly connected.

22. — rhomboidalis, *M'Coy.* Synopsis, pl. xxii, fig. 11, 1844; Dav., pl. xii, figs. 6, 7.

23. — glabra, *Martin*, sp. Pet. Derb., pl. xlviii, figs. 9, 10, 1809; and Dav., pl. xi, figs. 1—9, and pl. xii, figs. 1—5 = *obtusus*, Sow. = *oblatus*, Sow. = *linguifera, symmetrica,* and *decora*, Phillips.

24. — urii, *Fleming.* British Animals, p. 376, 1828; and Dav., pl. xii, figs. 13, 14.

25. — Carlukensis, *Dav.*, pl. xiii, fig. 14, 1857.

26. — lineata, *Martin*, sp. Pet. Derb., tab. xxxvi, fig. 3, 1809; and Dav., pl. xiii, figs. 1—13 = *imbricata*, Sow. = *reticulata*, M'Coy = *elliptica* and *mesoloba*, Phillips.

27. Spiriferina cristata, *Schloth.*, var. octoplicata, Sow., M. C., tab. dlxii, figs. 2—4, 1827; Dav., pl. vii, figs. 37—47.

28. ? insculpta, *Phillips.* Geol. York., pl. ix, figs. 2, 3, 1836; Dav., pl. vii, figs, 48, 55 = *crispus*, De Kon. (not of Linnæus) = *quinqueloba*, M'Coy.

29. ? minima, *Sowerby.* M. C., tab. ccclxxvii, fig. 1, 1822; Dav., pl. vii, figs. 56—59, uncertain species.

Doubtful Species.

? ? *partita*, *Portlock.* Report on the Geol. of the County of Londonderry, &c., pl. xxxviii, fig. 3, 1843; and Dav., pl. vii, figs. 60, 61.

    N.B. It has not yet been perfectly ascertained whether *Sp. insculpta* and *Sp. minima* have a perforated shell structure.

30. Cyrtina septosa, *Phillips.* Geol. York., pl. xi, fig. 7, 1836; Dav., pl. xiv, figs. 1—10, and pl. xv, figs. 1, 2.

31. — carbonarius, *M'Coy.* British Palæozoic Fossils, pl. iii D, figs. 12—18, 1855; and Dav., pl. xv, figs. 5—14.

— *dorsata*, *M'Coy.* Synopsis, pl. xxii, fig. 14, 1844; and Dav., pl. xv, figs. 3, 4.[1]

---

[1] The still imperfect knowledge we possess of the exact geological position or vertical range of several of the species of the Carboniferous system, renders every well-determined fact of considerable interest.—

Genus—ATHYRIS, *M'Coy*, = SPIRIGERA, *D'Orbigny*.

See article *Athyris*, 'Monograph of British Permian Brachiopoda,' Part IV, pp. 20—22, 1857.

ATHYRIS AMBIGUA, *Sowerby*. Pl. XV, figs. 16—22.

SPIRIFER AMBIGUUS, *Sowerby*. Min. Con., vol. iv, p. 105, tab. ccclxxvi, Nov., 1822 (*Atrypa* of the index).

TEREBRATULA AMBIGUA, *Phillips*. Geol. of York., p. 221, pl. xi, fig. 21, 1836.

? — PENTAEDRA, *Phillips*. Ibid., pl. xii, fig. 3.

ATRYPA SUBLOBATA, *Portlock*. Report on the Geology of the Coast of Londonderry, &c., p. 567, pl. xxxviii, fig. 2, 1843.

TEREBRATULA AMBIGUA, *De Koninck*. Animaux Fossiles de la Belgique, p. 296, pl. xx, fig. 2, 1843.

— — *M. V. K.* Geol. of Russia, vol. ii, pl. ix, fig. 12, 1845.

SPIRIGERA AMBIGUA, *D'Orbigny*. Prodrome, vol. i, p. 151, 1849.

ATHYRIS AMBIGUA, *M'Coy*. British Palæozoic Fossils, p. 432, 1855.

Spec. Char. More or less obscurely pentagonal, rather wider than long, moderately convex; beak not much produced, incurved; foramen small, circular, contiguous to the umbone of the opposite valve; a longitudinal, somewhat angular sinus extending from the extremity of the beak to the frontal margin. Dorsal valve almost evenly convex, or obscurely trilobed, the central lobe or fold being more often broad, and longitudinally divided by a narrow mesial groove; front deeply undulated. External surface smooth, marked only by a few concentric lines of growth. Shell structure not perforated. In the interior the spiral appendages are directed outwards, filling the larger portion of the shell. Dimensions very variable; three examples have measured—

Length 14, width 12, depth 8 lines.

„ 10 „ 11 „ 7 „

„ $10\frac{1}{2}$ „ $10\frac{1}{2}$ „ 8 „

Obs. This species varies considerably in the details of its external shape. It is sometimes nearly equally and evenly convex (especially in young shells), with hardly any definite mesial elevation in the dorsal valve (figs. 21, 22, 23), while at other times the appearance of the shell is obscurely trilobed, and when the furrow along the middle of the fold is strongly marked it sometimes resembles certain examples of the Jurassic *Terebratula quadri-*

Mr. J. H. Burrow, to whom I am greatly indebted for much valuable and liberal assistance, has informed me that he has found the following Spirifers in the *Lower Scar Limestone* of Settle, in Yorkshire: 1. *Sp. striata*; 2. *S. duplicicostata*; 3. *Sp. crassa*; 4. *Sp. planata*; 5. *Sp. fusiformis*; 6. *Sp. triangularis*; 7. *Sp. trigonalis*; 8. *Sp. bisulcata*; 9. *Sp. convoluta*; 10. *Sp. grandicostata*; 11. *Sp. cuspidata*; 12. *Sp. triradialis*; 13. *Sp. pinguis*; 14. *Sp. ovalis*; 15. *Sp. integricosta*; 16. *Sp. glabra*; 17. *Sp. lineata*; 18. *Sp. octoplicata*; 19. *Sp. insculpta*; 20. *Sp. Reedii* and *C. septosa*.

fida, Lamarck, (figs. 15, 16, 17). Sowerby states that "the produced beak and three-angu-
lar-sided front give the shell a five-angled contour, although the sides are rounded." In the
degree of convexity of its valves, different specimens vary to a very great extent, and it has
appeared to me probable that the *Terebratula ambigua* of Phillips, "pentagonal, depressed,
surface undulated, front and sides emarginate, perforation of the beak minute," (fig. 25,)
is only a more flattened condition of Sowerby's species. (?)

Atrypa sublobata, Portlock, has already been correctly identified with Sowerby's *Sp.
ambiguus,* for it is stated by Professor M'Coy, at p. 432 of his 'British Palæozoic Fossils,'
" I have at length succeeded in tracing, in the most gradual manner, the passages of all
the forms figured by General Portlock under the name of *Atrypa sublobata,* into each other,
and in the ordinary types of the present species. When decorticated, a few straight pallial
ridges are seen near the beak, radiating towards the front margin. General Portlock
notices the resemblance of some of the varieties to *S. unguiculus* of Sowerby, but the want
of area between the beaks and hinge-line separate the species." It is also possible that
Athyris trilobata, M'Coy, may require to be added to the synonyms of Sowerby's species,
but from such insufficient data as the simple dorsal valve represented in Tab. XX, fig. 21,
of the 'Synopsis,' it would be hardly safe to offer any decided opinion.

The presence of spiral appendages in this species did not escape the observing eye of
Sowerby, who in 1822 appears to have even hinted at the propriety of establishing a new
genus for its reception, for we find him stating, that "in general appearance it does not
agree with most species of *Spirifer,* but approaches nearer the smooth Terebratulæ; its
having a perforated beak, and little or no hinge-line, still further distinguishes it; but the
actual existence of spiral appendages seems to confirm it a Spirifer, *unless its combining
the characters of both genera should render it desirable to construct a new genus of it.* But
as the appendages within the Terebratulæ are very variable, it will be well to wait until more
of them are known."[1]

I have been informed that it was for the reception of *T. ambigua* and other similarly
organised forms, that in 1841 Professor Phillips created his *Cleiothyris,* which he described
with a "cardinal area obsolete, beak incurved over a minute perforation, which is often
obtect, or merely serves to receive the beak of the smaller valve;" but as the author inad-
vertently and unfortunately omitted to mention any known species as an example, Professor
M'Coy subsequently proposed the name *Athyris*[2] for similar kinds of shells.

---

[1] Professor de Koninck remarks, at p. 297 of his 'Animaux Fossiles de la Belgique,' "that the spiral
arms are formed of nine or ten coils, and are placed in an opposite direction; that it is solely on this
last character, which is nevertheless common to many species of Terebratulæ, that Mr. De Buch has relied
for expressing the opinion that *T. ambigua* and some others must be placed among the Spirifers, and
insinuates at the same time that the circular foramen might perhaps not be the work of nature. We can
affirm that this aperture is not due to chance, and that it is to be found on all well-preserved examples;
moreover, that Mr. Deshayes builds upon the characters of this species to combat the establishment of the
genus Spirifer, to which Sowerby had at first referred the species."

[2] Sowerby's *Spirifer ambiguus* has received no less than six different generic appellations—*Spirifer,*

Loc. Sowerby states he obtained his specimens from decomposed mountain limestone (rotten stone), near Bakewell. The shell is far from rare in the lower Carboniferous limestone of Derbyshire; the Isle of Man; Lowick, Northumberland; at Bolland, &c. In Scotland it is rather common in the Carboniferous limestone and shales of the Clydeside basin, at Carluke, Lowrieston,[1] at Westlothian; Beith, Ayrshire; and Berwick-on-Tweed. In Ireland it is mentioned as occurring at Millecent, Kilcummin, and Ballintrillick. On the Continent M. De Koninck states it to be rather rare at Visé, in Belgium; and M. de Verneuil and Count Keyserling mention Peredki and Valdai as Russian localities. In the United States it has been discovered at Chester, Illinois; and in other localities.

ATHYRIS LAMELLOSA, *L'Eveillé.* Pl. XVI, fig. 1, and Pl. XVII, fig. 6.

> SPIRIFER LAMELLOSUS, *Leveillé.* Mémoires de la Soc. Géol. de France, ii, p. 39, figs. 21—23, 1835.
> — SQUAMOSA, *Phillips.* Geol. of Yorks., ii, p. 220, pl. x, fig. 21, 1836.
> TEREBRATULA LAMELLOSA, *De Koninck.* An. Foss. de la Belgique, p. 299, pl. xx, fig. 5, *a, b, c,* 1843.

Spec. Char. Transversely elliptical, or obscurely pentagonal; valves moderately convex, somewhat depressed; cardinal line nearly straight. The mesial fold is of small elevation and at times slightly concave along its middle. In the ventral valve a sinus of moderate depth extends from the extremity of the small incurved beak to the frontal margin; foramen small, circular, and contiguous to the umbone of the opposite valve. Surfaces of both valves ornamented by from twelve to fifteen nearly parallel concentric lamelliform expansions. In the interior the spiral coils fill the greater portion of the shell.

Length 14, width 21, depth 8 (without the expansion).

Obs. This species was correctly described and illustrated by L'Eveillé and Professor De Koninck, who observe that its principal character resides in the presence of strong concentric lamelliform expansions; but this peculiarity is also common to other species of *Athyris,* such as *A. planosulcata,* to which L'Eveillé's species sometimes nearly approaches.

Terebratula, Atrypa, Cleiothyris, Athyris, and *Spirigera!* and it is to be regretted that those who were discussing Professor Phillips's name on his half-announced views, had not inquired from the author why he had conceived the group and name. On the Continent, D'Orbigny's term *Spirigera* is generally preferred to that of *Athyris,* and I would myself have adopted the French author's denomination, since it is freed from the incorrect derivation the term *Athyris* conveys, had I not found that English authors were so much disposed to prefer M'Coy's name on account of its priority of date.

[1] It may remain a matter of some uncertainty whether the representation given by David Ure, in pl. xvi, fig. 9, of his 'History of Rutherglen' (1893), was intended for *Ter. hastata* or for the shell under description; but from the greater abundance of Sowerby's *Ambiguus* in the district, I should almost feel disposed to refer the figure above mentioned to the last-named species.

Loc. In England *A. lamellosa* has been found in the Carboniferous limestone of Settle, Yorkshire; Dovedale, Derbyshire, &c.; Professor Phillips mentions Kendal and Florence court. It occurs also at Hook Point, in Ireland. I am not acquainted with any Scottish examples. Professor De Koninck states he has found but a single example in the Carboniferous limestone of Visé, but that it is more abundant in the clay of the same formation of Tournay, in Belgium. In America it has been found by Mr. Worthen in the mountain limestone of Keokuk, Iowa.

ATHYRIS PLANOSULCATA, *Phillips*. Pl. XVI, figs. 2—13, 15.

> SPIRIFERA PLANOSULCATA, *Phillips*. Geol. of York., vol. ii, p. 220, pl. x, fig. 15, 1836.
> TEREBRALUTA DE ROISSYI, *De Verneuil*. Bulletin de la Société Geol. de France, vol. xi, p. 259, pl. iii, fig. 1 *a*, 1840 (not *Sp. de Roissyi*, L'Eveillé, 1835).
> ATRYPA PLANOSULCATA, *J. de C. Sowerby*. Min. Con., vol. vii, p. 15, pl. dcxvii, fig. 2, 1840.
> — OBLONGA. Ibid., fig. 3.
> — PLANOSULCATA, *De Koninck*. An. Foss. de la Belgique, p. 301, pl. xxi, fig. 2, 1843.[1]
> ACTINOCONCHUS PARADOXUS, *M'Coy*. Synopsis of Carb. Foss. of Ireland, pl. xxi, fig. 6, 1844.
> ATRYPA OBTUSA, *M'Coy*. Ibid., pl. xxii, fig. 20.
> ATHYRIS PARADOXA, *M'Coy*. British Palæozoic Fossils, p. 436, 1855.

Spec. Char. Obscurely pentahedral or nearly orbicular; valves equally deep, and either moderately or evenly convex, without sinus or fold, or with a slight mesial depression towards the front in one or both valves. The beak is small or incurved, with a minute circular foramen placed close to the umbone of the opposite valve. Surface of both valves ornamented at intervals of less than a line, with numerous large, concentric, parallel, semicircular, lamelliform expansions, each plate being flat and longitudinally striated at about half a line apart. Interiorly the spiral appendages, which are each composed of from twelve to fifteen coils, fill the larger portion of the shell.

The following measurements of the same specimens, with and without their expansions, will convey some idea of the relative proportions.

Without expansions.	The same.	With expansions.
Length $11\frac{1}{2}$, width 11,		length 20, width $24\frac{1}{2}$ lines.
„ $8\frac{1}{2}$, „ 12,		„ 15, „ 21 „
„ 14, „ 15,		„ 25, „ 33 „
„ $19\frac{1}{2}$, „ 21, depth 13,	extent of expansions not known.	
„ 12, „ $10\frac{1}{2}$, „ 8,	ditto.	

---

[1] *Sp. fimbriata* and *S. expansa*, Phillips, mentioned by Professor De Koninck, will require to be excluded from the synonyms of *A. planosulcata*.

Obs. The true character of this species has not always been clearly understood, so that some confusion has resulted from incorrect identifications. Professor Phillips states that his shell is pentahedral, depressed, with the middle of each valve planosulcate; but although some specimens do agree with this description, as well as with the single illustration (fig. 2) published in the ' Geology of Yorkshire,' still the larger number are more or less circular or ovate marginally, with their valves equally deep and evenly convex (figs. 4, 8—10). Professor Phillips was not aware that in the perfect condition his shell was provided with numerous flat concentric plates, which were produced from each successive line of growth, and prolonged, in some specimens, nearly an inch from the surface of the shell (fig. 8), and it is to similar examples that Professor M'Coy applied the generic and specific demonstration of *Actinoconchus paradoxus* in 1844, and *Athyris paradoxa* in 1855.

In 1840 M. de Verneuil published a figure of the same shell, with its marginal expansions, which he had found at Visé, in Belgium, under the name of *Terebratula de Royssii*,[1] from not being then aware of the difference in character of the expansions in Phillips' and L'Eveillé's species.

These appendages in *A. planosulcata* have been described as continuous concentric plates, but those in *A. de Royssii* are in the shape of numerous concentric ridges, from each of which radiate closely-set fringes of elongated, somewhat flattened, spines.

In 1843 Professor de Koninck published a very good description of *A. planosulcata*, along with representations of the shell with and without its expansions.

It is certain, also, that *Atrypa oblonga*, of Sow., and *Atrypa obtusa*, of M'Coy, are only slight modifications in shape of Phillips's species, and in the work on ' British Palæozoic Fossils' (p. 436), Professor M'Coy still considers *A. planosulcata* as distinct from his *A. paradoxa*, but with which opinion I am obliged to dissent. The author, moreover, observes that, " When the extended, flattened lamellæ are broken off, as is the case in the greater number of specimens, they only leave traces of obscure lines of growth (about ten in two lines), so nearly obsolete in the rostral portion that it seems smooth, but showing by their thickness, and the extreme obtuseness of the edge in specimens approaching ten lines in length, that to be the ordinary adult size, although I have seen some rather larger not bearing such marks of age. It is only close to, or immediately at, the margin that we find these paradoxical, greatly extended, shelly, 'flat, radiated lamellæ, which, if perfect, would considerably exceed the length of the shell in width. I imagine that they are found at the margins, because there the corresponding lamellæ of the two valves would come in contact and support each other; but, by growth of the shell, they would become separated when a new edge was formed between them, leaving them erect, insulated, and liable to be broken off by the slightest accident. The same thing occurs in *Tridacna squamosa*, where the great scale-like laminæ near the beak are always effaced, while those near the margin are perfect."

[1] 'Bulletin Soc. Geol. de France,' vol. xi, pl. iii, fig. 1.

11

Loc. In England *A. planosulcata* has been collected in the lower dark Carboniferous limestone of Lowick, in Northumberland ; at Longnor, in Derbyshire ; at Bolland ; Settle ; and in several other localities. In Scotland it has been met with in Lanarkshire, but is most abundant in Ireland, whence Mr. Kelly furnishes us with the following localities : Blacklion, Millecent, Little Island, Milverton. On the Continent it has been found by M. de Verneuil and De Koninck in the Carboniferous limestone of Visé, in Belgium.

ATHYRIS EXPANSA, *Phillips.*   Pl. XVI, figs. 14, 16—18 ; Pl. XVII, figs 1.—5.

> SPIRIFERA EXPANSA, *Phillips.*   Geol. of Yorkshire, vol. ii, p. 220, pl. x, fig. 18.
> ATRYPA EXPANSA, *J. de C. Sow.*   Min. Con., pl. dcxvii, fig. 1 (the upper three large speci-
>   mens only), 1840.
> —   FIMBRIATA.   Ibid., fig. 4 (not *Sp. fimbriata*, Phillips).
> ATHYRIS EXPANSA, *M'Coy.*   British Palæozoic Fossils, p. 433, 1855.

Spec. Char. Transversely elliptical, always very much wider than long ; valves evenly and equally convex, often much depressed, with or without a gentle mesial depression in the ventral valve, no fold in the dorsal one ; beak small, incurved ; foramen circular and contiguous to the umbone of the opposite valve. External surface of both valves covered with fine, indistinct radiating lines or striæ, departing from the extremity of the beaks, and crossed by numerous concentric lines or ridges of growth. The spiral appendages occupy the larger portion of the interior. Dimensions very variable ; three specimens measured—

Length 21, width 27, depth 11, without the lamellar appendages.
     „   16,   „   22,   „   8,   ditto.
     „   10,   „   29,   „   6,   ditto.

Obs. This species appears to be more variable in shape than *A. planosulcata*, from which it may be usually distinguished by its very transversely elliptical form, some specimens being almost twice and a half as wide as long (figs. 17, 18) ; and is often found in different states of malformation (Pl. XVII, figs. 2—4). The external surface is marked by faint radiating lines, which, according to Mr. J. de C. Sowerby, are inter-sected by " broad striated imbricating fringes," and of which a representation is given in pl. dcxvi, fig. 1, of the ' Mineral Conchology.' It is probable that these lamelliform prolongations were very similar to those of *A. planosulcata ;* but as they were not present on any of the numerous individuals that came under my direct observation, I have contented myself with reproducing the statement made in the ' Mineral Con-chology.'

Ter. fimbriata (Phillips) has been classed by some palæontologists among the synonyms of *A. expansa,* but although the author of the ' Geology of Yorkshire ' has omitted to furnish us with a figure of his species, and that his description—" orbicular, depressed; beak of the lower valve prominent but small; surface strongly radiated and concentrically imbricated," might be applicable to this or to other species, I am able to assert, from the inspection of the original example in the author's possession, that *T. fimbriata*, Phillips, cannot be placed among the synonyms of *A. expansa*, but would be more properly located with those of *A. Royssii*.

I am also quite of Professors M'Coy and De Koninck's opinion, while stating that *A. fimbriata*, figured by Mr. J. de C. Sowerby in the 'Mineral Conchology,' has been drawn from a specimen of Phillips's *A. expansa*; and it is likewise certain that some examples labeled and described as *A concentrica* (Buch), by Professor M'Coy, belong to the species we are now describing. *Sp. expansa* appears to have been subject to much malformation[1] if we are to judge from the number of specimens in that condition which abound in certain localities.

Loc. Common at Kendal, Westmoreland; Settle, in Yorkshire; at Bolland, and in the lower Carboniferous limestone of Hittor-hill, and Longnor, in Derbyshire, &c. In Ireland Mr. Kelly mentions Bruckless, Drumdoe, and Milverton. I am not acquainted with any Scottish examples.

ATHYRIS SQUAMIGERA, *De Koninck* (?).   Plate XVIII, figs. 12, 13.

> MARTINIA PHALŒNA, *M'Coy.* Synopsis of the Carboniferous Fossils of Ireland, p. 140, 1844. (Not *Spirifera phalœna* of Phillips's 'Figures and Descriptions of Palæozoic Fossils,' p. 71, pl. xxviii, fig. 123, 1841.)
> TEREBRATULA SQUAMIGERA, *De Koninck.* Animaux Fossiles du terrain Carbonifere de la Belgique, p. 667, pl. lvi, fig. 7, 1851.

Spec. Char. Transversely oval, much broader than long; valves convex, sometimes gibbous; beak moderately produced, incurved, and truncated at its extremity by a small circular aperture. In the dorsal valve there exists a prominent mesial fold, and in the ventral one a sinus of variable depth, both commencing at a short distance from the extremity of the beaks. External surface ornamented with small imbricated striæ; interiorly there exists two spiral appendages, with their extremities directed outwards. Two specimens have measured—

Length 12, width 22, depth 9 lines.
    „   8, „ 16, „ 7 „

Obs. On comparing the specimen identified by Professor M'Coy as *Martinia phalœna* (our fig. 13) with Phillips's Devonian *Spirifera phalœna*, I was soon convinced that they

---

[1] Two of these specimens in the Museum of Practical Geology measure—
Length 23, breadth 33, depth 7 lines.
    „ 25, „ 30, „ 7 „

did not belong to the same species, nor even to the genera to which they had been referred. Both are undoubted *Athyris's*, but the Carboniferous shell has puzzled me much, and it was only after having consulted Professor de Koninck that I ventured doubtfully to suggest that M'Coy's *M. phalœna*, and some other similar specimens I had obtained from Ireland, might perhaps belong to Professor de Koninck's *M. squamigera*, the peculiar reticulated surface still preserved on some portions of M'Coy's specimen (in the collection of Sir R. Griffith) being very similar to that described by the Belgian palæontologist. *Athyris squamigera* is stated by Professor de Koninck to be nearly related to *A. Royssii*, but that it differs by its greater transversity, depth of sinus, and external sculpture.

Loc. *A. squamigera* is mentioned by Mr. Kelly to occur in the Carboniferous limestone and calcareous slate of Lisnapaste, Clonea, St. Doulough's, in Ireland. Sir R. Griffith's specimen is labeled Ballinacourty, Dungarvan, and I have another from Millecent, Ireland. I am not acquainted with any English or Scottish examples. It was found in the Carboniferous limestone beds of Tournay, in Belgium, by Professor de Koninck.

ATHYRIS ROYSSII, *L'Eveillé.* Plate XVIII, figs. 1—11.

SPIRIFER DE ROYSSII, *L'Eveillé.* Mémoirs de la Société Géologique de France, vol. ii, p. 39, pl. ii, figs. 18—20, 1835.
— GLABRISTRIA, *Phillips.* Geol. of Yorkshire, vol. ii, p. 220, pl. x, fig. 19, 1836.
— FIMBRIATA. Ibid., p. 220, not figured.
TEREBRATULA ROYSSII, *De Verneuil.* Bulletin de la Société Géologique de France, vol. xi, p. 259, pl. iii, fig. 1, *b, c, d* (not 1, *a,* and *e*), 1840.
— — *De Koninck.* Animaux Fossiles du terrain Carbonifere de la Belgique, p. 300, pl. xxi, fig. 1 (but not pl. xx, fig. 1), 1843.
ATHYRIS DEPRESSA, *M'Coy.* Synopsis of the Characters of the Carboniferous Fossils of Ireland, p. 147, pl. xviii, fig. 7, 1844.
— DE ROYSSII, *M'Coy.* British Palæozoic Fossils, p. 433, 1855.
— GLABRISTRIA, *M'Coy.* Ibid., p. 434.

Spec. Char. Circular, or transversely oval; subglobose; beak incurved, and truncated by a small circular foramen which is contiguous to the umbone of the dorsal valve. The valves are almost equally and uniformly convex up to a certain age, after which a broad mesial fold of greater or lesser elevation is gradually formed in the dorsal valve, and a corresponding sinus in the ventral one. The frontal margin is, therefore, either nearly straight or presents a greater or lesser curve; the external surface is regularly covered by numerous concentric scaly ridges, from each of which radiate closely-set fringes of elongated, somewhat flattened spines. In the interior, the hinge is strongly articulated, the dental or rostral plates in the ventral valve offering, by their position, much solidity to the beak of the ventral valve. The hinge-plate is perforated close to its summit by a minute circular aperture, and the spiral appendages for the support of the oral arms have their extremities directed outwards, and are united by a complicated system of lamellæ.

Dimensions and relative proportions very variable; two specimens have measured—

{ Without the spinose expansion, length 24, width 33, depth 15 lines.
{ With   ditto   „   32, „   39, „   17 „
{ Without   ditto   „   24, „   26, „   16 „
{ With   ditto   „   31, „   32, „   17½ „

Some examples have even slightly exceeded these dimensions, but the generality of specimens are much smaller.

Obs. This remarkable species has been often confounded with *Athyris plano-sulcata*, but from which it is easily distinguished by its spines, which are very different from the numerous large, concentric, lamelliform expansions of *A. plano-sulcata;* and it is from the generality of palæontologists having overlooked this circumstance, that they have so often confounded the two shells. So closely packed are the spinose ridges which invest the entire surface of *A. Royssii*, that no portion of the surface of the valve itself can be perceived, and I have counted as many as eighty of these pectinated fringes on one of the valves of a specimen which did not measure more than ten lines in length by thirteen in width. This arrangement will be easily understood by a glance at the enlarged representations (figs. 10 and 11), wherein a portion of the spines have been purposely omitted, so as to show the disposition of the pectinated expansions. The interior arrangements would appear to be exactly similar to those we have already described in *Athyris pectinifera*, and, indeed the two shells have been considered by several palæontologists as belonging to a single species.

As we have fully described and represented the interior dispositions in the Permian shell, and figured those also of another species of the same genus, *A. ambigua*, in Pl. XVII of the present monograph, it will not be necessary to repeat what has been already written.

In the limestone, on account of the hardness of the matrix, it is impossible to detach the specimen with its outer spinulous surface, and for this reason the shell, in that condition, was not recognised, and received from Professor Phillips the denomination of *Sp. glabristria*, while to the specimens from the shales which retained their spinous ridges the name of *Sp. fimbriata* was applied; but both will require to be added to the synonyms of L'Eveillés species, as well as that of *A. depressa* (M'Coy). Professor M'Coy observes, at p. 433 of his Cambridge work, that "The equal, thick, longitudinal spines, which fringe the narrow, concentric lamellæ of shale specimens (of the species under description), form an extremely marked character, and by their great strength and coarseness separate the species certainly from the Devonian *A. concentrica* of V. Buch." In Pl. XVIII, I have represented the shell under various aspects, and to which the reader is referred.

Loc. *A. Royssii* is a common shell in the Carboniferous limestone and shales of many localities. In England it was found at Bolland, in the Isle of Man, at Ulverstone, and near Settle, in Yorkshire, &c.   In Ireland, Mr. Kelly mentions Bundoran, Millecent, Little Island, Lisnaparte, Malahide, and Hook.   In Scotland it is found at Brokley, near Lesmahago, at Craiginglen, and at West Broadstone, near Beith, in Ayrshire.   On the Continent it was found at Tournay, and Pauquys, in Belgium, by Professor L. de Koninck, M. L'Eveillé, and others, &c.

ATHYRIS GLOBULARIS, *Phillips.*   Plate XVII, figs. 15—18.

> SPINIFERA GLOBULARIS, *Phillips.*   Geology of Yorkshire, vol. ii, p. 220, pl. x, fig. 22, 1836.
> ATHYRIS   —   *M'Coy.*   British Palæozoic Fossils, p. 434, 1855.

Spec. Char.   Subglobose; as wide, or wider than long; valves almost equally convex; beak moderately produced, incurved and truncated by a small circular foramen, which is contiguous to the umbone of the opposite valve.   The sinus in the ventral valve, and the fold in the dorsal one, commence to appear at a short distance from the beaks, and divide the shell into three lobes of almost equal breadth; the sinus is of moderate depth, while the fold is more or less elevated.   Surface smooth, marked only by a few concentric lines of growth.   Shell-structure not perforated.   In the interior, the spiral appendages for the support of the oral arms have their extremities directed outwards, and fill the larger portion of the shell.   Dimensions variable; two examples have measured—

Length 11, width 12, depth 9 lines.
  „   9, „ 9, „ 8 „

Obs.   A. globularis is closely related to both *A. ambigua*, Sowerby, and to *A. subtilata*, Hall.   From the first it may, however, be distinguished by its more globose and uniformly convex appearance, the absence of that narrow median groove or depression in the dorsal valve (so characteristic of Sowerby's shell), as well as by the more regular inflation of the lateral portions of the valves.   *A. ambigua* may be said to be obscurely divided into four lobes, while three would constitute the character of Phillips's species.   From *A. subtilata*, *A. globularis* is principally distinguished by its transverse shape and more clearly defined mesial fold, Professor Hall's shell being longitudinally ovoid, or much longer than wide.

Loc.   A. globularis occurs in the Carboniferous limestone of Bolland, Settle, in York-shire, and in several Derbyshire localities.   Professor M'Coy states it to be common in the Carboniferous strata of Glasgow, and in that of Craige, near Kilmarnock; rare in the Carboniferous limestone of Dalmellington, Ayrshire, but whence I have not seen any well-authenticated specimens.   On the Continent it has been found in the Carboniferous limestone of Visé, in Belgium, by Professor L. de Koninck.

ATHYRIS SUBTILITA, *Hall.*   Plate I, figs. 21, 22.   Plate XVII, figs. 8—10.

> ATHYRIS GREGARIA, *M'Coy.*   British Palæozoic Fossils, p. 435.

Obs.   At p. 18 of this monograph, I described the shell here named as *Terebratula* (?)

subtilita, Hall, the material then at my command not being sufficient to enable me to determine positively whether it was a *true Terebratula* or an *Athyris*. Many excellent examples having subsequently turned up, I was able to assure myself that it was with the last-named genus that Professor Hall's species must be located, and now hasten to place it among its congeners. I also assured myself that the shell described by Professor M'Coy, in his work on 'British Palæozoic Fossils,' under the denomination of *Athyris gregaria*, belonged to Professor Hall's *A. subtilita*, and not to the shell which the same author had originally so designated in his 'Synopsis of the Characters of Irish Carboniferous Fossils,' and to which the term *gregaria* belongs.

Athyris subtilita occurs also in the Carboniferous limestone of Tournay in Belgium, and fig. 1, pl. xx, of Professor de Koninck's work on 'Belgian Carboniferous Fossils,' is referable to this species. At p. 714, of his 'Iowa Report,' Professor Hall states " that *T. subtilita* has a very wide range, being known in the eastern Ohio, Indiana, Illinois, Iowa, Missouri, Kansas, Nebraska, and Pecos village, in New Mexico." The British localities have been already mentioned.

RETZIA RADIALIS, *Phillips.* Plate XVII, figs. 19—21.

> TEREBRATULA RADIALIS, *Phillips.* Geology of Yorkshire, vol. ii, p. 223, pl. xii, figs. 40, 41, 1836.
>
> TEREBRATULA MANTIÆE, *De Koninck.* Animaux fossiles Carbonifere de la Belgique, p. 287, pl. xix, fig. 4, *a, b, c, d,* 1843 (not *T. mantiæ* of Sowerby).
>
> ATRYPA RADIALIS, *M'Coy.* Synopsis of the Characters of the Carboniferous Fossils of Ireland, p. 156, 1844.
>
> RETZIA RADIALIS, *Morris.* A Catalogue of British Fossils, p. 145, 1854.
>
> SPIRIGERINA (?) RADIALIS, *M'Coy.* British Palæozoic Fossils, p. 438, 1855.

Spec. Char. Circular or longitudinally ovate; valves almost equally and moderately convex, without fold or sinus; margin of valves almost straight. Each valve is ornamented by about twenty small, rounded, radiating ribs, of which the central one in the dorsal valve is at times the largest, and to which, in the ventral valve, corresponds a deeper sulcus; but this difference is not always apparent, and in some specimens the dorsal valve appears for some distance divided by a median depression, from which arises a central rib. The beak is more or less produced and truncated by a circular foramen, which is more or less separated from the hinge-line by a small triangular area. Shell-structure punctate. Interiorly, two spiral appendages, with their extremities directed outwards, exist for the support of the oral arms. Two examples have measured—

Length 5, width 5, depth 3 lines.

„ 4, „ 4, „ 3 „

Obs. This little *Retzia* was in 1843 mistaken by Professor de Koninck for Sowerby's *T. mantiæ*, but the distinguished Belgian palæontologist subsequently discovered and admitted his mistake.   Professor Phillips describes his shell as follows: " Orbicular; no mesial fold; ridges equal, rounded, radiating."   This diagnosis will agree with many of the specimens, but the peculiarity presented by the central rib and corresponding sulcus in the opposite valve are characters not to be overlooked, as well as the small triangular area which is not represented in the figures of the shell given in the ' Geology of Yorkshire.'   In England, *R. radialis* has been obtained from the Carboniferous limestone of Bolland, also by Mr. Burrow in the lower scar limestone of Settle, Yorkshire; and Professor M'Coy mentions it from Derbyshire.   In Ireland it was found, according to Mr. Kelly, at Bruckless, Malahide, and Millecent, in the Carboniferous .limestone and calcareous slate.   In Scotland it has been discovered near Lesmahago by Dr. Sliman. In Belgium it has been collected at Visé and Tournay by Professor de Koninck; and by the authors of the ' Geology of Russia' in the Carboniferous limestone of Zaraisk, in the government of Riaisan, Russia.

RETZIA ULOTRIX, *De Koninck*.   Plate XVIII, figs. 14, 15.

TEREBRATULA (CRISPATA) ULOTRIX, *De Koninck*.   Animaux fossiles de la Belgique, p. 292, pl. xix, fig. 5, 1843.

Spec. Char.   Somewhat circular or slightly transverse; moderately convex, without fold or sinus; each valve is ornamented with from seven to nine angular ribs, of which the three central ones are the largest and most prominent; the beak of the ventral valve is somewhat produced, straight, but slightly incurved, and truncated by a small circular foramen, a flattened triangular area existing between it and the hinge-line.   Shell-structure punctate.   Interiorly there exists two spiral appendages, with their extremities directed outwards for the support of the oral arms.

Dimensions variable; two British examples have measured—

Length 6, width 6,   depth 3 lines.

,,   2, ,, $2\frac{1}{3}$, ,,   1 ,,

Obs.   This little *Retzia* was first described by Professor de Koninck under the designation of *T. crispata*, but in the explanation of his plate the name was altered to that of *T. ulotrix*, from the author having found that the first denomination had been already employed.   I am acquainted with only two English examples of this shell, one, from Bolland, I found in the British Museum; the other, from the Carboniferous limestone of Wetton, in Derbyshire, is in the Museum of Practical Geology; and upon sending drawings to Professor de Koninck, I was assured that they were referable to his species. This shell is stated to be likewise very rare in Belgium, and is found in the Carboniferous limestone of Tournay, where it would appear to have attained somewhat larger dimensions, the specimen figured by Professor de Koninck measuring—

Length $6\frac{1}{2}$, width $7\frac{1}{2}$, depth 4 lines.

R. ulotrix is easily distinguishable from *R. radialis* by its fewer and larger ribs, as well as by other peculiarities.

In recapitulation it may be mentioned that the most active research has not enabled me to recognise more than seven or eight species of *Athyris*, and two of *Retzia*, in the British Carboniferous deposits. These may be arranged in the following order, and those that are most closely related have been connected with a brace:

ATHYRIS.
1. Athyris planosulcata, *Phillips.*
2. — lamellosa, *L'Eveillé.*
3. — expansa, *Phillips.*
4. — Royssii, *L'Eveillé.*
5. — squamigera (?) *De Koninck.*
6. — ambigua, *Sowerby.*
7. — globularis, *Phillips.*
8. — subtilita, *Hall.*

RETZIA.
1. Retzia radialis, *Phillips.*
2. — ulotrix, *De Koninck.*

Family—RHYNCHONELLIDÆ.

The only two genera belonging to this family that have been hitherto discovered in British Carboniferous strata are *Rhynchonella*, Fischer ('Introduction,' p. 93), and *Camarophoria*, King ('Introduction,' p. 96, and 'Permian Monograph,' p. 23). The genus *Pentamerus* has been recorded by Professor M'Coy, but the species so referred to belongs to *Cyrtina*, and not *Pentamerus*, as will be found explained at p. 66 of the present Monograph.

After a very lengthened and difficult study of the various species of *Rhynchonella* and of *Camarophoria* stated to have been found in the Carboniferous rocks of Great Britain, I have ventured to refer the whole to the following species:

RHYNCHONELLA.
1. Rhynchonella reniformis, *Sowerby.*
2. — cordiformis, *Sowerby* (still doubtful).
3. — acuminata, *Martin.*
4. — pugnus, *Martin.*
5. — pleurodon, *Phillips.*
6. — flexistria, *Phillips.*
7. — angulata, *Linnæus.*
8. — trilatera, *De Koninck.*

CAMAROPHORIA.
{
1. Camarophoria crumena, *Martin.*
2.   —   globularis, *Phillips.*
3.   —   (?) laticliva, *M'Coy.*
4.   —   isorhyncha, *M'Coy.*
}

Doubtful species of RHYNCHONELLA or CAMAROPHORIA.
{
1. Rhynchonella or Camarophoria (?) proava, *Phillips.*
2.   —   semisulcata, *M'Coy.*
3.   —   gregaria, *M'Coy.*
4.   —   nana, *M'Coy.*
}

So that out of upwards of thirty so-termed species recorded by various palæontologists, or geologists, as having been found in the Carboniferous rocks of Great Britain, but seven of *Rhynchonella*, and three or four of *Camarophoria*, appear to me as worthy of being retained; all the others, with the exception of three or four still doubtful forms, are either synonyms or names of species that have not in reality been hitherto discovered in Great Britain. The extent of variations in shape and even character assumed by certain species is quite perplexing; for instance, *Rh. acuminata* is at times entirely smooth, or is covered more or less with ribs; sometimes it is a heart-shaped shell, while at other times it is completely depressed and almost flattened; still, every passage connecting these extreme cases can be easily recognised, and it should warn the palæontologist of the necessity of founding his appreciations, not from the study of one or two specimens only, but from the inspection of a large series of its varieties.

Notwithstanding all the care and trouble I have taken in the endeavour to determine as correctly as possible the limits of the various species, I cannot pretend to have always succeeded, and would strenuously advise the reader to study the matter independently, and not to blindly adopt the views here recorded, as certain of them may eventually prove erroneous.

In order to facilitate research, I have carefully selected and represented all the more important variations assumed by each species, as I have always considered the illustration of a typical form only, to be insufficient as it does not generally convey to the mind the true average characters of the species (if I may be permitted the use of such an expression).

RHYNCHONELLA RENIFORMIS, *Sowerby.*   Pl. XIX, figs. 1—7.

TEREBRATULA RENIFORMIS, *Sowerby.*   Min. Con., pl. ccccxcvi, figs. 1—4, Sept. 1825.
    —   —   *Phillips.* Geology of Yorkshire, vol. ii, p. 222, pl. xii, figs. 13—15, 1836.
HEMITHYRIS   —   *M'Coy.* British Palæozoic Fossils, p. 441, 1855.
RHYNCHONELLA —   *Morris.* A Catalogue of British Fossils, p. 146. 1854.

Spec. Char. Transversely reniform, wider than long; dorsal valve more or less elevated or gibbous; profile much arched, especially at the umbone, which projects further than the extremity of the beak; externally the valve is divided into three portions or lobes, of which the central one forms a mesial fold of but small elevation, and composed of three, four, or five ribs, these becoming obsolete close to the umbone; the lateral portions or lobes are inflated and smooth. In the ventral valve the beak is small and much adpressed, the sinus deep and wide, with two or three longitudinal ribs along the middle; the lateral portions of the valve are smooth, and more or less inflated, usually extending lower down than the margin, and to meet which they are abruptly bent. The entire external surface of shell is covered with minute, longitudinal, radiating striæ. Dimensions and proportions very variable; three specimens have measured—

Length 14, width 18, depth 15 lines.

„ $13\frac{1}{2}$, „ $16\frac{1}{2}$, „ 16 „

„ 12, „ 14, „ 13 „

Obs. Some difference in opinion has been expressed as to the specific value of *Terebratula reniformis;*[1] thus, in 1843, it was placed by Professor de Koninck among the synonyms of *R. acuminata*, but from which the same author removed it at a subsequent period.

In the same year (1843) Sowerby's *T. cordiformis* had also been considered by Professor de Koninck as a variety of *R. acuminata*, and my distinguished friend still adheres to the same opinion, although the generality of palæontologists maintain the three as separate species. How far this last view is correct I am not yet perfectly prepared to decide, notwithstanding the long and careful examination I made of the original types, along with many other specimens of the same species.

It is very difficult, nay, quite impossible, to express by words alone those minute differences which at times distinguish two closely allied species; I have, therefore, spared no trouble in carefully selecting and representing all those specimens which may enable the reader to form an opinion for himself, and especially so as I have still some misgivings as to the exact position *R. cordiformis* should occupy; that is to say, whether it should be viewed as a distinct species, or as a variety of *R. reniformis* or *R. acuminata*.

The general characters and facies of *Rh. acuminata* and *R. reniformis* appear to me sufficiently defined, and not to be confounded; the remarkable inflation of the umbone of

---

[1] I here reproduce the original description :

"TEREBRATULA RENIFORMIS, *Sow.* Vol. v, p. 496, tab. ccccxcvi, figs. 1—4.

"*Spec. Char.* Reniform; middle furnished with three or four longitudinal, rounded ridges, terminated by acute plaits in the much-elevated margin in the front; sides inflated below the entire edges; a much-rounded, almost two-lobed, inflated, and more or less depressed, shell. The form of the ridges and intermediate furrow is very remarkable, being rounded, while the corresponding notches in the margin are acute-angular. The peculiar form of the sides of the beaked valve, which are inflated so as to hang below the edges, will distinguish all the varieties of this species. Very abundant in the mountain limestone of Dublin and Cork."

the dorsal valve, which generally projects beyond the level of the beak, the curious convexity of the smooth lateral portions of both valves, and especially of the ventral one, which present the unusual appearance of hanging below the margin (as noticed by Sowerby), are not, to my knowledge, characters ever seen in *R. acuminata*. The last-named shell is also generally much more elevated in front, and the frontal commissure of the valves is peculiar, and different from that of *R. reniformis*.

The shell under description is usually transverse, with three or four ribs on the fold, and two or three in the sinus; five on the fold, and four in the sinus, are of much rarer occurrence. In the 'Geology of Yorkshire,' Professor Phillips represents two specimens, without mesial fold or ribs, as varieties of *R. reniformis*, but it appears doubtful to me whether these very exceptional specimens are properly defined.

Loc. *Rh. reniformis* is not a rare fossil in the Carboniferous limestone of Bolland, of Settle, in Yorkshire, at Twiston, in Lancashire, and in the Isle of Man, &c. In Ireland, Sowerby mentions Dublin and Cork, and Mr. Kelly adds Lisnapaste, Millecent, and Little Island. I am not certain whether the species really occurs in Scotland, at least I have not seen any well-authenticated examples.

RHYNCHONELLA CORDIFORMIS, *Sowerby.* Plates VIII, IX, X.

TEREBRATULA CORDIFORMIS, *Sowerby.*		Min. Con., vol. v, p. 495, tab. ccccxcv, fig. 2, 1825
ATRYPA	—	*M'Coy.* Synopsis of the Carboniferous Fossils of Ireland, p. 152, 1844.
RHYNCHONELLA	—	*Morris.* A Catalogue of British Fossils, p. 446, 1854.

Spec. Char. Obscurely pentagonal, either slightly wider than long or longer than wide; dorsal valve gibbous, front much elevated, with three or five ribs on the mesial fold, which become obsolete before reaching the extremity of the umbone; profile arched; the lateral portions of the valve are convex and tumid. In the ventral valve the beak is small and often adpressed; the sinus is wide and deep, with from two to four ribs along its middle; the lateral portions of the valve are smooth, and do not hang below the margin of the opposite one. Dimensions variable; three examples have measured—

Length 13½, width 16½, depth 16 lines. (Sowerby's type.)
   „   12, „ 14, „ 13 „
   „   9, „ 10, „   9 „

Obs. It is with much doubt and many misgivings that I here provisionally introduce *Ter. cordiformis* as a separate species, and in this respect I have followed both Professors M'Coy and Morris. Sowerby describes his shell as " heart-shaped; front much elevated, with a deep sinus in the margin; sides rather convex, sharp edged; middle ornamented with several acute furrows reaching to the beaks;" that it " differs from *Ter. acuminata* in being much more tumid in the middle, and in having three or more sharp angular furrows, extending along the middle of the large marginal sinus almost to the beaks, and is very variable in magnitude." A single specimen is represented by the author, and of this I have given four carefully drawn representations (fig. 8). In his ' Synopsis,' Professor M'Coy states that " *R. cordiformis* is chiefly distinguished from one of the varieties of *A. acuminata* by being more convex or tumid at the sides, and in the length and distinctness of the mesial plaits." Professor de Koninck, on the contrary, believes *R. cordiformis*, or at least the specimen described as such by Sowerby (fig. 8 of our plate), to be nothing more than a simple variety of *Rh. acuminata;* and although I am not yet perfectly convinced of the fact, it is very possible that my Belgian friend may be correct in his interpretation. In British collections, however, the shells which we usually designate *R. cordiformis*, and of which one or two examples so labeled exist in Sowerby's cabinet, are similar to those I have represented by figs. 9 and 10, and which are supposed to belong, rightly or wrongly, to the same species as Sowerby's type. These Professor de Koninck considers quite distinct from either *R. acuminata* or *R. reniformis*, since he sends me an Irish specimen marked " nov. sp.," and on the label points out the differences he had observed.

I must admit that I am still much puzzled how to decide; for I once even supposed that the last-named shells might, perhaps, constitute a variety of *R. reniformis*, and that Sowerby himself may have had a similar idea, from the fact that he figures among his specimens of *R. reniformis,* in tab. ccccxcvi of the 'Mineral Conchology,' a specimen very similar to the two I have represented (figs. 9 and 10), but without reference, as if he afterwards doubted the identity. If it should be determined that the typical example of *R. cordiformis* must be referred to *R. acuminata,* and that the other two (figs. 9 and 10) are distinct, I think the name *Cordiformis* might be advantageously retained for these last, as they have been so long known under that designation among British collections.

Loc. The shells here described are generally found in the Carboniferous limestone, associated with *R. reniformis.* Mr. Burrow has obtained it at Settle, in Yorkshire, and I have specimens from Bolland. In Ireland Messrs. Kelly and Morris mention Millecent, Little Island, Cork, and Ardconnaught. I am not acquainted with any Scottish specimen, but Professor de Koninck has figured the shell from Belgium.

RHYNCHONELLA ACUMINATA, *Martin* (sp). Plate XX, figs. 1—13; Plate XXI, figs. 1—20.

CONCHYLIOLITHUS ANOMITES ACUMINATUS, *Martin.* Petrif. Derbs., pl. xxxii, figs. 7, 8; and pl. xxxiii, figs. 5, 6, 1809.
TEREBRATULA ACUMINATA, *Sowerby.* Min. Con., tab. cccxxiv, fig. 1, Jan. 1822.
— — var. *sulcata, Sowerby.* Ibid., tab. ccccxcv, fig. 3, Sept. 1825.
— PLATYLOBA, *Sowerby.* Ibid., tab. ccccxcvi, figs. 5, 6, 1825.
— ACUMINATA, *V. Buch.* Ueber Terebratula, p. 33, 1834; and Mem. Soc. Geol. de France, vol. iii, p. 131, pl. xiv, fig. 1, 1838.

TEREBRATULA ACUMINATA, *Phillips.* Geology of Yorkshire, vol. ii, p. 222, pl. xii, figs. 4—9, 1836.

— MESOGONIA. Ibid., pl. xii, figs. 10—12.

— ACUMINATA, *De Koninck.* Animaux fossiles de la Belgique, p. 278, pl. xvii, fig. 3, *a, b, c, d, f* (not the other figures), 1843.

ATRYPA ACUMINATA, *M'Coy.* Synopsis of the Characters of the Carboniferous Fossils of Ireland, p. 151; also woodcut, fig. 32, 1844.

— — *D'Orbigny.* Prodrome de Paléontologie Stratigraphique, vol. i, p. 146, 1849.

RHYNCHONELLA ACUMINATA, *Morris.* A Catalogue of British Fossils, p. 146, 1854.

HEMITHYRIS — *M'Coy.* British Palæozoic Fossils, p. 381, 1858.

Spec. Char. Very variable in shape and character; more or less trigonal, heart-shaped, or obscurely pentagonal; generally wider than long. Dorsal valve convex, often gibbous, and much elevated at its frontal margin, the lateral portions sloping rapidly on either side; a distinct mesial fold is rarely perceptible. Beak in ventral valve small, incurved; foramen minute, situated under the extremity of the beak, and margined by a deltidium. Sinus either concave or angular, of great width, and of variable depth; originating at a short distance from the extremity of the beak, it extends to the front, where it forms either a regular curve or is prolonged in the shape of an acute cuneiform extremity. Externally the surface is smooth, finely striated, or more or less strongly plaited.

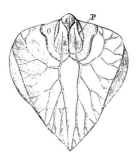 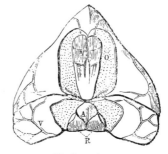

Ventral aspect.   Umbonal aspect.

Rh. acuminata, internal casts.

A. Adductor, or occlusor. R. Cardinal, or divaricator. P. Pedicle, or ventral adjustor, muscular impressions. V. Vascular markings. O. Ovarian spaces.

In the interior of the dorsal valve the hinge-plate is divided, and to these are attached a pair of short, curved, shelly processes, for the support of the oral arms. On the bottom of the valve may be seen the quadruple impressions produced by the adductor or posterior and anterior occlusor muscle of Hancock; and in the ventral valve the adductor (or occlusor), cardinal (or divaricator), and pedicle (or ventral adjustor) muscular scars may be distinctly traced. The ovarian and vascular spaces and impressions are also clearly defined in the interior of both valves. Dimensions very variable; four examples have offered the following proportions:

Length 25, width 26, depth 19 lines.

    ,,   15, ,, 17, ,, 22 ,,

    ,,   19, ,, 20, ,, 11 ,,

    ,,   15, ,, 19, ,, $7\frac{1}{2}$,,

Obs. At page 222 of his 'Geology of Yorkshire,' Professor Phillips has stated that the varieties of *T. acuminata* are almost innumerable, and that a specific character was at that time impossible. If, therefore, such was the case in 1836, I do not believe that matters have materially improved since that period. It is at all times most difficult, nay impossible, to convey in a diagnosis a sufficiently accurate description of a species, and especially of so variable a one as that now before us; and as figures always convey a more accurate idea of any object than words alone are able to express, I have represented in Plates XX and XXI all the principal and more important variations, both of specimen and age, assumed by this species, and which have been selected with great care from among some hundred specimens I was able to assemble from various collections and localities. Every intermediate link can be found connecting such extreme cases as those, Pl. XX, fig. 1, and Pl. XXI, fig. 11, and the want of space alone has prevented their being all here represented. The first notice I find of this remarkable shell consists of two very good figures of the typical variety, published by Andreae, in his 'Lettres écrites de la Suisse,' pl. xiv, 1763, and which illustrations very closely agree with those published by Martin forty-four years later. Several authors have referred fig. 1 of pl. ccxlvi, of the 'Encyclopédie Méthodique,' published by Bruguière, in 1788, to *T. acuminata*, and it may have been drawn from an example of the species, but it bears also much resemblance to some specimens of De Verneuil and Keyserling's *Rh. Meyendorfii*, and is not so good or so characteristic a representation as that by Andreae, above mentioned. It was, however, only in 1809 that the shell received a specific denomination, and that of *acuminata* was bestowed upon it by Martin,[1] who gave us at the same time a good description and correct figure of two of its varieties, and which agree also very closely with those I have represented in Pl. XX, fig. 2, and in Pl. XXI, fig 1. These will require to be looked upon as the typical shape of the species.

In 1822 James Sowerby describes and figures the same form, and in 1825 he represented two other of its varieties, which he designated by the names *sulcata* and *plicata*, and to a third he applied the specific denomination of *platyloba*. In 1836 Professor

---

[1] Martin describes his species in the following words:

"Conchyliolithus anomites (*acuminatus*) cordiformis lævis, sinu margine longissimo cuneato, s. p.

" A petrified shell; the original an *Anomia*. Perforate; both valves convex, cordiform or heart-shaped, and smooth, the surface being destitute of striæ, furrows, or tubercles (the larger specimens, under a glass, appear to be marked with very minute, close, and equal striæ). The hinge curved, and close. Foramen very minute, under the apex of the beak, which is small, sharp-pointed, and incumbent. The larger or beaked valve hollowed at the back into a single wave, ending in a very long, sharp-pointed, cuneiform, or wedge-shaped sinus at the margin. This very curious Anomite is not uncommon near *Bakewell* and *Buxton*, in the limestone strata; it varies considerably in size."

Phillips judiciously united into one species *T. acuminata* and *platyloba*, as well as the above-named varieties, but was perhaps mistaken while forming a new and separate species, *T. mesogonia*, for what seems to be a small variety of the typical shape of Martin's shell, and which the distinguished author of the 'Geology of Yorkshire' himself states to be " a miniature copy of the first var. of *T. acuminata*." Such would appear to be the named varieties and synonyms of the species under description, but it is necessary here also to record that several Palæontologists, and especially Professor M'Coy, would go still further, by adding *Rh. pugnus*, of Martin, to the varieties of *R. acuminata*; and although I am ready to admit that some extreme and exceptional shapes of both may be difficult to determine, and might lead to the idea of a passage, still the generality of specimens of both species appear to me so distinct that I would not feel myself justified in subscribing to the opinion advocated by the distinguished Irish professor.

In his great work on 'Belgian Carboniferous Fossils,' published in 1843, Professor de Koninck had assembled, as synonyms of *Rh. acuminata*, several species which he subsequently admitted to be distinct, at p. 664 of the supplement to that work, published in 1851, and among other observations he states it to be his decided opinion that *R. pugnus* is specifically distinct from *R. acuminata*.

Long and careful observations, based upon the study of many specimens, has induced some palæontologists as well as myself to consider that, although the typical form of the species under description may be heart-shaped, smooth, with a long, sharpened, pointed, cuneiform sinus (as described by Martin and others), the greater number of specimens did not present that character, the sinus being more uniformly rounded in front, and regularly or irregularly marked with rudimentary or prominent ribs, and of these Sowerby has formed his varieties *sulcata*, *plicata*, and *platyloba*. In Pl. XXI, fig. 10, a specimen is represented in which the ribs extend from the frontal margin to nearly the extremity of the umbone, and if the reader will again look at Pl. XXI, he will find that in figs. 1, 2, and 3, the fold, or what corresponds to it, is smooth and acute, while in fig. 4 it is divided into two, in fig. 5 into three; fig. 6 shows five ribs, and so on until the fold and sinus become regularly and numerously plaited, as in figs. 7, 10, and 11. The great elevation of the front is also at times gradually replaced by a more gentle curve, and in the young state especially the whole shell very often appears to be but slightly convex, and is even depressed, as in figs. 14—20 of the same plate. In the generality of specimens we find not a vestige of a rib on the lateral portions of the valves, but in other rarer examples these are sometimes present near the margin, and it is specimens so constructed that approach most in shape to *Rh. pugnus*. All this extreme variability, offered by a single species, shows how cautious we should be in our appreciations of specific characters, and especially so when we are unavoidably deprived of all anatomical assistance.

Mr. Hancock has stated in his admirable 'Memoir on the Anatomy of the Brachiopoda,' published in the 'Philosophical Transactions of the Royal Society' for 1858, that two forms, externally almost undistinguishable, have been found by him to possess animals so different in their anatomical details that they could not, in his opinion, be confounded under the same specific denomination ; and it has been shown, by other similar investigations, that shells presenting a wide range of external modifications of detail possessed an animal exactly similar. This demonstrates the almost impossibility of very often determining among the extinct forms what in reality belongs to the same species ; and our appreciations must therefore be always more or less uncertain, however much zeal we may display in our attempt to dive into the veiled secrets of nature.

Professor M'Coy was mistaken while stating, at p. 151 of his ' Synopsis of the Carboniferous Fossils of Ireland,' that the internal structure of *R. acuminata* was similar to that of *T. hastata*, the one being a *Rhynchonella*, the other a *Terebratula ;* but his imperfect woodcut representation (fig. 32 of the 'Synopsis') would denote that the Professor had observed the two small curved supports of the oral arms in *Rhynchonella acuminata*. In Pl. XX, fig. 6, the reader will find a more complete representation of the interior of the dorsal valve, and the two woodcuts given above show the shape of the muscular and vascular impressions.

Loc. *R. acuminata* and its varieties abound in the Carboniferous limestone of many English and Irish localities, but no well authenticated Scottish examples have hitherto come under my observation. Martin mentions Bakewell and Buxton, in Derbyshire. It is very common about Clitheroe and Bolland, in Lancashire, and at Park Hill, Longnor, Derbyshire ; at Settle and Malham Moor, in Yorkshire ; in the Isle of Man, &c.

In Ireland, Mr. Kelly mentions Mullaghfin, Millecent, Little Island.

On the Continent *R. acuminata* was found at Visé, in Belgium, by Professor L. de Koninck ; at Hausdorf, by M. De Semenow ; and at Cosatchi-Datchi (Oural), by M. de Verneuil and Count Keyserling, &c.

RHYNCHONELLA PUGNUS, *Martin* (sp.). Pl. XXII, figs. 1—15.

CONCHYLIOLITHUS ANOMITES PUGNUS, *Martin*. Petrificata Derbiensia, tab. xxii, figs. 4, 5, 1809.
TEREBRATULA PUGNUS, *Sowerby*. Min. Con., tab. ccccxxv, figs. 1—6, 1825.
  —   — *Phillips*. Geology of Yorkshire, vol. ii, p. 222, pl. xii, figs. 17 (but fig. 16 should also be included), 1836.
  — SULCIROSTRIS, *Phillips* (?). Geology of Yorkshire, vol. ii, p. 222, pl. xii, figs. 31, 32, 1836.
ATRYPA PUGNUS, *M'Coy*. Synopsis of the Characters of the Carboniferous Fossils of Ireland, p. 156, 1844.
  — LATICLIVA, *M'Coy* (?). Ibid., p. 154, pl. xxii, fig. 16.
TEREBRATULA PUGNUS, *De Verneuil* and *Keyserling*. Geology of Russia, vol. ii, p. 78, pl. x, fig. 1, 1845.
HEMITHYRIS ACUMINATA, var. *pugnus*, *M'Coy*. British Palæozoic Fossils, p. 338, 1852.

Spec. Char. Shell very variable in shape, transversely ovate or oblato-deltoidal; wider than long; dorsal valve gibbous, most elevated near the front, evenly convex at the umbone; mesial fold large, and more or less prominent. Ventral valve less convex than the opposite one, with a sinus of moderate depth, commencing at a short distance from the beak, and extending to the front. Beak small, much incurved, and contiguous to the umbone; foramen minute, placed under the extremity of the beak, and but rarely visible in full-grown shells. Each valve is ornamented with from nine to fourteen ribs, which become obsolete as they approach the beak and umbone; from three to six occupy the fold and sinus.

Three examples measured:

Length 12, width 17, depth 12 lines. (Martin's type.)

„ 24, „ 24, „ 20 „

„ 11, „ 15, „ 10½ „

Obs. This, and *R. acuminata*, are certainly the *Rhynchonellæ* at present known in the Carboniferous period which appear to have attained the largest proportions, and although believed by the generality of palæontologists to be specifically different, and in general easily distinguishable, it cannot be denied that there are exceptional examples, which from their intermediate shape would go some length to support the opinion expressed by Professor M'Coy, viz., that *R. pugnus* might be nothing more than a variety of *Rh. acuminata*.[1]

In 1809 Martin published two somewhat enlarged representations of what we must consider to be an abnormal shape and injured specimen of the species under description; nor did Sowerby neglect to mention that had he not possessed the individual specimen (which I have also seen and drawn) he would have felt some uncertainty as to its identity with the specimens he subsequently figured in tab. ccccxcvii of the 'Mineral Conchology.'

In very large, full-grown examples, such as in figs. 3—6 of my plate, the posterior half or upwards of each valve is evenly convex, and exhibits no trace of ribs, but these are always present round the margin and on the anterior portion of the valves. This is also the case with the generality of young shells; but in the larger number of middle-sized specimens the mesial ribs, or those which compose the fold and sinus, extend to more than

---

[1] 'British Palæozoic Fossils,' p. 381. Professor M'Coy proposes to subdivide his *Hemithyris acuminata* into four named varieties, viz.:

"1st var., — *acuminata* (Martin). Front sinus very high, acutely angular, few or no traces of mesial plaits, nor lateral marginal plaits, except in very large specimens.

"2d var., — *platyloba* (Sow.). Transversely ovate, plaits obtuse.

"3d var., — *pugnus* (Martin). Rhomboidal tumid; three to six mesial, and three or no lateral, short, strong plaits.

"4th var., — *mesogonia* (Phill.). Form and other characters exactly as in the type, var. *acuminata*, but the width less than one inch."

half the distance, and sometimes even reach nearly to the extremity of the beak and umbone. This character is well exhibited in Martin's type, and may be seen likewise in figs. 9 and 16 of my plate. The specimen represented in the ' Petrificata Derbiensia ' exhibits five large rounded ribs on the mesial fold of the dorsal valve, and four in the sinus of the ventral one, these last being indented or grooved along their centre to the distance of a couple of lines from the margin; but the shorter ribs on the lateral portions of the valves are not so marked, nor do they extend to any great distance from the margin. Now if we examine a large number of specimens of *R. pugnus*, it will be seen that in the generality of young and middle-sized examples the ribs are acute, and rarely present the indentation along the middle; this last-named peculiarity is common, but not general, to a smaller species, *R. pleurodon*, and I must hasten to observe that although by far the larger number of specimens of the last-named shell appear to be well distinguished from *R. pugnus* by the more numerous ribs which cover the entire surface of the valves; in some exceptional cases they become likewise gradually obsolete as they approach the beak and umbone, and in which condition are less easily separable from Martin's shell.

The prevailing number of ribs on the mesial fold in *R. pleurodon* is five, but that number is the exception in *R. pugnus*, where three, four, and six on the fold; and two, three, and five in the sinus would appear to be the usual number. The ribs are likewise sharper in young specimens, or races from certain localities than from others, and in this condition approach most to *R. pleurodon*. The surface of the valves, when perfectly preserved, appear to be finely striated, as is also the case with *R. acuminata* and some other species.

In young specimens of the shell under description the beak is acute, and but slightly incurved, and in this state the small circular foramen surrounded by its deltidium may be distinctly observed, but in more aged individuals, from the beak becoming much incurved over the umbone of the opposite valve, the aperture can no longer be seen. Some specimens are also much more transverse than others, and but few appear to have attained the dimensions of the larger ones represented in our plate. From *R. acuminata* and *R. reniformis*, Martin's shell may be distinguished by its lateral ribs.

The interior arrangements and impressions are those of the genus *Rhynchonella*. In the dorsal valve the shelly processes (for the support of the spirally coiled brachial appendages) consist of two short, flattened, grooved lamellæ, separate and moderately curved upwards (Pl. XXII, fig. 13), being prolongations of a deeply divided hinge-plate. Four muscular impressions produced by the adductor or occlusor muscle are visible on the bottom of the valve. In the ventral valve the hinge-teeth are supported by dental plates; scars formed by the adductor or occlusor; cardinal or devaricator; and ventral adjustor muscles may be distinctly recognised, as well as the vascular markings. Professor M'Coy appears to have been the first naturalist who observed in this species the slender curved lamellæ above described; for, at p. 156 of his ' Synopsis,' we find a woodcut representation which, although very defective and imperfect in drawing, conveys some idea of the arrangement.

The intimate shell structure does not differ in any respect from that which is common to its congeners.

In 1825, Sowerby represented several examples or variations in shape of the shell under description, observing, at the same time, that out of several hundred specimens, hardly two could be found alike; he mentions, likewise, that *Ter. lateralis* ('Min. Con.,' tab. lxxxiii, fig. 1), may probably be a variety or shape of *R. pugnus;* but as the original specimen could no longer be found in the Sowerby collection, and as the figure is insufficient, all that can be said is, that whether the specimen be referable to *R. pugnus,* or to *R. angulata,* it will require to be added to the synonyms.[1]

In 1836, Professor Phillips described and figured a Rhynchonella by the name of *T. sulcirostris.* "Rhomboideo-deltoidal; edge sharp, plaits obtuse, mesial plaits five to nine; upper valve (our dorsal) sulcate towards the beak"[2] (fig. 16, of my plate); and I have felt much puzzled how to deal with this species (?), for I could not satisfactorily recognise the original specimen, in the Gilbertsonian collection, all those bearing the name *sulcirostris* appearing to be nothing more than variations in shape of *R. pugnus,* in which the ribs on the fold and sinus have extended to nearly the extremity of the beak, and thus agreeing also with what we perceive to have been the case with Martin's type. *R. sulcirostris* appears to be one of those undecided shapes, bearing some resemblance to *R. pleurodon,* of which Mr. Morris has supposed it a synonym; but the shortness of the lateral ribs, as depicted in the figure, would assimilate it more closely to certain varieties of *R. pugnus,* than to *R. pleurodon,* which does not present that appearance. It appears to me, however, almost certain that the shell in question is a true *Rhynchonella,* and not a *Camarophoria,* as described by Professor M'Coy, at p. 446 of his 'British Palæozoic Fossils,' and it is evident that the last-named author must have had before him some other species while drawing up his description; but as he does not furnish us with a representation of his shell, it is impossible for me to offer any further opinion on the subject. It must also be remarked that the large specimen drawn in pl. xii, fig. 16, of the 'Geology of Yorkshire,' would be more properly located with *R. pugnus,* than with *R. pleurodon.*

The description and figure Professor M'Coy gives us of his *Atrypa laticliva,* at p. 154

---

[1] I have reproduced Sowerby's description of *R. lateralis,* under *R. angulata,* and the figure will be found in Pl. XXV, fig. 14.

[2] Phillips's 'Geology of Yorkshire,' vol. ii, p. 222.

[3] "Transversely rhomboidal, length two thirds the width, gibbous; middle of the shell elevated in front, with three obtuse ribs, reaching nearly to the beak; between the mesial elevation and the sides is a smooth space, equal in breadth to the mesial elevation; sides small, compressed, with three obtuse ridges reaching half way to the beak. This species is remarkable for the small number of ribs, and the broad space between those of the mesial elevation and the sides; length 6 lines, width 9 lines."

of the ' Synopsis,' has induced both Professor De Koninck and myself to suppose that the author had before him some specimen of *R. pugnus;* anyhow it cannot belong to the same species as that which Professor M'Coy describes and figures in 1855, under the denomination of *Camarophoria laticliva.*[1]

In 1843 Professor De Koninck had united *R. pugnus,* and one or two more species, under the single specific denomination of *Ter. acuminata;* but at p. 664 of his work on ' Belgian Carboniferous Fossils,' he has corrected the oversight, and referred figs. 3, 3 *h* and *i,* of his pl. xviii, to the shell we are now describing.

Loc. Martin states that the species is common in Carboniferous limestone at Castleton, Hope, and Little-Longstone, in Derbyshire. Gilbertson's largest examples appear to have been obtained at Linton, in the Craven district. It is also abundant at Whitewell, Widgill, and Twiston, all situated at a few miles from Clithero. It does not appear to be rare at Kendal, nor in the Isle of Man, &c.

In Scotland it occurs in several localities of the Lanarkshire and Stirlingshire Carboniferous basins, such as at Hill-head, Campsie, &c. In Ireland, Mr. Kelly mentions Ardagh, Millecent, Little Island, &c.

It is also a common fossil in many foreign Carboniferous districts; it occurs at Visé, in Belgium, as well as on the oriental side of the Oural, to the east of Miash, &c.

RHYNCHONELLA PLEURODON, *Phillips,* Sp.   Pl. XXIII, figs. 1—15, 16—22?

> TEREBRATULA MANTIÆ, *Sowerby.* Min. Con., tab. cclxxvii, fig. 1, May, 1821.
> —   PLEURODON, *Phillips.* Geology of Yorkshire, vol. ii, p. 222, pl. xii, figs. 25—30 (but not 16), 1836.
> —   VENTILABRUM, Ibid., p. 223, pl. xii, figs. 36, 38, and 39, 1836.
> —   PENTATOMA, *De Koninck.* Animaux fossiles de la Belgique, p. 289, pl. xix, fig. 2 (not *T. pentatoma,* Fischer) in p. 664; this mistaken identification is corrected to *T. pleurodon,* of Phillips, 1843, and 1851.
> ATRYPA PLEURODON, *M'Coy.* Synopsis of the Carb. Foss. of Ireland, p. 155, 1844.
> —   TRIPLEX, *M'Coy.* Ibid., p. 157, pl. xxii, fig. 17, 1844.
> TEREBRATULA PLEURODON, *De Verneuil* and *Keyserling.* Geol. of Russia, vol. ii, p. 79, pl. x, fig. 2, 1845.
> —   DAVREUXIANA, *De Koninck.* Animaux foss. de la Belgique, p. 664, 1851.
> HEMITHYRIS PLEURODON, *M'Coy.* British Pal. Fossils, p. 382 and 441, 1855.

Spec. Char. Transversely oval, rarely longer than wide, very variable in shape;

---

[1] ' British Palæozoic Fossils,' p. 444, pl. iii D, figs. 20, 21. I have reproduced in my plate the author's figures, published in 1844 and 1855, so as to enable the reader to form his own opinion upon the subject.

valves more or less convex, sometimes very gibbous, beak moderately produced, incurved, and exhibiting a small circular foramen under its angular extremity, which is surrounded and slightly separated from the hinge-line by a deltidium; medial fold large, almost square, and most elevated close to the front, when it is suddenly deflected, so as to meet the corresponding margin of the opposite valve; sinus in ventral valve of moderate depth; ribs numerous and angular, extending over the whole surface, and numbering from ten to about twenty-four in each valve, of which three to nine compose the fold, and two to eight the sinus; but five on the fold and four in the sinus is the general number. The plaits that cover the lateral portions of the dorsal valve are very much curved, while those of the ventral one are nearly straight, with their extremities bent upwards. The ribs are longitudinally grooved along their median portion to some distance from the margin. Dimensions very variable; three examples have measured:

Length 10, width 14, depth 11 lines.
„ 9, „ 13, „ 10 „
„ 6½, „ 11, „ 5 „

Obs. This Rhynchonella is very abundant in the Carboniferous rocks, but varies so much in shape, according to age and specimen, that several so termed species have evidently been manufactured out of what I conceive to be mere differences in shape, race, and even malformation of a single species.

R. pleurodon does not appear to have ever attained very large proportions, and is distinguishable from *R. pugnus*, *R. acuminata*, and *R. reniformis*, by the angular ribs which cover the entire surface of its valves. When young, with dimensions of from two, three, and sometimes more lines in length, both valves are at times much compressed, the fold and sinus being but little elevated or depressed beyond the regular convexity of the valves, a small longitudinal depression extending likewise along the median portion of the umbone of the dorsal valve. In the fry the shell is at times somewhat triangular, the width being equal to the length, but with growth the shape becomes more transverse, and rapidly increases in depth or convexity, the fold and sinus assuming likewise their characteristic appearance. Many young specimens of undoubted *R. pleurodon* possess but three ribs on the mesial fold, and two in the sinus, but in the larger number of full-grown shells five were prevalent, although we sometimes pick up specimens with as many as from six to nine ribs on the fold; but these are exceptions, and of much less common occurrence.

It may now be desirable to examine what are those so-termed species which I have been tempted to consider as simple variations in shape or synonyms of *R. pleurodon*.

First.—Rh. (*Terebratula*) *Mantiæ*, Sowerby, pl. xxiii, figs. 15, 15 *a, b, c* ('Min. Con.,' tab. cclxxvii, fig. 1, May, 1821).

Sowerby founded his *T. Mantiæ* from the inspection of a single specimen, said to have been collected in Ireland by a Mrs. Mant, but unfortunately the figure published in the 'Mineral Conchology' does not convey an accurate idea of the shell, and I am not therefore surprised that palæontologists should have felt embarrassed, and have identified with the imperfect illustration shells of an entirely different character. Having obtained the loan of the original example, through the kindness of its possessor, Mr. J. de C. Sowerby, I have been able to study and illustrate its characters with sufficient care. It is always very hazardous to establish species from the inspection of a single specimen, and especially among so variable a group of shells as that of the *Rhynchonellæ*. Sowerby's *T. Mantiæ* appears to me to be nothing more than an accidentally elongated malformation of *R. pleurodon*, wherein the mesial fold (composed of five ribs) had become unsymmetrical, from being twisted more to one side than to the other, an occurrence not uncommon to many species of *Rhnychonella*, which in their normal state have the fold and sinus in the middle. I purposely avoided referring, in the list of references, to the works of those authors who have alluded to *T. Mantiæ*, as they were evidently unacquainted with the characters of Sowerby's shell, which measures nine lines in length, eight in width, and six in depth. Notwithstanding the priority of date of *R. Mantiæ* over *R. pleurodon*, I believe all palæontologists will prefer retaining Phillips's denomination for the species; especially so as Sowerby's one has unfortunately been the cause of so much misapprehension, being founded on a malformation. At page 146 of the first volume of D'Orbigny's 'Prodrome,' *T. proava* of Phillips is considered a synonym of *T. Mantiæ*, but I believe the French author's guess to be incorrect.

In 1843, Professor De Koninck described and figured a totally different shell under the name of *T. Mantiæ*, and to which he had likewise added *T. radialis* of Phillips, as synonym, but the distinguished Belgian author has since then recognised his mistake. At page 437 of M'Coy's work on 'British Palæozoic Fossils,' a shell is described under the generic and specific appellation of *Spirigerina* (?) *Mantiæ*, Sow., sp., but as no illustration is appended, all that can be said is, that the shell so described cannot have belonged to Sowerby's species, because *T. Mantiæ* possesses all the exterior appearances and characters of a true *Rhynchonella*, and none of those of a *Spirigerina*, nor does it present " a large, high, flat, cardinal area," or any appearance of the punctured shell-tissue, described in the work above quoted. It is, therefore, certain that Professor M'Coy must have had some other species before him while drawing up his description.

Secondly.—*Rhynchonella* (Ter.) *ventilabrum*, Phillips (pl. xxiii, figs. 13, 14), ('Geology of Yorkshire,' vol. ii, p. 223, pl. xii, figs. 36, 38, 39, 1836).

Professor Phillips states that he " is not certain whether his species be distinct from *T. sulcirostris*, that it has no mesial elevation, the ribs rounded, and vanishing towards the margin." In Pl. XXIII, I have reproduced the author's figures, but feel but little doubt while placing *R. ventilabrum* among the variations of form, or synonyms of *R. pleurodon*, for the typical example (fig. 13) exactly resembles certain specimens of the last-named species, in which the mesial fold is indistinctly marked; in fig. 14 it may, however, be clearly perceived. *R. sulcirostris* has been with some doubt considered a synonym of *R. pugnus*, from the shortness of its lateral ribs, and I regret not having been able to discover the original examples in the Gilbertsonian collection ; the conclusions here recorded are therefore based upon an examination of the figure only, for the few words of description which accompany it do not materially assist.

Thirdly.—Rhynchonella Davreuxiana, De Koninck (Pl. XXIII, figs. 18—20), (De Koninck, 'Animaux Fossiles qui se trouvent dans le terrain Carbonifere de la Belgique.' Supplement, p. 664, 1851).

Notwithstanding the superior authority of my distinguished friend, Professor De Koninck, I cannot help feeling impressed with the idea that the form above named is only a small thick variety, or local race of *R. pleurodon* (?). None of the specimens I have been able to examine (and which were all obtained at Gilling, by Mr. E. Wood) did exceed $3\frac{1}{2}$ lines in length, 4 in width, and $2\frac{1}{2}$ in depth. In the larger number of specimens, three and four ribs were prevalent upon the fold, but in some others five could be counted ; the ribs generally extend from the beak and umbone to the margin, but sometimes, as was the case with certain young shells of *R. pleurodon*, the ribs became obsolete close to the extremity of the beaks. Should my present views relative to this shell be considered erroneous, palæontologists will do right in adopting Professor De Koninck's specific denomination, and I may observe that the specimens of *R. Davreuxiana* figured in my plate, were so determined by the Belgian author himself.

Fourthly.—Rhynchonella triplex, M'Coy, sp. (pl. xxiii, figs. 16—18) ; *Atrypa triplex*, M'Coy, 'Synopsis of the Carb. Foss. of Ireland,' p. 157 (pl. xxii, fig. 17, 1844).

Professor M'Coy describes his shell as " transversely oval, gibbous ; beaks very small, pointed ; surface with nine short, angular ribs, which reach but half-way to the beak ; front elevated with three of the ridges ; the three ridges on each side slightly larger than the mesial ones. This pretty little shell is remarkable for its three equal lobes, of three ridges each ; it is distinguished from the *A. raricosta*, Phillips, by the ridges extending only half-way to the beaks on the ventral valve, and its very small size. Length, two and a half lines ; width, three lines." In our Plate XXIII, fig. 17 is copied from the ' Synopsis.'

No locality is given, but Mr. Kelly supplies the deficiency, by informing us that the shell was found in yellow or reddish sandstone, forming the base of the Carboniferous system, at Kildress, in Tyrone ; and he kindly furnished me with specimens of the sandstone, which I found to contain, in addition to casts of *Spiriferina octoplicata, Athyris ambigua, St. crenistria,* and *Rh. pleurodon,* many examples of M'Coy's *Atrypa triplex ;* and although some few of the specimens of the last did present exactly the characters described by the author, still it was easy to perceive that this little *Rhynchonella* varied as well as all other species of the genus in the number of its ribs ; and on the same slab could be seen specimens which connected by gradual passages *R. triplex* to *R. pleurodon,* proper, and of which I believe it to be the fry. All the fossils above enumerated as occurring in the sandstone of Kildress are in the state of casts, and the ribs in some of the examples referable to *R. triplex,* extend much further along the surface of the valves than was indicated by Professor M'Coy. We have already had occasion to remark that in certain young shells of *R. pleurodon* a small portion of the beak and umbone was sometimes smooth, as is seen in the small *R. triplex* from Kildress. In the shales, on the upper portion of the Carboniferous limestone at Settle, in the parish of Carluke, and elsewhere, we find small specimens of a *Rhynchonella* with nine ribs, and which agree with the characters assigned by M'Coy to his species; but here also we observe that they pass, by insensible gradation, into *R. pleurodon,* and in some the sinus and fold is unsymmetrical or twisted, as was the case with the shell upon which Sowerby founded his *Terebratula Mantiæ.*

The specimen represented and described by General Portlock in pl. xxxviii, fig. 4, of his excellent 'Report on the Geology of Londonderry, Tyrone, and Fermanagh,' as *Terebratula* (?) *ferita,* (?), V. Buch, and which was found also in the red or yellow sandstone of Kildress, is evidently a young example of *R. pleurodon;* Von Buch's species being quite distinct, and belonging to another genus (*Retzia*).

It is well known, as I have already so often had occasion to remark, that the number of ribs varies exceedingly according to age and specimen in almost every known species of *Rhynchonella,* so that no definite number can be made use of as an unvariable character, and no name given to a species should be founded upon the number of ribs possessed by a single specimen.

Loc. In England *R. pleurodon* is abundant in the Carboniferous limestones and shales at Bolland, Settle, Kirby Lonsdale, Orton, and in several other Yorkshire and Derbyshire localities, &c. In Scotland it is found in similar beds in various localities in Lanarkshire, near Carluke, Campsie, also in Westlothian. In Ireland it occurs in the red sandstone of Kildress, and in the limestone and shales of other localities. It is also a common fossil in other Carboniferous districts of the world. At Visé, in Belgium, by Professor De Koninck. In Russia it is described from Archangleskoi, Cosatchi-datchi, Sterlitamak, &c. In Australia it has been recently discovered in beds of the Carboniferous period at Bundaba, Port Stephen; and is common in the Carboniferous rocks of America, &c.

RHYNCHONELLA FLEXISTRIA, *Phillips* (sp.)   Pl. XXIV, figs. 1—8.

> TEREBRATULA FLEXISTRIA, *Phillips.* Geol. York., vol. ii, p. 222, pl. xii, figs, 33 and 34, 1836.
> —   TUMIDA. Ibid., fig. 35.
> HEMITHYRIS HETEROPLYCHA, *M'Coy.* Annals and Mag. of Nat. Hist., 2d series, vol. x, and British Palæozoic Fossils, p. 440, pl. iii D, fig. 19, 1855.
> —   FLEXISTRIA. British Palæozoic Fossils, p. 439.

Spec. Char. Shell oblate, or transversely ovate; dorsal valve more convex than the ventral one; beak small, much incurved, sometimes inconspicuous, on account of the gibbosity of the dorsal valve; sinus moderately deep, fold more or less prominent. The surface of each valve is ornamented by from fifteen to forty ribs, those on the fold and sinus being generally larger and wider than those that cover the lateral portions of the valves; the ribs also become more numerous by intercalation or bifurcation at various distances from the beak and umbone, and especially so on the lateral portions of the valves, where they are generally much curved and smaller. Dimensions very variable; three specimens measured—

Length 10, width 13, depth 8 lines.
    ,,   8, ,,   10,   ,, 7 ,,
    ,,   7, ,,   9,   ,, 6 ,,

Obs. After a long and minute examination of the typical and many other examples of *Terebratula flexistria* and *T. tumida* (Phillips),[1] as well as of *Hemithyris heteroplycha* (M'Coy), I arrived at the conclusion that they were all variations in shape of a single species, for which the designation of *flexistria* has been retained. At p. 222 of the 'Geology of Yorkshire' (vol. ii), Professor Phillips describes his two shells with the following words: "*Terebratula flexistria.* Oblate, depressed, mesial elevation rounded; lower valve smaller, flatter, with inconspicuous beak; many obtuse striæ, *much curved on the sides.*"

"*Terebratula tumida.* Oblate, tumid, lower valve flatter, with inconspicuous beak; striæ strong and rounded on the middle, smaller and *curved on the sides.*" Both are stated to be from Bolland, and I have reproduced in Pl. XXIV the author's original figures (1 and 7), also those given by Professor M'Coy of his *H. heteroplycha* (fig. 5). Every intermediate passage or gradation of shape and character can be found, whereby the three shells above mentioned are intimately connected; in some the ribs on the fold and sinus are few in number, large, simple, and varying from three to six; those on the lateral portions of the valves being smaller and almost entirely simple; while in other examples, some or all the ribs have bifurcated or augmented by the means of intercalations at various distances from the beak and umbone, the relative disproportion in size of the ribs on the fold and lateral portions of the valves not being so apparent. The beak likewise is inconspicuous in some specimens, while in others it is slightly produced above the umbone of the dorsal valve. In Pl. XXIV, I have endeavoured to represent all these appearances or variations of detail, to which the reader can refer. The umbone varies also much in degree of convexity, being uniformly rounded in some specimens, while there exists in others a slight median depression.

Rh. flexistria and its varieties may be distinguished from *Rh. pleurodon* (to which some examples approach) by their general shape, disposition, and character of their dichotomising

---

[1] The original examples of Professor Phillips's species form part of the Gilbertsonian Collection in the British Museum. Professor M'Coy's *H. heteroplycha* is preserved in the Geological Museum at Cambridge.

ribs, which do not appear to have been grooved along the middle, as is the case with those of the last-named species.

Loc. *R. flexistria* is not a very common species ; it occurs in the Carboniferous limestone of Bolland, Clitheroe, and in several Derbyshire localities, &c. In Ireland, Mr. Kelly mentions Millecent, Rahoran, and Knockninny. I am not acquainted with any Scottish specimen, nor is it known to me from any Continental locality, although it is no doubt to be found in several.

RHYNCHONELLA ANGULATA, *Linnæus* (sp.) Pl. XIX, figs. 11—16.

> ANOMIA ANGULATA, *Linnæus*. Systema Naturæ, i, pars ii, p. 1154, 1767.
> TEREBRATULA EXCAVATA, *Phillips*. Geol. Yorks., vol. ii, p. 223, pl. xii, fig. 24, 1836.
> — ANGULATA, *De Koninck* Animaux foss. de la Belgique, vol. i, p. 284, pl. xix, fig. 1, 1843.
> RHYNCHONELLA ANGULATA, *D'Orbigny*. Prodrome, vol. i, p. 146.[1]
> HEMITHYRIS ANGULATA, *M'Coy*. British Palæozoic Fossils, p. 439, 1855.
> ANOMIA ANGULATA, *Hanley*. Ipsa Linnæi Conchylia, p. 133, 1855.

Spec. Char. Elongated, sub-trigonal or cuneiform, rarely as wide or wider than long, the greatest breadth at a short distance from the front ; valves sometimes much flattened, but assuming every degree of convexity, and even gibbosity. The lateral portions of the beaks in both valves are much depressed, forming large, broad, flattened spaces, which extend to upwards of half the length of the shell ; the beak of the ventral valve is moderately produced, tapering to a point, nearly straight, or but feebly incurved ; foramen small, margined by a deltidium. The dorsal valve is convex and sharply incurved at the umbone, almost straight towards the front, the lateral portions of the valve sloping rapidly on either side. The surface is smooth close to the umbone, but soon becomes ornamented by from six to nine large angular plaits, of which two, three, or four compose the mesial fold, while one, two, or three strongly curved plaits occupy each side of the lateral portions of the valve. The ventral valve, when viewed in profile, presents a very convex curve from the beak to the extremity of the sinus, this last being in general rather shallow, and composed of one, two, or three angular ribs. Measurements taken from three individuals have presented—

Length 12, width 9½, depth 7 lines.
 „ 10, „ 6 „ 7½ „
 „ 10, „ 11 „ 8 „

Obs. This remarkable but variable shell may always be distinguished from other

---

[1] This appears to be the only species belonging to the genus *Rhynchonella* D'Orbigny would admit in the Carboniferous period, all the other species of this genus are placed by him in the genus *Atrypa !*

species of *Rhynchonella* found in the Carboniferous system, by the compression or excavation of the lateral portions of its beaks, which give to the shell that peculiar sub-trigonal or cuneiform appearance which no doubt prompted the specific designation of *angulata*, and at a later period that of *excavata*, by Professor Phillips. The ribs are likewise characteristic of the species, but not to such an extent as the lateral compressions of the beaks.

Rh. angulata bears some resemblance to certain individuals of the Silurian *Rh. cuneata* of Dalman, but from which it may be distinguished by a difference in the character of its ribs. *R. lateralis* of Sowerby has been (by some authors) added to the synonyms of *angulata*; but if we are to judge from the description and figure published in the 'Mineral Conchology' (tab. lxxxiii, fig. 1), the identification must be looked upon as uncertain, for the Sowerby representation does not exhibit those flattened or depressed spaces on either side of the beaks which are observable in the Linnean species; and this view is further strengthened by Sowerby's own remark, that " *Ter. lateralis* is probably a variety of *R. pugnus*, although it has only three plaits" ('M. C.,' vol. v, p. 155).[1] In any case, whether *R. lateralis* be or not a synonym of the Linnean species, or a variety of *R. pugnus*, the name will require to be erased from the nomenclature, on account of Linnæus's species, as well as that of Martin, claiming priority of date over the Sowerby one. I must also differ with the statement made by Professor De Koninck, at p. 285 of his excellent work on the 'Carboniferous Fossils of Belgium,' that Bruguière has figured the species in the 'Encyclopédie méthodique,' or that it is the same as the one to which Lamarck gave the name of *T. plicata* at a later period; for although Bruguière's representation ('Ency. méth., pl. ccxliii, fig. 11, and pl. ccxli, fig. 1, 1789) might bear a somewhat obscure resemblance to certain examples of *R. angulata*, it represents a much larger shell, with a beak strongly incurved, and very different to that of Linnæus. Besides, I believe that Bruguière's figure represents a liasic shell found in the north of Italy, and the same as that termed *Terebratula plicata* by Lamarck, at p. 254 of the sixth vol. of his work, and wherein he refers to the figure of Bruguière above quoted. When writing my report on Lamarck's Fossil Terebratulæ, which was published in the fifth vol. of the 'Annals and Mag. of Natural History' (pl. xix, fig. 39, 1850), I was able to draw the original specimen upon which the author had founded his species. I am therefore

---

[1] To my regret the original example could no longer be found in the Sowerby collection, but I have reproduced the original representation (Pl. xxv, fig. 14), and here append the description taken from p. 189 of vol. i of the 'Mineral Conchology,' so that every one may form his own opinion.

" *Spec. Char.* Oval, broader than long, gibbous; middle of the front much elevated, with three deep, but short plaits; sides with two plaits each, much below the middle.

"The three plaits in the middle of the imperforated valve, though not continued far into the shell, produce three very deep angular notches, which are filled by as many sharp teeth on the edge of the other valve, which is not so much plaited, and is altogether flatter. The length of the edge between the central plaits and the lateral ones is remarkable. This is found in the limestone rock near Dublin, and in the black rock near Cork. The stone is generally a compact darkish marble, fetid when scraped."

acquainted with but a single certain synonym, that of *excavata*, published by Professor Phillips in 1836.

The original example of *R. angulata* is still preserved in the Linnean collection, where it has been compared by myself, as well as by several of our London Palæontologists. I am also of the opinion expressed by the late Mr. Sharpe, that *R. angulata*, of Sowerby, found in the Jurassic strata of England, is specifically distinct from that of Linnæus, and must receive another denomination.

Loc. *R. angulata* does not appear to be a very abundant fossil; it is found in the Lower Scar limestone of Yorkshire and of the Isle of Man. In Ireland it is stated by Mr. Kelly to occur at Ardagh, and abundantly at Ardelogh; but I am not acquainted with any Scottish example. On the Continent it has been collected by Professor De Koninck from the Carboniferous limestone of Visé, in Belgium.

RHYNCHONELLA (?) TRILATERA, *De Koninck.* Pl. XXIV, figs. 23—26.

> TEREBRATULA TRILATERA, *De Koninck.* Animaux fossiles de la Belgique, p. 292, pl. xix, fig. 7, 1843.
> RHYNCHONELLA TRILATERA, *Morris.* Catalogue of British Fossils, 2d ed., p. 148, 1854.

Spec. Char. Shell sub-triangular or sub-quadrate, with the angles rounded; valves almost equally deep, convex, and flattened along the middle; a slight median depression or groove, originating at the extremity of the beaks, in either valve, extends at times to the frontal margin. The beak of the ventral valve is acute, very small, and generally closely adpressed to the umbone of the opposite one; the lateral portions of the beaks in either valve are flattened or compressed, while the frontal margin is nearly straight, from the total absence of all sinus or fold. From sixteen to eighteen angular ribs ornament the surface of either valve, while the five or six central are generally much larger than lateral ones. Interior unknown. Two examples measured—

Length $4\frac{1}{2}$, width 5, depth 3 lines.

    ,,   5,   ,,   5, ,, $2\frac{1}{2}$,,

Obs. This remarkable species, of which no British specimen has till now been figured, does not appear to have ever attained much larger proportions than those above given, and is easily distinguishable from its congeners by the peculiar shape and character of its shell. It varies somewhat in the degree of convexity or flatness of its valves, and the median groove or depression is more or less apparent. The dorsal valve is more often flatter than the ventral one; the ribs also either diminishing regularly in width from the centre, or having the middle ones (comparatively) much larger than those that cover the lateral portions of the valves.

None of the interior characters being at present known, the shell is provisionally located with *Rhynchonella*, whence it may be hereafter removed, should the necessity for so doing become apparent. Seven British examples alone have come under my notice, and of these four will be found represented in our plate. Three good specimens from the Carboniferous limestone of Alstonfield, in Derbyshire, may be seen in the Museum of Practical Geology; while the remaining four, which form part of the Gilbertsonian Collection in the British Museum, were probably derived from a different locality, as they are more triangular and flattened than the Derbyshire examples above mentioned. Professor De Koninck, who first described and illustrated the species, mentions that it is among the rarer species from the Carboniferous limestone of Visé, in Belgium.

Doubtful Species.

A. Rhynchonella (?) nana, *M'Coy.* Pl. XXV, fig. 15.

Atrypa nana, *M'Coy.* Synopsis of the Characters of the Carboniferous Fossils of Ireland, p. 155, pl. xxii, fig. 19, 1844.

"*Spec. Char.* Orbicular, or slightly ovate, compressed, beak pointed; surface radiated with about ten straight, equal, obscurely angular ridges, none of which reach the beak. This little species resembles *T. radialis* in form, but is much flatter, and has considerably fewer, larger, and more angular ribs, all of which disappear before reaching the beak, as in the *A. subdentata* and *rotunda*, Sow., from both of which it is distinct by its greater flatness, and more numerous radiating ridges; it is also much smaller than any of those shells. Length two lines, width two lines." (M'Coy's 'Synopsis,' p. 155.)

Obs. As all the endeavours I have made to procure the sight of a specimen of this shell have proved unsuccessful, the only thing I can do is to reproduce the author's description and figures. These are not, however, sufficiently detailed or precise to permit of any comment as to the probable generic or specific value of the species; and it is even possible and probable that *A. nana* may be but the fry or young of some other of the described species, such as of *R. pleurodon* (?), of which some young examples of similar dimensions would closely resemble. The species cannot, therefore, be admitted upon such vague and uncertain grounds, and is only here appended for the sake of reference.

Mr. Kelly informs me that the *A. nana* is common at Rahoran, in the county of Tyrone, and that it occurs in a Calciferous slate, which is situated between the limestone and a reddish sandstone, which forms the basis of the Carboniferous system in Ireland. It is to be regretted that of a shell stated "so common," not a specimen could be procured for the present monograph.

B. Rhynchonella (?) semisulcata, *M'Coy*.   Pl. XXV, fig. 13.

> Atrypa semisulcata, *M'Coy*.   Synopsis of the Characters of the Carboniferous Fossils of Ireland, p. 157, pl. xxii, fig. 15, 1844.

" *Spec. Char.* Orbicular, depressed; beak small; mesial fold broad, flat, with about five strong, rounded, radiating ridges, continued to the beak; sides smooth, or very finely striated longitudinally. This species is very remarkable, from the coarsely ridged mesial fold, and nearly smooth sides. Length seven lines, width eight lines." (M'Coy.)

Obs. From not having been able to procure determinable specimens of this shell, I cannot speak as to its generic or specific claims. Through the kindness of Sir R. Griffith I have been able to examine the original example from which M'Coy's lower figure was drawn; but this is simply a flattened, obscure impression on a slab of what Sir R. Griffith calls " black calp slate," from Walterstown Skreen, and not very correctly represented either in the plate. Mr. Kelly informs me that the shell has been found in what he considers to be " coal shales," at Culkagh, Ballintree, and Walterstown, in Ireland.

C. Rhynchonella or Camarophoria (?) proava, *Phillips*.   Pl. XXV, fig. 10.

> Terebratula proava, *Phillips*.   Geology of Yorkshire, vol. ii, p. 223, pl. xii, fig. 37, 1836.

" *Spec. Char.* Beak produced; radiations obtuse; mesial fold square. Strongly allied to the Oolitic species *T. obsoleta* and *T. socialis*, &c." (Phillips.)   (Carboniferous limestone) Bolland.

Obs. The original specimen may be seen at the British Museum; it forms part of the Gilbertsonian Collection, but several other shells which do not belong to the species are evidently attached to the same tablet.

The specimen on which Professor Phillips created his species is of an elongated oval shape, with nineteen or twenty small ribs on each valve, but which become obsolete at the beaks; the mesial fold is of but small elevation above the general convexity of the valve, and is ornamented by five ribs; the sinus is shallow. Length eight, breadth seven, depth five lines.

Notwithstanding the examination I was able to make of the original specimen, I felt uncertain as to its being a good species. It might possibly be a *Camarophoria*, and perhaps even an abnormal shape of *Camarophoria crumena* (?). Until more material shall have been procured, it will be safer to consider *T. proava* as one of the doubtful or uncertain species.

D. Rhynchonella (?) gregaria, *M'Coy*.   Pl. XV, figs. 27, 28.

Atrypa gregaria, *M'Coy*.   Synopsis of the Carboniferous Fossils of Ireland, p. 153, pl.
xxii, fig. 18, 1844. (Not *Atrypa gregaria*, M'Coy, British Palæozoic
Fossils, p. 435, tab. iii D, fig. 20, 1855).

"*Spec. Char.*   Trigonal; beak of the dorsal valve (*ventral*, Owen) very large, produced, incurved; dorsal valve flattened; sides abruptly rounded; a very wide, but shallow, mesial depression, slightly produced in front; ventral valve (*dorsal*, Owen) an equilateral triangle; all the angles rounded very convex; front margin raised to a broad sinus; surface smooth." "This curious species is very remarkable in form, wholly unlike any of the other Palæozoic species, approaching in size and shape to the *Terebratula lineolata*, Phil., of the Speeton Clay; it is the only true *Atrypa* I know of, resembling in this respect several of the *Terebratulæ* of the more recent formations.   Length seven lines, width six lines, depth three lines and a half."   (M'Coy, 'Synopsis,' p. 153.)

Obs.   Of this shell I am acquainted with but the single ventral valve, that was kindly lent to me by Sir R. Griffith, and which I found to agree with the same valve that is figured by Professor M'Coy in his 'Synopsis;' but all my efforts to procure a bivalve example or the dorsal valve have proved unsuccessful.   On such insufficient material I could not venture to describe the species, and have, therefore, reproduced that given by Professor M'Coy in 1844; for I cannot admit that the shell subsequently published as *Athyris gregaria* in the work on 'British Palæozoic Fossils' by the same author, could belong to the same species as that of the 'Synopsis.'[1]   I am also uncertain as to its genus, and although placed in Pl. XV among the *Athyrises* with a point of doubt, I now prefer to locate it provisionally among the *Rhynchonellæ*, on the simple authority of M'Coy's figures, which bear more resemblance to a *Rhynchonella* than to an *Athyris*.   In my plate I have reproduced Professor M'Coy's three original figures, and have added a drawing of the ventral valve that was kindly lent me by Sir R. Griffith.

Mr. Kelly mentions that the species was derived from the Calcareous slate of Ballinglen, Kilbride, and White River, in Ireland; but the specimen lent me by Sir R. Griffith had all the appearance of having been taken from the Carboniferous limestone.   I am not acquainted with any English or Scotch specimens.

---

[1] It belongs to Professor Hall's *Athyris subtilita*.

CAMAROPHORIA CRUMENA, *Martin* (sp). Pl. XXV, figs. 3—9.

CONCHYLIOLITHUS ANOMITES CRUMENA, *Martin.* Petrificata Derbiensia, pl. xxxvi, fig. 4, 1809.

TEREBRATULA SCHLOTHEIMI, *Von Buch.* Ueber Terebratula, p. 39, pl. ii, fig. 32, 1834.

CAMAROPHORIA SCHLOTHEIMI, *King.* A Monograph of the Permian Fossils of England, p. 118, pl. vii, figs. 10—21, 1850. Dav. Mon. of British Permian Brachiopoda, part iv, p. 25, pl. ii, figs. 16—27, 1857, &c.

Spec. Char. Shell obscurely rhomboidal or deltoid, with marginal expansions; generally wider than long; greatest breadth towards the anterior portion of the shell. Beak of ventral valve small, moderately prominent, and incurved; foramen minute; sinus varying in depth and width according to age and specimen, flattened along the middle. Dorsal valve more convex than the opposite one, arched in profile; marginal expansions slightly bent upwards; the mesial fold commences at a short distance from the umbone, and varies in width and elevation according to the number of ribs which cover its surface; the lateral portions of the valve slope rapidly downwards. The ribs which ornament the valves generally commence at about the middle of the shell, and extend to the margin; in number they vary from thirteen to twenty-four in each valve, of which from three to six occupy the fold, while two to five ornament the sinus. The ribs on the lateral portions of the shell are sometimes strongly marked, while at other times they are but indistinctly defined. A large specimen measured, without its marginal expansions—length nine, breadth eleven and a half, depth six lines; but the dimensions were generally smaller.

Obs. Several palæontologists have alluded to the presence of *Camarophoria Schlotheimi* in the rocks of the Carboniferous period; thus M. de Verneuil and Count Keyserling state, in their great work on Russia (1845), that the Permian species was found by them in the Carboniferous limestone of Mount Chéractau, near Sterlitamack; at Sarana, on the Ufa; and at Cosatchi Datchi, to the east of Miask (Oural). In the second edition of Morris's 'Catalogue,' *C. Schlotheimi* is also recorded from the Carboniferous limestone of Derbyshire. At p. 119 of his 'Monograph of English Permian Fossils,' Professor King states that " *Camarophoria Schlotheimi* closely resembles the *Cam. crumena* of Martin, which appears only to differ from the former in being narrower and more accuminated behind; occasionally, however, a variety of the present species occurs, which can scarcely be distinguished from *C. crumena;* in short, both species apparently merge into each other so completely, that many would be inclined to consider them as specifically inseparable." In a note at the bottom of the page the same author has added, that having examined (what I believe he erroneously takes to be) Martin's original specimen of *Crumena,* in Mr. J. de C. Sowerby's Collection, he found that it belonged to the genus *Camarophoria;* and moreover that a tablet in the Gilbertsonian

Collection in the British Museum (erroneously labelled *T. plicatella*, Dalman[1]), and mounting nine specimens, with three to five ribs in the sinus, are undoubtedly *Camarophorias*, and that his notes state that they are identical with *C. Schlotheimi*. With all this evidence before me, I considered it necessary to ascertain what was really the *Anomites crumena* of Martin, and whether the Permian *C. Schlotheimi* does really occur in the Carboniferous limestone; and it was not until after much comparison and investigation that I became convinced that not only were the Carboniferous specimens alluded to by Professor King and others specifically identical with the Permian *Camarophoria*, but that it was impossible to distinguish the last from *A. crumena* of Martin.[2]

The term *Schlotheimi*, given to the species in 1834, by Baron V. Buch, must, therefore (much to my regret), be added to the synonyms of Martin's *A. crumena*, as we cannot preserve two names for the same species. *C. crumena* partakes of the general misfortune (as far as palæontologists are concerned) of presenting endless variations in shape and number of ribs; thus we find specimens with three, four, five, and six ribs on the mesial fold, and those on the lateral portions of the shell vary also in number, width, and degree of projection. In many examples the posterior portion of the shell is almost smooth, the ribs occupying only the anterior half; but in other specimens the ribs commence to be visible from the extremity of the beak and umbone, and extend uninterruptedly to the margin. Martin's figure portrays one of the shapes presented by this species, and I have

---

[1] Fig. 9 represents the largest specimen in question.

[2] Mr. Kirkby states that after having carefully examined the specimens and figures of *Camarophoria crumena* I sent him, with some hundreds of *C. Schlotheimi*, he could not find a single character distinguishing the one from the other. In both the series of specimens we have the same variation in length and width, the same difference in the elevation of the mesial fold, and the same latitude in the number of ribs ornamenting that fold; and that he should certainly consider both the Carboniferous and Permian specimens to belong to one species; that had he found the Carboniferous specimen in a Permian locality, and especially in the compact limestone, he would have referred them to *C. Schlotheimi*, without any doubt whatever.

As *C. crumena* is a rather important species, I have reproduced Martin's original figure (Pl. XXV, fig. 3), and will now transcribe the description he has given of his specimen:

" *Conchyliolithus anomites crumena*, scrotiformis, sulcis longitudinalibus subobsolescentibus, margine sino 3-plicato s.p.

" A fossil shell. The original an *anomia*. Perforate, valves convex, purse-like; or bellied and gradually increasing in size, from the beaks to the opposite extremity. Hinge curved, compact. Foramen oblong, very minute (rarely visible), situate under the apex of the larger beak, which is sharp-pointed and incumbent. The surface of the shell longitudinally furrowed; the furrows few, not more than ten or twelve, deep at the margin, but gradually becoming indistinct as they approach the beaks. The three central furrows form, in the smaller valve, a convex wave, answered in the other valve by a concave one; both terminating in a deep, three-plaited sinus at the margin. *Loc.* Winster."

The representation by Sowerby, tab. lxxxiii, fig. 3, of the ' Mineral Conchology,' does not convey an accurate idea of the species under description; and I do not believe that Martin's specimen could have been made use of for that illustration, as supposed by Professor King. Some of his figures, likewise, appear to have been taken from a Jurassic Rhynchonella.

endeavoured to represent some of the others. The marginal expansions are rarely obtained, on account of the hardness of the limestone matrix; but small portions were distinctly visible on several of the specimens that came under my observation, and I must refer the reader to p. 26, and Pl. II, of my 'Permian Monograph,' wherein he will find perfect exteriors and interiors of the species both represented and described. None of the Carboniferous specimens showed perfectly the interior arrangement, but the median septum and conjoined dental or rostral plates of the ventral valve, as well as portions of the septum which supports the spatula-shaped process in the dorsal one, could be distinctly seen in several examples.

Loc. Martin mentions that his shell was not common at Winster and Cromford, but it does not appear to be very rare in the Lower Scar limestone of Settle, in Yorkshire, whence many examples have been obtained by Mr. Burrow; and it is probable that Gilbertson's specimens in the British Museum were derived from some locality in the same county. It has been found also in the Carboniferous limestone of Dovedale, in Derbyshire, and in West Lothian, Scotland, by the late Dr. Fleming. Professor de Koninck has obtained specimens exactly similar to our own in the Carboniferous limestone of Visé, near Liège, in Belgium; and several Russian localities have been already mentioned.

CAMAROPHORIA GLOBULINA, *Phillips.* Pl. XXIV, figs. 9—22.

> TEREBRATULA GLOBULINA, *Phillips.* Encyl. Met. Geol., vol. iv, pl. iii, fig. 3, 1834.
> — RHOMBOIDEA, *Phillips.* Geol. York., vol. ii, p. 222, pl. xii, figs. 18—20, 1836.
> — SEMINULA, *Phillips.* Ibid., figs. 21—23.
> RHYNCHONELLA RHOMBOIDEA, *Morris.* A Catalogue of British Fossils, p. 147, 1854.
> HEMITHYRIS LONGA, *M'Coy.* British Pal. Foss., p. 440, pl. iii D, fig. 24, 1855.
> CAMAROPHORIA GLOBULINA, *Dav.* Mon. Br. Permian Fossils, p. 27, pl. ii, figs. 28—31, 1858.

Spec. Char. Shell small, globular or rhomboidal, either transverse or slightly elongated, entirely smooth up to a certain age. Dorsal valve rather more convex than the ventral one, with a fold of moderate elevation, which, commencing at about the middle of the shell, remains smooth or becomes divided into two or three (rarely four) angular ribs; the lateral portions of the valve are entirely smooth, or possess one or two short ribs on either side, close to the margin. The dorsal valve presents a rather deepened sinus, which, commencing at about the middle of the shell, extends to the front, with one or two short ribs along its middle; the beak is small and incurved, but showing between its extremity and the hinge line a small foramenal aperture. No marginal expansions; shell structure impunctate. Dimensions variable; three examples measured—

Length 5½, width 6, depth 4 lines.
,, 5 ,, 4½, ,, 3 ,,
,, 2¼ ,, 2, ,, 1½ ,,

Obs. After a lengthened comparison of numerous specimens of Phillips's *Terebratula rhomboidea* and *T. seminula*, it appeared to me evident that the last was nothing more than a young state of the first, and that neither could be distinguished from the Permian *Camarophoria globulina* of Phillips; the resemblance was indeed so great, that having mixed several specimens of each, it was with some difficulty that they could be afterwards separated. I was not able to expose the interior, but the single longitudinal median line which can be observed through the transparency of the shell, and which extends for a short distance along the back of the beak, leaves no doubt as to the species belonging to the genus *Camarophoria*, and not to that of *Terebratula* or *Rhynchonella*, as was hitherto supposed.

All the description given by Professor Phillips of his *T. rhomboidea* and *T. seminula* is, that the first has " no lateral plaits and perforation minute," while the second has " one lateral plait," and the " perforation is (also) minute;" but although some specimens of *T. rhomboidea* show no lateral ribs, other examples present one or two. The fold is also sometimes entirely smooth, one of Professor Phillips's figures denoting that the author was aware of the fact; and it is to this last variety that Professor M'Coy, at a subsequent period, applied the specific denomination of *Hemithyris longa*. We now, therefore, dispense with three so-termed species, and are able to add another to those that are common to the Carboniferous and Permian periods. In Pl. XXIV, I have represented the original figures, as well as all the principal variations of shape hitherto discovered.

Loc. *Camarophoria globulina* has been found in the Carboniferous limestone of Bolland and Settle, in Yorkshire; at Longnor, in Derbyshire, &c. Mr. Kelly mentions Cookstown, Blacklion, and Howth, as Irish localities. I am not acquainted with any Scottish example. It occurs also in the Permian shell limestone of Tunstall and Humbleton Hills, &c.

CAMAROPHORIA (?) LATICLIVA, *M'Coy*. Pl. XXV, figs. 11, 12.

> CAMAROPHORIA LATICLIVA, *M'Coy*. British Paleozoic Fossils, p. 444, pl. iii D, figs. 20, 21,
> 1855 (not *Atrypa laticliva*, M'Coy, Carb. Foss. of Ireland,
> p. 22, fig. 16, 1844).

Spec. Char. Transversely rhomboidal, wide, very deeply trilobed, rounded; both valves convex; mesial fold elevated, sinus deep; beak small, incurved, or adpressed to the umbone of the dorsal valve; foramen small, triangular. Each valve is ornamented with about twelve or thirteen ribs, of which a few have obscurely dichotomised; they originate at a short distance from the extremity of the beaks, and proceed from thence to the margin; three or four compose the fold, two or three the sinus, with wide flattened spaces on either side. One example measured—

Length 10, width 13½, depth 7 lines.

Obs. The above description has been abridged, or partly copied, from that published by Professor M'Coy, in his work on 'British Palæozoic Fossils,' because, from not having been able to obtain specimens of the species, I cannot express an opinion of my own as to its specific value. It is, however, more than probable that the shell which was briefly described, and vaguely represented, by M'Coy, in his 'Synopsis,' under the name of *Atrypa laticliva*, is specifically distinct from the one he subsequently described and figured as *Camarophoria laticliva;* and although I have no means of ascertaining whether the Professor is right while attributing the shell under description to the genus *Camarophoria*, I have reproduced the name and figures under his responsibility alone.

Professor M'Coy states, moreover, that " this species is very remarkable for its few ribs, deep trilobation (accounting for the wide, steep, smooth spaces on each side of the mesial ridges), and the very small beak. The smooth rostral space, free from ridges, extending nearly to the beak, clearly separates this species from *H. (R.) pleurodon;* and their great size and small number are distinctive characters from *C. ventilabrum.* Some very large specimens from Derbyshire, nearly ten lines long, show an obscure duplication of some of the great ridges near the margin. The very small, depressed sides, wide space free of ridges defining the mesial elevation, and the extremely sharp angular definition of the ridges, separate the species from the obtuse *H. (Rh.) pugnus.* The mesial septum in the (ventral) valve, in specimens of the above size, is from three to four lines long; the divarication forming the chamber in the beak is very distinct, but remarkably narrow, being upwards of a line long, but considerably less than half a line wide.

" *Position and locality.* Carboniferous limestone, Derbyshire; and Lowich, Northumberland."

CAMAROPHORIA ISORYNCHA, *M'Coy.* Pl. XXV, figs. 1, 2.

ATRYPA ISORHYNCHA, *M'Coy.* Synopsis of the Characters of the Carboniferous Fossils of Ireland, p. 154, pl. xviii, fig. 8, 1844.
CAMAROPHORIA ISORHYNCHA, *M'Coy.* British Palæozic Fossils, p. 444, 1855.

Spec. Char. Obscurely subcuboidal, globose; length, width, and depth, nearly equal; dorsal valve very gibbous, presenting in profile a rising curve, which is most elevated close to the front, where it becomes suddenly deflected (at almost right angles), so as to meet the margin of the opposite valve. The mesial fold is generally composed of four ribs, but does not project very much above the uniform gibbosity of the valve; the lateral portions of the beak and umbone are compressed, flattened, or slightly concave. The ventral valve is of small depth, and slightly convex to within a short distance of the front, when it becomes suddenly bent so as to meet the margin of the opposite valve; sinus wide and shallow; beak obtuse, not much produced or incurved. The surface of each valve is ornamented with about twenty obtusely angular ribs, which are strongly arched on the lateral portions of the shell.

Length 12½, width 13, depth 13 lines (Sir R. Griffith's specimen).

Obs. This species is stated by Professor M'Coy to be common in red arenaceous limestone, underlying the main limestone of the country, at Cookstown, Tyrone, in Ireland; and the only specimen I have been able to examine is one so labelled by M'Coy, and which forms part of Sir R. Griffith's collection. With such scanty material before me, I cannot venture upon any observations relative to its variations in form, age, and character; but the species seem well distinguished from other Carboniferous *Rhynchonellæ* or *Camarophorias*, by its general appearance. In the 'Synopsis,' Professor M'Coy describes the shell as an *Atrypa*, but at p. 444 of the work on the species of palæozoic fossils in the Cambridge Museum, the author states it to be a *Camarophoria*.

"The mesial septum in the dorsal valve strong, and about half the length of the flattened rostral portion; the mesial septum in the ventral one is nearly as long, and diverges to form a large wide chamber at the beak."

Sir R. Griffith's specimen does not show the interior, but on the external surface of each valve a single median line may be perceived, extending from the extremity of the beak and umbone to about a third of the length of the shell, which would indicate septa which probably supported processes peculiar to *Camarophoria*. We are also informed by Professor M'Coy, that the shell occurs in brownish siliceous Carboniferous rocks, near Shap Toll Bar, in England; and Professor De Koninck believes that the species is to be found in the Carboniferous beds of Tournay, in Belgium. No Scotch example has been recorded.

Family—STROPHOMENIDÆ.

This family comprises several genera and sub-genera, of which *Strophomena*, *Strepto-rhynchus*, and *Orthis* alone have been found to be represented in British Carboniferous strata. It must however be remarked, that although the families *Strophomenidæ* and *Productidæ* have been the subject of long and patient research, and that much progress has been made towards their elucidation, a great deal still remains to be done before the species of the first will have been grouped together with that degree of precision which is required in similar investigations; for it is certain that when we examine an extensive suit of the species which are at present located in the genera *Strophomena* and *Leptæna*, we perceive much interior dissimilarity, and it would be desirable to group more closely together those that resemble each other, and to separate those that are dissimilar. This, however, is a subject for further consideration, and no good would accrue to science by the hasty fabrication of a number of new genera on half worked-out material and ideas.[1]

Palæontologists appear to have generally agreed as to the propriety of maintaining the two families above mentioned, but it is still uncertain whether the characters by which they have been distinguished are of the importance at one time imagined. It was once supposed that external spines were peculiar to the *Productidæ*, and always absent in the *Strophomenidæ*; but it is now well known that certain species of *Orthis* were as thickly covered with short slender spines as any of the *Productidæ*—e. g., *P. punctatus*. In external shape also many *Strophomenidæ* differ but little from certain *Productidæ*, but interiorly the absence of the so-termed reniform impressions in all the *Strophomenidæ* hitherto discovered would appear to be the most constant character by which the two families can be distinguished.

In the *Strophomenidæ*, as well as in the *Productidæ*, no very prominent calcified processes have been hitherto detected, for the support of those beautifully fringed appendages which exist on either side of the mouth of the animal, and to which the (improper) designation of "oral arms," or "brachial appendages," has been given by the greater number of naturalists, and which are now believed to have subserved at once the function of gills and of sustentation, and to prove which, as observed by Mr. Hancock, "it is only necessary to refer to the manner in which the blood circles round the arms, and is carried to the cirri, but more particularly to its circulating through these latter organs, and to return direct from them to the heart."

Genus STROPHOMENA, *Rafinesque*, 1820.

Of this genus but a single species is known from the British Carboniferous strata, viz.:

STROPHOMENA RHOMBOIDALIS, *Wahlenberg*, Var. ANALOGA, *Phillips*. Pl. XXVIII, figs. 1, 2.

> ANOMITES RHOMBOIDALIS, *Wahlenberg*. Acta. Soc. Ups., vol. iii, p. 65, No. 7, 1821.
> PRODUCTA DEPRESSA, *Sowerby*. Min. Conch., pl. cccclix, fig. 3, 1823.
> —— RUGOSA, *Hisinger*. Vetensk. Acad. Hand. for är. 1826.

---

[1] It must be obvious to all, that the present work, whose publication will unavoidably have extended over a number of years, can be fairly viewed but as a continual attempt to work out a great difficulty. Availing myself, as I have constantly done, of every new discovery made by myself or by other competent observers, and imbued with no preconceived idea, I have continually modified my views as science has progressed; and this I must plead as a valid excuse for the changes (contradictory, perhaps) which may be noticed here and there in the many pages of which the monograph is composed. It is, however, my intention (should I ever be able to complete my arduous undertaking) to correct and co-ordinate the whole in the concluding pages.

Leptæna rugosa, *Dalman.* Kongl. Vetensk. Handl. for är. 1827, pl. cvi, fig. 1.
Productus quadrangularis, *Steininger.* Bemerk. über die Verstein. des Eifels, p. 35, 1831.
Strophomena pileopisis, *Dumont.* Const. Géol. de la Province de Liège, p. 354, 1832.
— marsupit. Ibid.
Productus elegans, *Steininger.* Mém. Soc. Géol. de France, vol. i, p. 361, 1834.
Strophomena rugosa, *Bronn.* Lethæa, Geog. i, p. 87, pl. ii, fig. 8, 1835.
Producta analoga, *Phillips.* Geol. of Yorkshire, vol. ii, p. 215, pl. vii, fig. 10, 1836.
Orthis rugosa, *V. Buch.* Ueber Delthysis, p. 70, 1837.
Leptena tenuistriata, *Sowerby.* Sil. Syst., tab. xxii, fig. 2 a, 1838.
— distorta, *J. Sowerby.* Min. Conch., vol. vii, pl. dcxv, fig. 2, 1840.
— nodulosa, *Phillips.* Palæozoic Fossils of Cornwall, pl. xxiv, fig. 95, 1841.
— depressa, *De Koninck.* Description des animaux fossiles de la Belgique, pl. xii, fig. 3, 1843.
Leptagonia rugosa, *M'Coy.* Synopsis of the Characters of the Carb. Foss. of Ireland, p. 118, 1844 (also 1855).
— multirugata, *M'Coy.* Pl. xviii, fig. 12, 1844.
Strophomena rhomboidalis, var. Analoga, *Dav.* Carboniferous System in Scotland. The Geologist, vol. iii, p. 102, pl. i, figs. 26—33, 1860.

Sp. ch. Shell more or less transversely semicircular or sub-quadrate; valves geniculated: hinge line straight, and as long as the greatest width of the shell, with rounded cardinal angles, which are at times prolonged in the shape of expanded wings. The ventral valve is slightly convex at the beak, from whence it becomes flattened to a certain distance and age, when the valve is suddenly bent downwards at almost right angles. The frontal margin is undulated, concave near the cardinal angles; it afterwards bulges out laterally, to form in front a slight outward curve. On the flattened portion of the disc, there exists a variable number of slightly undulating and occasionally interrupted concentric wrinkles, which turn outwardly towards the cardinal angles, and thus follow the marginal curves. The entire surface is also covered with numerous radiating, thread-like striæ; and a small circular foramen is generally observable close to the extremity of the beak, and up to a certain age, but which becomes obliterated or cicatrised in the adult. The dorsal valve is concave, and usually follows the curves of the opposite one, and is similarly wrinkled and striated. In the interior of the ventral valve, two diverging teeth articulate with corresponding sockets in the opposite valve. The muscular impressions (in this valve) are margined by a semicircular ridge, continued from the base of the teeth, and curving on either side so as to produce a saucer-shaped depression; the adductor or occlusor leaves a scar on either side close to a small median ridge, the cardinal or divaricator muscle filling on either side the anterior portion of the cavity; the ventral adjustor and pedicle muscles do not appear to have produced any very definite impressions, but it is highly probable that an attachment for these muscles existed in the posterior portion of the saucer-shaped depression above described, from the fact that a small circular peduncular foramen is also observable at a small distance from the extremity of the beak, and which denotes that a pedicle muscle must have existed, although the foramen became closed as soon as the animal found it could dispense with the moorings required during the early stages of its development.[1]

In the interior of the dorsal valve the cardinal process is divided into two lobes, and not connate with the diverging socket-ridges. From the base of this a slight median ridge runs down and separates the two pairs of adductor or occlusor scars, which are bordered by prominent ridges. The vascular impressions consist of large primary vessels which run at once direct from near the centre of the valve, to a short distance of the frontal margin, when they become reflected on either side to surround the ovarian spaces, and, near the margin, some of the vessels bifurcate several times.

Dimensions very variable: a large example measured, length 17, breadth 32 lines.

Obs. As may be observed from the synonyms and references given (and which could have been considerably increased in number), this species has already been located in at least six genera, and has received twelve or more specific denominations. It is not, however, surprising that before the class had been sufficiently investigated, and while efforts where being made to value and define the different interior characters, as a means of classification, that geologists and palæontologists should have felt uncertain where to locate this and other species, so that the different generic denominations may thus be easily and naturally accounted for. We next come to the specific denominations; and here we open out a field susceptible of much divergence in opinion, for the preconceived opinions of some palæontologists and geologists make them averse to recognise the existence of the same species in so extensive a vertical range as the *Silurian, Devonian*, and *Carboniferous* periods; and although the greater number of palæontologists have admitted, in their various works, that it was scarcely possible to separate the specimens of the shell as found in the three great divisions of the Palæozoic period, they still preferred to retain for each a distinct specific denomination: and this will explain the principal cause of the many names the shell we are now describing has received at different periods and from different hands.

In 1836, while describing the Carboniferous shell, Professor Phillips observes:— " I cannot distinguish this from the Dudley species, *L. depressa;*" and the same author begins the description of his *Prod. analoga* by stating that, " Perhaps this is really different from the last (*L. depressa*), but it is difficult to fix on characters." In 1843, Professor de Koninck observed, that it was impossible to distinguish the Silurian *Stroph. rhomboidalis* from the Carboniferous *Lept. analoga*, and he united the two under a single denomination. In p. 389 of his ' British Palæozoic Fossils,' while alluding to *Lept. analoga*, Professor M'Coy remarks, that " the shell in the Middle and Upper Palæozoic rocks can scarcely be considered as a distinct species from *L. depressa* of the Lower Palæozoic strata, and of which it is only a variety, not separable when the extremes are compared, but generally

[1] According to Darwin, a muscle does not exist simply because it was created, but because a necessity for it existed originally, and having once come to be of general use, traces of it might remain (as a rudimentary organ) in after-times, when the creatures had ceased to use it.

recognisable as a distinct variety much in the same way as the Devonian variety *aspera*,
of the Silurian *Spirigerina reticularis*, is usually distinguishable from the older varieties.
The general characters of the species are exactly those of *Lept. depressa*, and the
description given in the Lower Palæozoic section of this species will suit the present one in
nearly all respects. The differences which strike an observer are, that the transverse
wrinkles in the Lower Palæozoic *L. depressa* are usually from twelve to fifteen in number,
but in this variety they are from fifteen to twenty, and the longitudinal striæ are, on an
average, considerably coarser in *L. analoga* than in *L. depressa.*" The last two
observations are not, however, of much importance; for I possess examples of the Silurian
and Devonian shell with as many wrinkles as may be counted upon the generality of
Carboniferous specimens, nor does there exist that great difference in the quality of the
longitudinal striæ. Professor M'Coy then goes on to observe, that as to the varieties
named *Lept. distorta* by Sowerby (Carboniferous), and *Lept. nodulosa* by Phillips (Devonian),
he has in the mountain limestone traced all the intermediate varieties to the more regular
types, and no doubt exists in his mind as to their specific identity. It is therefore
evident that, in external shape and character, it is hardly possible to distinguish the
Silurian, Devonian, and Carboniferous shell, and that those little differences that are
occasionally observable, such as number of wrinkles and coarseness of longitudinal striæ,
may be due to local conditions which often slightly modify the size and appearance of the
same species.

Now let us cast a glance at the interior of the Silurian and Carboniferous shell
In the interior of the valves no very important differences can be traced, and the only
peculiarity I can perceive is, that the ridges margining the muscular scars are generally
more elevated or produced (though not always so) in the Carboniferous specimens; and it
appears to me, therefore, that by retaining the term *analoga* as a varietal designation
for the Carboniferous form, we shall be following the views advocated by the greater
number of palæontologists.

The variety *distorta*, which has prevailed in certain localities, can be viewed in no
other light but as that of a malformation, in which the dorsal valve is generally more or less
convex, and bending more or less rapidly inwards before becoming again inflected to
follow closely the deflected prolongation of the ventral valve.

Loc. In England, the Carboniferous variety is common in the Lower Carboniferous
limestone of Longnor, as well as in other Derbyshire localities: it has been collected
at Bolland and the Craven and Wensleydale districts, in Yorkshire; in Carboniferous shales
at Redesdale, Northumberland; also in the Middle Carboniferous limestone of Ronaldsway and Poolwash, Isle of Man.

In Scotland, it is found at Gare, in Lanarkshire, in limestone, at 239 fathoms below
the " Ell Coal," and 343 at Waygateshaw; in Renfrewshire, near Thornliebank; in
Stirlingshire, in the Campsie Main limestone and Corrie Burn beds.

In Ireland, it appears to occur in all the Lower Carboniferous strata, as well as in the

Mountain limestone division. It is not rare in the Carboniferous slate, in the parish of Ballyseedy, county Kerry. Mr. Kelly mentions Ring, Currens, Millecent, and Ballina county, among the Irish localities; the variety " *distorta* " being very abundant, especially in the county of Kildare.

It is also a common species in the Carboniferous rocks of many foreign countries, very large and fine examples occurring at Visé and Tournay, in Belgium.

The Silurian and Devonian localities will be mentioned in the monographs treating of the form belonging to those epochs.

Sub-Genus—STREPTORHYNCHUS, *King.* 1850.

The shells composing this sub-genus are closely related to *Strophomena*; they are usually semicircular-convex or concavo-convex, externally striated and interstriated; the ventral valve possessing a prolonged and oftentimes bent or twisted beak. (See Part IV, p. 29, of the present volume.)

Many are the so-termed species that have been described as occurring in the Carboniferous rocks of this and other countries, but so exceedingly variable are these shells, and so intimately do they all appear connected and linked together by intermediate and insensible graduations of shape, that it becomes most puzzling and difficult to determine how far we may be permitted to limit the extent of variation, or to determine what shapes ought to be separated or combined under a single species. Darwin considers "the term species as one arbitrarily given for the sake of convenience to a set of individuals closely resembling each other, and it does not essentially differ from the term variety, which is given to less distinct and more fluctuating forms; that the term variety again, in comparison with mere differential differences, is also applied arbitrarily and for mere convenience sake." And further on the same author observes, that " no one can draw any clear distinction between individual differences and slight varieties, or between individual differences or more plainly marked varieties and sub-species and species." [1] And how often are we not too prone to solve a difficulty in the way of identification, by at once cutting the Gordian knot, and arbitrarily fabricating a new species, without seeking to determine or to trace the connection of the specimen with some other form, of which it may be but a variety or mere difference in shape.

Hundreds of British and foreign specimens of Carboniferous *Streptorhynchus* have been assembled and carefully examined, but after much research and uncertainty from not finding characters of sufficient permanence to warrant the establishing of distinct species, I resolved (provisionally so, at least) to retain but one, of which *S. crenistria* (Phillips) may be considered the type, and to describe under separate heads, but with

[1] Darwin 'On the Origin of Species.'

varietal designations only, those few forms whose connection with the typical shape could not be entirely established; and by so doing I hope to concur with the views of those palæontologists who are not disposed to go so far as I feel inclined to do. No trouble has been spared in the selection and illustration of the most important and characteristic forms assumed by the species and its varieties, and many more intermediate shapes could have been figured, had space permitted.

STREPTORHYNCHUS CRENISTRIA, *Phillips,* sp. Pl. XXVI, fig. 1, Pl. XXVII, figs. 1—5 and 10? Pl. XXX, figs. 14—16.

SPIRIFER CRENISTRIA, *Phillips.* Geol. of Yorkshire, vol. ii, pl. ix, fig. 6, 1836.
— SENILIS. Ib., fig. 5.
LEPTŒNA ANOMALA, *J. de C. Sow.* Min. Con., tab. dcxv, figs. 1[b] (but not 1[a,d,c]), 1840.
ORTHIS UMBRACULUM, var., *Portlock.* Report on the Geology of the County of Londonderry, Tyrone, and Fermanagh, pl. xxxvii, fig. 5, 1843.
— — *De Koninck.* Animaux Fossiles du Terrain Carbonifère de la Belgique, p. 222, pl. xiii, figs. 4—7, 1843.
— QUADRATA? *M'Coy.* Synopsis of the Characters of the Carb. Fossils of Ireland, pl. xx, fig. 18, 1844.
— BECHEI. Ib., pl. xxii, fig. 3.
— COMATA. Ib., pl. xxii, fig. 5.
— CADUCA. Ib., pl. xxii, fig. 6.
LEPTŒNA SHARPEI, *Morris.* Catalogue, p. 138, 1854.
STROPHALOSIA STRIATA, *Morris.* Catalogue, p. 155, 1854, (part). Min. Con., tab. dcxv, fig. 1[a] only.
LEPTŒNA CRENISTRIA and L. SENILIS, *M'Coy.* British Palæozoic Fossils, pp. 450 and 452, 1855.
ORTHIS KEOKUK, *Hall.* Iowa Report, pl. xix, fig. 5, 1858.
— ROBUSTA. Ib., pl. xxviii, fig. 5.
STREPTORHYNCHUS CRENISTRIA, *Dav.* Mon. of Scottish Carb. Brach., p. 32, pl. i, figs. 16—22, 1860.

Spec. Char.—Very variable in shape, transversely or longitudially semicircular; hinge-line straight, slightly exceeding or somewhat shorter than the greatest width of the shell; cardinal angles rounded or prolonged with acute terminations; ventral area variable in width, flat, and divided by a fissure covered with a pseudo-deltidium; dorsal area linear. Ventral valve variable in its curves, slightly convex at the beak, flat in the middle, and partly concave near and at the margin, or more or less regularly convex throughout; beak straight, produced, and often twisted or irregularly inclined to one side. Dorsal valve moderately or extremely convex. Surface of both valves covered with numerous strong, radiating, rounded striæ with flattened interspaces of variable width, partly occupied by one or two smaller striæ, the ribs and interspaces being, at the same time, closely intersected by fine concentric lines or striæ, giving to the longitudinal ones a crenulated appearance.

In the interior of the ventral valve, a strong hinge-tooth is situated on either side at the base of the fissure, and strengthened by small dental or rostral plates ; the muscular impressions form a saucer-shaped depression, partially surrounded by a slightly elevated ridge ; the adductor or occlusor (A) occupies the central portion, and forms two small elongated depressions, separated by a slightly elevated mesial ridge, and on either side are the larger scars (R), apparently composed of two parts, the anterior or central being due to the divaricator, while the other or outer one would be produced by the ventral adjustor ?   In the interior of the dorsal valve, the cardinal process (J, to which were attached the divaricator muscular fibres, Pl. 27, fig. 6, 7) is composed of two testaceous projections ; the socket plates are large, and partially united to the lower portion of the cardinal process. Under these, on the bottom of the valve, may be seen the quadruple impressions left by the occlusor, and which occupy above one third of the length of the valve, and are arranged in pairs divided by a short rounded median ridge.

Dimensions and relative proportions very variable ; some examples have attained or exceeded three lines in length, by four in width, the depth varying also from a few lines to about two inches.

Obs.   Several palæontologists have already coincided in the belief that *S. senilis* is nothing more than a different state of *S. crenistria*, viz., in which the shell has attained an excessive depth or degree of convexity, accompanied by a very large area.   Professor M‘Coy justly states *S. Sharpei* to be an undoubted synonym ; and he might have said as much of *O. Bechei*, for the so-termed species has been made out of a crushed imperfect cast or impression of the same derived from the Carboniferous slate of Whiling Bay, Younghall, Ireland, as may be seen from an inspection of the original specimen in Sir R. Griffith's collection.   *O. caduca* is a small flattened valve of *S. crenistria*, or of var. *arachnoidea ?*, the original specimen in Sir R. Griffith's possession being derived from a black Carboniferous slate at Rahoran, Fivemiletown, Ireland.   *O. comata* is founded on a fragment of dorsal valve ; for the original specimen in Sir R. Griffith's collection shows the cardinal process, although not represented in the ' Synopsis ;' the striæ are also not so close and regular as in the Irish author's enlarged illustrations, but agree more with those of *S. crenistria*, to which the specimen should be referred; and I cannot help feeling somewhat surprised that any author could fabricate a species upon such insufficient material as the third part of single valve must naturally be.   In his work on the ' Cambridge Palæozoic Fossils,' while describing *S. crenistria*, Professor M‘Coy justly observes that " the striation is very variable, according to the state of preservation : in some the principal striæ, being nearly a line apart, and the intervening flat spaces having very distinct, longitudinal, fine lines, internally punctured, the middle one largest and crossed by fine, close, deep, irregularly transverse wrinkles ; in others, the intervening striæ nearly equal the principal ones in size, and the shell appears more closely and coarsely striated sub-alternately : but both extremes may be seen on the one specimen, and the differences are clearly the result of the loss of one or more layers of shell." I have also had the advantage of being able to examine typical examples of Professor Hall's *O. Keokuk* and *O. robusta*, which were kindly presented to me by Mr. Worthen, and am thus enabled to affirm that these American shells cannot be specifically distinguished from British varieties of *S. crenistria*.

For a long time I felt puzzled how to deal with a certain curious fragmentary ventral valve, figured by Mr. J. de C. Sowerby, in the 'Mineral Conchology,' under the designation of *Leptæna anomala* (tab. 615, fig. 1, *b*), and which I have also represented in Pl. XXX, fig. 15. This shell was subsequently referred by Messrs. Salter and Morris to the genus *Strophalosia*, and united with it under the same denomination was *Productus striatus;* but here a double error has been committed, for the shell in question is neither a *Strophalosia*, nor does it belong to *Productus striatus*, which is an entirely different species, although so confounded likewise by Mr. J. de C. Sowerby. Having received from Mr. Burrow some curiously shaped examples of *Strept. crenistria*, (Pl. XXX. fig. 14), I at once perceived that *Lept. anomala* was no other than a malformation of Phillips's *S. senilis*, and consequently a *Streptorhynchus*, while the *Mytilus striatus* of Fischer belongs to the genus *Productus*. *O. quadrata* was created for a very small imperfect ventral valve of a *Streptorhynchus*, which looks very like a young *S. crenistria ;* and the shell, such as it is, belongs to Sir R. Griffith, and is said to have been found in calp at Ballintrillick, Bundoran, Ireland. Its striation is exactly similar to that of *crenistria*, and to which it is referred at least provisionally, the material being too imperfect to admit of its being regarded as a well made-out species. Such are some of the undoubted synonyms of Phillips's species, and to which I would have added others, but for the reasons already given.

S. crenistria appears to have had a very extended vertical range, and is recurrent from the Devonian if not Silurian period? In the Carboniferous rocks, it is found in all the stages, from the lowest beds (such as the Lower Carboniferous Red and Yellow Sandstone of Kildress) up to the highest beds above the Mountain or Carboniferous Limestone. It is also a far-spread species, having been found in various parts of Europe, America, Asia, and Australia, everywhere assuming the same shape and variety.

In England, it occurs at Bolland, Kendal, Settle, the Isle of Man, &c. ; in Scotland, at Bowertrapping, three miles south of Dalry, and in many localities in Lanarkshire, Renfrewshire, Dumbartonshire, Ayrshire, &c. It was figured by David Ure, in his ' History of Rutherglen and Kilbride (pl. xiv, fig. 19), as far back as 1793. In Ireland, it occurs in many localities; among which may be mentioned Hook, Bundoran, Ballyduff, Millecent, &c.

In Belgium, it is found at Visé and Tournay, &c. In America at Keokuk, Iowa;

Warsaw, Nauvoo, St. Clair county, Illinois, &c. In India, at Moosakhail, &c., in the Punjaub. It was found also in Spitzbergen, and in Australia, &c., &c.

VAR. A.—STREPTORHYNCHUS ARACHNOIDEA, *Phillips*, sp. Pl. XXV, fig. 19—21, Pl. XXVI, fig. 2, 3, 4 (lower figs.), 5, 6.

SPIRIFER ARACHNOIDEA, *Phillips*. Geol. Yorks., vol. ii, pl. xi, fig. 4, 1836.
ORTHESINA PORTLOCKIANA, *Semenow*. Ueber die Fossilien des Schlesischen Kohlen-kalkes, pl. ii, fig. 1, 1854.
ORTHIS ARACHNOIDEA, *De Verneuil*. Geol. of Russia, vol. ii, pl. x, fig. 18, 1845.

Professor Phillips describes his *S. arachnoidea* as "very depressed, truncato-orbicular, hinge-line wide as the shell; striæ fine, sharp, and continually subdivided; upper valve convex, as in *S. resupinata*." This description would certainly suit many specimens of true *S. crenistria*; but the figure in the 'Geology of Yorkshire' would not indicate a shell with a convex valve, such as that possessed by the Martins species, and was evidently intended for those very depressed varieties in which the valves were almost flat or but very slightly convex, one of them being even at times a little concave from the middle to the margin, as will be found represented in our Pl. XXV, fig. 19—21; Pl. XXVI, fig. 2, 3, 4. Such, at any rate, is the shell we can conceive as having been intended for *S. arachnoidea*. The area in this is narrow, while the striation is entirely similar to what we have already described. This variety or variation in shape of *S. crenistria* generally occurs in the same localities where the more convex forms are met with.

Professor Phillips mentions Stradon, Haltwhistle, Allenheads, near Heskel Newmarket. It is found also at Rutcheugh, and other places in Northumberland; in Scotland, in the localities already enumerated; while in Ireland, it occurs at Curragh, Ardmore, Kildare, Kildress, &c., &c.

VAR. B.—S. KELLII, *M'Coy*. Pl. XXVII, fig. 8.

ORTHIS KELLII, *M'Coy*. Synopsis of the Characters of the Carboniferous Fossils of Ireland, p. 124, pl. xxii, fig. 4, 1844.

In shape, this species or variety? is marginally semicircular and slightly indented in front, the hinge line being a little shorter than the greatest width of the shell. Ventral area narrow, and divided by a triangular fissure covered by a pseudo-deltidium. The ventral valve is slightly concave, the beak alone presenting a small convexity; while the dorsal one is moderately convex, with a mesial furrow or depression. The surface of both valves is closely covered with numerous radiating, imbricated striæ, a smaller one or two intervening between the larger, as in *S. crenistria*. The typical specimen measured, length 19½, width 28½, depth 5 lines.

Professor M'Coy has stated that his species " seems to be constantly distinguished from *O. crenistria*, by its deep rounded mesial furrow, which indents the frontal margin ;" but it is not always more convex than *S. crenistria*, nor are the angles in the last always acute, for many examples had them rounded, as in *S. Kellii*. It is quite true that the striation is so variable a character as to be of little use in specific distinctions, in the British species of *Streptorhynchus ;* but, at the same time, I cannot allow that, in all the examples of *S. Kellii* that have come under my observation, the striation was finer, closer, or more equal than is commonly the case with *S. crenistria*, although such is the case in many individuals.

In England, *S. Kellii* has been found in the Black Rock, Clifton. In Scotland, it occurs in Ayrshire. In Ireland, at Monaghan; Annghilla, three miles south-west of Ballyganty in Tyrone, &c.

VAR. C.—S. CYLINDRICA, *M'Coy.*   Pl. XXVII, fig. 9.

ORTHIS CYLINDRICA, *M'Coy.*   Synopsis of the Characters of the Carboniferous Fossils of Ireland, p. 123, pl. xxii, fig. 1, 1844.

Almost circular, hinge-line shorter than the greatest width of the shell, with rounded cardinal angles: area triangular, not very wide; fissure covered by a pseudo-detidium. Ventral valve convex at the beak, concave from near the middle to the margin; dorsal valve very convex, almost geniculated. Surface covered with numerous radiating striæ, of which two or three smaller ones intervene between the larger ones. Of this beautiful species or variety (?) I have never been able to obtain the sight of any other besides the type, which measures eighteen lines in length, by twenty-one and a half in breadth, and ten in depth.

Professor M'Coy observes that, in this species, the striation is similar in character to that of *O. crenistria*, *O. Kellii*, and *L. euglypha*, but that the form at once distinguishes it from every known species.

It is at all times hazardous to make a new species upon the inspection of single examples; and as my knowledge in the present instance does not extend beyond the one figured in the 'Synopsis,' I dare not venture to conjecture too closely as to its probable affinities or variations in shape. The original example in Sir R. Griffith's collection is said to have been found in Arenaceous Limestone at Castle Espie, Comber, Ireland.

Var. D.—S. radialis, *Phillips.* Pl. XXV, figs. 16—18.

Spirifer radialis, *Phillips.* Geol. of Yorks., vol. ii, pl. xi, fig. 5, 1836 ; and *Dav.*, Mon. of Scottish Carb. Brachiopoda, pl. i, figs. 21, 25, 1860.

Shell variable in shape ; transversely or longitudinally semicircular ; hinge-line usually shorter than the greatest width of the shell. Dorsal area of moderate width ; fissure closed by a pseudo-deltidium. Valves moderately convex, the dorsal one being the deepest, with sometimes a slight depression near the front of the ventral one. Surface covered with strong radiating striæ, with intervening smaller ones, crossed by imbricated lamellæ and deep concentric lines or interruptions of growth.

This species or variety (?) does not appear to have attained the large dimensions of *S. crenistria*, but the following average measurements may be given :

Length 17, width 21, depth 7 lines.
„ 16, „ 15, „ 6 „

The interior details do not appear to differ from those of *S. crenistria ;* but on a young specimen from Gare there appeared to exist a small circular foramen at the extremity of the beak, similar to what we see in some species of *Strophomena ;* the aperture does not, however, exist in full-grown individuals, and thus shows that the animal could dispense with its temporary moorage.

The striæ differ also much in different specimens, from the larger ones being either closer or more widely separated, two or four smaller ones occupying the interspaces. It must be, however, observed that it is difficult at times to distinguish certain individuals of *S. radialis* from others of *crenistria.*

Professor Phillips mentions Florence Court, near Enniskillen, as the locality whence his type was obtained ; but the figure in the ' Geology of Yorkshire' is very incomplete, and drawn from a crushed specimen. Good examples have been collected at Whatley, near Frome, in Somerset ; at Gare and Middleholm, in Lanarkshire, Scotland ; and at Middleton, in Ireland.

Genus—ORTHIS.

The genus Orthis forms a well-characterised group, especially numerous and abundant in the Silurian and Devonian systems, is considerably reduced during the Carboniferous period, to appear no longer (?) in subsequent stages. Four species alone have been hitherto discovered in the Carboniferous rocks of Great Britain.

ORTHIS RESUPINATA, *Martin*, sp. Pl. XXIX, figs. 1—6. Pl. XXX, figs. 1—5.

> CONCHYLIOLITHUS ANOMITES RESUPINATUS, *Martin*. Petrif. Derb., tab. xlix, figs. 13, 14, 1809.
>
> TEREBRATULA RESUPINATA, *Sow.* Min. Con., tab. 325, Feb., 1822.
>
> SPIRIFERA RESUPINATA, *Phillips.* Geol. Yorks., vol. ii, pl. xi, fig. 1, 1836.
>
> —   CONNIVENS. Ib., fig. 2.
>
> —   RESUPINATA, *V. Buch.* Mém. Soc. Géol. de France, vol. iv, pl. x, fig. 32, 1840.
>
> ORTHIS RESUPINATA, *De Koninck.* Déscription des Animaux Fossiles de la Belgique, pl. xiii, fig. 9, 1843.
>
> ATRYPA GIBBERA, *Portlock.* Report on the Geology of Londonderry, Tyrone, and Fermanagh, pl. xxxviii, fig. 1, 1843.
>
> ORTHIS LATISSIMA, *M'Coy.* Synopsis of the Carb. Limest. Fossils of Ireland, pl. xx, fig. 20, 1844.
>
> —· GIBBERA. Ib., pl. xviii, fig. 9.
>
> —   RESUPINATA, *Dav.* 'Introduction,' pl. vii, fig. 135, 1853.
>
> —   —   *M'Coy.* British Palæozoic Fossils, p. 449, 1855.
>
> —   —   *Dav.* Scottish Carb. Brach., p. 28, pl. i, figs. 11, 12, 13, 1860.

Spec. Char. Shell transversely oval or elliptical; valves convex, sometimes gibbous; hinge-line straight, much shorter than the greatest width of the shell, with rounded cardinal angles. Dorsal valve generally the deepest, regularly and evenly convex, or slightly flattened and depressed along the middle to the front, area narrow. Ventral valve variable in its curves, moderately convex throughout, or at the rostral portion only, becoming flattened or slightly concave near the lateral and frontal margins. The frontal margin presents a uniform or undulated curve, indenting to a lesser or greater extent the margin of the opposite valve. Beak small and moderately incurved; area triangular, with an open fissure. Exteriorly, the valves are closely covered with numerous fine, thread-like, rounded, radiating striæ, which increase in number by interstriation and bifurcation at variable distances from the beaks, and at intervals the striæ themselves augment in thickness and prominence, producing small, hollow, thread-like, tubular spines, which become more numerous towards the margin. The intimate shell structure is perforated by innumerable canals, of which the exterior orifices, in the shape of minute punctures, cover the entire surface of the valves.

In the interior of the ventral valve, the dental plates extend to some distance along the bottom of the shell, and between these a small rounded or angular ridge divides the muscular scars, which thus form two elongated depressions margined on their outer sides by the prolonged basis of the dental plates. The adductor or occlusor leaves a small, not always clearly defined impression on either side of the mesial ridge, and it is probable that the larger impressions, termed divaricator (R), in our figures of this species (as well as of *O. Michelini*), is apparently composed of two parts, the anterior or central being the cardinal or divaricator, while the other, the posterior or lateral (which is parallel), may belong to the ventral adjustor?

In the dorsal valve, the fissure is almost entirely occupied by a moderately produced shelly prominence or cardinal process, to which were no doubt affixed the divaricator muscular fibres. The inner socket walls are somewhat prolonged under the shape of projecting laminæ, and to the extremity of which free spiral arms may perhaps have been attached (?); while under this shelly process a longitudinal ridge, with a wide, flattened space on either side, separates the quadruple impressions of the adductor or occlusor muscles; these last producing two oval-shaped depressions, placed obliquely one above the other, and separated by lateral elevations branching from the central ridge. Vascular impressions and ovarian markings are at times clearly observable in the interior of both valves.

Dimensions and degree of convexity very variable two extremes have measured—

Length, 32; width, 41; depth, 21 lines (the largest example I have seen).

„ 16; „ $17\frac{1}{2}$ „ 16 „

Obs. *O. resupinata* is a common, well-known, and widely spread Carboniferous shell. Its exterior appearance has been often described and illustrated, but not always properly understood. In shape it is subject to considerable variation, and hence the specific denominations of *resupinata, connivens, gibbera*, and *latissima*, which appear to Mr. Morris and myself to have been applied to what we must regard as different states of a single species; and in my plates will be found represented some of the most remarkable shapes. The striation is also finer in some specimens than in others; and certain examples show, more than others do, the small drop-like elevations from which originate the little thread-like spines, which are generally broken close to the surface of the valves in a large number of specimens.

Interiorly, some unimportant variation in detail may likewise be occasionally observed and which I have endeavoured to illustrate.

In England, *O. resupinata* abounds in the Carboniferous limestone and shales of many localities, such as at Greenhow Hill; Hawes, Otterburn; at Dovedale, in Derbyshire; the Isle of Man; Lowick, Northumberland; Settle, Malham Moor, Withgell, Yorkshire; at Ulverston, Bolland, &c.

In Ireland, it is found at Tyrone, St. John's Point, Dunkineely, Cornacarrow, Enniskillen; Cruiceroth, near Drogheda, County Meath; Little Island; Cornagrade, Bundoran, Cookstown, Millecent, &c.

In Scotland, it occurs at Gare, Raes Gill, Middleholm and Brockley, near Lesmahago, Capel Rig, East Kilbride, &c., in Lanarkshire; in Ayrshire, at Auchenskeigh, near Dalry, West Broadstone, Beith, &c; in Dumbartonshire, at Castlecary; in Stirlingshire, at Balglass Burn, as well as in the Campsie Main Limestone and Corrieburn beds. It occurs also in Midlothian and other Scottish counties.

On the Continent, it is not rare at Visé and Tournay, in Belgium, &c.

ORTHIS KEYSERLINGIANA, *De Koninck*.   Pl. XXVIII, fig. 14.

> ORTHIS KEYSERLINGIANA, *De Koninck*.   Déscription des Animaux Fossiles qui se trou-
> vent dans le Terrain Carbonifère de Belgique,
> p. 230, pl. xiii, fig. 12, 1843.

Spec. Char.   Shell transversely oval or sub-quadrate, with rounded angles; ventral valve convex, and much deeper than the dorsal one, with a deep longitudinal median sinus or groove, which extends from the extremity of the incurved umbonal beak to the front, this valve being likewise the longest and largest, on account of the umbonal beak being more elevated than the level of the beak and area of the dorsal one; area narrow. Dorsal valve straight, flattened, and sometimes slightly concave from the extremity of the beak to the front, the lateral portions sloping gently on either side; area triangular, flattened, and bent backwards; beak small, angular, slightly convex, and on a level with the area; fissure triangular, open; margin moderately flexuous. Exteriorly, each valve is covered with numerous thread-like, rounded, radiating striæ, which increase in number by the means of numerous interstriations and bifurcations from the extremity of the beak to the front; the striæ at intervals augmenting in thickness as well as projection, and giving rise to small thread-like spines (broken close to the surface in the generality of specimens). Numerous concentric lines of growth are also observable on the valves, which are likewise perforated by minute tubuli. Interior unknown.

Two specimens have measured—

Length 11, width 12½, depth 6½ lines.

    „   8,   „   10,   „   6

Obs. This interesting and well-marked species was first brought to my notice as British by Professor de Koninck, who had received a specimen from England, but without locality; and it was only after some research that I learnt that Mr. Burrow had found several examples in the Carboniferous limestone of Settle, in Yorkshire, this being also the only British locality, with which I am at present acquainted, wherein the shell has been discovered. *O. Keyserlingiana* cannot be confounded with *O. resupinata*, on account of its peculiar and different shape, although the striation is very similar in both. In Belgium, it has been found occasionally at Visé.

ORTHIS MICHELINI, *L. Eveillé*.   Plate XXX, figs. 6—12.

> TEREBRATULA MICHELINI, *L'Eveillé*.   Mém. de la Soc. Géol. de France, vol. ii, p. 39,
> pl. ii, figs. 14—17, 1835.
> SPIRIFERA FILIARIA, *Phillips*.   Geol. of Yorks., vol. ii, pl. xi, fig. 3, 1836.

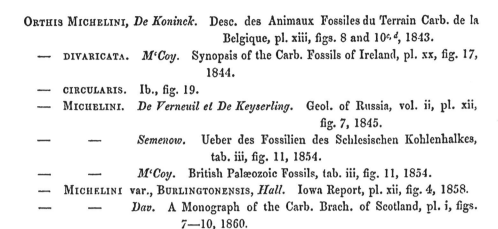

Orthis Michelini, *De Koninck.* Desc. des Animaux Fossiles du Terrain Carb. de la Belgique, pl. xiii, figs. 8 and 10*ᶜ*ᵈ, 1843.

— divaricata. *M'Coy.* Synopsis of the Carb. Fossils of Ireland, pl. xx, fig. 17, 1844.

— circularis. Ib., fig. 19.

— Michelini. *De Verneuil et De Keyserling.* Geol. of Russia, vol. ii, pl. xii, fig. 7, 1845.

— — *Semenow.* Ueber des Fossilien des Schlesischen Kohlenhalkes, tab. iii, fig. 11, 1854.

— — *M'Coy.* British Palæozoic Fossils, tab. iii, fig. 11, 1854.

— Michelini var., Burlingtonensis, *Hall.* Iowa Report, pl. xii, fig. 4, 1858.

— — *Dav.* A Monograph of the Carb. Brach. of Scotland, pl. i, figs. 7—10, 1860.

Spec. Char. Shell depressed, almost circular or subtrigonal, as wide or wider than long, with greater width near the frontal margin, this last assuming a gentle outward or slightly inward curve; hinge-line much attenuated, not above one third the width of the shell, sometimes less; ventral area small, but rather larger than that of the opposite valve, and divided by a fissure which is almost entirely filled up by the cardinal process of the dorsal valve. Ventral valve either moderately convex, with a slight depression commencing about the middle and extending to the front, or flattened throughout with the exception of the beak, which is small, slightly incurved, and prominent. Dorsal valve, moderately and uniformly convex, or gently depressed towards the front. Surface of both valves closely covered with thread-like, radiating, rounded striæ, which increase in number by means of numerous interstriations and bifurcations, while from all the little ribs hair-like hollow spines project and become more closely packed towards the margins. Intimate shell structure perforated by small canals, of which the external orifices, under the shape of punctures, cover the entire surface of the valves.

In the interior of the dorsal valve, the cardinal process is situated between two projecting laminæ (which perhaps afforded attachment to the spiral arms) while under this shelly process a narrow median longitudinal ridge separates the quadruple impressions of the adductor or occlusor muscle; these last producing on either side two oval-shaped depressions, one placed above the other, and separated by lateral elevations branching from the central ridge.

In the interior of the ventral valve, the dental plates extend to some distance along the bottom of the shell, and between these are situated the muscular impressions. At the base of the fissure, and between the dental plates, may be observed a horizontally striated impression, which may perhaps be due to the pedicle muscle; under this a smaller median oval depression was produced by the occlusor, while the larger elongated depressions on each side of this are apparently formed of two parts, the anterior or one close to the occlusor being the cardinal or divaricator, while the other, the posterior or lateral, which is parallel, may belong to the ventral adjustor? Ovarian and vascular impressions are also visible.

Dimensions variable; three examples have measured—
Length 15, width 16½, depth 5 lines.
„   15, „ 15,   „   6 „
„   11, „ 11,   „   6 „

Obs. ·*O. Michelini* is a common and far-spread Carboniferous species, always recognizable, and easily distinguished from *O. resupinata* and *O. Keyserlingiana* both by external shape and by interior differences, which last are evinced in the narrowness of the median ridge, and less inclined slope of the adductor or occlusor muscular impressions; while in the dorsal valve the space occupied by the adductor, divaricator, and ventral adjustor muscles is wider than in *O. resupinata.* There is also a singular impression, marked N in fig. 9, which is with some uncertainty attributed by Mr. Hancock to the pedicle muscle. The striation in *O. Michelini* differs also from that of *O. resupinata* and *O. Keyserlingiana* by the absence of that peculiar dilatation of some of the striæ before the occurrence of a spine-like projection, and the remarkable abrupt diminution of the swollen ridge beyond the spine, which produces those elongated drop-like mark, so visible over the surface of the last-named species. Professor De Koninck, who in 1843 first noticed the spiny investment of the shell under description, expressed the opinion that the dorsal valve was alone so adorned; but a careful examination of many specimens has convinced me that spines existed on both, although perhaps more numerously spread on the dorsal valve. The spines were likewise more abundant on some examples than on others, nor do they seem to have anywhere exceeded about a quarter of an inch in length.

Orthis Michelini appears also to have varied less in shape than the preceding species; the shortness of the hinge-line is, however, in some specimens very remarkable.

O. Michelini abounded where it occurred, and especially so in certain shales at the top of the Mountain Limestone, such as at Clattering Dykes, on the middle of Malham Moor, in Yorkshire; nor is it rare in the Mountain or Carboniferous Limestone of Settle, Bolland, Fountain Fell, the Isle of Man, &c.

In Ireland, it occurs in the Carboniferous Limestone and shales at Lisnapaste, Little Island, Millecent, Ballyduff, Bruckless, Malahide, Old Leighlin, &c. It may also be collected by millions (in the condition of internal casts and impressions) in the calciferous states of Ballyseedy, County Kerry.

In Scotland it is also very plentiful in certain localities, such as at Gateside, near Beith, where detached valves occur by thousands in the same stratum of shale whence David Ure collected his specimens of the species.[1] In Lanarkshire, it occurs at Langshaw Burn, Brockley, and Middleholm, near Lesmahago, Auchentibber, Calderside, and Phillipshill, High Blantyre; Capel Rig, East Kilbride, &c. In Renfrewshire, at Orchard Quarry Thornliebank; Barrwood and Howood, near Paisley. In Ayrshire, at Roughwood, West Broadstone, and Treehorn, near Beith, &c. In Stirlingshire it is found in several stages; and occurs also in Dumbartonshire, Fifeshire, &c.

[1] David Ure appears to have been the first naturalist who figured this shell in Great Britain; for figs. 13 and 14 of the fourteenth plate of the 'History of Rutherglen,' published in 1793, evidently belong to the species under description.

In Belgium, it is common in the limestone and shales of Tournay; but rarer at Visé. In Russia, at Cosatchi-datchi, &c. While in America it has been collected at Burlington, Iowa; Quincy, Illinois; Hannibal, Missouri, &c.

ORTHIS? ANTIQUATA, *Phillips.* Plate XXVIII, fig. 15.

> TEREBRATULA ANTIQUATA, *Phillips.* Geol. Yorks., vol. ii, pl. xi, fig. 20, 1836.
> RHYNCHONELLA ANTIQUATA, *Morris.* 'Catalogue,' p. 146, 1854.

Spec. Char. Somewhat elongated oval; hinge-line straight, shorter than the greatest width of the shell; ventral valve deep and convex, but flattened towards the front; beak small, incurved; area narrow, with an open triangular fissure (?). Dorsal valve shallow, and divided into three lobes, of which the two largest diverge from the umbone to the lateral margins, the third or the less produced forming a small median elevation, originating near the middle, and extending to the front, where the valve assumes a somewhat triundate margin and appearance. Surface smooth; length 4, width 3, depth 2½ lines.

Obs. Of this little shell I have seen but one example—that described and figured by Professor Phillips, and which is stated to have been obtained at Bolland, and now forming part of the Gilbertsonian collection in the British Museum. All endeavours to procure the sight of another example have proved unavailing; so that my observations have been confined to the examination and description of the original type, which looks more like a small *Orthis* than a *Terebratula* or *Rhynchonella*, and it is therefore here at least provisionally located under the genus *Orthis*. Professor Phillips describes his species in the following words:—" a very singular, small Brachiopodous shell (perhaps *Producta*) of an oval figure; lower valve convex; upper plane, with two diverging convexities; hinge-line straight." Upon such insufficient material, the species itself can only be admitted as provisional.

Family—PRODUCTIDÆ.

The PRODUCTIDÆ have been divided into four groups, viz., *Productus, Aulosteges, Strophalosia,* and *Chonetes;* but they all appear to bear so natural and indeed so intimate a relation towards each other, that it is very questionable whether the last three should be considered more than simple sub-genera or modifications of PRODUCTUS; and this is also the opinion of Professor de Koninck, to whom science is indebted for a valuable monograph of the many forms of which this family is composed.

All the species at present known are restricted to the limits of a portion of the Palæozoic period. No British examples of Carboniferous *Strophalosia* or *Aulosteges* have been hitherto recorded; but a few forms of those sub-genera occur in equivalent rocks of other countries, such as Belgium and India. In all British species of *Productus* and *Chonetes*, the shell is more or less concavo-convex, oval, semi-oval, or angular, and generally auriculated; the hinge-line is straight and strong, with or without teeth and sockets for the articulation of the valves. All well-authenticated species of *Productus*, hitherto examined, have shown themselves to be edentulous, but whether this character was general and without exception may remain a question for further consideration; anyhow, the dorsal valve must have turned on its long hinge-line with as much precision as in *Chonetes*, which possessed regularly articulating teeth. It has been often asserted and believed that *Productus* might be distinguished from its sub-genera by the total absence of an area; and although this is the prevalent character of the genus, still in certain species, such as *P. sinuatus*, a perfectly developed area is generally present in the ventral valve. There exists also an occasional tendency to the formation of hinge-area in several species, as may be seen, for example, in the remarkable example of *P. semireticulatus* of which a representation is given in Pl. XLIII, fig. 5.

All species of *Chonetes* at present known have, in addition to the regular articulation, an area in each valve, this being larger in the ventral than in the dorsal one, which is also divided by a fissure, more or less arched over by a pseudo-deltidium, the cardinal process of the opposite valve filling up and effectually closing any portion that might have otherwise remained uncovered.

The external surface in *Productus* varies according to the species; in some, it is almost smooth, in others longitudinally and finely striated or coarsely costated, as well as intersected by numerous concentric wrinkles or lines of growth. All the species appear to have been furnished with tubular spines; in some forms they are small, delicate, and so closely packed as to conceal every portion of the shell, with the exception of the area; while in others they are irregularly scattered, and chiefly confined to the auriculate portions of the valves. In certain species, the spines exceeded by four or five times the length of the shell; and while some were almost as delicate as the hair of one's head, others exceeded a line diameter; the dimensions of the shell having nothing to do with that of the spine, for in some small species these were few and large, while the reverse has occasionally been found to be the case with species of the largest dimensions.

Chonetes differs also somewhat from *Productus* by the manner in which its spines are disposed along the cardinal edge, these last sloping outwardly, and increasing in length as they approach the extremities of the hinge-line; but in many species, in addition to these, there existed on the surface of the valves small spines, disposed as in *Productus*.

The intimate shell-structure has been described by Dr. Carpenter to be perforated, and that, where the shell is furnished with spines, the perforations are continued into them, and that such passages are of more than average dimensions. In *Productus* the internal surface of the dorsal valve is more or less convex, and presents in the middle of the hinge-line, a prominent bilobed or trilobed projection or cardinal process (J),[1] its upper surface is often striated, and affords attachment to the cardinal or divaricator muscles (R).   Under this a narrow longitudinal ridge generally extends to about half (or more) of the length of the valve, and on either side are seen the ramified or dendritic impressions, which we consider to be attributable to the adductor or posterior and anterior occlusor muscle (A), but which are often situated so close to each other, on either side of the mesial ridge, as to render the quadruple attachment not so distinct as could be desired; they are, however, well defined in *P. longispinus*, and in some other species.   Outside, and in front of these, are the two " reniform impressions" (x).   Their surface is generally smooth, and bordered by ridges, which after dividing the occlusor muscles proceed in an outward, oblique, or almost horizontal direction; then turning abruptly backwards, they terminate at a short distance from their origin.   There exists also, in many species, but not in all, two prominences (w), one on each side of the median ridge, and close to the base of the muscular scars; they are very apparent in *Productus*, but not observable in *Chonetes*.

The internal surface of the valves (in all the family) is covered with innumerable granulations, of which some have been thought by Mr. Hancock to have been " probably produced by the muscular bands, which retracted the margin of the mantle." I will now describe the internal appearances observable on the concave surface of the larger or ventral valve.   A narrow mesial ridge, originating under the extremity of the beak separates two large, elongated, ramified or dendritic impressions (A), which have in all probability been produced by the adductor or occlusor muscle, although otherwise referred by some Palæontologists.   In advance of the larger scars, we sometimes perceive smaller impressions closely connected with the larger ones above described (E), and which were in all probability produced by a portion of the occlusors themselves.   On either side of the adductor or immediately under there exist two deep, longitudinally striated, sub-quadrate impressions, which are due to the cardinal or divaricator muscles (R).   I have in vain, hitherto, sought for impressions referable to the adjustor muscles; but as no peduncular aperture existed, such muscles may be supposed not to have existed: however, as the valves in the species hitherto known of *Productus* possessed no teeth or sockets, and therefore are not strongly articulated, as in the *Terebratulidæ*, it is not impossible that the adjustors may have been so arranged as to keep the valves adjusted to each other, and that they thus acted as a substitute for a hinge.

The only point remaining to be here noticed in connection with the interior of this valve are the deep concave, often sub-spinal, depressions (L), visible in some species, such as *P. giganteus*, *P. humerosus*, &c., and which were hollows no doubt occupied by the spiral arms, for it would seem impossible to conjecture otherwise, how they had originated. Similar hollows could not of course be expected to be present in those species in which

[1] See Pl. XXXVII, figs. 1—4.

the shell did not possess a sufficient thickness, as they never influenced the regular curve or convexity of the exterior of the valve. In all the PRODUCTIDÆ we therefore find muscles destined to open and close the valves, as well as evidence that they possessed spiral arms, and which have been justly, I think, supposed to be brachial appendages, which subserved at once the function of gills and of sustentation.

In *Chonetes*, the muscular impressions are somewhat similar to those of *Productus*, but of more often comparatively smaller proportions, as may be perceived by a glance at the figures I have given of both.

The determination and arrangement of the British Carboniferous species of *Productus* and *Chonetes* has demanded a lengthened examination, for much confusion still existed among the synonyms, notwithstanding the able and valuable researches of Professor de Koninck. Through the kindness of several friends, I have been enabled to study the original types of the larger number of species, as well as a very extensive series of specimens; and although I have rejected a great number of so-termed species, and been able to add several others new to Great Britian, I should still urge further research upon these very interesting shells, as it has not been always possible to determine the specific claims of some among them with sufficient accuracy.

After much examination, and in some cases much uncertainty, the following twenty-eight species have been retained, and which may be arranged in the following three groups.[1]

1. *Striati, or Semireticulati* . . . . .

1.	*Productus striatus,*	Fischer.
2.	— *giganteus,*	Martin.
		var. *hemisphæricus,* Sowerby.
3.	— *latissimus,*	J. Sowerby.
4.	— *Cora,*	D'Orbigny.
5.	— *semireticulatus,*	Martin.
6.	— *costatus,*	Sowerby.
		var. ? *muricatus,* Phillips.
7.	— *longispinus,*	Sowerby.
8.	— *sinuatus,*	De Koninck.
9.	— *humerosus,*	Sowerby.
10.	— *Margaritaceus,*	Phillips.
11.	— *undatus,*	Defrance.
12.	— *arcuarius,*	De Koninck.
13.	— *carbonarius,*	De Koninck.
14.	— *ermineus,*	De Koninck.
15.	— *proboscideus,*	De Verneuil.
16.	— *Wrightii,*	Davidson.
17.	— *tessellatus,*	De Koninck.

[1] Those groups and species most nearly related have been connected by the means of a brace. It is difficult, with words, to explain the exact contour, or shape, of such variable shells as the Producti; but

		18.	*Productus scabriculus,* Martin.
		19.	— *pustulosus,* Phillips.
		20.	— *fimbriatus,* Sowerby.
2. *Spinosi*			var. *laciniatus,* M'Coy.
		21.	— *punctatus,* Martin.
			var. *elegans,* M'Coy.
		22.	— *aculeatus,* Martin.
		23.	— *Youngianus,* Davidson.
		24.	— *Keyserlingianus,* De Koninck.
		25.	— *spinulosus,* J. Sowerby.
		26.	— *mesolobus,* Phillips.
3. *Sublævi*		27.	— *plicatilis,* Sowerby.
		28.	— *sublævis,* De Koninck.

PRODUCTUS STRIATUS, *Fischer.* Plate XXXIV, figs. 1,—5.

> PRODUCTUS COMOIDES, *Dillwyn.* An Index to the Hist. Conch. of Lister, p. 24 (not of Sow.), 1823.
>
> MYTILUS STRIATUS, *Fischer.* Oryct. du Gouv. de Moscou, p. 181, pl. xix, fig. 4, 1830.
>
> PINNA INFLATA, *Phillips.* Geol. of Yorkshire, vol. ii, pl. vi, fig. 1, 1836.
>
> PECTEN TENUISSIMUS, *Eichwald.* Bullet. Scient. de l'Acad. de St. Petersburg, vii, p. 86, 1840.
>
> LIMA WALDAICA, *V. Buch.* Kersten's Archiv fur Min. Geogn., &c., p. 60, 1840.
>
> LEPTŒNA ANOMALA, *J. de C. Sow.* Min. Conch., vol. vii, pl. 615, fig. 1, *a, c, d* (not 1 *b,* which belongs *Streptorhynchus crenistria,* var. *senilis*), 1841·
>
> PRODUCTUS LIMÆFORMIS, *V. Buch.* Abhand. der Königl. Akad. der Wissens. zu Berlin, aus dem Jahre, Theil. i, p. 22, pl. i, figs. 4, 5, and 6, 1841.
>
> — STRIATUS, *De Kon.* Desc. des Anim. Foss. du Syst. Carb. de Belgique, pl. vi, fig. 10, and pl. viii bis, fig. 4, 1843.
>
> — — *De Verneuil.* Russia and the Ural Mountains, vol. ii, p. 254, pl. xvii, fig. 1, 1845.
>
> — — *De Keyserling.* Reise in das Petschoraland, pl. iv, fig. 8, 8a, and 8b, and pl. v, fig. 1, 1846.
>
> — — *De Kon.* Mon. du genre Productus, pl. i, fig. 1, *a—d,* 1847.
>
> STROPHALOSIA STRIATA, *Morris.* Catalogue, p. 155, 1854.
>
> PRODUCTUS STRIATUS, *M'Coy.* British Pal. Fossils, p. 473, 1855.

Spec. Char. Shell thin, exceedingly variable in shape, and often irregular; usually much elongated, triangular, with acuminated beak and rounded front; or broad, transversely semicircular; hinge-line always shorter than the width of the shell, sometimes exceedingly so. Ventral valve moderately convex, rarely gibbous, without sinus; beak pointed, rarely

a glance at the figures will, I trust, remove all uncertainty. In order not to unnecessarily burden the list of references, I have many times purposely omitted to allude to those catalogues or works in which the species are named, but not illustrated.

projecting beyond the hinge-line; ears small, flattened, and from which project numerous long slender spines; surface covered with fine, thread-like, waving, radiating striæ, which increase in number at variable distances from the beak by the interpolation of other striæ. Spines few in number, and irregularly scattered over the surface. Dorsal valve concave, following the curves of the opposite one, and similarly ornamented.

Dimensions very variable; three British specimens have afforded the following measurements:

Length 39, width 20 lines.
  „   41,   „   35 „
  „   20,   „   26 „

Obs. This species is very remarkable, on account of the various forms it is capable of assuming, and has consequently not only been attributed to seven different genera, but has received as many specific denominations. The shell and its history have been minutely described by Professor de Koninck, as well as by some other authors, but it was Lister who, in 1688, first noticed the shell, ' Hist. sive Synops. Conch.,' pl. 468, fig. 27 ; although it received its first specific denomination from Fischer de Waldheim in 1830, or shortly after. The shell is sometimes so very irregular in its shape as to be hardly recognisable, and although Professor M'Coy states that he has seen some Irish specimens one foot in length by five inches in width, I have never observed any examples exceeding four inches in length. The space between the valves is also remarkably small, and in no case have I ever noticed an area in either valve. In 1840, Mr. J. de C. Sowerby confounded some specimens of the shell under description with a distorted imperfect valve of *Streptorhynchus crenistria*, var. *senilis*, and consequently erroneously described the species as possessing a large triangular hinge area. Mr. Morris has also fallen into the same mistake, for he classes Fischer's species in the sub-genus *Strophalosia ;* the reader is however referred to p. 126 of the present monograph for further information upon the subject. The spines are very numerous and closely packed upon the ears, Professor de Koninck having counted as many as twenty-four on each of the auriculate expansions in certain specimens, but that they are exceedingly rare on the surface of the shell. I have never seen any perfect interiors of either valve, but fragments have shown that the details are very similar to those of other species of the genus.

In England it occurs in the lower scar limestone of Settle, in Yorkshire; the lower carboniferous limestone, Park Hill, Longnor, in Derbyshire; the dark middle limestone of Lowick, Northumberland, as well as of the Isle of Man, &c. No specimen has been hitherto found in Scotland, and although Ardagh in Ireland is mentioned in Mr. Kelly's list, I have seen no specimen. In Belgium it occurs at Visé. In Russia, in the valleys of Stolobenka and Prikcha (Valdai) in the Petschora, near Zvenigorod, &c. In the Punjaub (India), &c., &c.

PRODUCTUS GIGANTEUS, *Martin.* Plate XXXVII, figs. 1—4; Plate XXXVIII, fig. 1; Plate XXXIX, figs. 1—5; Plate XL, figs. 1—3.

ANOMITES GIGANTEUS, *Martin.* Petrif. Derb., pl. xv, fig. 1, 1809.
— CRASSUS, *Martin.* Ib., pl. xvi, fig. 2, 1809.
PRODUCTUS GIGANTEUS, *J. Sow.* Min. Con., vol. iv, pl. 320, Jan., 1822.
— HEMISPHÆRICUS, *J. Sow.* Ib., pl. 561, 1827.
PRODUCTA GIGANTEA, *Phillips.* Geol. of York., pl. viii, fig. 5, 1836.
— AURITA, *Phillips.* Ib., pl. vii, fig. 6, 1836.
— EDELBURGENSIS, *Phillips.* Ib., pl. vii, fig. 5, 1836.
LEPTÆNA VARIABILIS, *Fischer.* Oryct. du Gouv. de Moscou, p. 144, pl. xxi, 1837.
PRODUCTUS GIGAS oder GIGANTEUS, *Von Buch.* Abhandl. der Königl. Akad. der Wissens. zu Berlin, aus dem Jahre, p. 19, 1841.
— COMOIDES, *Von Buch.* Ib., pl. i, figs. 1—3 (not Sow.).
PRODUCTA HEMISPHÆRICA, *D'Archiac* and *De Vern.* Trans. of the Geol. Soc. of London, 2d series, vol. vi, p. 397, 1842.
PRODUCTUS GIGANTEUS, *De Koninck.* Animaux Foss. des Terr. Carb. de Belgique, pl. vii, fig. 1, 1843.
— COMOIDES, *De Koninck.* Ib., pl. vii bis, fig. 1 (not Sow).
LEPTÆNA MAXIMA, *M'Coy.* A Synopsis of the Characters of the Carb. Limest. of Ireland, pl. xix, fig. 12, 1844.
PRODUCTUS STRIATUS, *E. Roberts.* Atlas Géologique des Voyages en Scandinavie et de la Commission Scientifique du Nord, pl. xx, fig. K, 1845 (non Fischer).
— GIGANTEUS, *De Verneuil.* Russia and the Ural Mount., vol. ii, pl. xvi, fig. 12, and pl. xvii, fig. 2, 1845; and *P. Edelburgensis*, ib., pl. xviii, fig. 2.
— — *De Koninck.* Mon. du genre Productus, pl. ii, fig. 1; pl. iii, fig 1; pl. iv, fig. 1; and pl. ix, fig. 8, 1847.
— — *M'Coy.* British Palæozoic Fossils, p. 463, 1855.
— — *Dav.* The Geologist, vol. ii, pl. iii, 1859; and Scottish Carb. Mon., pl. v, figs. 1—4, 1861.

Spec. Char. Shell large, varying in shape and striation according to age and specimen; somewhat transversely oval; hinge-line straight, exceeding the width of the shell, and sometimes (though not commonly) possessing a narrow rudimentary hinge area in the larger valve. Ventral valve greatly thickened, especially towards the middle, less so at the beak, and near the margin; more or less gibbous, and often much dilated at the ears, which are semi-cylindrically enrolled, and more or less defined. Beak moderately developed, incurved, and overhanging the hinge-line near its extremity only. Exterior evenly convex, or more or less deeply and irregularly furrowed, the surface being covered with a vast number of longitudinal flexuous striæ, which vary according to age and specimen, three or more usually occupying the width of a line towards the middle or margin of the valves. The striæ are also at times confluent, bifurcating, or suddenly disappearing and again reappearing and increasing in number towards the margin of the valve. They are likewise contiguous or separated by sulci or interspaces of variable width. At irregular distances the striæ give rise to a few short, cylindrical, hollow spines, which are more numerous upon the auriculate portions of the valve. Dorsal valve thin, concave, following the curves of the opposite one, and similarly sculptured, the visceral portion and ears being also sometimes concentrically wrinkled, while the entire surface is crossed by minute concentric lines of growth. In the interior of the ventral valve the divaricator muscular scars are immediately under and outside of the adductor or occlusor ones, and lower down towards the centre of the valves there are two deep subspiral depressions. In the interior of the dorsal valve the cardinal process is trilobed and V shaped, under which a narrow longitudinal ridge extends to about half the length of the valve, and on either side are situated the ramified dendritic impressions of the adductor or occlusor muscle, while outside and in front of these are the two "reniform impressions." The internal surface is covered with innumerable asperities, and the shell structure perforated by canals. Dimensions variable, four British examples measured—

Length 6 inches 2 lines, width 11 inches 6 lines.

    „   3   „ 9 „   „ 6   „   6 „

    „   4   „ 3 „   „ 5   „ 10 „

    „   6   „ 5 „   „ 7   „   8 „

Obs. Of the many species of which the genus *Productus* is composed, this is certainly the largest and most remarkable, some examples having attained eight inches in length by twelve in breadth, but it is also most variable in its shape, so much so, indeed, that one is at times seriously puzzled, and tempted to fabricate more than one so-termed species out of its variations. The shell has been several times minutely described since its first discovery, but Palæontologists are far from unanimous with reference to some of its supposed synonyms. Martin's figure of *Anomites giganteus* represents the adult typical shape of the species, with its developed auriculate expansions, while the same author's *A. crassus* illustrates an adult individual of the same, but in which the ears are broken or undeveloped, and it is quite easy to find in the same quarry and bed specimens graduating from the one into the other. In one, Plate XXXVIII, is represented, of the natural size, a very remarkable example of the typical form, with unusually expanded ears, while in Plate XXXIX (fig. 3) will be found a reduced illustration of a still more circular specimen than that upon which Martin's *P. crassus* was founded. The greater number of Palæontologists have considered *P. hemisphæricus* a synonym of *P. giganteus*, and I am inclined to believe that they are substantially correct, but it must be remembered that in this case the name has been, at any rate, applied to two conditions of the species. James Sowerby's original illustrations in Tab. 328 of the 'Mineral Conchology,' which we consider the typical ones, appear to represent what I take to be a local modification, which had better be described separately with the varietal designation "*hemisphæricus*," while the figures of *P. hemisphæricus* in Tab. 561 of the same work, published some five years later by Mr. J. de C. Sowerby, represent the true adult condition of that shape of *P. giganteus* to which Martin had applied the denomination "*crassus.*" *P. auritus* of Phillips is another undoubted synonym, while *P. Edelburgensis* of the same author represents those examples of *P. giganteus* in which the valves are more coarsely striated, with interspaces of almost equal or greater width, and of which a figure will be found in Pl. XL, fig. 2, of the present work. The striation is, however, sometimes very variable and irregular, even on the surface of a same specimen, and I have at present before me a large example in which the striæ are small, regular, and contiguous for about one inch and a half's distance from the extremity of the beak, then comes a band of about one inch and a quarter in breadth, wherein the ribs become suddenly reduced to nearly half their original number, from many having become obliterated, and thus having left interspaces of irregular width between the remaining striæ ; again, for the last two inches, the striæ become suddenly smaller, twice as numerous, irregularly twisted, unequal in their respective widths, and assuming every kind of modification until they reach the margin. In this example, and in many similar ones, we have therefore combined the striation of both the typical *P. giganteus* and its modification *Edelburgensis.* The term, *P. maxima,* was given by M'Coy to that modification of *P. giganteus* in which the valves are uniformly convex and concave, without longitudinal furrows, while the so-termed *P. personatus* was established on what is believed by some to be the internal cast of a circular example of *P. giganteus,* but of which I am not yet perfectly satisfied. We now come to another form, which has by some been considered a synonym, by others a variety, and by many a separate species, viz., *P. latissimus,* Sow., and I confess that it is not easy to satisfactorily determine which of the interpretations comes nearest the truth. It has, however, appeared to me that if *P. latissimus* is not a distinct species, it is certainly a well-marked variety, and had better, at least for the present, be separately described. *Prod. Scoticus,* Sow., and *P. pugilis,* Phillips, have been classed by Professor de Koninck among the synonyms of *P. giganteus;* but this is a mistake, for both those shells, of which I have been able to study the original type, are synonyms of *Prod. semireticulatus,* as will be hereafter shown.

In the young state *P. giganteus* varies almost as much as it does in the adult; it is at times very slightly convex, while the number of striæ are also far less numerous, these last increasing in number as the shell becomes older, by the means of interstriation or bifurcation. A very great disproportion in the respective thickness of the valves is also usually observable, that of the ventral one being in certain examples five or six times greater than that of the dorsal valve. The spines also were far more numerous in certain specimens than in others, and in some specimens a row of short ones projected from close to the cardinal edge of the ventral valve, but never as represented by Von Buch in Pl. I, fig. 1, of his monograph,[1] such a restoration is founded on a supposition which no example I have seen has ever warranted, but they occur sometimes as represented in his fig. 3.

Loc. Productus giganteus is common in the English grey lower carboniferous limestone

---

[1] 'Ueber Productus oder Leptaena Gelesen in der Akademie der Wissenschaften,' 1841.

of Derbyshire; it occurs near Richmond and Thornton Wensleydale, in Yorkshire; while Addleburgh, Bolland, Fountainsfell, Ulverston, Aldstone Moor, Hawes, Askrigg, Dentdale, Rockeley, &c., are given by Professor Phillips as the localities in the carboniferous limestone from whence his specimens were derived. It has also been found in the dark carboniferous limestone of Lowick, Northumberland, as well as in the Isle of Man. In Scotland it characterises some of the lower stages of the carboniferous system, wherein Brachiopoda have been found; thus at Braidwood Gill, in Lanarkshire, it is found for the first time at 397 fathoms below the horizon of the "Ell coal." In Stirlingshire it occurs in the Mill-Burn beds, Campsie. In the island of Arran, and in red limestone at Closeburn, in Dumfriesshire; in Edinburghshire, at Joppa; in Haddingtonshire, at Cat Craig, near Dunbar; in Peebleshire, at Carlops, &c. In Ireland, Mr. Kelly mentions that it occurs in the carboniferous limestone of Cookstown, Millecent, Tankardstown, Tornaroan, Drumreagh Etra, Castle Espie, Armagh, Little Island, &c. On the Continent the species occurs at Visé, Chokier and Temploux being, according to Professor de Koninck, the only localities in Belgium where the shell has been found. It has been also obtained from several Russian localities, such as Karova, Zerovskoi in the Oka of Mydynsk, on the Valdaï, at Peredki, in the basin of the Donetz, &c. In the Oural, at Kamensk and Bielobac in the river Isset, &c. In Silesia it occurs at Altwasser, &c. In Carinthia, in the neighbourhood of Bleiberg and Ratingen, and it has also been discovered at the top of Mount Misery, in Bear Island, along with *P. striatus* and *Punctatus*,[1] &c.

Var.? HEMISPHÆRICUS. Pl. XL, figs. 4—9.

> PRODUCTUS HEMISPHÆRICUS, *James Sowerby.* Min. Conch., tab. 328, Feb., 1822 (not tab. 561).
> — — *M'Coy.* British Palæozoic Fossils, p. 464, 1855.

Shell hemispherical or transversely oval, hinge-line usually exceeding the width of the shell. Ventral valve evenly convex, ears more or less semicylindrically enrolled, and sloping more or less gradually into the gibbous body. Surface covered with numerous small, rounded, radiating, contiguous striæ, which increase in number by bifurcation as well as by the occasional formation of new striæ between those already existing. The spines are very few on the body of the shell, but a close row of small curved ones line and curve over the cardinal edge. Dorsal valve very concave, following the curves of the opposite one, and similarly ornamented. A few obscure concentric wrinkles may be observed, on the ears of certain specimens. Interior similar to that of *P. giganteus* proper, dimensions variable. Two specimens have measured—

Length 17, width 26 lines.
„ 15, „ 17 „

[1] Further details with reference to the foreign localities wherein this widely spread species has been discovered, will be found in Professor de Koninck's very valuable 'Monographe du genre Productus,' 1847.

Obs. Much difference of opinion has been expressed relative to the shell under description, and it was not until after much examination and hesitation that I determined to provisionally retain the shells first figured by James Sowerby, in tab. 328 of the ‘Min. Conch.,’ as a variety of *P. giganteus*—a view entertained by Prof. de Koninck, when publishing his monograph in 1847. Prof. M‘Coy, describes them as belonging to another and distinct species ; so that the subject may still remain an open question, although I am myself inclined to consider *P. hemisphæricus* as nothing more than a local variety of *Productus giganteus?* Mr. E. Wood, who has had occasion to collect *in situ* some hundreds of specimens of this and the typical shape, assures me that the one never occurs in the same bed or zone along with the other; that the bed in which the *P. hemisphæricus* occurs runs for miles along Warfdale in Yorkshire, and that the layer is covered and filled with closely packed specimens of this one fossil, which is also always exactly the same, and much under the true *P. giganteus* bed. My reasons for supposing it a variety of *P. giganteus* are based on the following consideration, viz., that I have several times seen large examples of *P. giganteus* evenly convex, and closely resembling some of the smaller specimens of *P. hemisphæricus* above described,[1] and it is well known that the longitudinal furrows which cover the valves of many examples of Martin’s shell are not always present; that the striation of *P. hemisphæricus* agrees likewise with that of many specimens of *P. giganteus*, and that the small curved cardinal spines so constantly present in the Warfdale specimens are also observable in many undoubted large examples of Martin’s shell; and, lastly, that the interior is smaller in both. I have also had the advantage of being able to study James Sowerby’s original specimens of *P. hemisphæricus*, and found them to agree with those from the gray carboniferous limestone of Warfdale, in Yorkshire.

PRODUCTUS LATISSIMUS, *J. Sowerby.* Plate XXXV, figs. 1—4.

PRODUCTUS LATISSIMUS, *J. Sow.* Min. Conch., pl. 330, Feb., 1822.
— — *Phillips.* Geol. of Yorks., vol. ii, pl. viii, fig. 1, 1836.
— — *De Koninck.* Monographie du genre Productus, pl. ii, fig. 2, and pl. iii, fig. 2, 1847.
— — *Dav.* Scottish Carb. Brach., pl. ii, figs. 8, 9, 1861.

Spec. Char. Shell thin, transversely elliptical or spindle-shaped, with a long, straight hinge-line, the breadth of the shell being more than twice the length. Ventral valve very much vaulted and convex, with a slight mesial longitudinal depression, the gibbosity forming in profile more than a semicircle, while the passage from the body of the valve into the lateral expansions is usually so gradual as to become insensible. The small flexuous striæ which cover the surface augment by the means of numerous interca-

[1] In his work on ‘British Palæozoic Fossils,’ Professor M‘Coy has considered *P. aurita* to be a synonym of *P. hemisphæricus ;* and I feel convinced, that any one who examines the Gilbertson specimens of the first (now in the British Museum) will class it among the synonyms of *P. giganteus.*

lations, and from which at intervals project short slender spines. Dorsal valve very concave, following the curves of the opposite valve, and similarly sculptured. In the interior of the ventral valve, the muscular impressions are located in the rostral portion of shell close to the extremity of the beak.

Dimensions variable: a large example measured, length 2 inches, width $4\frac{1}{2}$ inches; a small one, 2 lines by 4.

Obs. At p. 463 of his 'British Palæozoic Fossils,' Prof. M'Coy mentions that "the magnificent suite of specimens of *P. giganteus* now in the University Collection enables him to state positively that the distinctions relied on by M. de Verneuil in the 'Geology of Russia,' and M. de Koninck in his 'Monograph of *Productus*,' for separating *P. latissimus* from *P. giganteus*, do not really exist; several specimens in the collection showing in one individual the gradual conoidal passage of the ear into the body of the shell (as in *P. latissimus*), on the one side, and a narrow sub-cylindrical ear, projecting abruptly from the side of the defined gibbous body of the shell (as in the *P. giganteus*), on the other: that the other characters mentioned as distinctive by them and the older authors, such as thinness of the shell, and few or no longitudinal folds, greater depression of the ears, &c., are characters of the young shell, and particularly the entering valve: further, both forms (contrary to what those authors suppose) occur together in abundance in the same bed in Northumberland." This is not, however, the opinion of the larger number of palæontologists, who cannot recognise in the peculiar spindle shape of *P. latissimus* a simple variation in form of *P. giganteus;* and I must confess that the examinations I have made of many examples of both would not enable me to arrive at so positive an opinion as that of the distinguished author above quoted, although I would not dispute the possibility of its being correct. At all ages and dimensions, from that of four lines in breadth to four inches and a half, the shell appears to retain its very transverse appearance; and its interior dispositions, although similar in character to those of *giganteus*, appear to me somewhat different in their minor details, and, notwithstanding what has been asserted by M'Coy, the shell seems to me much thinner than that of Martin's species.

In England, *P. latissimus* is common in the limestone and shales of various localities. Sowerby mentions that it occurs in a cherty limestone at Tyddmaur Farm, in Anglesea; it is not uncommon at Settle in Yorkshire, Fountains Fell, Kirby Lonsdale, in Northumberland, &c. In Scotland, it is one of the most characteristic species in some of the lower stages of the Carboniferous system: it is found in Lanarkshire at two different levels; thus at Belston Burn it occurs at 265 fathoms below "Ell Coal," and 391 at Braidwood Gill, also at Brockley, near Lesmahago; in Renfrewshire, at Arden Quarry, Thornliebank; in Ayrshire, at Roughwood and West Broadstone Beith, &c.; in Stirlingshire, in the Craigenglen (Campsie) beds. In Ireland, it is stated by Mr. Kelly to occur in the calciferous slate and carboniferous limestone of Lisnapaste, Killymeal, and Raheendoran. On the Continent, it has been found at Visé in Belgium, and in several Russian localities.

PRODUCTUS HUMEROSUS, *Sow.* Pl. XXXVI, figs. 1, 2.

> PRODUCTUS HUMEROSUS, *Sow.* Min. Conch., t. 322, January, 1822.
> — HORRIDUS, *De Koninck.* Mon. du genre Productus, 1847 (not of Sow. ?)
> — ACULEATUS (*Schlotheim*), *Von Buch.* Uber Productus oder Leptaena Akademie der Wissenschaften, p. 35, 1841.

Of this Productus internal casts alone have been hitherto found, so that no description of the exterior can be given. The length generally slightly exceeds the breadth, while the hinge-line is usually somewhat shorter than the width of the shell. The ventral valve was gibbous and of great thickness, with perhaps a slight longitudinal depression or sinus? while the dorsal one was concave, light, and thin. The interior details are sharply sculptured upon the internal casts. The ventral valve in the cast is very remarkable on account of two large projecting conical protuberances situated about the middle of the shell, with their extremities directed towards the beak, and which in the enormously thickened shell formed very deep cavities, which no doubt afforded accommodation to the spirally coiled? oral arms. The muscular impressions form also in the cast a remarkable protuberance, and in the shell itself occupy the rostral or incurved portion under the cavity of the beak. The adductor or occlusor produces two radiating dendritic impressions, longitudinally divided by a narrow ridge, and on either side there is (in the shell) a deep, strongly grooved or striated sub-quadrate impression, which is due to the cardinal or divaricator muscle, the remainder of the inner surface of the valve being covered with rugosities. In the interior of the dorsal valve, the quadruple impressions of the adductor, as well as the reniform impression, do not differ from those of the generality of Producta. The absence of the shell itself renders specific determination uncertain, and the name is consequently only provisionally retained. In their monographs, Baron von Buch and Prof. L. de Koninck have placed *P. humerosus* among the synonyms of the Permian *P. horridus;* and although I would not deny the possibility of this view being correct, still as none of the internal casts of the Permian shell, that have come under my observation, have exhibited those enormously developed and peculiar conical protuberances present in the Carboniferous casts, and as the muscular impressions in the ventral valve differ also somewhat in their details, I have preferred (at least provisionally) retaining the denomination under which the Carboniferous Productus is generally known.

The internal casts of *P. humerosus* are from the Magnesian Limestone of the Carboniferous series of Breedon in Leicestershire, and a fine series of specimens may be seen in the Collection of the Geological Society and British Museum.

PRODUCTUS CORA, *D'Orbigny.*   Pl. XXXVI, fig. 4, and Pl. XLII, fig. 9.

> PRODUCTUS CORA, *D'Orb.*   Paléontologie du Voyage dans l'Amérique Méridionale, pl. v, fig. 8, 9, and 10, 1842.
> — COMOIDES, *De Koninck.* Descrip. des Anim. Fossiles du Terrain Carb. de la Belgique, pl. xi, fig. 2 *ᵃ, ᵇ* and fig. 5 *ᵃ, ᵇ* (not of Sow.), 1843.
> — CORRUGATA, *M'Coy.* Synopsis of the Carb. Limest. Fossils of Ireland, pl. xxvi, fig. 13, 1844.
> — NEFFEDIEVI, *De Verneuil.* Russia and the Oural Mountains, vol. ii, pl. xviii, fig. 11, 1845.
> — CORA. *De Koninck.* Monographie du genre Productus, pl. iv, fig. 4, and pl. v, fig. 2, 1847.
> — CORRUGATA, *M'Coy.* British Pal. Foss., p. 459, 1855.
> — PILEIFORMIS, *M'Chesney.* Desc. of New Species of Fossils from the Palæozoic Rocks of the Western States of America, p. 40, 1859.
> — CORA, *Dav.* Mon. of Scottish Carb. Brach., pl. iv, fig. 13, 1860.

Spec. Char. Shell thin, longitudinally oval, or semi-cylindrical, usually longer than wide; hinge-line about as long as the breadth of the shell; ventral valve gibbous, evenly convex, or slightly flattened along the middle: beak wide and incurved; ears small, and generally crossed by four or five deep undulating folds or large wrinkles, which extend to some distance over the lateral portions of the valves. Surface covered with numerous longitudinal, straight or slightly flexuous, narrow, thread-like, rounded striæ, with sulci or interspaces of rather less width, while smaller striæ are often intercallated between the larger ones, the ribs being also regularly crossed by small concentric lines. Spines few in number, sparingly scattered over the surface, but more numerous on the ears. Dorsal valve concave, following the curves of the opposite one, and similarly ornamented.

Dimensions variable, some examples have attained or exceeded two and a half inches in length, by three and a half in width.

Obs. This *Productus* is well characterised and distinguishable from its congeners both by shape and sculpture; but the four or five large concentric wrinkles which usually cover the ears and lateral portions of the valves are much less developed in certain examples than in others. In the dorsal valve, the visceral portion of some specimens is also entirely crossed by a variable number of concentric folds or wrinkles (as represented in Pl. XLII, fig. 9), but which do not exist in the generality of specimens. The shell varies likewise very much in shape, being at times almost oval, while in other examples the margin becomes considerably expanded. Prof. de Koninck, who has had the advantage of being able to examine D'Orbigny's original specimens, positively asserts that they cannot be distinguished from the European examples we are now describing, and that consequently Prof. M'Coy's *P. corrugata* will require to be added to the synonyms; and I believe that my Belgian friend is likewise correct when he refers the Russian *P. Neffedievi* to D'Orbigny's species.

In England, *P. cora* has been found in the carboniferous limestone of Kendal, Lowick, Settle, Poolwash (Isle of Man). In Scotland, in the Mill Burn and Balgrochan beds, as well as Campsie Main limestone, Stirlingshire; at Arden Quarry, near Thornliebank, in Renfrewshire; West Broadstone, near Beith, Ayrshire. In Ireland, at Larganmore, Millecent, Milverton, Skerries, Little Island, &c. In Belgium, at Visé, Tournay, &c. In Russia, at Cosatchi-Datchi, near Sterlitamak, &c. In America, it was found by D'Orbigny in a blue compact limestone above Patapatani, in one of the islands of the lake Titicaca, and in a gray limestone at Yarbichambi. It has been also obtained from the Mountain Limestone of Chester, Illinois, and by Dr. Fleming and Mr. Purdon in the Punjaub (India), &c. &c.

PRODUCTUS SEMIRETICULATUS.   Pl. XLIII, figs. 1—11, and Pl. XLIV, figs. 1—4.

> ANOMITES SEMIRETICULATUS, *Martin.* Petrif. Derb., pl. xxxii, figs. 1, 2, and pl. xxxiii. fig. 4, 1809.
> — PRODUCTUS, *Martin.* Ib., pl. xxii, figs. 1—3.
> — SCOTICUS, *Sow.* Min. Con., pl. lxix, fig. 3, Oct., 1814.
> — MARTINI, *Sow.* Ib., pl. 317, figs. 2—4, Dec., 1821.
> — ANTIQUATUS, *Sow.* Ib., figs. 1, 5, and 6.
> — CONCINNUS, *Sow.* Ib., pl. 318, fig. 1, 1821.
> — MARTINI and ANTIQUATUS, *Phil.* Geol. York., pl. vii, fig. 2, 1836.
> — PUGILIS, *Phil.* Ib., pl. viii, fig. 6.
> LEPTŒNA ANTIQUATA, *Fischer.* Oryct. du Gouv. de Moscou, pl. xxvi, figs. 4, 5, 1837.
> — TUBULIFERA, *Fischer.* Ib., pl. xxvi, fig. 1 (not Deshayes).
> PRODUCTUS MARTINI, *De Kon.* Desc. des Animaux Foss. du Terrain Carb. de la Belg.. pl. vii, fig. 2, 1843.
> — INCA, *D'Orb.* Paléont. du Voyage dans l'Amérique Mér., pl. iv, figs. 1—3, 1844 (according to M. de Koninck).
> — PERUVIANUS, *D'Orb.* Ib., fig. 4.
> — FLEXISTRIA, *M'Coy.* Synopsis of the Carb. Limest. Fossils of Ireland, pl. xvii, fig. 1, 1844?
> — SEMIRETICULATUS, *De Kon.* Mon. du genre Productus, pl. viii, fig. 1; pl. ix, fig. 1; pl. x, fig. 1, 1847.
> — — and MARTINI, *M'Coy.* Brit. Palæozoic Fossils, pp. 467 and 471, 1855.
> — — *Dav.* Mon. of the Scottish Carb. Brach., pl. iv, figs. 1—5, 7, and 12, 1860.
> — — *Salter.* Quarterly Journal of the Geol. Soc., pl. vi, fig. 1, 1861.

Spec. Char. Very variable in shape, transversely oval, sub-cylindrical or elongated; hinge-line as long or somewhat shorter than the width of the shell; ventral valve gibbous and variably vaulted, with a shallow longitudinal median sinus or depression; auriculate expansions moderately developed; beak wide, incurved, usually covered with irregular, concentric, undulating wrinkles, larger and deeper upon the ears, while the entire surface of the shell is ornamented by many radiating, longitudinal, rounded striæ, which become more numerous towards the margin from bifurcation and interstriation, and from which project, at variable intervals, tubular spines of sometimes considerable length. Dorsal valve moderately concave, following the curves of the opposite one, and similarly sculptured. Dimensions variable, some examples having attained three inches in length, by four in breadth.

Obs. This species has varied very much in its general shape, and I am disposed to coincide with Prof. de Koninck, while considering *P. semireticulatus, P. antiquatus, P. Martini, P. concinnus,* and, I will add, *P. Scoticus,* as simple variations in shape of a single species, and for which the term *semireticulatus* is here retained. I am also quite ready to admit that, if we examine only typical examples of *P. semireticulatus* and *P. Martini,* a certain degree of difference is perceptible, on account of the profile of the first being simply semicircularly curved, while in the second the valves are geniculated; but these two extreme shapes are intimately connected by insensible gradation, and indeed very often to such an extent that it would be impossible to say to which in particular certain specimens should be referred. In Mr. Salter's opinion, *P. costatus* should also be considered a variety of the species under description;[1] and although that learned palæontologist may be quite correct in his supposition, I do not feel myself at present sufficiently justified in uniting the two under a single denomination. *P. sulcatus* has, I fear, without sufficient caution been located, by myself and others, among the synonyms of *P. semireticulatus,* but I am now inclined to consider it a variety of *P. costatus,* for reasons which will be given hereafter. *P. Scoticus,* of which I have seen the original type, appears undistinguishable from *P. semireticulatus;* and several other so-termed species must likewise be placed under the same denomination.

The width of the striæ, as well as the interspaces between them, vary also according to the specimen, two or more usually occupying the breadth of a line. The larger number are simple, but others bifurcate here and there, and especially so on the lateral portions of the shell; two or more in rarer cases will also sometimes unite towards the margin so as to form a single rib, while others are due to intercalation. Several ribs at times cluster together so as to produce an elevation, and thus give to the frontal portion of the shell a somewhat grooved or undulated appearance; and this was particularly the case with the crushed specimen from which Phillips founded his *P. pugilis,* the original example being still preserved in the author's collection. In the species under description, the spines are likewise often more numerous and longer in certain specimens than in others, but always most so upon the auriculate portions of the valve, where they do not appear to have been generally disposed in a double row, or rather to have protruded from an elevated ridge, as in *P. costatus,* and its variety, *P. sulcatus.*

[1] 'Quarterly Journal of the Geol. Soc.,' vol. xvi, p. 441.

The interior need hardly be described, as the details which we have carefully drawn do not differ materially from those already noticed in other Producta ; so that all we require to notice is that, in the ventral valve, the adductor or occlusor muscular impressions are situated almost on a level, or are longitudinally parallel, with the divaricator scars, and consequently much lower in the ventral valve than in *P. giganteus* and some other species. In the dorsal valve, the occlusor impressions are often beautifully sculptured, and the cardinal process tri-lobed. The ventral valve is more thickened than the dorsal one, and possesses sometimes, though rarely, a well-defined area and fissure covered by a pseudo-deltidium, this character being beautifully displayed upon a specimen in the British Museum, and of which an illustration will be found in my plate. Since several palæontologists appear, however, desirous of maintaining the term *Martini* as at least a varietal designation, and although I cannot draw any line of distinction between the two, it may be as well to mention that figs. 6 to 11 of my Pl. XLIII represent the typical shapes of *P. Martini*, while the other figures would represent *P. semireticulatus*. In its typical condition, *P. Martini* may be distinguished by the great length and sometimes irregularity of its anterior prolongation, the visceral portion alone being regularly arched, the shell becoming afterwards suddenly bent downwards in an almost straight line, giving to some specimens a peculiarly elongated and geniculated appearance, as seen in figs. 6 and 7. The sudden bend of the ventral valve makes it sometimes, in the fossil state, liable to fracture where the sudden bend in the valve takes place, as seen in figs. 9 and 10. The lateral portions of the valve are likewise much dilated, while numerous spines, clustered together, project from the auriculate portions of the valve. This is also the variety to which Martin, in 1809, applied the specific denomination of *Anomites productus ;* and Sowerby's *P. concinnus* is evidently nothing more than a smaller form or synonym of the same. *Productus semireticulatus* is one of the most abundant and far-spread species of the genus, and is found in almost every locality where Carboniferous Brachiopoda occur.

In England it is found in the Carboniferous Limestone and shales of Bolland, Settle, Kirkby Lonsdale, Coverdale, as well as in various Derbyshire localities ; at Poolwash, in the Isle of Man ; Clifton, near Bristol, &c. In Scotland, it is plentiful in Lanarkshire, Renfrewshire, Ayrshire, Dumbartonshire, Stirlingshire, Fifeshire, Edinburghshire, Dumfries-shire, Peebleshire, Linlithgowshire, and Berwickshire. In Ireland, at Millecent, Lisnapaste, Tankardstown, Tornaroan, Cookstown, &c. In Belgium, it occurs in the lower stages of the Carboniferous Limestone of Visé, De Chokier and De Lives, near Namur, and in various other localities, as well as in the Middle Limestone of Tournay. M. de Koninck states he has found it likewise in the following German localities :—Crumford, at Ratingen, Altwasser, &c. In Russia, M. de Verneuil and Count Keyserling obtained it in several localities such as the neighbourhood of Moscou, on the Dwina, the Oka, the Pinega, and Petschora, as well as in the Timans Mountains, near the Glacial Sea, also in the Oural Mountains, at Sterlitamack, &c. M. de Tchihatchef found it also in the Altai Mountains. It has also been found in Spitzbergen and Australia ; in the Punjaub, and in various

American localities, such as at the Island of Quehaja, near the Lake Titicaca, at the foot of the Bolivian Andes; at Quincy, Illinois; Keokuk, Iowa, &c.

PRODUCTUS COSTATUS, *J. de C. Sow.*   Pl. XXXII, figs. 2—9.

> PRODUCTA COSTATA, *J. de C. Sow.*   Min. Conch., pl. 560, fig. 1, May, 1827.
> —   SULCATUS, *Sow.*   Ib., pl. 319, Jan., 1822.
> —   COSTATA (et SULCATA, *Sow.*), *Phillips.*   Geol. of York., vol. ii, pl. vii, fig. 2, 1836.
> —   COSTELLATUS, *M'Coy.*   Synopsis, pl. xx, fig. 15, 1844.
> —   COSTATUS, *De Verneuil.*   Russia and the Ural Mountains, vol. ii, pl. xv, fig. 13 [a, b] 1845.
> —   —   *De Koninck.*   Mon. du genre Productus, pl. viii, fig. 3, and pl. x, fig. 3, 1847.
> —   —   *Dav.*   Mon. of Scottish Carb. Brach., pl. ii, figs. 22—24, 1860.

Spec. Char.   Shell very variable in shape, transversely semi-cylindrical, wider than long; hinge-line about as long as the width of the shell.   Ventral valve gibbous, very much vaulted; abruptly arched, or obscurely geniculated; beak incurved, but not overlying the hinge-line except at its attenuated extremity, a median longitudinal sinus or depression dividing the valve to a greater or less extent into two lobes; ears more or less developed, sloping abruptly from the visceral portion, with a strong, rugged, semicircular ridge on either side, obliquely placed to the hinge-line, and from which project several long, cylindrical, hollow spines, similar to those situated close to the cardinal edge.   Surface covered with a variable number of strong, longitudinal rounded ribs of unequal width, and which become more numerous towards the margin from occasional bifurcation or intercalation, while the whole visceral portion is crossed by numerous regular concentric wrinkles, producing reticulate tuberculations.   The spines are long, but variable in number, projecting here and there from the ribs.   Dorsal valve somewhat geniculated, following the curves of the opposite valve, a slight median elevation corresponding to the sinus of the ventral valve; the visceral portion is usually somewhat flattened, while the anterior portion of the valve becomes more or less abruptly bent upwards, the sculpture being similar to that of the opposite valve.

Dimensions variable: a typical specimen measured, length 18, width 23 lines.

Obs.   So variable do the shells composing this species appear to be, that it is very puzzling to know how to dispose of certain shapes which, although individually somewhat different from the typical form, appear, nevertheless, linked to them by insensible gradation.   Sowerby states his species to be "transversely oblong, with an angular depression in the middle; costæ few, broad, decussated at their upper part, compressed upon the deflected front, each side furnished with two or three spines and a small tube." This description will suit typical specimens, such as the one the author had at the time before him, but it would scarcely be equally applicable to all the shells which compose the species. Thus, the median depression is extremely shallow, and even absent in certain examples; the characteristic rugged, semicircular ridge, with its strong, elongated spines, present on typical specimens, is at times very little developed, and even absolutely wanting in certain examples, while the number of ribs will vary from eighteen to fifty, and are more often rounded than flattened, the spines being likewise more numerous on some examples than on others. When publishing my monograph of Scottish Carboniferous Brachiopoda I erroneously placed Sowerby's *P. sulcatus* among the synonyms of *P. semi-reticulatus*, but a more attentive examination of the subject has convinced me that Phillips was correct when, in 1836, he placed *P. sulcatus* among the synonyms of the species under description.[1] M'Coy's *P. costellatus* is evidently a synonym of *P. costatus*, but it is more difficult to determine whether *P. muricatus* and another closely allied form from Corrieburn, in Stirlingshire, should be considered as varieties of Sowerby's *P. costatus*. Having obtained the loan of the original example of *P. muricatus* preserved in the museum at York (fig. 11 of my plate), I found that it was imperfect, and much more circular than the typical and usual shapes of *P. costatus*; but it presented, however, the same median depression, and although the tubular spines projecting from the convex surface of the ribs were more numerous, still intermediate forms seemed to unite Phillips's species to *P. costatus* proper. I am therefore inclined to believe that *P. muricatus* is only a variety of the shell under description, or a race peculiar to certain localities; but as I may be mistaken, and in order to allow the subject to remain an open question, I will describe Phillips's shell under a separate or varietal denomination.

Var.? P. MURICATUS, *Phillips.* Pl. XXXII, figs. 10—14.

PRODUCTA MURICATA, *Phillips.* Geol. of York., pl. viii, fig. 3.

Shell circular, about as long as wide; hinge-line somewhat shorter than the greatest width of the shell; ventral valve regularly arched, so as to present in profile nearly a half circle, uniformly convex, or slightly longitudinally depressed along the middle; beak large, wide, rounded and incurved, but not overlaying the hinge-line except at its attenuated extremity; ears very small; surface covered with numerous rounded ribs, of unequal width, augmenting in number here and there by bifurcation or intercalation, while two or three will sometimes unite into one; long, slender spines project at intervals from the ribs, and form also a row close to the cardinal edge. Dorsal valve regularly concave and similarly sculptured. Proportions variable. Two specimens have measured—

Length 11, width 10, greatest depth 5 lines.
„ $10\frac{1}{2}$ „ 11   „   „ 5 „

[1] As already mentioned under *P. semireticulatus*, Mr. Salter is inclined to consider the last-named shell and *P. costatus* as varieties of a single species.

To the typical forms of this variety we have united certain specimens from Corrieburn, of which the figs. 13 and 14 of our plate are representations; but it is desirable, at the same time, to mention that the Corrieburn specimens are remarkable on account of the narrowness or smallness of their ribs, none much exceeding half the width of those of Phillips's type; and as the general form of the shell in both, as well as the arrangement of the spines, is similar, I have not ventured to apply to it a separate specific denomination, although it may possibly be distinct. One of the principal differences observable in *P. muricatus* lies in the ventral valve being regularly convex, and the dorsal one regularly concave, while in all well-shaped examples of *P. costatus* the valves are generally somewhat obscurely geniculated. The interior of *P. muricatus* is still unknown, but the dorsal valve of *P. costatus* has been found several times, and does not differ in its details from what we observe in other *Producta*.

P. costatus has been obtained from the Carboniferous limestone of several English localities, such as Settle and Richmond, in Yorkshire; at Bolland; in the dark Carboniferous limestone of Lowick, Northumberland, &c. In Scotland it has been found at Hillhead, and Brockley, in Lanarkshire; Barrhead, in Renfrewshire; Roughwood and West Broadstone, &c. In Ireland, in the upper limestone of Old Leighlin.

Prof. de Koninck states in his 'Monographie' that he has not discovered the species in Belgium, but I possess a specimen from Visé identical in shape and character with many of our British forms. In Russia it occurs in the Carboniferous limestone of the neighbourhood of Sloboda, Government of Toula, also of Botcharova, on the Volga, &c. In the limestone of St. Louis, in the Missouri, America, very abundant and unusually large; in the Punjaub, &c.

The variety *P. muricatus* is stated by Phillips to occur at Harelaw and Kirby Lonsdale. In Scotland it has been found at Cessnock, and Gateside Beith, in Ayrshire, at Corrieburn, in Stirlingshire, and Castlecary, in Dumbartonshire.

PRODUCTUS LONGISPINUS, *Sowerby.* Pl. XXXV, figs. 5—17.

> ANOMIÆ ECHINATÆ (pars),*Ure.* History of Rutherglen, p. 314, pl. xv, figs. 3, 4, 1793.
> PRODUCTUS LONGISPINUS, *Sow.* Min. Conch., vol. i, p. 154, pl. lxviii, fig. 1, October, 1814.
> — FLEMINGII, *Sow.* Ib., fig. 2.
> — SPINOSUS, *Sow.* Ib., pl. lxix, fig. 2, 1814.
> — LOBATUS, *Sow.* Ib., pl. 318, figs, 2—6, 1821.
> — ELEGANS, *Davreux.* Const. Géol. de la Prov. de Liege, p. 272, 1833 (according to De Koninck).
> — SETOSA, *Phillips.* Geol. of Yorkshire, vol. ii, pl. viii, figs. 9 and 17, 1836.
> — LOBATUS, *V. Buch.* Verhandl. der Königl. Akad. der Wissens. zu Berlin, aus dem jahre, Theil. i, p. 32, pl. ii, fig. 17, 1841.
> — CAPACII, *D'Orb.* Paléont. du Voyage dans l'Amérique Méridionale, pl. iii, figs. 24—26, 1843.

PRODUCTUS LONGISPINUS, *De Koninck.* Descript. des Animaux foss. du terr. Carb. de Belgique, p. 187, pl. xii, fig. 11*ᵃ,ᵇ*, and pl. xii*ᵇⁱˢ*, fig. 2, 1843. *P. Flemingii,* Mon. du genre Productus, pl. x, fig. 2, 1847.

— LOBATUS, *De Verneuil.* Russia and the Ural Mountains, vol. ii, pl. xvi, fig. 3; pl. xviii, fig. 8, 1845.

— TUBARIUS, *De Keyserling.* Reise in das Petschora Land, p. 208, pl. iv, fig. 6, 1846.

— LONGISPINUS, *Dav.* Introduction to British Fossil Brach., pl. ix, fig. 221, 1853.

— WABASHENSIS, *Norwood* and *Pratten.* Notice of Producti in the Western States and Territories, Journal of the Academy of Nat. Sciences of Philadelphia, pl. i, fig. 6, 1854.

— SPLENDENS, *Norwood* and *Pratten.* Ib., pl. i, fig. 5.

— FLEMINGII, *M'Coy.* British Palæozoic Fossils, p. 461, 1855.

— LONGISPINUS, *Dav.* Mon. Scottish Carb. Brach., pl. ii, figs. 10—19, 1860.

Spec. Char. Shell very variable in shape, rather small, subcylindrical, usually slightly transverse, rarely longer than wide: hinge-line as long, or a little shorter than the greatest width of the shell; frontal margin rounded or more or less indented. Ventral valve gibbous, evenly convex, or more or less divided into two lobes by a longitudinal sinus of variable depth, commencing at a short distance from the extremity of the beak; auriculate expansions small; beak incurved, but rarely protruding much beyond the hinge-line. Dorsal valve moderately and uniformly concave, following the curves of the opposite valve, or with a small mesial fold slightly developed towards the front. Surface of each valve covered with numerous small, longitudinal, rounded striæ, tolerably regular in their course and respective widths, but augmenting in number by occasional bifurcation and intercalation. Visceral portions crossed by small, concentric, undulating wrinkles, more developed upon the auriculate expansions; long tubular spines project from some of the ribs at irregular intervals, more numerous on the ears. Dimensions variable; two British examples measured—

Length 9, width 10 lines.
„ 8, „ 7 „

Obs. At least nine or ten so-termed species have been fabricated out of variations in shape of this common, far-spread, very variable species; but as the study of the types themselves and of a multitude of specimens has clearly shown that every variety or variation is intimately connected by intermediate gradation and inseparable links, we should not be justified in retaining, even as varietal denominations, any of the names enumerated among the synonyms. In Great Britain the shell does not appear to have ever attained much larger proportions than those given above; and as the term *longispinus* is the best known and first recorded in the 'Mineral Conchology,' it is retained from the species. *P. Flemingii* was badly drawn from a very imperfect specimen, nor does Sowerby's figure even do justice to the original specimen, which, along with *P. longispinus* and *spinosus,* may still be seen in Dr. Fleming's collection. *P. spinosus* is another condition or variation of the shell under description, while *P. lobatus* represents those specimens in which the median sulcus is very much developed, while every connecting link can be found in the same bed and quarry between the evenly convex specimens and those with a deepened median sinus. The striation is sometimes very fine in some examples, coarse in others, so that on two shells of the same dimensions fifty ribs will be counted on the ventral valve of the one, thirty on the other; the hinge line is also in many examples shorter than the width of the shell, longer in others. The long, slender, tubular spines which are scattered over its surfaces vary likewise considerably in number in different individuals, and Sowerby was almost correct while mentioning that "two principal ones are nearly straight and cylindrical, attached to the convex side of the convex valve, and extending in line parallel to the hinge." These two spines are rather constant in the majority of specimens I have examined, and are often seen as represented in Sowerby's figure (5 of my plate), but not always parallel to the hinge; the other spines are more irregularly scattered over the surface, while some are two or even three times as long as the length of the entire valve. Many of the most marked modifications in shape assumed by this species have been carefully represented in my plate, and these will suffice to show how large a range of variation must necessarily be given so as not to violate the laws of nature. Beautifully perfect interiors of both valves are not rare in certain localities. On the concave surface of the ventral one two elongated contiguous dendritic adductor or occlusor impressions project at times considerably above the level of the valve, and immediately under, but outside, of these may be seen the two large longitudinally striated subquadrate impressions, attributable to the divaricator muscle, while a glance at my figures of this and the corresponding valve of *P. semireticulatus* will show the difference in position occupied by these muscles in the two species. The occlusors in the last-named shell are almost upon a level and longitudinally parallel with the divaricator, while in *P. longispinus* the divaricators commence only at or close to the base of the occlusors. A difference in the arrangement of these muscles occurs likewise in *P. punctatus*, and denotes that the three species might be distinguished alone by the details connected with their interior arrangements. In the interior of the dorsal valve the cardinal process is proportionally large and trilobed, under which a median longitudinal ridge extends to a little more than half the length of the valve, and becomes much elevated and thickened towards its extremity; on either side may be seen a pair of dendritic scars formed by the adductor or occlusor muscle; the reniform impressions are also well defined and often much raised, and the surface of the valve is covered near its margin with numerous spinulose asperities; minute canals traversing the valves are also clearly visible in the shape of punctures, especially upon specimens that have been slightly weathered.

In the 'Geology of Yorkshire,' Prof. Phillips figures two variations in shape of the shell under description by the name of *setosa*; his fig. 9 (15 of my plate) is a large, but not unusual, shape of *P. longispinus*, while his fig. 17 (16 of my plate) is a rarer condition of the same shell, but of which many examples have been found in the same quarries along with every intermediate passage form. Possessing examples of Norwood and Pratten's *P. Wabashensis* and *P. splendens*, I could not recognise the smallest difference between them and the generality of our Scotch examples of Sowerby's shell, and I have but little doubt that the list of synonyms might be further increased had we the means of comparing certain so-termed species with our British forms. I consider that Prof. M'Coy is, however, mistaken while locating *P. pugilis*, Phillips, and his own *P. costellatus* among the synonyms of the shell under description. In England *P. longispinus* is common in the Carboniferous limestone and shales of several Derbyshire and Yorkshire localities, and at Lowick, Northumberland, Poolwash, Isle of Man, &c.

In Scotland it is exceedingly abundant in various counties, and occurs in several stages. At Braidwood, in Lanarkshire, it occurs at 337 fathoms lower than the "Ell Coal," and a little lower again at Hallcraig, Raes Gill, Langshaw, Hillhead, Kilcadzow, &c., in the Parish of Carluke; also in the same county at Kersgill and Brockley, near Lesmahago, Auchentibber, and Calderside, High Blantyre, Capel Rig, East Kilbride,[1] &c. It occurs also in Renfrewshire, Dumbartonshire, Ayrshire, Stirlingshire, Edinburghshire, Haddingtonshire, Fifeshire, &c.

In Ireland it occurs in the Calciferous slate and Carboniferous limestone of many localities, such as Lisnapaste, Howth, Ballyduff, Mohill, Hook, Bundoran, Millicent, Culkagh, &c.

In Belgium, at Visé, Tournay, &c., and, indeed, is found in almost any Carboniferous district where Brachiopoda have been found. In Russia, at Karova, Government of Kalouga, &c. Yarbichambi, on the edge of the Lake Titicaca, on the Bolevian table-land of the Andes, and Sangamon, County Illinois, in America, &c. In the Carboniferous limestone of the Punjaub in India. Carrocreek in Tasmania, &c.

PRODUCTUS SINUATUS, *De Koninck.* Pl. XXXIII, figs. 8—11.

LEPTŒNA SINUATA, *De Koninck.* Description des Animaux fossiles du Terrain Carb. de
la Belgique. Supplément, p. 654, pl. lvi, fig. 2, 1851.
— — *M'Coy.* British Palæozoic Fossils, p. 453, 1855.
PRODUCTUS SINUATUS, *Dav.* The Geologist, vol. iv, p. 48, Feb., 1861.

Spec. Char. Shell transverse, longer than wide; obscurely subtetragonal; hinge-line about as long as the greatest width of the shell; cardinal angles rounded. Ventral valve geniculated from the anterior half, being bent at right angles to the posterior or visceral portion; ears large, reflexed, curving away from the hinge-line; visceral disc somewhat flattened, but divided by a longitudinal sinus commencing at a short distance

[1] In his 'History of Rutherglen and East Kilbride,' published in 1793, David Ure gives two figures of the shell under description.

from the small, slightly convex beak, and extending to the front, dividing the valve into two convex lobes. Cardinal area well defined, subparallel and narrow, divided by a small fissure covered by a pseudo deltidium. Surface ornamented by numerous small, rounded, longitudinal striæ, sometimes bifurcating or increasing in number by intercalation at various distances from the beak, and crossed by numerous small, concentric lines or wrinkles, while about fourteen long, vertical, slender, tubular spines are arranged in two rows close to the cardinal edge. The dorsal valve follows the curves of the opposite one, but has a mesial elevation in lieu of a sinus, and is similarly sculptured. Dimensions variable; length 7 inches, width 12 lines.

Obs. Of this remarkable species a few imperfect examples were discovered for the first time by Prof. de Koninck in the Carboniferous limestone of Visé, in Belgium, and who, misled by the well-defined ventral area, supposed his species to be referable to the genus *Leptæna*. Prof. de Koninck's figures do not represent the perfect condition of the shell, for none of his examples retained the peculiarly extended and reflexed ears which I have drawn with great care from some very perfect specimens discovered by Mr. Burrow in the Lower Scar limestone of Settle, in Yorkshire, and from which my description is taken. When it so happens that, from fracture, the auriculate expansions are absent and that the area is not exposed, some examples in this condition might be mistaken for certain deeply sinuated specimens of *P. longispinus*, but the last-named shell is usually more regularly vaulted and rarely geniculated to the extent observable in *P. sinuatus*. Interiorly, both species differ in the position which the adductor muscular impressions in the ventral valve occupy relative to the divaricator ones; those in *P. sinuatus* being longitudinally parallel or on a level with the divaricators, while in *P. longispinus* the last-named, muscular scars commence under and outside of the adductor impressions, so that in Sowerby's shell the four muscular scars in the ventral valve occupy a larger space than in *P. sinuatus*, where they are all on a level or parallel to each other, and occupy a small saucer-shaped space in the rostral portion of the valve. In the dorsal valve the muscular and reniform impressions do not appear to differ much in detail from those of *P. longispinus* or of the generality of other *Producta*, and in any case the interior markings denote with certainty that *P. sinuatus* is a true *Productus*, and could in no case be classed with *Leptæna*. None of the specimens or internal casts exhibited evidence of teeth, so that it is probable that the valves were unarticulated, as they appear to be in all known species of the genus.

Prof. M'Coy mentions that the shell under description is rare in the Carboniferous limestone of Derbyshire, and it has been recently discovered at Bowertrapping, near Dalry, Ayrshire, Scotland, and which I was happy to recognise among some duplicates forwarded to me by Mr. Young.

PRODUCTUS MARGARITACEUS, *Phillips*.   Pl. XLIV, figs. 5—8.

> PRODUCTA MARGARITACEA, *Phillips*.   Geol. of Yorksh., vol. ii, p. 215, pl. viii, fig. 8, 1836.
> —   PECTINOÏDES, *Phillips*.   Ib., pl. vii, fig. 11.
> —   MARGARITACEUS, *De Kon.*   Desc. des Anim. foss. du Terr. Carb. de Belgique, pl. vii, fig. 3; pl. viii ᵇⁱˢ, fig. 5, 1843.   Mon. du genre Productus, pl. iv, fig. 3, 1847.
> —   —   *De Keyserling.*   Wissensch. Beobacht. auf einer Reise in das Petschora Land., p. 210, pl. iv, fig. 7, 1846.
> —   —   *M'Coy.*   British Palæozoic Fossils, p. 466, 1855.

Spec. Char.   Shell thin, somewhat circular or transversely semicircular, rarely longer than wide; hinge-line rather less than the width of the shell.   Ventral valve regularly convex, without sinus; beak small, incurved; ears narrow and but slightly distinguishable from the general convexity of the valve; surface ornamented with numerous thick, obtusely rounded or flattened ribs, separated by shallow sulci, the costæ often bifurcating near the margin, while the whole surface is closely crossed by concentric imbricating striæ, which at intervals produce strongly marked foliaceous interruptions or lines of growth.   The spines are few in number, but four or five strong ones project from each of the auriculate expansions, while two or three more are sometimes irregularly scattered over the surface.   Dorsal valve concave, following the curves of the ventral one, and similarly ornamented.   Interior details imperfectly known.   Dimensions variable; two British examples have measured—

Length 15, width 18 lines.
    „   17,   „   16 „

Obs.   This is a well-marked species, easily distinguishable from its congeners, both by shape and sculpture.   Its shell is likewise remarkably thin and delicate, the ears very brittle and often broken, while the space occupied by the animal between the valves is very small.   In shape it is more often transverse, and to this variety the name *margaritacea* had been applied, while to the rarer or elongated form that of *pectinoides* was given; but as every gradation in shape between the two can be readily found, a single specific denomination can alone be retained.

P. margaritaceus does not appear to be a very common species, and is confined to the middle and lower stages of the Carboniferous system.   In England it occurs at Bolland, Kendal, Settle, in Yorkshire, as well as in some Derbyshire localities, &c.   In Ireland it may be obtained at Florence-court, Millecent, Kildare.   Mr. Kelly mentions also Lisnapaste, Bundoran, and one or two more places where it is said to occur.

On the Continent it is not abundant in the Carboniferous limestone of Visé and Tournay in Belgium, and has been found at Ratingen, D'Altwasser, in Silesia, and in one or two Russian localities.

PRODUCTUS ARCUARIUS, *De Koninck*. Pl. XXXIV, fig. 17.

> PRODUCTUS ARCUARIUS, *De Kon.* Descript. du Animaux foss. du terrain Carb. de la Belgique, p. 171, pl. xii, fig. 10 a, b, 1843, and Mon. du genre Productus, pl. iv, fig. 2.

Spec. Char. Shell small, transversely oval; hinge-line much shorter than the greatest width of the shell. Ventral valve almost hemispherical and regularly vaulted up to a certain distance or age, when it is again continued under the shape of a wide, prominent border, separated from the rostral half by a groove, which, originating on either side of the beak, curves round the valve. The margin is also sometimes bent up at right angles, and thus forms a flattened rim all round the shell. Beak and auricular expansions small. Surface ornamented by numerous radiating, thread-like striæ, which become more numerous by interstriation, and are here and there swollen out, giving rise to a variable number of slender, tubulated spines, which are more numerous on the ear-shaped expansions. Dorsal valve concave, following the curves of the ventral one, and similarly ornamented. Interior unknown. Dimensions variable; a British example measured—length $5\frac{1}{2}$, width 8 lines.

Obs. This little shell does not appear very rare in the Carboniferous limestone of Settle, in Yorkshire, where it was discovered for the first time in Britain by Mr. Burrow. Having submitted specimens to Prof. de Koninck, he declared them to be specifically identical with those from Visé, in Belgium. The concentric groove which divides the ventral valve into two portions is not, however, always so distinctly marked in the British specimens that have come under my observation as in certain Belgian individuals figured by Prof. de Koninck.

PRODUCTUS CARBONARIUS, *De Koninck*. Pl. XXXIV, fig. 6.

> PRODUCTUS CARBONARIUS, *De Koninck.* Descrip. des Animaux foss. der terr. Carb. de la Belgique, pl. xii bis, fig. 1, 1843, and Mon. du genre Productus, pl. x, fig. 4, 1847.
> — — *De Verneuil.* Russia and Oural Mountains, pl. xvi, fig. 2, 1845.
> — — *Dav.* A Mon. of the Carb. Brachiopoda of Scotland, pl. iv, fig. 14, 1860.

Spec. Char. Shell of median size, about as wide as long, or slightly transverse; ventral valve very gibbous, evenly rounded, and without sinus; auriculate expansions small; hinge-line nearly as wide as the greatest width of the shell. Surface ornamented with very fine, thread-like, radiating striæ, from which project at short intervals numerous slender spines, the rib becoming thickened at the spot where the spine originates; beak small, much incurved, and covered with a few slight concentric wrinkles. Dorsal valve following the curves of the opposite one, and similarly ornamented. Interior unknown. A British specimen measured—length $9\frac{1}{2}$, width 11, lines.

Obs. I am acquainted with but a single British example of this species, stated to have been found in Carboniferous limestone north of Glasgow (?) and now preserved in the museum of Practical Geology. The identification is given on the authority of Prof. de Koninck, and it would be very desirable that more specimens should be obtained and examined before its specific claims be definitely recognised. Our Scottish example agrees, however, very nearly with the figure published by Prof. de Koninck of a specimen found in a calcareous nodule at Chokier, in Belgium, and M. de Verneuil states he has obtained the species in a Carboniferous limestone in the valley of Prikcha (Valdai), Russia.

PRODUCTUS UNDATUS, *Defrance.* Pl. XXXIV, figs. 7—13.

> PRODUCTUS UNDATUS, *Defrance.* Dic. des Sc. nat., vol. xliii, p. 354, 1826.
> — — *De Koninck.* Desc. des Animaux foss. du Terrain Carb. de Belgique, p. 156, pl. xii, fig. 2, 1843; and Mon. du genre Productus, pl. v, fig. 3.
> — TORTILIS, *M'Coy.* Synopsis of the Carb. Foss. of Ireland, tab. xx, fig. 14, 1844.
> — UNDATUS, *De Verneuil.* Russia and the Ural Mountains, vol. ii, p. 261, pl. xv, fig. 15, 1843.
> — — *Dav.* Mon. Scottish Carb. Brach., p. 41, pl. iv, figs. 15—17, 1860.

Spec. Char. Shell somewhat sub-orbicular or slightly transverse; hinge-line rather less than the width of the shell. Ventral valve regularly vaulted, very convex, without sinus; beak small, rounded, incurved, not extending much beyond the hinge-line; auriculate expansions small. Surface covered with numerous irregular or interrupted sub-parallel, undulating, concentric folds or wrinkles, which become wider and more produced with age, and having their narrow, almost perpendicular, side directed towards the beak; the valve is, moreover, ornamented by numerous minute, rounded, thread-like striæ, separated by narrow sulci, and of which from five to six may be counted in the thickness of a line, and swelling out at intervals, they give rise to a slender spine. Dorsal valve concave, following the curves of the opposite one, and similarly ornamented. Interior unknown. Dimensions variable; two British examples have measured—

Length 14, width 13, lines.
    „   12,   „   13,   „

Obs. This remarkable species, although not common, is by no means very rare in the Carboniferous limestone of Great Britain, and may be easily distinguished from other British Producta by that curious terrace-shaped or crumpled-like appearance it presents, and which is caused by the numerous concentric wrinkles which cover the surface of both valves. In some specimens these ridges are very regular and uninterruptedly continuous, but in the larger number of individuals some of them are here and there interrupted in their course and absorbed, while two at other times unite during their passage along the middle. The radiating thread-like striæ bear much resemblance to those of *P. Cora*, and increase in number here and there by interstriation.

In England, *P. undatus* occurs in the lower scar limestone of Settle, in Yorkshire, in the lower Carboniferous limestone of Lowick, Northumberland, middle limestone of Poolwash, Isle of Man, and upper Carboniferous limestone of Derbyshire. In Scotland, it occurs in limestone at Gare and Headsmuir, at about 240 fathoms below the horizon of the "Ell Coal." In Stirlingshire, in the Campsie main limestone, and at Castlecarry, in Dumbartonshire. In Ireland, at Tullynagaigy, Fermanaugh, and at Little Island, near Cork.

On the Continent, it occurs in the limestone of Visé, in Belgium, and at D'Unja, near Kosimof, as well as at Nikoulin and Karova, in Russia. M. de Koninck mentions also that a specimen is preserved in the Museum of Paris, found in Tasmania.

PRODUCTUS WRIGHTII, *Dav*. Pl. XXXIII, figs. 6, 7.

Spec. Char. Shell small, transversely oval; hinge-line rather less than the width of the shell. Ventral valve convex and regularly vaulted; beak and auriculate expansions small. Surface covered with numerous smooth, regular or irregularly interrupted, undulating, concentric folds or wrinkles, from which, at short intervals, project scattered spines; the margin of the valve is, moreover, bent up at right angles, forming a wide, flattened, but ribbed boarder or frill round the shell. Dorsal valve and interior unknown. The largest specimen hitherto discovered measures—length $3\frac{1}{2}$, width 5 lines, of which the bent-up margin is from 1 to $1\frac{1}{2}$ line in width.

Obs. Of this interesting little species two examples were discovered by Mr. J. Wright in the Carboniferous limestone of Middleton, near Cork, in Ireland, and have been considered by Prof. de Koninck, as well as by myself, a new species (?) The concentric folds or wrinkles resemble much those of *P. undatus* in their terrace-like arrangement, but are distinguished by the total absence of those longitudinal, thread-like striæ so beautifully disposed in Defrance's species; the wide fringe or frill which surrounds the shell is also peculiar.

The species is named after Mr. Joseph Wright, to whom the author of this monograph is indebted for much valuable material and information relative to the species from the neighbourhood of Cork, in Ireland.

PRODUCTUS PROBOSCIDEUS, *De Verneuil.* Pl. XXXIII, figs. 1—4.

PRODUCTUS PROBOSCIDEUS, *De Verneuil.* Bulletin de la Soc. Geol. de France, vol. xi,
p. 259, pl. iii, fig. 3, 1840.
CLAVAGELLA PRISCA, *Goldfuss.* Petref. Germ., vol. ii, p. 285, pl. clx, fig. 17, 1841.
PRODUCTUS PROBOSCIDEUS, *V. Buch.* Abhand. der. K. Akad. der Wissens. zu Berlin,
Erster Theil., p. 40, 1841.
— — *De Koninck.* Descript. des Animaux foss. du Terrain Carb.
de Belgique, p. 11, fig. 4, 1843; and Mon. du genre
Productus, pl. vi, fig. 4 [a, b, c, d], 1847.

Spec. Char. Shell of moderate size, and very variable in shape, on account of the singular prolongation of its ventral valve. Dorsal valve small, gently and regularly concave; slightly transverse or elongated, with a straight hinge-line always shorter than the greatest width of the shell. Surface covered with fine, radiating, thread-like striæ, intersected by numerous concentric wrinkles. The ventral valve is composed of two well-marked parts; the first (which corresponds with almost the entire dorsal valve) is marginally somewhat circular, moderately convex, and about as wide as long, the beak being very small and but slightly incurved over the hinge-line of the opposite valve. Auriculate expansions small. The second portion is composed of a general prolongation of the margin of the same valve, which commences by the lateral portions extending on either side, and, by gradually forcing open the smaller valve, unite round its margin, to be prolonged afterwards in the shape of a cylindrical tube of lesser or greater length. This tube is concentrically wrinkled, and irregular in width and direction, while the surface of the entire valve is minutely striated, and numerous long, slender spines rise and project forwards from the lateral expansions of the ventral valve. Interior unknown. Proportions very variable; two British examples have measured

Ventral valve, entire length, with tube, 13½, width 5, lines.
Dorsal valve „ „ 4, „ 4½, „
Another example measured—
Ventral valve, entire length, with tube, 11, width 4½, lines.
Dorsal valve „ „ 3, „ 3½ „

Obs. This Productus is one of the most remarkable of the genus, and its discovery by Mr. Burrow in the Carboniferous limestone of Settle, in Yorkshire, was the more interesting from the fact that authentic examples of the species had not until then been found in any other locality than Visé, in Belgium.[1] The shell has been fully described and

[1] Another form, (?) nearly related by shape to that under description, has been described and figured by Messrs. Norwood and Pratten from the Carboniferous rocks of Graysville, Illinois, America, under the denomination of *P. clavus,* but which is stated by these palæontologists to "differ from *P. proboscideus,* with which alone it can be confounded, in the longitudinal ribs not extending over the visceral portion, and in the prolongation not showing the numerous and well-marked transverse folds of that species. The

illustrated by Messrs. de Verneuil and de Koninck, who state that, although the tube was generally simple, it sometimes bifurcated, and then two independent tubes were produced. Its direction is generally that of the longitudinal axis of the shell, but it is often irregularly twisted to the one or the other side, and may likewise remain straight for some time before suddenly bending and assuming another direction, as represented in one of the specimens figured in my plate. The tube has also attained at times considerable length, one Belgian example figured by Prof. de Koninck being nearly two inches in length, with not more than four lines in width, and although the specimen was broken and imperfect at its extremity, the width had not altered in all its length. M. de Koninck has also observed that he has never been able to procure a single specimen in which the tube was complete, and therefore concludes that its extremity was open, and served for the passage of muscular fibres, by which the animal attached itself to submarine objects. But, as I have already had occasion to observe at p. 119 of my " General Introduction," I feel compelled to differ with my learned friend in this last assumption. M. d'Orbigny believed the prolongation to be due to malformation produced by accidental circumstances, connected with the supposed constrained position in which the animal lived, which forced the mantle to prolong its edges so as to reach the surface of the sea-bed; but in the localities in which the form is found a vast number of other species of the genus occur, which do not present this peculiarity, so that we must regard the structure as *normal*, although M. d'Orbigny's explanation of its function is probably correct; and in any case I cannot concur in the supposition that the shell may have been fixed to submarine bodies by the means of muscular fibres issuing from the open extremity of the tube. The prolongation of the ventral valve beyond that of the dorsal one is not a feature peculiar to the shell under consideration, for it has been noticed already in two or three more species of the same genus, as well as in a Liassic form of Thecidium, where the ventral valve is prolonged much beyond that of the dorsal one.

The two parts of which the ventral valve is composed are clearly defined by the means of a groove.

PRODUCTUS ERMINEUS, *De Koninck*. Pl. XXXIII, fig. 5.

> PRODUCTUS ERMINEUS, *De Koninck*. Descript. des Anim. foss. du Terr. Carb. de Belgique, p. 181, pl. x, fig. 5, 1843; and Mon. du genre Productus, pl. vi, fig. 5, and pl. xviii, fig. 1.

Spec. Char. Shell of moderate dimensions, longer than wide; hinge-line much

longitudinal ribs of *P. proboscideus* are also much finer, numbering about fifty in the space of ten millemètres at the interior border of the visceral part. *P. clavus* shows no trace of sinus." It must, however, be remembered that the American specimen is stated to have been in so imperfect a state of preservation that for some time the authors hesitated in including it in their memoir, and it is still possible that *P. clavus* may, after all, belong to M. de Verneuil's curious species (?).

shorter than the greatest width of the shell. Ventral valve gently convex from the extremity of the beak to the front, the lateral portions being much extended and bent downwards at almost right angles to the plane of the valve; beak and auriculate expansions small, with a ridge commencing at their extremity, extending to some distance on either side, thus dividing the visceral from the labial extensions of the valve. The surface is covered with numerous thread-like, radiating striæ, which increase in number by interstriation at various distances from the beak, from which project a few scattered spines; it is also ornamented by many concentric, undulating wrinkles, which become larger or wider as they recede from the extremity of the beak. The dorsal valve is concave, following closely the curves of the opposite one, and is similarly ornamented. Interior like that of other Producta. A British example measured—

Length 10, width 9½, lines.

Obs. This species differs from *P. proboscideus*, by its dorsal valve following exactly the curves of the ventral one, and by the last-named valve not being prolonged in the form of a cylindrical tube. It occurs in the Carboniferous limestone of Settle, in Yorkshire, where it was found for the first time in England by Mr. Burrow, and is there associated with *P. proboscideus*, as well as at Visé, in Belgium, the only two localities in which the species has been hitherto discovered.

PRODUCTUS TESSELLATUS, *De Koninck.* Pl. XXXIII, figs. 24, 25; Pl. XXXIV, fig. 14.

> PRODUCTUS MURICATUS, *De Koninck.* Descript. des Animaux foss. du Terr. Carb. de Belgique, p. 192, pl. ix, fig. 2; pl. 13 bis, fig. 5 (not Phillips), 1843.
> PRODUCTUS TESSELLATUS, *De Koninck.* Monographie du genre Productus, Mémoires de la Soc. Royale des Sciences de Liége, vol. iv, pl. xiv, fig. 2, 1847.
> — — *Morris.* Catalogue, p. 145, 1854.

Spec. Char. Shell small, transversely oval, a little wider than long; hinge-line much shorter than the width of the shell. Ventral valve evenly convex and regularly vaulted, without sinus; beak small, prominent, and incurved, not projecting beyond the hinge-line; auriculate expansions very slightly developed. Surface ornamented by from fifteen to twenty round, salient costæ, from which project a few scattered spines, and intersected by about as many concentric grooves; margin bent upwards, wide, flat, or slightly concave, and marked by numerous rounded, bifurcated, or trifurcated ribs, with dorsal valve gently concave, and ornamented as in the ventral one. Interior unknown. Dimensions variable; two British examples have measured—

Length 11, width 12, lines.

,, 6, ,, 8½, ,,

The flattened border is nearly one third of the entire length of the valve.

Obs. This elegant shell appears rare in British Carboniferous rocks, but is easily distinguished from other forms of the genus by shape and sculpture, as well as by the flattened, ribbed border which encircles the valves. In England it has been found by Mr. Burrow in the Calcareous limestone of Settle, in Yorkshire, by the Rev. T. G. Cumming in the Isle of Man, while in Ireland it occurs in the Calcareous limestone of Kildare. On the Continent it is not very common at Visé, in Belgium.

PRODUCTUS ACULEATUS, *Martin.* Pl. XXXIII, figs. 16—20.

> ANOMITES ACULEATUS, *Martin.* Petrif. Derbiensis, p. 8, pl. xxxvii, figs. 9, 10, 1809, (not Schlotheim).
> PRODUCTUS ACULEATUS, *Sow.* Min. Con., tab. lxviii, fig. 4, Oct., 1814.
> PRODUCTA LAXISPINA, *Phillips.* Geol. of York., pl. viii, fig. 13, 1836.
> PRODUCTUS SPINULOSA, *Phillips.* Ib., vol. ii, pl. vii, fig. 14 (not of Sow.), 1836.
> —　GRYPHOIDES, *De Koninck.* Desc. des Anim. foss. du Terrain Carb. de la Belgique, pl. ix, fig. 1, and pl. xii, fig. 12, 1843 (?)
> —　ACULEATUS, *De Koninck.* Mon. du genre Productus, pl. xvi, fig. 6, 1847.
> —　—　*M'Coy.* British Palæozoic Fossils, p. 458, 1855.
> —　—　*Dav.* Mon. of Scottish Carb. Brach., pl. ii, fig. 20, 1860.

Spec. Char. Shell of moderate dimensions, ovate or semicircular, slightly transverse or elongated; hinge-line less than the width of the shell. Ventral valve regularly vaulted, gibbous, without sinus; beak much incurved and overlying the hinge-line at its attenuated extremity; auriculate expansions small and thin. Surface in the young shell covered with a variable number of irregularly scattered, elongated tubercules, from which project small, curved spines, but with age the tubercules are at times so close and elongated on the anterior half of the valve that they become transformed into longitudinal ribs; the surface is moreover intersected by regular or undulating laminar, concentric lines. Dorsal valve very concave, following closely the curves of the opposite one, and similarly ornamented. Interior unknown. Dimensions variable; three British examples have measured—

Length　6, width　$6\frac{1}{2}$, lines (Martin's type).

„　11, „　12, 　„

„　11, „　13, 　„

Obs. In general appearance this *Productus* varies considerably, according to age and specimen, Martin's type (of which I have given a carefully enlarged representation) having been described and drawn from a young shell, which was almost smooth, from the very few, irregularly scattered, tubular spines which covered its surface;[1] but in other specimens, with

---

[1] Martin describes his species—"A fossil shell. Original an *Anomia*, imperforate, with one of the valves concave, the other convex and gibbous. Hinge close, straight, but less extended than in most other *Anomitæ* of the same division (Syst. G. B., 'Anomitæ,' *b, b*). The convex valve prickly; the

age the tubercules became closer and closer, until converted into ribs, the posterior half of the valves usually preserving the characters of the young shell. Martin describes the spines as pointed backwards towards the beak, but this observation has already been shown by Prof. de Koninck and M'Coy to have been erroneous, their direction being similar to that of other Producta. *P. laxispina* of Phillips has been drawn from a large example of Martin's species, and the *P. spinulosa*, Phillips (but not of Sow.), would appear to be likewise referable to the same species.

P. aculeatus is a common shell in many localities. In England, it occurs in Carboniferous limestone at Bakewell, Buxton, and Chrome Hill, Longnor, in Derbyshire; at Settle, in Yorkshire; Bolland, in Westmoreland; Lowick, Northumberland; in the middle limestone of Poolwash, Isle of Man, &c. In Scotland, at Calderside; High Blantyre, in Lanarkshire; Orchard Quarry, Thornliebank, Renfrewshire; at West Broadstone, Beith, in Ayrshire; as well as at Craigie, near Kilmarnock, and Auchenskey, near Dalry, in the same county. In Ireland, in limestone at Lisnapaste, Howth, Old Leighlin, Little Island, and Windmill.

On the Continent, it occurs at Visé, in Belgium. In Russia it has been found in several localities, such as in the neighbourhood of Buregi, Government of Nowgorod, as well as at Cosatchi-Datchi, on the Oriental side of the Oural, to the east of Minsk, &c.

PRODUCTUS YOUNGIANUS, *Dav.* Pl. XXXIII, figs. 21—23.

> PRODUCTUS YOUNGIANUS, *Dav.* Mon. of Scottish Carb. Brach., pl. ii, fig. 26, and pl. v, fig. 7, 1861.

Spec. Char. Shell longitudinally oval; hinge-line shorter than the width of the shell. Ventral valve regularly arched, convex, and without sinus; auriculate expansions very small; beak incurved and comparatively large, not overlying the hinge-line except quite at its attenuated extremity. Dorsal valve very concave, following closely the curves of the opposite one. Surface in both valves ornamented with numerous small, rounded ribs, of which a certain number are due to intercalation, and from which, at short distances, project slender, tubular spines, more widely scattered in some specimens than in others; the valves are also covered with irregular, undulating, squamiform, concentric expansions, which overlap each other very numerously in some examples, sparingly in others. In the interior the muscular and reniform expansions are similar to those of other *Producta*. Dimensions variable; two British examples have measured—

Length 13, width 9, lines.
  „   9, „ 6½, „

Obs. This shell appears distinguishable from *P. aculeatus* by its more regularly oval

prickles few, scattered, very minute, short, appressed, or squeezed flat to the surface, and pointing backwards or towards the beak. The beak small and hooked. . . . . . . It is found near Bakewell and Buxton."

and elongated appearance, its ribs being likewise more numerous and regular, both in the young and adult condition, than what is usually observable in Martin's species. In this opinion I am supported by Prof. de Koninck and Mr. Young, but am still somewhat uncertain whether the shell under description is in reality more than a variety of Martin's species.

In Scotland, *P. Youngianus* occurs abundantly in a white friable shale above a coralline bed (*Lithodendron fasciculatum*, Fleming, *Lithostrotion Martini*, M. Edwards). It is found also at Brockley, near Lesmahago, in Lanarkshire, as well as in Renfrewshire, Ayrshire, and Fifeshire.

In England in Carboniferous limestone at Llangollen, in Wales.

PRODUCTUS PUSTULOSUS, *Phillips.* Pl. XLI, figs. 1—6; Pl. XLII, figs. 1—4.

> PYXIS TRANSVERSIM STRIATA, *Chemnitz.* Martini's Neues Syst. Conch. Cabinet,
> vol. vii, p. 301, pl. lxiii, figs. 605 and 606, and
> vol. viii, fig. 69, 1784.
> PRODUCTA PUSTULOSA, *Phillips.* Geol. of Yorks., vol. ii, p. vii, fig. 15, 1836.
> — RUGATA, *Phillips.* Ib., fig. 16.
> — OVALIS, *Phillips.* Ib., pl. viii, fig. 14.
> PRODUCTUS PUSTULOSUS, *De Koninck.* Desc. des Anim. foss. du Terr. Carb. de
> Belgique, pl. xii [bis,] fig. 3, 1843.
> — PUNCTATUS, *Kon.* (non Martin). Ib., pl. ix, fig. 6.
> — PUSTULOSUS, *Kon.* Mon. du genre Productus, pl. xiii, fig. 1, and pl. xvi,
> figs. 8, 9, 1847.
> — PYXIDIFORMIS, *Kon.* Ib., pl. xi, fig. 7; pl. xii, fig. 1; pl. xvi, fig. 2.

Spec. Char. Shell thin, rotundato-quadrate, wider than long, rarely oval and longer than wide; hinge-line shorter than the greatest width of the shell. Ventral valve regularly arched and gibbous, divided longitudinally by a sinus of greater or lesser depth and width; beak moderately developed and incurved, not overlying the hinge-line except at its attenuated extremity; ears wide, nearly rectangular, and flattened; margin slightly indented in front. External sculpture varying somewhat in appearance in different specimens, but usually covered with numerous continuous or interrupted transverse, undulating wrinkles, while numerous pustules or elongated tubercules, bearing slender, tubular spines, are subquincuncially arranged or disposed in irregular rows over the entire surface. Dorsal valve slightly concave, with a small mesial elevation or fold, while numerous transverse wrinkles, tubercule-pits, and short spines, cover the surface. Dimensions variable; three examples measured—

Length 44, width 46, greatest depth between valves 17, lines.
„ 28, „ 39, „ „ „ 10, „
„ 24, „ 20, „ „ „ 8, „

Obs. After a lengthened comparison and study of more than one hundred specimens of the shell under description, there can exist but little doubt that *P. rugata*, *P. ovalis*, and *P. pyxidiformis*, are merely different states of *P. pustulosus*, and I quite coincide with Professor M'Coy when stating that " scarcely any two examples of *P. pustulosus* agree in the strength or directness of the transverse ridges ; in specimens perfectly typical in this respect near the beak, the ridges will be often found indistinct, undulated, and interrupted on other parts, and when this is the case the tubercules generally increase in size and become quincuncially arranged. I have traced the passage from the most regularly wrinkled type (like Koninck's figure, op. cit., t. xii, fig. 4), with the spines on the summits of the transverse ridges, through those in which the spines do not coincide with the (still well-marked) ridges (like his t. xvi, fig. 9), to those in which the wrinkles gradually become irregular, interrupted, and nearly obsolete, as in *P. pyxidiformis*, by the most imperceptible gradations. In all these varieties the isolation of the tubercules, instead of their being mere inflations of distinct, longitudinal striæ or ridges, distinguishes the species from the true *P. scabricula*." I have never yet obtained a perfect specimen of the shell under description, for in all the British examples the spines were broken close to their base.

The interior has been sometimes obtained, and of which figures are given in my plate. In the umbonal portion of the ventral valve the occlusor or adductor scars are placed on either side of a small ridge between and on a level with the cardinal or divaricator impressions. In the dorsal valve the adductor and riniform impressions do not differ much in detail from what we find in the generality of *Producta*. *P. pustulosus* is not rare in the Carboniferous limestone of many English localities, such as Bolland, Settle, Kendal, the Isle of Man, in Derbyshire, &c. In Scotland it has been stated to occur at Cat Craig, near Dunbar. In Ireland in the Calciferous slate and Carboniferous limestone of Bundoran, Ballyduff, Carrigaline, Lisnapaste, Millecent, Tankardstown, Florence Court, Little Island, shores of Lough Gill, valley of the Maine, Hook, St. Doolas, near Dublin, &c.

On the Continent it occurs at Visé and Tournay, &c., in Belgium ; Ratingen (Prussia) ; and it has also been found in America.

PRODUCTUS SCABRICULUS, *Martin.* Pl. XLII, figs. 5—8.

ANOMITES SCABRICULUS, *Martin.* Petrif. Derb., p. 8, pl. xxxvi, fig. 5, 1809.

PRODUCTUS — *Sow.* Min. Conch., t. lxix, fig. 1, Oct., 1814.

PRODUCTA SCABRICULA, *Phill.* Geol. of York., vol. ii, pl. viii, fig. 2, 1836.

— QUINCUNCIALIS, *Phill.* Ib., pl. vii, fig. 8.

— SCABRICULUS, *De Koninck.* Desc. des Anim. foss. du Terrain Carb. de Belg., pl. xi, fig. 3 *a, b* (?), 1843 ; and Mon. du genre Productus, pl. xi, fig. 6, 1847.

— CORBIS, *Potiez* et *Michaud.* Galer. des Mollusques du Mus. de Douai, vol. ii, pl. xli, fig. 2, 1844.

— SCABRICULUS, *De Verneuil.* Russia and the Ural Mount., pl. xvi, fig. 5 ; and pl. xviii, fig. 5, 1845.

— — *M'Coy,* British Pal. Fossils, p. 470, 1855.

— — *Dav.* Scottish Carb. Brach., pl. iv, fig. 18, 1861.

Spec. Char. Marginally rotundato-quadrate, generally wider than long, lateral margins sub-parallel; rounded or slightly indented in front; hinge-line rather less than the greatest width of the shell. Ventral valve convex, with a wide, shallow, median depression or sinus; beak incurved, not overlying the hinge-line, except at its attenuated extremity; ears small, flattened. Surface covered with numerous sub-regular striæ, swelling out at close intervals in the shape of oblong tubercules, arranged somewhat irregularly in quincunx, and from each of which rise slender, curved spines, of rather less than half an inch in length; feeble concentric wrinkles sometimes traverse the valve, and are especially marked on the ears. Dorsal valve concave, near the margin, with a slight median elevation commencing not far from the front; surface covered with numerous concentric wrinkles, tubercule-pits, and short, depressed, slender spines. Dimensions variable; two British examples measured—

Length 28½, width 33½ lines.

,, 12, ,, 12 ,,

Obs. Variations in shape of this shell have received different names, and it has been sometimes suggested that it and the preceding species should be united.[1] Professor de Koninck mentions that it bears much resemblance with the young state of his *P. pyxidiformis*, and that it approaches likewise to *P. Humboldtii*, and I am ready to admit that a certain external resemblance does sometimes exist among certain examples of these species. In true *P. scabriculus* the swollen-out, alternating, elongated tubercules are connected by a continued ridge, so that the shell has the appearance of being ribbed, while in *P. pustulosus* the tubercules are more often isolated. The interior, however, in the dorsal valve presents a difference in the shape of the median ridge, which extends from under the cardinal process and divides the adductor impression. In *P. pustulosus* this ridge is simple, while in *P. scabriculus* it is composed of two ridges, which converge and unite at some distance from their origin into a single ridge (fig. 8). I have observed this character in several examples, and believe it constant.

P. scabriculus is common in the Carboniferous limestone of the Craven district, and Settle, in Yorkshire; near Bolland; in dark Carboniferous limestone at Lowick, Northumberland; Martin found it in the limestone of Tideswell, in Derbyshire. It is not uncommon at Coalbrook Dale; near Bristol, &c. In Scotland, it is plentiful in ironstone at Jock's Burn, Braidwood, Brockley, and many other Lanarkshire localities. In Stirlingshire it occurs in several stages, such as the Craigenglen beds, Campsie main-limestone and ironstone, and at Corrieburn. It has also been found in Renfrewshire, Dumbartonshire, Ayrshire, the Lothians, and Fifeshire. In Ireland, Mr. Kelly mentions that it occurs in the Calciferous slate and Carboniferous limestone of Lisnapaste, Millecent, Little Island, and of, no doubt, many other localities. It is also a common fossil on the Continent, having been found at Visé, in Belgium; Peredki (Valdaï), Sloboda, &c., in Russia. Specimens have also been collected in America, &c.

---

[1] Explanations to accompany sheets 102 and 112 of the maps of the Geological Survey of Ireland, p. 16.

PRODUCTUS FIMBRIATUS, *J. de C. Sowerby.* Pl. XXXIII, figs. 12—15, and Pl. XLIV, fig. 15.

ANOMITES PUNCTATUS, *Martin.* Petrif. Derb., pl. xxxvii, figs. 7, 8 (fig. 6 exclusa), 1809.

PRODUCTUS FIMBRIATUS, *J. de C. Sow.* Min. Conch., pl. 459, fig. 1, July, 1823.

STROPHOMENA MARSUPIT, *Davreux.* Const. Geogn. de la province de Liege, pl. iv, fig. 2 A, B, 1833.

PRODUCTA FIMBRIATA, *Phillips.* Geol. of York., vol. ii, pl. viii, figs. 11, 12, 1836.

— — *V. Buch.* Abhandl. der K. Akad. der Wissens. zu Berlin, Theil i, p. 27, pl. ii, figs. 21—23, 1841.

— — *De Koninck.* Desc. des Animaux foss. du Terr. Carb. de Belgique, pl. x, fig. 3 $^{a, b, c, d,}$ 1843; and Mon. du genre Productus, pl. xii, fig. 3 $^{a, b, c,}$ 1847.

? PRODUCTA LACINIATA, *M'Coy.* Synopsis of the Carboniferous Fossils of Ireland, pl. xx, fig. 12, 1844.

PRODUCTA FIMBRIATA, *M'Coy.* British Palæozoic Fossils, p. 461, 1855.

— — *Dav.* Mon. of Scottish Carb. Brach., pl. ii, fig. 27, 1860.

Spec. Char. Longitudinally oval or ovate; hinge-line a little shorter than the width of the shell. Ventral valve uniformly convex, gibbous, and greatly arched in profile; beak much incurved, overlying the hinge-line at its attenuated extremity; ears small and but slightly marked. Surface regularly traversed by numerous sub-regular, concentric, prominent bands or ridges, with flattened, intervening spaces; a row of elongated tubercules covering each ridge, and from which project long, tubular, cylindrical spines. Dorsal valve nearly flat or moderately concave, traversed by numerous concentric ridges, with concave interspaces; a row of short, adpressed spines projecting from every ridge. Dimensions variable; two specimens have measured—

Length 18, width 17, greatest depth between valves 9 lines.

„ 15, „ 12, „ „ „ 7 „

Obs. This species does not attain the proportions of *P. punctatus,* is less variable in its shapes, possesses no sinus, and is especially distinguished by the single row of cylindrical spines which rise from each row of elongated tubercules, as above described. With *P. laciniata,* M'Coy (Pl. XLIV, fig. 15), I am not sufficiently acquainted. It has been supposed by some a synonym of *P. punctatus,* but the study of two Irish examples leads me almost to agree with M'Coy while stating that it bears more resemblance to *P. fimbriatus,* of which it is possibly only a variety or variation in shape, more transverse, with a greater number of concentric ridges, smaller and more numerous elongated tubercules or cylindrical spines.[1] This I must, however, leave as an open question for the present. The interior of the dorsal valve of *P. fimbriatus* alone is at present

[1] At p. 110 of the 'Synopsis,' M'Coy describes his *P. laciniata* as "nearly semicircular, length one sixth less than the width; hinge-line rather less than the width of the shell; moderately convex; beak small, prominent; ten or twelve rounded, concentric wrinkles, fringed on their marginal declivity with close, regular, lengthened spine-bases. Length eleven lines; width one inch one line." Fig. 15 of our 44th plate is taken from the original specimen.

known, and appears to present certain peculiarities which have been carefully represented in the figures of my plate. Thus, the adductor muscular impressions are very much produced, and form two pairs of contiguous projections, separated by a short median ridge which extends to nearly the centre of the valve, where it becomes much elevated, with a depression or pit on either side, margined by a vertical wall, as in *Strophomena analoga* and some other species. The reniform impressions are also much more oblique in their direction than is usual to the species of this genus. In the greater number of specimens the spiniferous ridges are separated by almost smooth spaces, but in some rarer examples the tubercules or spine-bases are so much elongated as to give the surface somewhat the appearance of certain examples of *P. scabriculus.*

P. fimbriatus is not uncommon in the Carboniferous limestone of Derbyshire, of Lowick, Northumberland; limestone and shale of Settle, in Yorkshire; Bolland; Poolwash, Isle of Man, &c. In Scotland at Hillhead, at Middleholm, Lanarkshire; Gateside and West Broadstone Beith, Ayrshire; in Fifeshire, &c. In Ireland, at Little Island, Bundoran, Tornaroan, &c. In Belgium it is common in the lower stage of the Carboniferous system of Visé, rare in that of Chockier; and has been found also at Sterlitamak, in Russia, &c.

PRODUCTUS PUNCTATUS, *Martin.* Pl. XLIV, figs. 9—16.

ANOMITES PUNCTATUS, *Martin.* Petrif. Derb., pl. xxxvii, fig. 6 (7, 8 exclusa), 1809.
TRIGONIA RUGOSA, *Parkinson.* Organ. Remains, vol. iii, pl. xii, fig. 11, 1811.
ANOMITES THECARIUS, *Schloth.* Nachtr. zur Petrefactenk., pl. xiv, fig. 1, 1822.
PRODUCTUS PUNCTATUS, *J. Sow.* Min. Conch., t. 323, Jan., 1822.
PRODUCTA PUNCTATA, *Phillips.* Geol. of York., vol. ii, pl. viii, fig. 10, 1836.
LEPTŒNA SULCATA, *Fischer.* Oryct. du Gouv. de Moscou, pl. xxiii, fig. 2 (non Sow.), 1837.
PRODUCTUS PUNCTATUS, *V. Buch.* Abhandl. der K. Akad. der Wissens. zu Berlin, Thiel. i, pl. ii, figs. 10, 11, 1841.
— — *De Kon.* Desc. des Animaux foss. du Terr. Carb. de Belgique, pl. viii, fig. 4; pl. x, fig. 2; pl. ix, fig. 4, 1843; and Mon. du genre Productus, pl. xii, fig. 2, 1847.
— CONCENTRICUS, *Potiez et Michaud.* Galer. des Mollusques du Mus. de Douai, vol. ii, p. 25, pl. xli, fig. 1, 1844.
— PUNCTATUS, *De Vern.* Russia and the Ural Mountains, vol. ii, pl. xvi, fig. 11, 1845.
— TUBULOSPINA, *M'Chesney?* Desc. of New Species of Fossils from the Palæozoic Rocks of the Western States, p. 37, 1859.
— PUNCTATUS, *Dav.* Scottish Carb. Mon., pl. iv, fig. 20, 1860.

Spec. Char. Shell thin, variable in shape; transversely rotundato-quadrate, or slightly elongated oval; hinge-line shorter than the width of the shell. Ventral valve convex, sometimes gibbous, with a shallow, longitudinal, mesial depression or sinus, commencing at a short distance from the extremity of the beak and extending to the front; beak incurved, overlying the hinge-line at its attenuated extremity; ears flattened, but slightly defined.

Surface covered with numerous sub-regular, concentric bands or ridges, which increase in number and breadth as they recede from the extremity of the beak, but in adult shells becoming again closer as they approach the margin; these bands (in the ventral valve) are slightly raised towards their lower margin, and are abruptly separated from each other by a narrow, smooth space, after which there exists a tolerably regular row of lengthened tubercules or slender, shining, tubular spines, and again below these the remaining space is filled up by irregularly scattered, but closely packed, smaller spines, all overlapping one another, and lying close to the valve. Dorsal valve moderately concave, with a slight mesial elevation, and ornamented as in the dorsal one, but the bands are slightly concave. In the interior of the ventral valve the adductor or occlusor muscular impressions extend much lower down in the shell than do those attributable to the divaricators. In the dorsal valve the muscular and reniform impressions differ but slightly from those of other Producta. Two specimens have measured—

Length 28, width 31, greatest depth between valves 12 lines.
„ 29, „ 23, „ „ „ 11 „

Obs. This common and very characteristic species appears to have been described and figured for the first time by David Ure, who mentions that " both valves are covered with small spines resembling hair, and so numerous that a largish example contains upwards of ten thousand, and lie so closely together that the surface of the shell is entirely concealed from view,"[1] but it is rare to find the specimens in that condition, the valves being usually deprived of their spiny investment. Martin appears, however, to have been the first naturalist who applied to the shell a specific denomination, but he confounded under the same name the *P. fimbriatus* of Sowerby, a closely allied, but easily distinguishable species.

Productus elegans has been generally considered a synonym or a young condition of the shell under description; but as I am still uncertain whether it be really so or a distinct species, as believed by M'Coy, it will therefore be preferable to describe it separately, but, perhaps, as a variety of *P. punctatus*, to which it bears much resemblance in the arrangement of its spines. *P. punctatus* is characterised in all well-shaped examples by a median depression or sinus, but as this commences usually at a short distance from the extremity of the beak it has been supposed, correctly or erroneously, that in the young state the ventral valve may have been convex and without sinus (?).

Var? ELEGANS, *M'Coy.* Pl. XLIV, fig. 15.

PRODUCTUS ELEGANS, *M'Coy.* Synopsis Carb. Foss. Ireland, pl. xviii, fig. 13, 1844; and British Palæozoic Fossils, p. 460, pl. iii H, fig. 4.

Shell rather small, longitudinally ovate; hinge-line a little shorter than the width of the valve. Ventral valve regularly arched and gibbous, without sinus. Surface crossed

[1] 'History of Rutherglen and East Kilbride,' pl. xv, fig. 1, 1793.

by from ten to sixteen regular broad, sub-equal, transverse ridges, obtusely angulated in the middle, each having three or four rows of spines disposed exactly as in *P. punctatus* proper. Dorsal valve slightly concave, and ornamented as in the opposite one. Length 8, width 7 lines.

Professor M'Coy states, in his work on the Palæozoic fossils, that " this species is intermediate in all its characters between *P. punctata* and *P. fimbriata*, but is perfectly distinct from both as a species. It agrees with the former in the *numerous*, instead of single rows of spinules on each concentric band, while it differs from it, and agrees with *P. fimbriata* in the elongato-ovate form, absence of mesial furrow, and very great gibbosity of the ventral valve, differing, however, from it in the smaller size, greater number of concentric bands in a given space at the same distance from the beak, and in having several rows of minute punctures, instead of a *single* row of elongate tubercules on each band."

Both *P. punctatus* and *P. elegans* occur in many of the same localities. In England they are abundant in almost every locality where Carboniferous Brachiopoda have been found, such as in Derbyshire; abundantly also in the Craven district; near Bolland; in shale and limestone at Settle, in Yorkshire; in dark Carboniferous limestone at Lowick, Northumberland; at Kendal, Westmoreland; Poolwash and Ronaldsway, Isle of Man, &c. In Scotland it occurs abundantly in the Carboniferous limestone and shale of Lanarkshire, Renfrewshire, Ayrshire, Buteshire, Dumbartonshire, Stirlingshire, Haddingtonshire, and Fifeshire. In Ireland, in Calciferous slate and limestone at Lisnapaste, Millecent, Tankardstown, Bruckless, Cornacarrow, &c. On the Continent of Europe it is also common at Visé, Lives, Namur, &c., in Belgium; in the valley of Prikcha, at Kosatchi-Datchi, &c., in Russia. In America it abounds in several localities, such as Zanesville (Ohio), Eddyville (Kentucky), and in coal measures throughout the western states, &c.

PRODUCTUS KEYSERLINGIANUS, *De Kon.* Pl. XXXIV, figs. 15, 16.

PRODUCTUS ACULEATUS, *De Koninck.* Descrip. des Anim. foss. du Terrain Carb. de Belgique, pl. x, fig. 8 *a, b, c* (not Martin), 1843.

— KEYSERLINGIANUS, *De Koninck.* Mon. du genre Productus, pl. xiv, fig. 6 *a—d*, 1847.

— — *M'Coy*, British Carb. Foss., p. 466, 1855.

Spec. Char. Shell small, subrectangular, or rotundato-quadrate; hinge-line slightly exceeding the average breadth of the shell. Ventral valve very gibbous, sometimes rather geniculated, and feebly depressed along the middle; beak small, and not overlying the hinge-line except at its attenuated extremity; ears small, slightly produced, and well defined. Surface traversed by numerous concentric lines of growth, as well as by a few wrinkles on the ears; a variable number of comparatively large spine-tubercles being likewise arranged somewhat in quincunx over the visceral portion of the valve. Dorsal valve very slightly concave, traversed by small wrinkles and minute concentric lines of growth, a few elongated tubercule-pits being also irregularly scattered over the visceral portion of the valve. Length 5, breadth 6, greatest depth between the valves 2 lines.

Obs. As observed by Prof. de Koninck, this species is distinguished from *P. aculeatus* by its smaller dimensions, more transverse shape, slight median depression, and lesser concavity of its smaller valve. The spiny tubercules are not scattered over the entire surface of the ventral valve, as in Martin's species, but are restricted to the visceral portion of the valve. The interior is very similar to that of other *Producta*. *P. Keyserlingianus* is not a rare fossil in the Carboniferous limestone of Settle, in Yorkshire, but much less so in that of Derbyshire. I am not acquainted with any specimens from either Scotland or Ireland.

On the Continent it occurs at Visé, in Belgium, as well as at Likwin, Government of Kalouga, and of Cosatchi-Datchi, Ural.

PRODUCTUS SPINULOSUS, *J. Sowerby.* Pl. XXXIV, figs. 18—21.

PRODUCTUS SPINULOSUS, *J. Sow.* Min. Conch., pl. lxviii, fig. 3, Oct., 1814.
PRODUCTA GRANULOSA, *Phillips.* Geol. of Yorks., pl. vii, fig. 15, 1836.
— CANCRINI, *De Kon.* Desc. des Anim. foss. du Terrain Carb. de Belgique, pl. ix, fig. 3 (not of De Verneuil), 1843.
— PAPILLATUS, *De Kon.* Ib., pl. x, fig. 6.
— GRANULOSUS, *De Kon.* Mon. du genre Productus, pl. xvi, fig. 7, 1847.
— — *M'Coy.* British Palæozoic Fossils, p. 472, 1855.
— — *Dav.* Scottish Carb. Brach., pl. iv, figs. 22—24, 1861.

Spec. Char. Shell transversely semicular; hinge-line a little shorter than the greatest width of the shell. Ventral valve regularly convex and evenly arched, without sinus; beak incurved, but not overlying the hinge-line, except at its attenuated extremity; ears flattened, with a few concentric wrinkles; surface covered with sub-regular, small, slightly elongated tubercules, irregularly or quincuncially arranged, each tubercule producing a slender spine. The tubercules are more often about half a line or so apart near the middle of the shell, but closer as they approach the beak. Dorsal valve deeply and evenly concave, and covered with small tubercule-pits, arranged somewhat in quincunx. Interior unknown. Length 7, width 9 lines.

Obs. This species does not appear to have attained proportions very much larger than those above recorded, and is easily distinguished from all other British Producta by shape and sculpture. To those examples in which the tubercules or spines were regularly arranged in quincunx the term *spinulosus* has been applied, while to those in which the spines or tubercules were less regularly disposed the term *granulosus* was given. I quite coincide, however, in the opinion expressed by Prof. M'Coy, that the two are simply different conditions of the same species, and should be therefore united. Prof. M'Coy is also of opinion that *P. Koninckianus*, De Verneuil, is distinct, and it is probable that *P. spinu-*

losus, De Koninck (but not of Sow.), is likewise a different species, for in addition to the tubercules, the shell is represented as being finely longitudinally striated, a character never observable in the many British examples of *P. spinulosus,* Sow., and *P. granulosus,* Phillips, that have come under my observation. I have never yet seen a well-preserved exterior of the smaller valve, but believe it to have possessed spines. In England *P. spinulosus* occurs in the Carboniferous limestone of Bolland; Settle, in Yorkshire; the Isle of Man, and in several Derbyshire localities. In Scotland it has been collected at Nellfield and Hillhead at 375 fathoms below "Ell Coal;" occurs likewise at Brockley, near Lesmahago, in Lanarkshire. In Ayrshire, at West Broadstone and Auchenskeigh, near Dalry. At Cat-Craig, near Dunbar, &c. In Ireland Mr. Kelly mentions Knockinny, Killukin, Millecent, and I possess a specimen from Little Island, near Cork, for which I am indebted to Mr. J. Wright. On the Continent it occurs at Visé, in Belgium, and has been found in Russia.

PRODUCTUS PLICATILIS, *J. Sow.* Pl. XXXI, figs. 3—5.

> PRODUCTUS PLICATILIS, *J. Sowerby.* Min. Conch., pl. 459, fig. 2, July, 1823.
> — — *Phillips.* Geol. of York., vol. ii, pl. viii, fig. 4, 1836.
> LEPTŒNA POLYMORPHA, *Muenster.* Verzeichniss der in der Kreis-Natur. Samml. zu Bayr. befindl. Petref., p. 45, 1840 (according to Prof. de Koninck).
> PRODUCTUS PLICATILIS, *De Kon.* Desc. des Anim. foss. du Terr. Carb. de Belgique, pl. xii, fig. 7, 1843; and Monographie du genre Productus, pl. v, fig. 6, 1836.
> — — *M'Coy.* British Pal. Foss., p. 168, 1855.

Spec. Char. Shell thin, transversely oblong, slightly indented in front; hinge-line about as long as the greatest width of the shell; ventral valve gibbous, and more or less geniculated; the visceral disc is semicircular and very slightly convex, while the anterior prolongation is abruptly arched; a median depression or sinus commences likewise at a short distance from the beak and extends to the front. Beak small, not overlying the hinge-line; ears flattened. Exteriorly, the visceral disc or posterior half of the valve is traversed by numerous regular or irregular, undulating, concentric wrinkles, while the entire surface is covered with fine, thread-like, longitudinal striæ. Spines long, slender, few in number, and rising here and there from the surface of the valve. Dorsal valve concave, with a small mesial elevation or fold towards the front. Sculpture similar to that of the opposite valve. Dimensions variable; two examples have measured—

Length 16, width 23, greatest depth between valves 7 lines.

„ 10, „ 13, „ „ 4 „

Obs. This Productus is well distinguished from its congeners both by shape and sculpture. It bears some slight resemblance to *Strophomena analoga,* but (as justly observed by Profrs. de Koninck and M'Coy) does not possess the area of the last-named species, and its interior details are those of *Productus,* as may be seen by a glance at fig. 5

of our plate, which represents an internal cast of the ventral valve, on which the adductor and divaricator impressions are well marked. The interior of the dorsal valve does not appear to have been hitherto discovered, at least no example has come under my observation. From *P. sub-lœvis* the form under description is distinguished by its transverse shape and much smaller dimensions, &c.

In England it is found in the Carboniferous limestone of Settle, in Yorkshire; the gray or lower limestone of Longnor and Castleton, in Derbyshire. In Ireland Mr. Kelly mentions that it occurs in the limestone of Armagh, Salmon, and Little Island. No specimen from Scotland has been hitherto produced, but it is very common in the limestone of Visé, in Belgium, as well as in the dark limestone of Hof, in Bavaria, and Falkenberg, in Silesia. It has also been stated to occur in Russia, but was not met with there by the authors of the 'Travels in Russia and Ural.'

PRODUCTUS SUB-LÆVIS, *De Koninck*. Pl. XXXI, figs. 1, 2; Pl. XXXII, fig. 1; and Pl. LI, figs. 1, 2.

> PRODUCTUS SUB-LÆVIS, *De Koninck*. Descript. des Anim. foss. du Terr. Carb. de Belgique, pl. x, fig. 1, 1843.
>
> STROPHOMENA ANTIQUATA, *Potiez et Mich*. Galer. des Moll. de Douai, vol. ii, pl. xlii, fig. 5, 1844 (according to De Koninck).
>
> PRODUCTUS SUB-LÆVIS, *De Keyserling*. Reise in das Petschora Land, pl. v, figs. 3 and 3ᵃ, 1846.
>
> — CHRISTIANI, *De Koninck*. Monographie du genre Productus, pl. xvii, fig. 3, 1847.

Spec. Char. Shell elongated oval, sub-quadrangular, and longer than wide; hinge-line about as long as the greatest width of the shell. Ventral valve gibbous, much vaulted upon itself at the beak, and longitudinally divided by a median groove or ridge, which commences to appear at a short distance from the beak, and extends to the front, while the lateral portions of the valve are more or less deeply furrowed; ears moderately developed, sub-cylindrically coiled, and sharply defined from the body of the shell by a row of tubular spines. Externally the visceral portion of the valve is traversed by numerous concentric, feebly marked wrinkles, while the entire surface is longitudinally and finely striated. Dorsal valve concave, following the curves of the opposite valve, and similarly sculptured. Interior unknown. Dimensions variable; two British examples have measured—

Length 4 inches 8 lines, width 4 inches 1 line.
„   3   „   9   „   „   3   „   10   „

Obs. This remarkable Productus varies much in appearance, according to age and specimen. When the external coat or surface is well preserved, it is very finely striated, as well as crossed by many concentric lines of growth, but when the outer surface has been removed by fossilisation the shell appears to be almost entirely smooth. Some examples

23

are likewise more deeply longitudinally grooved than are others, while in the ventral valve of certain examples there exists a median depression, which in others is converted into a projecting ridge.

It has appeared to me, after the attentive examination of about a dozen specimens, that Prof. de Koninck's *P. sub-lævis* and *P. Christiani* are only different conditions of a same and single species, and should therefore be united. In England *P. sub-lævis* occurs in the Carboniferous limestone of Clitheroe, in Lancashire, where it has been several times obtained by Mr. J. Parker, Curator of the Manchester Museum. It has also been stated to have been found in Wales, but no other British locality than that above given has been correctly ascertained. On the Continent it has been found by M. de Koninck in the Carboniferous limestone of Visé, in Belgium, and of Glageon, near Avesnes, in France, and appears to have sometimes attained rather large dimensions, although nowhere hitherto found in any profusion.

PRODUCTUS MESOLOBUS, *Phillips*. Pl. XXXI, figs. 6—9.

PRODUCTA MESOLOBA, *Phillips*. Geol. of Yorks., vol. ii, pl. vii, figs. 12, 13, 1836.
PRODUCTUS MESOLOBUS, *De Kon.* Desc. des Anim. foss. du Terr. Carb. de Belgique, pl. xii, fig. 8, 1843; and Monographie du genre Productus, pl. xvii, fig. 2, 1847.
— — *De Verneuil.* Russia and the Ural Mount., vol. ii, pl. xvi, fig. 8, 1845.
— — *M'Coy.* British Pal. Fossils, p. 468, 1855.
— — *Dav.* Mon. of Scottish Carb. Brach., pl. xxi, 1861.

Spec. Char. Shell thin, transversely oblong, and generally rather wider than long; hinge-line as long, and sometimes longer, than the average width of the shell; ears or cardinal extremities wide, and at times much extended. Ventral valve gibbous at the beak, and somewhat geniculated towards the front, with a wide, flattened, or slightly concave sinus, interrupted in the middle by a narrow median rib, and on either side of the sinus a rib is present, while another intervenes between these and the cardinal angles. The beak is proportionately large, and not overlying the hinge-line, except quite at its attenuated extremity. Exteriorly the visceral disc is covered with numerous concentric wrinkles, and a few tubular spines project from the five ribs. Dorsal valve concave, with a narrow median groove and two slightly marked lateral ones, while the surface is covered with concentric wrinkles and lines of growth. Interior unknown. Dimensions variable; three British examples have measured—

Length 11, width 19 lines.
    „   9,   „   12 „
    „   15,   „   14 „

Obs. Some foreign specimens have exceeded the proportions here given, and although possessing some of the characters of both *P. plicatilis* and *P. sub-lævis*, can always be distinguished by its narrow median ridge, deeper sinus, and lateral ribs.

In England it is found in the Carboniferous limestone of Settle, in Yorkshire, the dark limestone of Kendal, Poolwash, Isle of Man, as well as in the gray limestone of Derbyshire. In Scotland it has been found at 375 fathoms below "Ell Coal." At Braidwood, in Lanarkshire, also at Brockley, near Lesmahago; in Stirlingshire, in the Glarat lime works, or Campsie main limestone. In Ireland at Cornacarrow, Millecent, Little Island. On the Continent it occurs in the limestone of Visé and Tournay, in Belgium. In Russia, at Ilinsk, on the Tchusovaya, Ural Mountains.

Sub-genus—CHONETES, *Fischer.*[1]

We have already alluded to the close relationship which exists between *Chonetes* and *Productus*, and must now refer to the great difficulties in the way of a correct and definite determination of its species, which appears to have been unnecessarily multiplied, and at times been fabricated out of undeterminable or uncharacterised specimens or fragments. The confusion in which I found the species was so great that all my many efforts, consultations, and researches, have proved ineffectual to satisfactorily solve the difficulty with reference to some few of Prof. M'Coy's Irish so-termed species; but in order that the reader may form his own opinion, or be better enabled to continue the research, I have given figures of all the uncertain species (?), and, when desirable, reproduced the original descriptions. I have not been able to determine satisfactorily more than five or six species, viz.—1. *Chonetes comoides*; 2. *C. papilionacea*; 3. *P. Dalmaniana*; 4. *C. Hardrensis*; 5. *C. Buchiana*; and 6. *C. polita.* Those termed *C. laguessiana*, De Kon., *C. papyracea*, *C. crassistria*, *C. tuberculata*, *C. sub-minima*, *C. gibberula*, *C. sulcata*, *C. volva*, *C. perlata*, and *C. serrata*, M'Coy, may still demand further research, although I have myself but little doubt that the whole number are merely different conditions or synonyms of some of the five or six species above recorded; and even out of this number I do not feel entirely confident with reference to *C. Dalmaniana*. In his monograph of the sub-genus *Chonetes*, Prof. de Koninck states that "the slight difference which exists between the form and the exterior sculpture of the various species renders their classification much more difficult than that of *Productus*, and that he has taken as basis for his arrangement the number and

[1] Prof. M'Coy considers *Chonetes* to be simply a sub-genus of Leptœna, removing it, as well as *Strophalosia* (King) and *Aulosteges* (Helmersen), from the family PRODUCTIDÆ ('British Pal. Foss.,' Cambridge Museum,' p. 211), and placing them as a sub-genus among his *Orthisidæ*; but this view appears to me far from correct, as I have endeavoured to demonstrate in p. 112 of my 'General Introduction,' when showing that the reniform impressions are the same in *Chonetes* as in *Productus*, while a completely different arrangement prevails among the *Strophomenidæ*, of which Leptœna constitutes only a section.

nature of the longitudinal ribs which ornament the surface of all the species, with the exception of a single one, wherein these are replaced by concentric plaits." In Great Britain this last-named shell has not been hitherto discovered, but another, which is entirely smooth, has been obtained, so that the British species can be divided into two groups, which we will distinguish as follows, and to which we have added the probable synonyms.

STRIATÆ.	*C. comoides*, Sowerby.
	C. papilionacea, Phillips = *C. multidentata*, M'Coy = *C. papyracea*, M'Coy?
	? *C. Dalmaniana*, De Koninck.
	C. Hardrensis, Phillips = *C. sub-minima* = *C. gibberula* = *C. sulcata ?* = *C. volva?* = *C. perlata ?* M'Coy = *C. laguessiana ?* De Koninck.
	C. Buchiana, De Koninck = *C. crassistria*, M'Coy.
LÆVES.	*C. polita*, M'Coy.
Uncertain species?	*C. tuberculata* and *C. serrata*, M'Coy.

Although my distinguished friend has considered the arrangement, the number, and nature of the longitudinal striæ as the best means or character to be used in the discrimination of the species, I have not always found them sufficient, on account of the extraordinary variability in this particular which almost every specimen has presented not only at different stages of growth, but in shells of similar dimensions, and I believe that we will find the interior details in each of the species to be somewhat different, and a character of greater importance even than that of the exterior, as may be noticed in those of which we have been so fortunate as to procure specimens; and, indeed, in so difficult a matter every point, both exterior and interior, will require to be carefully examined and compared before a correct determination can be arrived at, and surely the difficulty will be great indeed, and even insurmountable, in those cases where the so-termed species have been fabricated from a crushed or imperfect fragment of a single valve !

CHONETES COMOIDES.   Pl. XLV, fig. 7 (1 to 6 ?).

PRODUCTUS COMOIDES, *J. Sow.* Min. Conch., pl. 329, April, 1816.
CHONETES   —   *De Keyserling.* Reise in das Petschora-land, pl. vi, fig. 1, 1846.
  —   —   *De Koninck.* Monographie du genre Chonetes, pl. xix, fig. 1, 1847.
  —   —   *Dav.* Quarterly Journal of the Geol. Soc., vol. x, pl. viii, fig. 1, and 2—8? 1853.

Spec. Char. Shell large, transversely semicircular, concavo-convex; hinge line straight, as long or a little shorter than the greatest width of the shell; valves strongly articulated by the means of teeth and sockets. Ventral valve convex, beak more or less developed and incurved, but not overlying the cardinal edge. Dorsal valve concave, following the curves of the other. Each valve is provided with a subparallel area of greater or lesser width, but always narrower in the smaller valve. In the middle of this last there exists a produced, trilobed, cardinal process, which enters and almost fills a corresponding triangular fissure in the ventral area, and which was probably partly covered or arched over by a pseudo-deltidium. The external surface of the shell is finely striated. In the interior of the greatly thickened ventral valve, and under the extremity of the beak, at the base of the fissure between the projecting teeth, originates a large, deep, pyriform muscular cavity, extending to beyond half the length of the valve, and occupying upwards of a third of its inner surface, its greatest breadth being towards the centre of the shell. In this depression are situated four elongated muscular impressions, and these are separated to a greater or lesser extent by three longitudinal ridges, the central one of which is shorter than the others, and assumes the character of a mesial septum; the two smaller scars, situated on either side near the central ridge, are due to the adductor, the outer and larger ones to the cardinal or divaricator muscles, the muscular scars not being equally deep in every example. Of the interior of the dorsal valve nothing is at present known. Dimensions variable; two British specimens have measured—

Length 2 inches 7½ lines, width 3 inches 10 lines.

„ 3 „ 5 „ „ 6 „ 2 „.

Obs. Sowerby's specimens of this remarkable species are incomplete, and his descriptions and illustrations consequently so, but he did not fail to mention that the shell is " very thick and rough within," this allusion having reference to the ventral valve only, for, as I have elsewhere shown, the convex valve was often four or five times as thick as the concave or dorsal one. Sowerby's figures do not show the fissure which exists in the area of the ventral valve, nor do they exhibit the area of the dorsal one, which is so well displayed in a specimen preserved in the Bristol Institution Museum (Pl. XLV, fig. 7). No spines could be detected on any of the few examples that have come under my observation, although some small circular holes could be perceived here and there, close to the cardinal edge, in one of the specimens; but it must likewise be remembered that the cardinal spines were very small in some other species of *Chonetes*, and especially so in the large *C. papilionacea*, and may, consequently, not have existed, or been destroyed, on the few fossil examples we possess.

It has been questioned whether the shell under description should be located with *Chonetes* or with *Productus*, and that we are not acquainted with the interior of the smaller and most important valve; still the area in both, and the strongly articulated hinge, would be an anomaly in *Productus*, which, on the contrary, is the constant character of *Chonetes*, and this alone would, at least provisionally, induce me to leave *C. ? comoides* in the last-named sub-genus. It has also been doubted whether the shells, fig. 1—6 of my Pl. XLV, do really belong to Sowerby's species; and here, again, I must repeat what I said in 1853, viz., " that I do not feel convinced that sufficient grounds exist for the establishing of two distinct species (an opinion in which I was then supported by Messrs. Salter and Woodward), the original type not appearing to us to differ materially in its convexity from several of those represented in my plate." The area, I am ready to admit, is certainly wider in both Mr. Sowerby's specimens than in those belonging to Mr. Ormerod (figs. 1—6), but the area generally varies much in its width in specimens of a same species, as I have already often had occasion to notice. Mr. Sowerby's second example (not figured in the ' Min. Conch.') is an incomplete interior of the *ventral valve,* in which, from the shell being young and shallow, the muscular impressions could not be as deep or as indented as in adult, very convex and thickened valves, such as is the original specimen figured in the ' Mineral Conchology,' or those illustrated in my plate (figs. 3, 4); nor could I perceive that the interior of Sowerby's specimen varied in any *essential* particular from those more perfect examples I had examined. In any case, fig. 7 must be regarded as the typical shape of *C. comoides,* of which the other specimens figured by me are probably variations.

Chonetes comoides does not appear to be a very common species, and all the specimens hitherto procured are from the Carboniferous limestone. It has been found in England and in Ireland in the following localities:—Llangaveni and Beaumaris, in Anglesea; Llanymynech and Tidenham Chase, in Gloucestershire; Chepstow, Treflach Wood, south-west of Oswestry; Bundoran, County Donegal; and Lough Erne, Fermanagh. On the Continent it appears to be equally rare; Prof. de Koninck mentions having found a single specimen at Visé, in Belgium, and that he possesses another example from Sablé, in France; that Count Keyserling has found it in the Carboniferous limestone at the banks' of the River Ylytsch, in the Ural, and that the Museum of St. Petersburgh possesses examples from the neighbourhood of Switschei, a village on the banks of the Ugra, in the Government of Kaluga, in Russia.

CHONETES PAPILIONACEA, *Phillips.* Pl. XLVI, figs. 3—6.

PECTINITES FLABELLIFORMIS, *Lister.* Hist. Conch., lib., iii, pl. 475, fig. 31, 1688.
SPIRIFERA PAPILIONACEA, *Phillips.* Geol. of York., vol. ii, pl. ii, fig. 6, 1836.
CHONETES PAPILIONACEA, *De Koninck.* Description des Anim. foss. du Terr. Carb. de Belgique, pl. xiii, fig. 5, and pl. xiii bis, fig. 1, 1843; and Monog. du genre Chonetes, pl. xix, fig. 2, 1847.
— MULTIDENTATA, *M'Coy.* Synopsis of the Char. of the Carb. Fossils of Ireland, pl. xx, fig. 8, 1844.
— PAPYRACEA, *M'Coy?* Ibid., pl. xx, fig. 2.

Spec. Char. Shell thin, sometimes rather large, transversely semicircular, depressed, slightly concavo-convex, but almost flat when young. Hinge-line straight, and as long as the width of the shell. Ventral valve slightly convex at the beak, but much flattened at the sides, and especially so near the cardinal edge; beak small, and not protruding beyond the cardinal edge; dorsal valve very gently concave along the middle, lateral portions much flattened; a well-defined area is present in each valve, the ventral one, which is the widest, being divided in the middle by a triangular fissure, partially arched over by a pseudo-deltidium, the cardinal process of the opposite valve filling up the remaining open space. Surface of both valves covered with numerous fine, thread-like, straight or flexuous striæ, which become more numerous by means of bifurcation and interstriation at variable distances from the beak and umbo, the striæ and interspaces being closely crossed by numerous fine, concentric lines of growth, which produce a beautifully crenulated appearance. Small spines rise from the striæ at variable distances, and a row of from twenty to thirty short ones project from the cardinal edge. Valves strongly articulated by means of teeth and sockets. Interior details imperfectly known. Dimensions variable; two British examples have measured—

Length 2 inches, breadth 5 inches 3 lines, greatest width between the valves 3 lines.
„ 1 „ „ 1 „ 9 „ „ $1\frac{1}{2}$ „

Obs. This is perhaps the largest and most beautiful species of the sub-genus hitherto discovered, and is remarkable on account of its elegant shape and sculpture. *C. comoides* is the only species with which the shell under description might be compared, but from which it can be distinguished without much difficulty, *C. comoides* being a much more ponderous and gibbous shell, and whose beak appears to be generally more rounded and produced than is that of *C. papilionacea.* Although it received the very appropriate name of *papilionacea* from Phillips in 1836, it had been figured and shortly described by Lister in 1688 — " *Pectinites flabelliformis tuberculoso commissura seu pectinites semicircularis compresso minuta ad modum striatus.*"

C. multidentata, M'Coy, of which I have been able to examine the original specimen, is evidently a synonym of the shell under description, and I am inclined to believe that the fragment figured and described by the same author under the denomination of *C. papyracea* is also another synonym. In England *C. papilionacea* has been found in the Carboniferous limestone of Bolland, Otterburn, Kendal, Settle, Ronalds Way, Isle of Man, and Dalton, in Furness, Lancashire, &c. In Ireland at St. John's Point, Ballybodonnel, Dunkineely; Cheeverstown, County Dublin, &c. On the Continent it has been found by Prof. de Koninck in the Carboniferous limestone of Visé, Chokier, and Temploux, in Belgium; Karova, Government of Kalougarear, Moscou; and at Sablé, in France.

CHONETES DALMANIANA, *De Koninck.* Pl. XLVI, fig. 7.

> CHONETES DALMANIANA, *De Kon.* Desc. des Animaux foss. du Terrain Carb. de Belgique, pl. xiii, fig. 3; pl. xiii[bis], fig. 2, 1843; and Mon. du genre Chonetes, pl. xix, fig. 3, 1847.

Spec. Char. Shell thin, transversely semicircular, concavo-convex; hinge-line as wide as the greatest breadth of the shell. Ventral valve regularly, but moderately, convex; area narrow, sub-parallel, and divided in the middle by a small fissure, partly arched over by a pseudo-deltidium. Dorsal valve concave, following the curves of the opposite one; area narrow; surface of both valves covered with numerous fine, radiating striæ, which become more numerous as they proceed from the beak and umbo to the margin from bifurcation and interstriation. Spines few on the surface of the ventral valve, but a row of longer ones rise from and follow close to the cardinal edge. Interior unknown. Dimensions variable; a British example measured—length 8, breadth 14 lines.

Obs. I am but very imperfectly acquainted with this species. Certain fragmentary specimens from the lower Scar limestone of Settle, in Yorkshire, could not be distinguished from some small Belgian examples of *C. Dalmaniana* I had received from Prof. de Koninck, and for which reason the species (?) is, with some uncertainty, here provisionally introduced. The specimens I have been able to examine of the shell under description appear to me intermediate in shape and character between *C. papilionacea* and *C. Hardrensis.* Prof. de Koninck mentions that it can be distinguished from the young age of *C. papilionacea* by its greater curvity (or convexity) and longer cardinal spines; it appears, also, to attain larger dimensions than *C. Hardrensis,* which some examples very closely resemble. Prof. de Koninck has given as synonyms of his species *Leptæna (Chonetes) volva* and *L. (Chonetes) multidentata* of M'Coy, but I cannot coincide with this determination, for, as already stated, *C. multidentata,* M'Coy, evidently belongs to *C. papilionacea,* and I am almost disposed to look upon *C. volva* as a variety of *C. Hardrensis?*

CHONETES BUCHIANA, *De Koninck.* Pl. XLVII, figs. 1—7, and 28.

> CHONETES BUCHIANA, *De Kon.* Descrip. des Anim. foss. du Terr. Carb. de Belgique, pl. xiii, fig. 1, 1843.
> —   —   *Dav.* A Monograph of Scottish Carb. Brachiopoda, pl. ii, fig. 1, 1861.
> LEPTŒNA CRASSISTRIA, *M'Coy.* Synopsis of the Carb. Foss. of Ireland, tab. xx, fig. 10, 1844; and British Carb. Foss., pl. iii H, fig. 5, 1855.

Spec. Char. Shell marginally transversely semicircular, concavo-convex, about one third wider than long; hinge-line straight, and either a little shorter, with its cardinal angles rounded, or exceeding the width of the shell, with rectangular or slightly acute and extended terminations. Both valves are provided with narrow sub-parallel areas; the ventral one, which is the largest, being divided by a small fissure, partially covered with a pseudo-deltidium, while in the middle of the ventral one there exists a prominent, V-shaped cardinal process. The ventral valve is moderately convex, and flattened towards its auriculate cardinal extremities. The beak, which is small and incurved, does not overlie the hinge-line. Dorsal valve concave, following the curves of the opposite one. Exteriorly the surface of the ventral valve is ornamented with from twelve to thirty generally simple ribs, with wider or narrower interspaces, the lateral ribs being sometimes larger than those which occupy the middle of the shell; the dorsal valve is similarly ornamented. Spines short and not very numerous, on the surface of the valve, but a row rises and follows close to the cardinal edge. In the interior of the ventral valve there exists a tooth on each side of the fissure; the adductor muscular impressions are small, but prominent, and separated by a median ridge or septum, which extends to about one third of the length of the valve. Immediately under and outside of the adductor scars are two concave, sub-quadrate, longitudinally grooved impressions, attributable to the cardinal or divaricator muscles. In the interior of the dorsal valve, under the cardinal process, exists a mesial ridge or plate, which extends to nearly two thirds of the length of the valve, and on either side may be seen two well-defined muscular scars, which are produced by the quadruple attachment of the adductor or occlusor muscle, while outside and in front of these are situated the reniform impressions. The remaining surface not occupied by muscular impressions is in both valves covered with small asperities. Dimensions very variable; three British examples have measured—

Length 8, width 11, greatest depth between valves 1½ lines
„ 6, „ 9, „ „ 1 „
„ 7, „ 13, „ „ 2 „

Obs. This species is, in general, easily distinguished from its congeners by its strong, simple ribs; but although these are simple in the greater number of specimens, I have noticed a tendency in a few rare instances to bifurcation and even interstriation close to the margin, and especially so in one example from the gray Carboniferous limestone near Settle (fig. 7), and which I had at one time distinguished by the varietal designation of *interstriata*. The ribs are exceedingly variable in their number and strength in different specimens, and every number between twelve and thirty can be counted in different examples, so that the gradation is complete. It is highly probable, if not certain, that Prof. M'Coy's *C. crassistria* is nothing more than a variety of *C. Buchiana* with small ribs, and, indeed, closely agrees with some Belgian examples of this last I received from Prof. L. de Koninck.

In England the species under description occurs in the Carboniferous limestone of Settle, and again in shales at Malham Moor, in Yorkshire, the specimens in the limestone possessing, in general, a greater number and smaller ribs than in the variety which is met with in the shales. It has been found also at Rutcheugh, in Northumberland. In Scotland it occurs at Gare, in Lanarkshire, at 239 fathoms below "Ell Coal," and in black Carboniferous shales (*calp*) at Bundoran, in Ireland. On the Continent it was first discovered by Prof. de Koninck in the lower Carboniferous limestone of Visé, in Belgium, where it is very rare.

CHONETES HARDRENSIS, *Phillips.*   Pl. XLVII, figs. 12—16, 17, 18, and 25?

> PECTEN   -   -   -   -   *Ure.* Hist. of Rutherglen and East Kilbride, p. 317, pl. xvi,
> figs. 10, 11, 1793.
> CHONETES HARDRENSIS, *Phillips.* Figures and Descriptions of the Palæozoic Fossils
> of Cornwall and West Somerset, p. 138, pl. lx, fig. 104,
> 1841.
> LEPTŒNA (CHONETES) HARDRENSIS, *M'Coy.* British Pal. Foss., 454, 1855.
> CHONETES HARDRENSIS, *Dav.* Mon. of Scottish Carb. Brach., pl. ii, fig. 2, 1861.

Spec. Char. Shell marginally semicircular, wider than long, concavo-convex; hinge-line straight, and either a little shorter or somewhat longer than the width of the shell, with rounded or angular terminations; each valve is provided with a sub-parallel area, but which is widest in the ventral one, and divided in the middle by a small fissure, partially covered by a pseudo-deltidium; ventral valve moderately convex, sometimes slightly depressed along the middle and flattened towards its auriculate cardinal extremities; the beak, which is small and incurved, does not overlie the hinge-line, while the dorsal valve assumes in different specimens a greater or lesser degree of concavity, with, at times, a slight longitudinal elevation along the middle, and flatness near the cardinal extremities. The surface of both valves is covered with numerous thread-like, radiating, and often bifurcating striæ, which increase in number by the interpolation of striæ at various distances from the beak and umbo, so that as many as 120 striæ may in some examples be counted round the margin, while at irregular distances small spines rise from their rounded surface in addition to those on each side of the beak; in adult examples there exist along the cardinal edge from five to nine slanting, tubular spines, which become longer and larger as they approach the extremities of the cardinal edge. The valves are articulated by means of teeth and sockets, while the muscular and other impressions do not differ materially from those already described in the preceding species. Dimensions variable; an average-sized specimen has measured—length 7, width 11, greatest depth 1½ lines.

Obs. The determination of the present species has given me much trouble; and although I have spent much time in the endeavour to arrive at a satisfactory conclusion, it is not without some hesitation that the term *Hardrensis* is here provisionally retained; provisionally, because I am as yet unable to determine whether Phillips's Devonian shell is the same as that to which Schlotheim, in 1820, applied the denomination *sarcinulata,* as Prof. de Koninck's illustrations of this last differ so much from those given by Prof. Schnur and some other palæontologists. I am likewise uncertain whether J. de C. Sowerby's *Lept. sordida* (1840) be really a synonym of the last-named shell, or different from Phillips's *Hardrensis,* as has been stated to have been the case by some authors; and lastly, because my learned friend, Prof. de Koninck, who has paid so much attention to the species of the genus, maintains a different opinion to that here recorded, while not absolutely denying the possibility of mine being correct.

Geologists and palæontologists have for many years been in the habit of distinguishing the Chonetes we are at present describing by the name *Hardrensis*, and although Prof. de Koninck, in page 206 of his ' Monographie du genre Chonetes,' has referred this shell and the one figured by Ure in 1793 to *C. variolata*, he subsequently determined that our form could not be assimilated to D'Orbigny's species, and proposed that the Scottish shell, which occurs also at Visé, in Belgium, should be made a new species of under the designation of *C. alternata.*[1] Having received from Prof. Phillips the loan of his four best and figured examples of *C. Hardrensis*, and having compared these with our Scottish and other examples, the result was that I could perceive no difference in the shape, areas, and striation, so that I deemed it preferable to allow the *Chonetes* we are describing to retain the name *Hardrensis*. *C. Hardrensis* is certainly a very variable species, and this has, no doubt, induced palæontologists to consider some of its variations in shape to be distinct species. The striæ vary much in number and strength. In some specimens they are exceedingly numerous and fine, while in other examples they are less numerous and coarser, the shell differing also much according to age and locality. *C. Hardrensis* (as I understand it) occurs in the limestone and shales of many English, Scottish, and Irish localities. In England it is found at Settle, in Yorkshire; Sturaway, in Shropshire, Newton-on-the-More, Northumberland, &c. In Scotland it is met with at Gare, in Lanarkshire, at 239 fathoms below "Ell Coal," 343 at Raes Gill, 356 at Hillhead. It occurs also at Capelrig, East Kilbride, Auchentibber and Calderside, High Blantyre; Brockley, near Middleholm, Lesmahago; Robroyston, north of Glasgow. In Renfrewshire at Arden Quarry and Orchard Quarries, Thornliebank. In Stirlingshire in various stages, such as Craigenglen, Mill Burn, the Campsie main limestone, Corrieburn, &c. In Ayrshire at West Broadstone, Beith; Auchenskeigh, Dalry; Goldcraig, Kilwinning, &c. In Ireland it occurs in many localities in the counties of Dublin and Kildare, &c.

In the shale above the " Hosie limestone" at South Hill, Campsie, in Stirlingshire, we find millions of specimens of a small variety (?) of the shell under description (fig. 22), but it does not appear to have exceeded some two and a half lines in length by three in width, the generality of specimens being even smaller. Mr. Young, to whom we are indebted for

---

[1] Having been enabled to compare a great many specimens of *C. variolata*, D'Orb., = *C. granulifera*, Owen, with our Carboniferous *C. Hardrensis*, I could perceive little or no difference between many of the specimens, although the ribs are at times, perhaps, finer and more numerous in certain examples of the American *C. variolata* than in some specimens of Phillips's species, so that I do not consider Prof. de Koninck to have been much mistaken when he referred Ure's figures to D'Orbigny's species. I may likewise observe that the interior details are in both exactly similar. *Chonetes striatella*, Dalman, sp., from the Silurian limestone of Gotland, closely resembles some examples of *C. Hardrensis*; it is, however, a little more concavo-convex, and I could not trace the existence of spines on its surface.

the knowledge we possess relative to this shell, thought it might be, perhaps, different from *C. Hardrensis*, but I am inclined to look upon it as a simple local variation or smaller race of the last-named species. In the shales of Newton-on-the-More and Derwick, as well as in those of Rahoran, in Ireland, we find another variety, with very fine striæ, which Prof. de Koninck has identified as his *Chonetes laguessiana* (fig. 19 of my plate), 'Monographie du Genre Chonetes,' p. 191, pl. xx, fig. 6; but after a very careful examination of many specimens, I could not bring myself to consider it specifically distinct from *C. Hardrensis*. Several of Prof. M'Coy's so-termed species appear to me to be variations or synonyms of the shell under description, but as palæontologists may perhaps object to the view I have taken, it will be preferable as well as desirable to allude to them under separate heads.

A.—*Chonetes* (*Leptæna*) *gibberula*, M'Coy. Synopsis of the Characters of the Carb. Foss. of Ireland, pl. xx, fig. 11, 1843 (23 of my plate).

" Semicircular, length two thirds the width, very gibbous in the middle; ears acute, flattened; surface very finely and regularly striated longitudinally; length one and a half lines, width three lines.
Loc. " Calcareous slate of Lisnapaste, Ireland." M'Coy.

B.—*Chonetes* (*Leptæna*) *subminima*, M'Coy. British Pal. Fossils, p. 456. Pl. iii, fig. 31, 1855 (fig. 24 of my plate).

" Rotundato-quadrate; length three fourths or four fifths of the width; receiving (ventral) valve very gibbous in the middle; greatest depth a little behind the middle; hinge-line as long as the shell is wide, forming flattened ears, slightly acute from the sigmoid outline of the sides, having three or four moderately long, slender spines on each side of the beak, extending backwards, as usual, in the plane of the margin; front margin moderately convex. Surface uniformly covered with close, obtuse striæ, once or twice branched, but nearly uniform in size on all parts of the shell, and so fine that twelve at the margin only occupy half a line when decorticated, the impressed lines between the striæ of the surface being coarsely punctured, and the beak slit by very deep impressions of the mesial septum, extending half the length of the shell. Entering (dorsal) valve nearly as concave as the receiving one is convex; surface similar in both valves, the striæ being crossed by fine, close lines of growth. Average width one and a half line, the depth seems about half the width. Very abundant in a piece of the black upper Carboniferous limestone of Derbyshire." (M'Coy, p. 456.)
Obs. These two so-termed species appear to me to be nothing more than young shells of *C. Hardrensis*, and Prof. de Koninck has placed the first among the synonyms of *C. sulcata*, which I am likewise inclined to look upon as a synonym of *C. Hardrensis*. I

think, however, that my Belgian friend is not quite correct while uniting *Chonetes crassistria* to *C. gibberula* and *sulcata,* for reasons already stated.

c.—*Chonetes (Orthis) sulcata,* M'Coy. Synopsis of the Characters of the Carb. Foss. of Ireland, pl. xx, fig. 6 (fig. 20 of my plate).

"Semicircular, gibbous; ears flattened; surface with very coarse, rounded, frequently branched striæ; hinge-line exceeding the width of the shell, furnished with ten strong, conical spines." (M'Coy, p. 126.)

Loc. Arenaceous shale, Bruckless, Dunkineely.

Having had the advantage of being able to examine the original specimen in Sir R. Griffith's collection, I could not distinguish it from many specimens of *C. Hardrensis.* The figure in the 'Synopsis' is not quite correctly drawn; that in my plate was taken from the type.

d.—*Chonetes (Leptœna) volva,* M'Coy. Synopsis of the Characters of the Carb. Foss. of Ireland, pl. xviii, fig. 14 (fig. 21 of my plate).

"Semicircular, gibbous; ears involute, separated from the body of the shell by a shallow depression; hinge-line twice the length of the shell; furnished with twenty-four slender, hooked spines; surface covered with fine, flexuous striæ. Length eight lines, width one inch four lines." (M'Coy, p. 121.)

Loc. Lower limestone, Millecent, Clare, &c.

Obs. Through the kindness of Sir R. Griffith I have been enabled to examine the type and several other examples of this so-termed species, and I must confess that, although I am not quite prepared to positively assert that *C. volva* is a synonym of *C. Hardrensis,* I must observe that many of the specimens could not be distinguished from the last-named shell. The striæ are not simple, but arranged exactly as in *C. Hardrensis,* of which it is very possibly nothing more than a variety. In his 'Monographie,' Prof. de Koninck has placed *C. volva* among the synonyms of his *C. Dalmaniana,* to which he has added also *C. multidentata.* This last belongs to *C. papilionacea;* but I am convinced that, had my learned friend been able to study the original types of several of M'Coy's so-termed species of Chonetes, he would have arrived at different conclusions.

e.—*Chonetes (Leptœna) perlata,* M'Coy. Synopsis of the Characters of the Carb. Foss. of Ireland, pl. xx, fig. 9 (fig. 25 of my plate).

"Semicircular, nearly twice as wide as long, convex; front margin nearly straight; surface marked with very fine, longitudinal, slightly flexuous striæ; hinge-line with about eight long, slender spines, set nearly at right angles; length three lines, width five and a half lines." (M'Coy, p. 120.)

Loc. Carboniferous slate of Rahoran, Fivemile Town, Ireland.

Obs. Prof. M'Coy states that this species seems closely allied to that which he has figured under the name of *Leptœna serrata*, but is distinguished by its angular plaits and rounded extremities, the radiating ridges being striated transversely. I have examined the original example, and must say that it appeared to me to be nothing more than a small specimen or young shell of *C. Hardrensis*. Prof. M'Coy has exaggerated the inward curve of the spines, and the enlarged illustration of the striæ is not correct, being exactly similar in the specimen to what we find in *C. Hardrensis*.

CHONETES POLITA, *M'Coy*. Pl. XLVII, figs. 8—11.

LEPTŒNA (CHONETES) POLITA, *M'Coy*. British Palæozoic Fossils, p. 456, pl. iii D, fig. 30, 1855.

Spec. Char. Shell transversely semicircular, concavo-convex; hinge-line slightly longer than the width of the shell, with a narrow, sub-parallel area in either valve, the ventral one, which is the widest, being divided in the middle by a small fissure, arched over with a pseudo-deltidium, both valves being articulated by the means of teeth and sockets; ventral valve gibbous, much vaulted in the middle and at the beak, which last is small, and does not overlie the cardinal edge; dorsal valve concave, following the curves of the opposite one. Surface of both valves smooth, and marked only by fine, concentric lines of growth, a few scattered spines rising from the surface of the larger valve, and two or three longer ones project from the cardinal edge on either side of the beak. In the interior of the ventral valve the muscular impressions are very feebly marked, and are divided by a small mesial ridge or septum, while pustule-markings rise from the whole remaining surface. In the interior of the dorsal valve, under and a little lower down than the cardinal process, there exists four small impressions left by the adductor or occlusor muscle, while two prominent ridges divide the central pair, and are prolonged to about two thirds of the length of the valve, the remaining portion of the valve being covered with elongated pustular markings. Dimensions very variable; two specimens have measured—

Length $2\frac{1}{2}$, width 3 lines.

„ 2 „ 2 „

Obs. This remarkable little shell is at once distinguished from all its congeners by the smoothness of its valves, as well as by the two peculiarly prominent ridges in the interior of the dorsal valve, and which are likewise present in one or more Silurian species. Prof. M'Coy remarks that this form has much the shape of *C. volva*, but is not so wide, is more gibbous, and is the only Carboniferous species he knows that has a smooth surface.

In England it is stated by Prof. M'Coy to be rare in the dark Carboniferous limestone of Lowick, Northumberland, that it occurs at Mount Rath, in Ireland, and we are indebted to the great zeal of Mr. Young for his discovery in Scotland, and where it occurs in vast numbers in the Craigenglen beds in Stirlinghire, and, as far as their position is at present known, in the lowest portion of the marine limestone, or close upon the horizon of the Scabricula limestone.

We will now conclude this perhaps unsatisfactory account of the British Carboniferous *Chonetes* by reproducing the description given by Prof. M'Coy of two very doubtful so-termed species.

Chonetes (Leptæna) serrata, M'Coy, Synopsis of the Char. of the Carb. Foss. of Ireland, pl. xviii, fig. 10 (fig. 26 of my plate).

"Semicircular, convex; surface covered with numerous, rather coarse, branched ridges; hinge-line equal to twice the length, straight, furnished with twenty small, hooked spines; breadth ten lines, length five lines. One of the rarest fossils we have." (M'Coy, p. 121.)

Obs. I have seen the original undeterminable fragment upon which this so-termed species has been fabricated, and which was obtained from the lower limestone of Millecent, in Ireland. Prof. de Koninck has referred it to *C. variolata*, but I would not venture to emit any opinion based upon such insufficient material.

Chonetes (Leptæna) tuberculata, M'Coy. Synopsis of the Characters of the Carb. Limest. of Ireland, pl. xx, fig. 5 (fig. 27 of my plate).

"Semicircular, convex; length two thirds the width; surface with about forty-eight thick, rounded, dichotomous, smooth ribs, each bearing towards the margin a row of from five to eight round tubercles.

"This curious little species is nearly semicircular, the hinge-line being much shorter than the width of the shell; moderately convex; the radiating ribs are thick, smooth, and distinctly separated, branching as they approach the margin, where they bear a row of six or eight little round tubercles. Length five lines, width eight lines." ('Synopsis,' p. 121.)

Loc. The lower Carboniferous limestone of Millecent, Clare, Ireland.

Obs. I am acquainted with the original specimen, which consists of a single valve, in the collection of Sir R. Griffith, but upon such insufficient material would not venture to pass any decided opinion as to its specific claims.

Family—CRANIADÆ.

Genus—CRANIA, *Retzius*, 1781.

The shells composing this remarkable and widely spread genus vary much in shape, although not much difference has taken place in this respect in time, for some Palæozoic species can hardly be distinguished from more recent and even living types. They are all marginally more or less circular or sub-quadrate, rarely free, but generally attached to marine bodies by the beak (when such does exist) or by the entire surface of the lower or ventral valve ; and it is from this circumstance that the ventral or attached valve varies so much in shape and sculpture. The upper or dorsal valve is always more or less limpet shaped, with a sub-central vertex, the surface being smooth or variously sculptured by concentric or radiating striæ, or ribs, some also possessing a spiny investment. There exists no articulated hinge, the valves being kept in place by a peculiar disposition of the muscles ; and although the animal has not been hitherto completely investigated, we will give figures of the interior of the valves, for the sake of explaining the more recent but provisional interpretation and names that have been applied to the muscular impressions by Mr. Hancock. But we must hasten, at the same time, to observe that the interior appearance and shape of the muscular and other impressions are very different in detail in certain species, although very similar in others. The figures here given will, however, suffice to explain the general character.

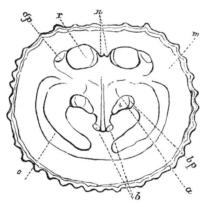
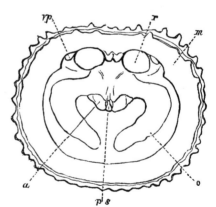

Dorsal or free valve.   Ventral or attached valve.

Crania Ignabergensis (var.), Cretaceous.

a. Occlusor (Hancock) = anterior adductors (Woodward).
r. Divaricator (H.) = posterior adductor (W.).
v, p. Ventral adjustor (H.).
d, p. Dorsal adjustor (H.) = protractor sliding muscle (W.).
p, s. [?] Anterior extremity of dorsal adjustor (H.).
b, p. Brachial muscle, posterior extremity (H.) = retractor sliding muscles (W.).
b. Brachial muscle, anterior extremity (H.) = retractor sliding muscles (W.).
n. Mesenteric muscle, destined probably to draw the alimentary tube backwards (? H.).
o. Ovarian (?), *m.* granulated margin.

Mr. Hancock, who, at my request, in May, 1859, examined the animal of three or four badly preserved specimens of *C. anomala* (the only examples then to be procured), has informed me that the impressions *a* are undoubtedly due to the occlusors, *r* to the divaricators, and that when the former muscles relax and the latter contract, the fluid in the perivisceral chamber will be forced forwards, and thus the valves will be opened a little in front, the action being the same as in *Lingula*; that *v,p*, is due to what may be termed the ventral adjustors; that these muscles form a scar close to the outer border of the divaricator in the ventral valve; the other extremities of this muscle converge and pass round the outer margin of the occlusor, to which they adhere, but Mr. Hancock could not exactly determine how they terminate; *d,p*, are considered due to the dorsal adjustors (?), one end of the muscle being attached to the dorsal valve, close to the outer border of the divaricators, the other most probably to the anterior process of the ventral valve. Although this could not be satisfactorily determined, from the very indifferent state of preservation of the specimens, at any rate the fibres of this extremity were firmly united to the inner border of the occlusors. The brachial muscle has both its extremities attached to the same valve (the dorsal), the anterior end to the ventral process, the dorsal close to the outer margin of the occlusor, with which it blends its fibres; that the arms are fixed to these muscles, which, perhaps, may be named the brachial. The mesenteric (*n*) is a flat, thin, membranaceous muscle, binding the dorsal mysentery to the process of the hinge-margin, to which, according to Mr. Woodward, the cardinal muscle is attached; but we may hope that, before long, Mr. Hancock will have been able to investigate anatomically some well-preserved examples, which may be dredged alive, close to some portions of our Scottish or Irish shores. The oval arms are thick, fleshy, and spirally coiled, the volutions are few, and directed vertically towards the cavity of the dorsal valve, somewhat as is seen in *Discina* and other genera. We may also notice that the brachial muscle is very closely united to the occlusor; that it is difficult to distinguish the two in the generality of specimens.

Dr. Carpenter has stated the structure of the shell in this genus to be widely different from that of Brachiopoda generally, but as still conformable to it in being penetrated by canals which are prolonged from the lining membrane of the shell, and which pass towards its external surface, these differing, however, from Terebratulæ in not arriving at that surface, and in breaking up into minute subdivisions as they approach it.

Although three so-termed species of British Carboniferous *Crania* will be here described, *C. quadrata* is the only well-determined species. Of *Crania? trigonalis* I have never seen any other than the original type, and a sight of its interior would be necessary prior to the species being definitely adopted. Of *Crania? (Patella) Ryckholtiana*, De Koninck = *C. vesicularis*, M'Coy, I am acquainted with but a single Irish specimen.

CRANIA QUADRATA, *M'Coy*. Pl. XLVIII, figs. 1—13.

ORBICULA QUADRATA, *M'Coy*. Synopsis of the Char. of the Carb. Foss. of Ireland, pl. xx, fig. 1, 1844.

CRANIA QUADRATA, *Dav*. Mon. of Scottish Carb. Brach., pl. v, figs. 12—21, 1860.

Spec. Char. Very variable in shape, on account of its mode of attachment, which is by the entire surface of its lower valve; when quite regular, is marginally sub-quadrate, almost circular, or slightly elongated, oval, the posterior edge being usually straight, or with a slight inward curve, while the shell is at the same time wider anteriorly than posteriorly. The upper or free valve is conical or limpet-like, the vertex being sub-central and closer to the posterior than to the anterior margin. Externally, the surface is marked with numerous but irregular concentric striæ or lines of growth, which give to the shell a somewhat roughened appearance. The interior of the attached or ventral valve is surrounded by a raised, thickened border, of moderate width, and upon it the tubular shell-structure is sometimes clearly discernible. In each corner of the disc, close to the posterior inner margin of the raised border, may be seen two somewhat circular, slightly convex, and prominent, but widely separated, muscular scars; while towards the centre of the disc two other prominent, but appropriate, muscular impressions exist, and which are, at the same time, somewhat hollowed out along their middle.[1]

The interior of the upper or free valve shows in each corner of the disc, close to the posterior inner margin of the border, a convex, oval-shaped, muscular prominence (and which is, according to Mr. Hancock, due to the divaricator muscle), while towards the centre of the valve are two prominent approximate impressions, which are referred by the same authority to the adductor or occlusor muscles. Dimensions very variable; a large example measured—length 7, width 7 lines. The generality of specimens are, however, much smaller.

Obs. The mode of existence peculiar to this as well as to other similarly constructed species is the cause of the great irregularity in shape assumed by the larger number of individuals, for it was the habit of the young of this as well as of other species of the genus to fix themselves as parasites to all kinds of marine objects, and they were sometimes so numerously and closely clustered together that their individual regular growth was prevented, from which it can be easily understood that in such cases the animal must have been compelled to develop itself in whatever direction it could find available space. When first formed, and up to a certain age, the shell of the attached valve was exceedingly thin, and adhered so closely to the surface of the object to which it was fixed as to have

[1] Mr. Hancock attributes the two first-mentioned scars to the divaricator, while the central pair are referred to the occlusor; the other muscular, ovarian, and vascular impressions which should exist in the interior were not sufficiently defined in the present species to admit of their being accurately described.

reproduced all the inequalities of its surface, but with age, and from the shell acquiring greater thickness, these inequalities were generally levelled. Nor is it an uncommon circumstance to find the roughness or sculpture of the object to which the lower valve adheres likewise reproduced upon the outer surface of the upper or unattached valve, in a similar manner to what we find to be the case with certain species of oyster. It is no easy matter to distinguish certain forms of Crania, as several species bear so close a resemblance to each other, and it was not until I had been able to study the original type of Prof. M'Coy's *Orbicula quadrata* (kindly communicated by Sir R. Griffith) that I could identify it with the same form which occurs so abundantly in the Carboniferous shales of Scotland.

In England but few examples have been hitherto discovered; one or two specimens were found by Mr. C. Moore, at Holwell, near Frome. In Scotland it occurs at Gare, in Lanarkshire, at 239 fathoms below "Ell Coal," 343 at Langshaw Burn, and 375 at Kilcadzow. It occurs also at Auchentibber and Calderside, High Blantyre; Capelrig, East Kilbride; Brockley, near Lesmahago; and Robroyston, north of Glasgow. In Ayrshire, at West Broadstone, Beith; Goldcraig, near Kilwinning; Cessnock, near Galston; and on the bank of the stream Pomillen, near Strathavon. In Renfrewshire, at Howood, near Paisley; and Orchard Quarry, Thornliebank. In Kircudbrightshire, in strata cropping out on the seashore, near Kircudbright. In Stirlingshire, in the Balglass Burn beds, and in those of the Campsie main limestone. In Ireland it was found in Carboniferous shale at Rahan's Bay, in Donegal, one mile south-west of Dunkineely village. On the Continent it occurs at Tournay, in Belgium; and at Tuscombia, Alabama, in America.

CRANIA RYCKHOLTIANA, *De Koninck.* Pl. XLVIII, figs. 15, 16, 17?

> PATELLA RYCKHOLTIANA, *De Kon.* Animaux Foss. de la Belgique, pl. xxiii, fig. 5, 1843.
> CRANIA VESICULARIS, *M'Coy.* Synopsis of the Char. of the Carb. Foss. of Ireland, pl. xx, fig. 3, 1844.

Spec. Char. Ovato-orbicular, nearly circular, or slightly longer than wide. Upper valve conical, limpet-like, the vertex being sub-central and closer to the posterior than to the anterior margin; surface smooth, marked only by a few concentric lines of growth; lower or attached valve unknown. Dimensions variable; two British examples have measured—

Length 8, width 10, depth 6 lines.
„ 12, „ 12, „ 4 „

Obs. It appears quite probable that the *Patella Ryckholtiana* and *Crania vesicularis* belong to the same species, but unfortunately, although several exteriors of the upper valve have been discovered, no example showing the interior has hitherto turned up. In England it has been found in the Carboniferous limestone of Castleton, in Derbyshire.

In Ireland it is mentioned from the limestone of Millecent, in Kildare. No Scottish specimen has hitherto been discovered.

CRANIA? TRIGONALIS, *M'Coy.*   Pl. XLVIII, fig. 14.

ORBICULA TRIGONALIS, *M'Coy.*   Synopsis of the Carb. Foss. of Ireland, pl. xx, fig. 2, 1844.

Spec. Char. " Conical, obovate, trigonal; anterior end narrow, rounded; posterior sub-truncate; surface irregular, marked with close, rounded, radiating ridges from the beak, which is small, deflexed, and little more than one fourth the length from the anterior margin. Length four and a half lines, width three lines." (' Synopsis,' p. 104.)

Obs. I have reproduced Prof. M'Coy's description, as I know so little about the shell, a single valve having been hitherto discovered, and which was kindly lent me by Sir R. Griffith; and although I would not dare to assert that it positively belongs to the genus *Crania*, it appears to me more probably so than to *Discina*, where placed by the author of the ' Synopsis.' *C.? trigonalis* was obtained from the Calciferous slate of Lisnapaste, in Ireland.

Family—DISCINIDÆ.

Genus—DISCINA, *Lamarck*, 1819.

The shells belonging to this genus are usually circular or longitudinally oval, the larger or imperforated valve being conical, or limpet-like, with the apex inclining towards the posterior margin. The lower valve is conical, opercular, flat, or partly convex, and perforated by a narrow, oval, longitudinal slit, which reaches to near the posterior margin, and which in recent species is placed in the middle of a depressed disc, the shell being always attached to marine bodies by means of a pedicle, and never by the substance of the shell, as in Crania. The valves are unarticulated, and kept in place by a particular disposition of muscles, the occlusor and divaricator impressions being somewhat similarly situated to those of Crania. Much has still to be done before the animal will have been completely or satisfactorily anatomically investigated. The so-termed oral arms have been described by Mr. S. P. Woodward, in his excellent ' Manual,' as being curved backwards, returning upon themselves, and ending in small spires, directed downwards towards the ventral valve, and the only process which could possibly have afforded support to the arms is developed from the centre of the ventral valve, as in Crania. In recent species the shell is stated by Dr. Carpenter to be horny and minutely punctate, the tubuli being generally arranged in fasciculi, so that their transverse sections present a series of dots. Dr. Gratiolet believes, however, that the shell is not entirely composed of a horny substance, but somewhat similar to that of

Lingula, although the calcareous element is enormously greater in the last-named genus. The chemical composition of the shell of Discina has been stated by Mr. S. Cloëz to be similar to that of Lingula, of which an analysis will be found further on.

Discina appears to have existed during almost the entire series of Palæozoic and Mesozoic periods up to the present day, and it is probable that the animal was not at any period the inhabitant of very deep water, for all the recent species of Lingula and Discina, or those species with a horny shell, have prevailed in the littoral zone, and do not appear to have descended deeper than about eighteen fathoms. The reader is referred for more ample details to Prof. Suess's excellent 'Memoir on the Habitat and Distribution of the Recent and Fossil Brachiopoda' recently published in Vienna.

One or two British species only have been hitherto found in the Carboniferous rocks.

DISCINA NITIDA, *Phillips.* Pl. XLVIII, figs. 18—25.

> ORBICULA NITIDA, *Phillips.* Geol. of York., vol. ii, p. 221, pl. ix, figs. 10—13, 1836.
> — CINCTA, *Portlock.* Report of the Geol. of Londonderry, &c., pl. xxxii, figs. 15, 16, 1843.
> DISCINA BULLA, *M'Coy.* British Palæozoic Fossils, pl. iii D, fig. 32, 1855.
> — NITIDA, *Dav.* Mon. of Scottish Carb. Brach., pl. v, figs. 22—29, 1860.

Spec. Char. Shell marginally circular or elongated oval, the posterior portion being rather narrower than the anterior one. The larger or free valve is conoidal or limpet-like, and more or less elevated, the pointed apex being situated at variable distances between the centre and the posterior margin, but it is not always the most elevated portion of the valve. The surface is covered with numerous small, irregular, concentric wrinkles or striæ. The smaller or lower valve is somewhat flattened or slightly concave towards its anterior margin, with an oval-shaped foramen, surrounded by an elevated, convex margin, which extends from near the centre of the valve to a variable distance from the posterior edge. This valve is likewise ornamented with numerous small, irregular, concentric ridges or wrinkles, with small, flattened interspaces. No interiors have been hitherto obtained. Dimensions variable; three examples have measured—

Length 10½, width 10, depth 6 lines.
  „ 7 „ 7½ „ 3 „
  „ 7 „ 5½ „ 4 „

Obs. After a lengthened examination of *Discina cincta,* as well as of *Discina bulla,* I could perceive no valid grounds for separating these two so-termed species from *D. nitida;* and any one possessing a sufficiently numerous series of specimens of the last-named form would, I think, soon perceive that Phillips's shell presented every degree of elevation—from that of an almost depressed shell to that extreme " inflated, bubble-like form" described by Prof. M'Coy. I am therefore quite disposed to concur in the Irish author's opinion when he considers *D. cincta* as nothing more than the perfect condition of *D. nitida;* for when the outer surface of the last-named shell is absent, which is often the case, the cast is generally almost smooth, or marked only with a few faint concentric and radiating lines, a circumstance which has apparently led some palæontologists to believe that Phillips's shell was smooth, while that of Portlock's was concentrically striated.

It is also highly probable that the Permian *Discina Koninckii* cannot be specifically separated from the Carboniferous *D. nitida.*

Discina nitida is a common shell in many localities. In England it is mentioned by Prof. Phillips to occur at Bowes, Pateley bridge, Lee Harelaw and Otterburn; also at Coalbrookdale. It has been found also at Lowick, Northumberland; in the upper part of the Carboniferous limestone of Derbyshire, &c. In Scotland it abounds at Belston Place Burn, in Lanarkshire, at 173 fathoms below "Ell Coal," 239 at Gare, 265 at Belston Burn, and 354 at Raes Gill, in the parish of Carluke. It is likewise found at Haw-hill, near Lesmahago; Auchentibber and Calderside, High Blantyre; and Capel-rig, East Kilbride. In Renfrewshire, at Arden Quarry, Thornliebank. In Stirlingshire, at Craigenglen, and in the Balgrochen Glen ironstone, &c. In Ayrshire, at Cragie, near Kilmarnock; Cessnock, parish of Galston; and Netherfield, near Strathavon. In Fifeshire, at Strathkenny, St. Andrew's, &c. In Haddingtonshire, at Cat Craig, near Dunbar. It occurs also in Edinburghshire, and along the Berwickshire coast, from the mouth of the Tweed to Ross. In Ireland it occurs at Benburb, Bundoran, Culkagh. In America it has been found in Pike and Adams County, Illinois, &c. &c.

Discina Davreuxiana, *De Koninck.*   Pl. XLVIII, fig. 26.

> Orbicula Davreuxiana, *De Koninck.*   Desc. des Animaux Foss. du Terrain Carb. de
> Belg., p. 306, pl. xxi, fig. 4, 1843.

Spec. Char. Shell marginally oval, longer than wide; larger or upper valve conoidal or limpet-like, the apex being situated between the centre and the posterior margin. Surface smooth, with concentric, rounded wrinkles. Lower valve unknown. Length 3, width $2\frac{1}{2}$, depth 1 line.

Obs. I am acquainted with but a single British example, which was found by Mr. Joseph Wright in the Carboniferous limestone of Little Island, near Cork, in Ireland. The specimen bears a close resemblance to the *O. Davreuxiana*, but I am still uncertain whether this last is a good species or only a variety of *D. nitida*, the material at my command not being sufficient to enable me to decide the question. In Belgium it occurs near Tournay.

Family—LINGULIDÆ.

Genus—LINGULA, *Bruguière* 1789.

The limit of variation among the shells composing the genus Lingula appears to be more restricted than what is prevalent among the generality of other genera and species of Brachiopoda. It is, therefore, very often no easy matter to distinguish and correctly determine some fossil species, even when occurring in different and often widely separate geological periods.

The shell of Lingula is thin, equilateral, usually longer than wide, and broader at the front than at the beaks, which are likewise more or less pointed, while the front is either nearly straight or with a slight inward or outward curve. The shell is also sub-equivalve ; the extremity of the beak of the dorsal valve being somewhat more elongated and pointed than that of the ventral one.[1] The external surface is also either nearly smooth or concentrically striated. The valves are usually moderately convex, and generally deepest or most elevated towards the beak, and become more flattened as they approach the front. The apex of the dorsal valve is likewise situated quite close to, but not contiguous with, the rounded margin of the beak, and by which character the valves can be readily distinguished both in the recent and fossil condition.

When alive, the valves of Lingula were slightly gaping at each end, contiguous only along the lateral margins ; but the animal could, at its will, by the action of certain muscles, close or draw together one or other extremity ; nor does there exist any articulation, the valves being kept in place by the means of a complicated system of muscles, to be hereafter described. The animal was also provided with a very long pedicle, of a peculiar construction, which was chiefly attached to the inner groove situated in the beak of the ventral valve; and when alive, did not inhabit great depths, most recent species having been found at low water, buried in sand.

The intimate shell-structure of Lingula has been described by Dr. Carpenter, and we will therefore only refer to Dr. Gratiolet and Mr. S. Cloëz's more recent observations. The first-named *savant* states that the shell is composed of two distinct elements, the one being horny, the other shelly. That they are disposed in layers, or thin laminæ, which succeed each other alternately from the convex surface of the valves, the outer or superficial one being horny ; that these layers have not the same thickness, the testaceous ones being the thickest on and near the visceral side, while the horny ones are more so towards the exterior surface ; and that while the horny layers are entirely formed of parallel fibres,

---

[1] Anatomists appear to differ as to the names by which the valves should be designated; it may therefore be as well to mention those that are synonyms. The shortest is the *dorsal valve* of Woodward, Hancock, &c. ; = *valve inférieure*, Gratiolet ; = *valve droite*, Vogt. The longest is the *ventral valve* of Woodward, Hancock, &c. ; = *valve supérieure*, Gratiolet ; = *valve gauche*, Vogt. Mr. Hancock is of opinion that if the names of the valves were to be changed, that they should be called *anterior* and *posterior*. In this monograph we will continue to make use of those first mentioned.

without trace of perforations, the testaceous ones are traversed by a multitude of minute canals, recalling those of the Terebratulidæ.[1]   Mr. S. Cloëz has likewise shown that the valves of Lingula, when dried at 100°, contain, for 100 parts—

Organic matter	45·20
Carbonate of lime	6·68
Phosphate of lime	42·29
— of magnesia	3·85
— of sesqui-oxide of iron	1·98
Silica	Traces.

This distinguished French chemist observes, at the same time, that this composition of the test of Lingula approximates to that which M. Chevreuil signalised in the scales of the Lepidostria, as well as to the test of insects as described some years ago by Hatchett.

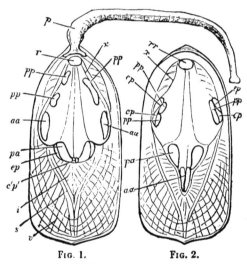

FIG. 1.   FIG. 2.

Lingula anatina (recent).

Fig. 1. *Ventral valve*, Woodward, Hancock, &c.; valve supérieure, Gratiolet.

Fig. 2. *Dorsal valve*, Woodward, Hancock, &c.; valve inférieure, Gratiolet.

p, a.   Posterior occlusor, Hancock; = anterior adductors, Woodward; = pré-adducteurs, Gratiolet; = untere schiese muskelbündel, Vogt; = muscles qui vont directement d'une coquille à l'autre, Cuvier.

a, a.   Anterior occlusor, Hancock; = anterior retractors, Woodward; = muscles obliques postero-antérieurs, Gratiolet; = mittlere schiese muskelbündel, Vogt.

r, r, r. Divaricator, Hancock; = posterior adductor, Woodward; = post-adducteurs, Gratiolet; = oberer schliessmuskel, Vogt.

p, p.   Posterior adjustor, Hancock; = posterior retractors, Woodward; = muscles croisé ou oblique transverse, Gratiolet.

e, p.   External adjustor, Hancock; = external protractors, Woodward; = surface sur laquelle se prolonge l'insertion du muscle oblique antero-postérieur paire externe, Gratiolet.

c, p, and *c', p'.* Central adjustor, Hancock; = central protractor, Woodward.   *c, p.* Muscles obliques antérieurs paire interne, Gratiolet; and *c', p'*, muscles obliques antérieurs paire interne et externe, Gratiolet.

x.   Line indicating the posterior parietals, Hancock; = peaussier verticaux, Gratiolet.

p.   Peduncular muscle, Hancock; = capsule of pedicle, Woodward; = muscle intérieur du pedoncule, Gratiolet.

i.   Impressions produced by the central branches of sinuses, Hancock.

s.   Impressions produced by the outer or lateral branches of sinuses.

v.   Impressions produced by pallial lobe.

' I have considered it desirable to reproduce these details, as they are important to British Palæontologists, and have been taken from the first portion of Dr. Gratiolet's recently published memoir on 'The Anatomy of *Lingula anatina*;' and I avail myself of the present opportunity to express my grateful

In the interior of the valves may be seen a number of muscular and other impressions with which the palæontologist should become acquainted, but it would be out of place were we to enter into a minute anatomical description of the animal itself in a work exclusively devoted to fossil species. We will therefore briefly place before the reader a few details only concerning those muscles which have left recognisable impressions in the interior of the valves. It must also be observed that, although Mr. Woodward, Mr. Hancock, Dr. Gratiolet, and others, agree as to the shape and position of the various muscles, they do not interpret the functions of some of these exactly in the same manner, and as a number of names have been applied to designate the same muscle, the first thing to do will be to place before the readers figures showing the position of the impressions and the synonymous terms that have been employed, and these will also save us the necessity of describing the scars, which a glance at the figures will explain far better than could be done by simple words.

The muscular system is much more complex in the unarticulated divisions of the Brachiopoda than in the articulated groups. While describing their shape and direction we cannot do better than to follow what has been said by Mr. Hancock in his admirable memoir upon the "Anatomy of the Brachiopoda," published in the 'Philosophical Transactions of the Royal Society,' 1858:

The *anterior occlusors* " are a pair of stout muscles, of about equal thickness throughout; they pass from the ventral valve, one at each side, in front of the visceral mass, and inclining forwards and inwards, they go to be attached to the sides of the central ridge of the dorsal valve, about one third of the length of the shell from the anterior margin. The dorsal extremities are compressed, and have their sides in contact."

The *posterior occlusors* " are rather stouter and much stronger, and go directly from valve to valve, parallel with each other. The ventral extremities are placed a little in advance of the corresponding terminations of the anterior pair, and the dorsal extremities of the former are situated a little behind those of the latter."

The *divaricator*, " though forming a single mass, is really two muscles combined. It is short and stout, and is situated at the posterior extremity of the perivisceral chamber, passing directly between the valves, and having its attachments immediately within the umbones. The extremities have a semicircular form, arched behind, and slightly bifid in front, indicating its double nature."

thanks to the distinguished French anatomist for the high honour he has conferred by dedicating to me the result of his admirable researches. For details concerning the animal of *Lingula* I must refer the reader to the following memoirs:—Cuvier, 'Mémoire sur l'Animal de la Lingula,' 1797 and 1802. Vogt, ' Anatomie der Lingula anatina,' 1845. Owen, "On the Anatomy of the Brachiopoda," 'Trans. of the Zool. Soc.,' 1835; as well as in Davidson's 'General Introduction,' chap. i, 1853. S. P. Woodward, 'Manual of the Mollusca,' 1854. But especially to the magnificent memoir by Hancock, "On the Organization of the Brachiopoda," 'Trans. Royal Soc.,' 1858. As well as to Gratiolet's most important and excellent memoir, " Études Anatomiques sur la Lingula anatina," in the 'Journal de Conchyliologie,' for January and April, 1860.

26

The *central adjustors*.—"This pair are attached to the ventral valve by fine points between the posterior occlusors in front; they are placed close together, one on each side of the median line. Sweeping round the inner border of these muscles, they diverge posteriorly, and increasing in size as they go, ascend towards the dorsal valve, to which they become adherent, one on each side, immediately within the parietes of the body."

The *external adjustors* "arise from the ventral valve, at the outside of the posterior occlusors, and in contact with them. They are at first pretty stout, but on passing outwards and backwards they enlarge a little, and ascending, are inserted into the dorsal valve, one on each side, immediately behind the central pair."

The *posterior adjustors* "are large and powerful muscles, and though they may be considered as a pair, they are asymmetrical, there being two on one side and only one on the other. As they pass across from valve to valve they intersect each other, the single one passing between the other two. The single one is as large as the other two both together, and is attached to the left side of the ventral valve, about midway between the divaricator and the anterior occlusor. From this point it passes diagonally upwards and forwards... and on reaching the opposite side of the dorsal valve has the other end inserted into the latter, immediately within the posterior terminations of the external and central adjustors of the same side. At the points of attachment the three muscles are pressed so close together that they appear at first sight as only one. The two opposite posterior adjustors take their origin from the right side of the ventral valve, considerably apart; but both of them close to the lateral parietes of the body, one only a little in advance of the divaricator, and the other a short distance further forward. They converge as they penetrate the visceral mass, and sloping forward, one on each side of the visceral muscle, with the alimentary tube above them, they ascend to their insertion into the left side of the dorsal valve, directly within those of the external and central adjustors. Therefore at this point there are the terminations of four muscles in close contact."

The *peduncular muscle* "has its insertion immediately within the umbo of the ventral valve, and close behind the divaricator."

x is a line indicating attachment of the posterior parietals.

There are some other muscles, but as they do not leave any impressions upon the surface of the shell, will not require to be recorded here; but now that the reader has had the names, shape, and direction of the muscles explained, it will be necessary to mention as briefly as possible what are their supposed functions, and for this purpose I will particularly mention the views of Mr. Hancock and Dr. Gratiolet, as they are the authors who have more recently examined the animal with the greatest attention. In order to avoid the possibility of error upon my part, I requested the last-named two anatomists to kindly transmit me their views, which I will now transcribe.

According to Mr. Hancock, the functions of the various muscles might be shortly described as follows :

"The *anterior* and *posterior occlusors* are mainly instrumental in closing the valves.

" The *divaricators* are the chief agents in opening them. When they contract, the umbonal regions of the valves are approximated, and thus pressing forward the fluid in the peri-visceral chamber, their anterior margin is separated.

" The primary function of the three pairs of *adjustors* is to keep the valves opposed to each other, or, in other words, to adjust them, and in this respect to compensate for the deficiency of the hinge and condyles. When in full action and in co-operation with the occlusors and divaricator, they likewise assist in closing the valves. The adjustors are the sliding muscles of those authors who believe in the sliding of the valves over each other. The anterior occlusors have had a similar function assigned to them.

" The *peduncular* muscle attaches the shell to the peduncle, and has probably the power of moving the former upon the latter."

Some doubt exists concerning the homology of the adjustor muscles (in Lingula), but Mr. Hancock has not expressed any strong opinion on the point; he thought it likely that the muscles so named in the articulated and unarticulated genera of Brachiopoda were probably homologous, but is ready to admit that he may be possibly mistaken, and, if so, he would not, however, be disposed to change the names, for in both divisions the function of these muscles is to adjust the valves. It is only necessary to keep in view that they are *not* homologous.

We will now give Dr. Gratiolet's description of the functions of the muscles, and for which I am indebted to the author himself, who has kindly therein distinguished the effects of *simultaneous action* and those of the *alternate action;* and for the assistance of the reader Mr. Hancock's names have been added within brackets.

1. MUSCLES PRÉADDUCTEURS (posterior adductor) and POST-ADDUCTEURS (divaricators).

    a. Simultaneous action.—The energetically drawing together the valves in their whole length.

    b. Alternative action.—When the préadducteurs contract themselves alone, they close the shell in front and make it gape behind. When the post-adducteurs contract alone, they close the shell almost completely behind and make it open in front.

2. MUSCLES PEAUSSIERS VERTICAUX (posterior parietal muscles).

    a. Simultaneous action.—They depress the body behind, cause the internal fluid to flow towards the arms, and consequently come strongly in aid of the action of the " muscles post-adducteurs."

3. MUSCLES OBLIQUES TRANSVERSALEMENT, *muscles croises* of Cuvier (posterior adjustors)

    a. Simultaneous action.—They energetically draw together the valves.

    b. Alternative action.—Taking for a fixed point the valve which I call the superior (ventral), the right muscle, causing a traction upon the opposite side of the inferior valve, makes it deviate a little to the right by a sliding, the extent of which I should not know, *a priori*, how to measure. The double-cross muscle of the left side, acting symmetrically, makes it deviate a little to the left.

4. MUSCLES OBLIQUES POSTERO-ANTÉRIEURS (anterior occlusor) *et* ANTERO-POSTÉRIEURS (central and external adjustors).

    a. Simultaneous action.—They draw the valves together energetically.

    b. Alternative action.—Supposing always the superior (ventral) valve as a fixed point the "muscles postero-antérieurs acting from behind forward upon the inferior (dorsal) valve, make it slide backward. The muscles " antero-postérieurs " acting from the front backward upon the inferior valve, make it slide forward.

    N.B.—If one admitted an alternative possible between the longitudinal oblique muscle of the right side and the left side, their movements would evidently come in aid of those of the cross muscles.

5. MUSCLES PÉDONCULAIRES (peduncular muscle) *et* MUSCLES MARGINAUX.

These muscles leave also their traces upon the shell. The first erect (adjust) the body upon the peduncle, and that in two ways—the first by a direct action, in the second place by causing the fluids which fill the internal cavity of the peduncle to ebb into the body. The second act exclusively upon the border of the great pallial lobes.

It will therefore be seen, from what has been stated, that, although anatomists agree as to the shape and position of the muscles, they entertain different views respecting some of their functions. Thus, Mr. Hancock objects entirely to the notion of the sliding of the valves in different directions over each other by the aid of the adjustors (protractor sliding muscles of Woodward[1]), a theory first propounded by Cuvier and Owen; while Dr. Gratiolet believes that the cross disposition of certain muscles, whether from behind forward or whether from right to left, would lead one to imagine a compensated antagonism from which equilibrium would result during the simultaneous contraction of all the elements; and that the oblique muscles transversely crossed of Cuvier, his "muscles obliques postero-antérieurs" and "antero-postérieurs," were employed in the sliding action of the valves. Mr. Hancock, on the other hand, observes that in *Crania*, where the muscular system is arranged after the plan of *Lingula*, there exists *no* sliding movement, and that Mr. Lucas Barrett, who has seen *Crania* alive, has distinctly stated that "the valves open by moving upon the straight side, as on a hinge, without sliding of the valves;" but it would be out of place and presumptuous were I to dwell any longer upon

---

[1] According to Mr. S. P. Woodward, *Lingula* would possess a pedicle muscle ; three adductor muscles, the posterior pair combined; two pairs of retractors, the posterior pair unsymmetrical, one of them dividing; and two posterior sliding muscles.

this controversed question, my object having been attained if I have been able to lay before the geological and palæontological reader the views of two such eminently distinguished anatomists as Mr. Hancock and Dr. Gratiolet, and no doubt time will prove which is the correct interpretation; for now that the question at issue has been made known, it will not be difficult for some observer who may happen to be where *Lingula* is found alive to notice whether or not the valves do slide upon one another. We will now conclude the little we had deemed necessary to say of the animal by observing that the so-termed oral arms are not supported, as in many of the articulated genera, by a more or less complicated system of lamellæ; that they are fleshy, with their spires directed towards each other.

After much examination I have reduced the so-termed species of British Carboniferous Lingulæ to four, viz., *L. squamiformis*, *L. mytiloides*, *L. Credneri* (which may possibly be a variety of *L. mytiloides*), and *L. Scotica*.

LINGULA SQUAMIFORMIS, *Phillips.* Pl. XLIX, figs. 1—10.

> LINGULA SQUAMIFORMIS, *Phillips.* Geol. of Yorkshire, vol. ii, pl. ix, fig. 14, 1836.
> — — *Portlock.* Report on the Geol. of Londonderry, &c., pl. xxxii, fig. 5, 1843.
> — — *M'Coy.* British Pal. Foss., p. 475, 1855.
> — — *Dav.* Mon. of Scottish Carb. Brach., pl. xi, fig. 14, 1861.

Spec. Char. Shell longitudinally oblong, one third or less longer than wide, with sub-parallel sides, the broadest towards the anterior extremity, the frontal margin assuming either a very slight inward or outward curve. The anterior portion is gradually curved on either side, the beak being rounded or but slightly angular at its extremity in the dorsal valve, with a thickened margin, tapering, pointed retrally at its termination in the ventral one, which is consequently so much longer than the opposite valve. The valves are slightly convex, but somewhat depressed along their middle. In the dorsal one there exists a small apex close to the rounded margin of the beak, and from which usually radiate three small, rounded ridges, separated by shallow sulci. The external surface in both valves is covered with numerous fine, concentric striæ, or lines of growth, giving to the shell a beautifully and delicately sculptured appearance, for the minute plications of growth succeed each other with much regularity, while some stronger lines or interruptions of growth are produced at variable distances. The internal muscular impressions are similar to those already described, the occlusor and external adjustors of Hancock being especially observable. Dimensions variable; two British examples have measured—

Length 19, width 13 lines.
    „   9   „ $6\frac{1}{2}$ „

Obs. This is a common species in certain Scottish strata and localities, and is found likewise in England and in Ireland, and can be distinguished from its congeners by shape and sculpture, although the Silurian *Lingula granulata* of Phillips approached nearest to it; this last is, however, usually less elongated, and does not present those radiatory ridges which are generally, but not always, observable in the Carboniferous species.

In his work on 'British Palæozoic Fossils,' Prof. M'Coy concludes his description of *L. squamiformis* by stating that "the wide, short, oblong form of this species easily distinguishes it from the others in the upper Palæozoic rocks. The more elongated, narrow, oblong species, well figured in Portlock's 'Geological Report,' tab. xxxii, fig. 5, under this name, might be called *L. Portlocki* (M'Coy). Its proportional width is only 55-100th. in the long, and 60-100th in the short valve." But specimens connecting the narrow and the wider varieties are so numerous that I could not admit the two extremes as distinct species, besides which, Phillips's original example is perfectly similar to many of the Scottish examples of the species, although not very correctly figured in the 'Geology of Yorkshire,' and from which circumstance may have led some to doubt the identity. Phillips's specimen (fig. 1 of my plate), which may be seen in the British Museum, consists of a shell and counterpart, or rather the shell is equally divided between the two sides of a split nodule, so that neither of them show the true structure. When the shell is removed the matrix shows regular, concentric striæ, similar to those above described, but elsewhere only fractured lines of laminæ and radiating striæ. The nodule is black, and the shell dark and pyritous. *L. squamiformis* has sometimes attained comparatively large dimensions; thus, in a coaly shale intercalated between bands of ironstone at one mile to the north of Glasgow millions of specimens may be seen in a crushed or distorted condition, but of which some examples, when perfect, measured about one inch and a half in length. Mr. Rodwell discovered also a specimen one inch two lines in length in a shale at about a mile to the east of Bally Castle (on the north coast of Antrim), but the shell does not usually attain such large proportions. In England it is stated by Prof. Phillips to occur at Bolland; it is found also at Lemmington, Northumberland. In Scotland it occurs at Raes Gill, at 341 fathoms below "Ell Coal," 343 at Hall Craig, 317 at Braidwood Gill, 354 at Langshaw Burn; it is found also at Hall Hill, near Lesmahago; in Renfrewshire at Orchard Quarry, Thornliebank; in Dumbartonshire at Netherwood, near Castlecary; in Stirlingshire in the Mill Burn beds, Campsie main limestone and Corrieburn beds. It is also found at Bishopsbriggs, three miles north of Glasgow; in Haddingtonshire at Cat Craig, near Dunbar; in Edinburghshire at Wardie (Western Breakwater, Granton); and occurs also in Fifeshire and the Berwickshire coast. In Ireland it it has been found to the east of Bally Castle (north coast of Antrim), Leam, Fermanagh, Enniskillen (in shale), &c.

LINGULA SCOTICA, *Davidson*. Pl. XLVIII, figs. 27, 28.

LINGULA SCOTICA, *Dav.* Mon. of Scottish Carb. Brach., pl. v, figs. 36, 37, 1860.

Spec. Char. Shell of an elongated triangular shape, tapering at the beak, slightly rounded laterally and in front. The valves are slightly convex, but much compressed, while the entire surface is covered with numerous minute, concentric striæ, with still wider, flattened interspaces. Interior unknown. Dimensions variable; two examples have measured—

Length 15, width $12\frac{1}{2}$ lines.

„ $6\frac{1}{2}$, „ 5 „

Obs. This remarkable species, which has at times exceeded the proportions above given, is easily distinguished by its triangular shape, tapering sides and beaks, as well as by the delicate and peculiar sculpture which adorns its surface. In shape it approaches to certain exceptional examples of Phillips's *Lingula cuneata*, but the Carboniferous and Silurian species cannot be confounded.[1]

L. Scotica has not been hitherto discovered either in England or in Ireland. In Scotland it occurs at Gare, in Lanarkshire, at 239 fathoms below the "Ell Coal," and from which locality it has been known to a friend in Carluke for upwards of thirty years. My attention was, however, first directed to the shell by Mr. Young, of the Hunterian Museum, Glasgow, who had been struck by its peculiar triangular appearance, and it was subsequently discovered at Robroyston, north of Glasgow, in beds upon a similar horizon to those of Gare, while the larger examples were procured by Dr. Slimon from Hall Hill, Auchenheath, and Coalburn, Lesmahago, about 300 fathoms below "Ell Coal."

LINGULA MYTILOIDES, *Sowerby*. Pl. XLVIII, figs. 29—36.

LINGULA MYTILOIDES, *Sow.* Min. Con., tab. xix, figs. 1, 2, 1812.
— ELLIPTICA, *Phillips.* Geol. of York., pl. xi, fig. 15, 1836.
— MARGINATA, *Phillips.* Ib., fig. 16, 1836.
— PARALLELA, *Phillips.* Ib., figs. 17—19, 1836.
— MYTILOIDES and L. PARALLELA, *Portlock.* Report on the Geol. of Londonderry, &c., pl. xxxii, figs. 6—9, 1843.
— — *De Koninck.* Desc. des Anim. Foss. de Belgique, pl. vi, fig. 9, 1843.
— — *Dav.* Mon. of Scottish Carb. Brachiopoda, pl. v, figs. 38, 43, 1860.

---

[1] I possess also an American Lingula from the Pottsdam Sandstone of the Falls of St. Croix, Minnesota, which is stated by Mr. Worthen to be the oldest American species of the genus. In shape it is very similar to *L. Scotica*, but differs from it in sculpture, as well as in the convexity of its valves.

Spec. Char. Shell very variable in shape, but usually more or less regularly elliptical or ovate, with its greatest width either towards the posterior or anterior extremity. Its sides are also sometimes nearly parallel and rounded in front, but both front and beaks are in some examples about equally and regularly elliptically attenuated. The valves are generally gently convex, and most elevated along the middle, where there exists likewise a flatness, which becomes gradually wider as it extends from the apex of the beak to the front, the lateral portions of the valves sloping rather abruptly on either side, while the surface is marked at intervals by a greater or smaller number of concentric ridges or lines of growth. Interior unknown. Dimensions variable; three British examples have measured—

Length 10, breadth 5½ lines.

    „   7   „   3½ „

    „   5   „   1½ „

Obs. After the examination of a considerable number of specimens, it has appeared to me that *L. parallela* and *L. elliptica* are only slight variations in shape of Sowerby's *L. mytiloides?* Some palæontologists will, however, probably differ with me in this conclusion, and may prefer retaining *L. mytiloides* and *L. parallela* as separate species, and I should be glad to adopt their views if they can point out the characters by which the two can be distinguished. It will, perhaps, be as well, therefore, to reproduce the original description and figures. Sowerby describes *L. mytiloides* as follows :

"Ovate, anterior end slightly truncated ; beak indistinct. Nearly an inch long, and three fifths wide ; the older shells are flattened towards the front, with rather a straightish edge. Shining, and of a grayish-blue colour.

"These are mostly found in pairs at Wolsingham, in the county of Durham, in a dark-coloured limestone. I am told they are sometimes larger than the figure. They are preserved so well that they have the appearance of a recent mussel."

Phillips describes his species as follows :

"*L. elliptica.*—Long, elliptical, acuminated retrally, surface with delicate, radiating, and concentric lines. Ashford, in Derbyshire.

"*L. marginata.*—Very oblong, with parallel sides, truncate in front, rounded retrally; edges of the valves turned up ; slight mesial ridge on a flat space ; small, oval hollow, fine radiating and concentric lines. Bowes.

"*L. parallela.*—Magnified views of a species which seems different from the last by its rounded front and equally convex surface. Northumberland."

Besides the English localities above given I have seen specimens from the Carboniferous shales of Brakewell and Ashford, in Derbyshire; Denwich, Northumberland. In Scotland it occurs abundantly in slaty ironstone in the parish of Carluke, in Lanarkshire, at 160 fathoms below the "Ell Coal," 239 at Gare, 300 at Mashock Burn, 237 at Raes Gill, Braidwood, and Langshaw Burn, 371 at Kilcadzow. It occurs likewise at Hall Hill, near Lesmahago, Capelrig, East Kilbride; Calderside and Auchentibber, High Blantyre;

Bishopbriggs and Robroyston, north of Glasgow. In Renfrewshire at Orchard Quarry, Thornliebank. In Ayrshire at West Broadstone, Beith. In Stirlingshire at Craigenglen and Corrieburn. In Fifeshire, at Craig Hartle, &c. It has also been found along the Berwickshire coast, and at Marshall Meadows, three miles north of Berwick. In Ireland it has been found in Tyrone, Desertcreat, also at Clogher (Pollock). On the Continent it has been described by Prof. de Koninck, from Visé, in Belgium.

LINGULA CREDNERI, *Geinitz.* Pl. XLVIII, figs. 38—40.

> LINGULA CREDNERI, *Geinitz.* Versteinerungen der Zechteingebirges, pl. iv, figs. 23—29, April, 1848.
> — — *M'Coy.* British Palæozoic Fossils, p. 474, 1855.
> — — *Kirkby.* Proceedings of the Geol. Soc., vol. xvi, p. 412, 1860.

Spec. Char. Shell small, oval, occasionally oblong, with the posterior end acuminated valves gently convex; marked with concentric lines of growth. Interior unknown. Length 3, width 2½ lines.

Obs. Is this shell specifically distinct from *L. mytiloides?* Might it not be a small variety of the last-named shell? Indeed, I would hardly venture to positively assert that it is specifically distinct. Until March, 1860, *L. Credneri* was known only from the " marl slate" and the lower beds of the compact limestone of the Permian system, but Mr. Kirkby informs us that the shell occurred to him as a Carboniferous species at Ryhope Winning, near Sunderland, in the summer of 1858; that he observed it first in a thin bed of dark shale, at a depth of 951 feet from the surface, or 592 feet from the base of the overlying Permian strata, though in this bed it was exceedingly rare; but that he found it more plentiful in a thick stratum of gray shale just above the bed already mentioned. That from the first (to use his own expression) he was struck with the resemblance of these *Lingulæ* to the Permian species, *L. Credneri,* and that his opinion was only strengthened by the acquisition of a full suit of specimens, and which he submitted to Mr. Hancock's and my own examinations.

There is no essential difference (he adds) between the Permian and Carboniferous specimens. The form of both is nearly oval; the Carboniferous specimens have the median elevation more prominent than those of the marl state, but that in this respect they only approach more closely to the Permian example from the Kupsfer Schiefer, the German equivalent of the marl slate, and that in no respect do the Permian examples differ from the Carboniferous specimens more widely than do individuals of the same series from each other.

LINGULA LATIOR, *M‘Coy.*   Pl. XLVIII, fig. 37.

LINGULA LATIOR, *M‘Coy.*   British Palæozoic Fossils, pl. iii D, fig. 33, 1855.

Spec. Char.   " Broad-ovate anteriorly, gradually acuminated posteriorly ; moderately convex towards the beak, very gradually flattened towards the margins ; sides meeting at the beak at an angle of about 75° ; front wide, semi-elliptically rounded ; greatest width at about the middle of the length, whence the posterior end is rapidly narrowed to the beak.   Surface with fine, sharply defined, strong, close, elevated, obtuse, concentric striæ, slightly irregular from occasional branchings and interruptions ; crossed in parts by longitudinal microscopic striæ.   Length $4\frac{1}{2}$ lines, proportional width $\frac{80}{100}$." (M‘Coy, p. 475.)

Obs.   I have never seen this so-termed species, although stated by Prof. M‘Coy to be not very uncommon in the black limestone over the main limestone of Derbyshire.   It is therefore provisionally given on the authority of Prof. M‘Coy, and until the species? may have been completely studied.

APPENDIX AND CONCLUSION.

Five years have elapsed since I first commenced my researches among the Carboniferous Brachiopoda in connexion with this Monograph, and although the time employed may be thought great, it must be remembered, in justice to the many gentlemen who have so zealously afforded me their valuable assistance, that the country and strata has during that interval been continually searched in order to obtain every specimen that might tend to complete the history of our species.

During this lengthened investigation, a vast number of specimens have turned up, and been attentively studied, so that it will be necessary in these supplementary pages to propose some few alterations to the published portions of the work, as well as to add further observations and new species which, having become subsequently known, could not be included in the regular succession of described species.

At the time I commenced my researches, about 260 so-termed British carboniferous species of Brachiopoda had been recorded by different palæontologists, but after a most searching investigation, I could not conscientiously admit more than about 100 of these, and in order to arrive at such a reduction, no small labour was required, nor was I unmindful of the danger Palæontologists should guard against in the breaking down of species, which if injudiciously done, would be as great an evil as that of uselessly multiplying them.[1]

---

[1] It may be as well to mention that in the second and improved edition of Prof. Morris's ' Catalogue,' published in 1854, 193 species are recorded, but of these about 93 only are retained on my lists.

In 1836, Prof. Phillips enumerated 100 species as having been found in England, of which about 52 are retained.

In his ' Synopsis of the Carboniferous Fossils of Ireland,' published in 1844, Prof. M'Coy described 230 so-termed species of Carboniferous Brachiopoda stated to have been found in Ireland, but he figured only 62. Of the 230, not more than 70 appear to me good species, about 61 are *Devonian* or *Silurian* names not known or *proved* to have been found in true Irish carboniferous strata, and about 117 are either synonyms, or are due to incorrect determinations.

In his memoir ' On the Localities of Fossils of the Carboniferous Limestone of Ireland,' Mr. Kelly enumerates no less than 240 so-termed Irish species ! the catalogue comprising the 230 described in the ' Synopsis,' and some others from Portlock's ' Report on the Geology of Londonderry, &c. ;' and if we add a few more subsequently discovered, about 250 species ! would be put down to Ireland, while my most

Before proceeding further it may be as well that we should inquire into what has led to this extraordinary multiplying of species? Has it not been caused by the uncertainty and difference in opinion that exists among naturalists with reference to what should constitute *genera* and *species*, as well as by the ignorance and precipitation with which we are so often apt to consider new, what may not be known to us?

Deshayes in his paper on the distribution of Acephalous Mollusca in the tertiary basin of Paris, observes:—"For us the *genus* is a creation of our own mind very happily conceived, so as to favour the grouping of those beings which have between them the largest number of common characters, than with any of those which are after them the most nearly related. That is a natural system, and consequently rational one; the genera represent equal degrees, and of comparable organization. That it is while, considering them in this manner, that in our actual researches they acquire the most interest. The fundamental basis of natural history, reposing on the exact and profound knowledge of the *species*, being that which emanates directly from the hands of the Creator, while the art of grouping those which we have recognised is human." Darwin considers the term species, on the contrary, as one arbitrarily given, for the sake of convenience, to a set of individuals closely resembling each other; and that it does not effectually differ from the term variety, which is given to its less distinct and more fluctuating forms. That the term variety, again, in comparison with mere differential differences, is also applied arbitrarily and for convenience sake; that no one can draw any clear distinction between individual differences and slight varieties, or between more plainly marked varieties and sub-species and species.

strenuous efforts have not shown the existence of more than about 80. Mr. Kelly, whose knowledge of Irish geology is equal to that of any other man, and who has visited almost every Irish fossiliferous locality, expresses himself averse to my rejecting so many Devonian and Silurian species said to have been found in his carboniferous strata and localities, but I may again, without hesitation, assert that the larger number, at any rate, are due to incorrect identification, for the examination of many of the original specimens in Sir Richard Griffiths' and other collections have convinced Prof. de Koninck, Mr. Salter, and myself of this important fact.

Mr. Kelly has, however, informed me by letter that a large portion of the doubtful fossils were got in localities of the calciferous slate, a band which lies under the limestone; that out of about 70 not proved to me, because I have not seen specimens, 22 were obtained at Lisnapaste and Donegal; that in these localities there is a great variety, and that they occur in black soft shale, as soft and as easily decomposed by exposure to the atmosphere as any that occurs in the coal-measures: that a lump of this black shale exposed to sun and rain for one summer, would slake or fall to pieces: and he therefore supposes that by far the larger number of Lisnapaste specimens that were originally in Sir R. Griffiths' collection were lost by their removal to the Great Exhibition held in Dublin in 1852, as those tender shales would not bear the agitation of carriage, and consequently mouldered away into very small fragments. That there are six or eight other localities in the calciferous slate in which similar shales occur with fossils, and that he finds, upon looking over his lists, that most of the Devonian species I object to were obtained in those localities. Along with Lisnapaste there is Larganmore, Bruckless, Kildress (the red shales near Cookstown in the old red series), Bundoran, Malahide, Curragh, &c.

It is quite certain that too little attention is given to the many modifications of which a species is susceptible; this fact has been clearly demonstrated by Dr. Carpenter in his admirable researches among the Foraminifera, and will be easily exemplified by the Brachiopoda, which are traceable through the entire series of fossiliferous rocks.

Two opinions appear to prevail at the present time on the origin of species. The greater number of naturalists believe in the creation of separate forms or species capable of producing varieties, but to how great an extent he who made them only knows. Darwin, on the contrary, supposes all species to have been derived from a common progenitor, but to be able to positively admit or refute such an idea, it would be necessary to possess a far more extended and minute knowledge of species, and the causes of their variation, than we at present possess; and although I could not conscientiously go the full length with Darwin, I heartily concur with Prof. Huxley, while observing that " all competent naturalists and physiologists, whatever their opinions as to the ultimate fate of the doctrines put forth, acknowledge that the work in which they are embodied is a solid contribution to knowledge." I will not therefore follow those, who blindly admit the theory, nor concur with those who unhesitatingly pronounce it a chimera, but will do my utmost to register the great facts as they stand, with such comment as I can give, and we may thus be led by degrees to a better understanding of many problems relating to species and their origin than we at present possess. Palæontologists should, above all, be zoologists, and as zoologists have little to do with geological divisions or systems; when they have to inquire into the resemblances and variations in species they should always endeavour to trace a species through its many modifications as far back as they can, or, in other words, to search for its probable progenitor, be it located in the Carboniferous, Devonian, Silurian, or in any other system of strata, as well as to follow or trace its recurrence in more recent periods, and I may boldly assert that when our knowledge of the Brachiopoda shall have extended, that the intimate connexion of many of the so-termed species will be discovered, and that a large proportion of them will be traced through their various modifications to a parent form in stages far more ancient than we are in many cases disposed to admit.

Before concluding, let me therefore recapitulate the result of five years' attentive study of the British Carboniferous species, and point out, however imperfectly, as far as our means will permit, those points which appear to have been clearly made out, as well as some of those I am necessarily compelled to leave unsettled, for I am far from believing that we have arrived at finite or satisfactory results, with reference to several of the species.

TEREBRATULA (pp. 11—18). Plate I and XLIX.

In 1857 I described *T. hastata, T. sacculus, T. Gillingensis,* and *T. vesicularis,* as distinct species, but the subsequent study of a very considerable number of specimens of each of the above-named forms, has shown so many intermediate shapes, that it must remain a question whether the last three are in reality more than varieties or simple modifications of the first in shape.

At page 12 I mentioned that Mr. De Verneuil, and several other experienced palæontologists, were of opinion that *T. sacculus* was only a variety of *T. hastata;* I also stated (p. 17) that *T. Gillingensis* had been supposed by some observers to be a variation in form or the young of *T. hastata*, and while describing *T. vesicularis* (p. 16), I did not omit to remind the reader that De Koninck's species was extremely variable both in shape and character, so much so, that, to my eyes, certain examples did appear undistinguishable from others of Martin's *T. sacculus*, and to which, Prof. De Koninck admits his shell to be very closely related. No one will, therefore, feel much surprised when I affirm that it is impossible to determine whether very many intermediate shapes or specimens should be referred more to one than another of the four above-named so-termed species, and that there is absolutely no line of demarcation between any of the four forms above recorded.

It would therefore not be surprising, if all the British Carboniferous Terebratulæ hitherto discovered, were to prove mere modifications of a single very variable species, capable of assuming different shapes (influenced no doubt from local circumstances), and not presenting a greater extent of modification than what is common to many other species. For example, is not *Terebratula plicata* and *T. fimbriata* entirely smooth up to a certain age, and indeed often so to an advanced stage, when they afterwards suddenly or by degrees become more or less regularly plicated during the remaining period of their growth? And many other examples could be given of still greater modifications.

T. hastata is the largest British Carboniferous *Terebratula* with which we are at present acquainted, some specimens having attained twenty-six and a half lines in length, nineteen and a half in width, and thirteen in depth (Pl. XLIX, fig. 11), which is certainly the full-grown condition of the species, but which, even under the most favorable conditions, was exceedingly variable, as may be seen by casting a glance at the numerous examples represented in Pl. I and XLIX of the present Monograph.

T. sacculus, in its typical shape, appears to be a thickened dwarfed condition of Sowerby's species, and although it has been urged that Martin's shell was never coloured (so far as known), while *T. hastata* was ornamented with purple-colour bands, it must be also remembered that the number of specimens discovered which have shown these remains of colour have been few; and that we are by no means certain that *T. sacculus* may not have been similarly ornamented. It is likewise not correct to say, that *T. hastata* always possessed sharp edges, for if we examine a large series of specimens we will soon perceive among them *many* as thick edged as any hitherto discovered of the so-termed *T. sacculus*.

If the reader will refer to Pl. VI of my Cretaceous Monograph, he will find therein many modifications of *T. biplicata* carefully represented, and will, I am sure, perceive as great a difference between certain specimens of this cretaceous shell as any he could produce between *T. hastata* and *T. sacculus*. Look, for example, at the deeply biplicated example (fig. 9) and to that without any biplication at all (fig. 11) of the plate above quoted; still both have been recognised by Palæontologists as belonging to a single species, and to be intimately connected by every degree of modification. I may also, while upon this subject, again remind the reader that out of many thousand specimens of *T. biplicata*, collected near Cambridge, but one (fig. 6) showing remains of colour has been hitherto procured, and whose markings very closely resembled those of our carboniferous *T. hastata*.

T. Gillingensis, in its extreme form (Pl. XLIX, figs. 19, 20), appears different enough from the usual shapes of *T. hastata*, but we must not consider extremes alone, but rather the character presented by the larger number of individuals, and then we will soon find every intermediate shape by which these extremes may be connected with Sowerby's species.

T. vesicularis also, with its deep triundate front, is certainly very peculiar, but this is not the common condition of the generality of specimens, which indubitably by gradation assume the characteristic shapes of *T. sacculus* and *T. hastata*. It is quite evident that in both of the last-named forms there exists at times a tendency to the production of a small central undulation or rib near the front of the dorsal valve, but the frontal margin of the shell may be, and indeed very often is, triundate, without necessitating the production of a median rib. The intimate connexion between *T. hastata* with its straight frontal margin, and the *T. vesicularis* shape with deep triundate or triplicate dorsal valve, or frontal margin, has been furnished by a small limestone quarry at Bowertrapping, near Dalry, in Ayrshire, and of which variety a series of specimens have been carefully represented in Pl. XLIX, figs. 21 to 26. In this locality the *T. vesicularis* shape has attained very large proportions, while in Yorkshire an exactly similar series has been found, but with much smaller proportions; still the Yorkshire *T. vesicularis* is a miniature fac-simile of the large Bowertrapping variety, and in both these cases these extremes merge into the common shape of *T. hastata* or of *T. sacculus*.—As to the other synonyms, I am still of opinion that *T. ficus*, M'Coy (p. 13), should be considered a very convex specimen of *T. hastata*, in which the frontal margin is slightly triundate; and in Pl. I will be found many examples of Sowerby's species with or without a mesial depression in either valve; and this leads me to observe that the Permian *T. elongata* is in all probability, and *T. sufflata* certainly, a recurrent form of *T. hastata* and *T. succulus*, and in proof of which I would beg the reader to cast his eye at Pl. LIV, figs. 1 and 2, 3 and 4, of this Monograph, and he will surely be struck by the close resemblance of the figures of *T. hastata* and *T. elongata* represented therein. The interior details, loops, &c., are exactly similar in all the forms of *Terebratula* here described, and their animal was no doubt so likewise.

One mistaken synonym must, however, be corrected.

At p. 1 *Atrypa virgoides*, M'Coy, was supposed to be a form of *T. hastata*, and to this species is certainly referable the *Seminula virgoides* represented at Pl. 3ᴰ, fig. 23, of M'Coy's 'British Palæozoic Fossils,' but the true *Atrypa virgoides*, described and figured in

1844 in the 'Synopsis of the Carboniferous Fossils of Ireland,' Pl. xxii, fig. 21, is nothing more or less than an elongated malformation of *Athyris plano-sulcata!* a discovery entirely due to my zealous friend Mr. J. Wright, of Cork, who having obtained the typical specimen (formerly in the collection of Dr. Haines), and a number of similar malformations found in the limestone near Cork, has left no possible doubt in the matter. (Pl. LI, 11 and 11*a*.)

It is well known that the carboniferous fossils of the South of Ireland are usually much distorted by pressure or cleavage, and that the mere form is of but little specific value; and in illustration of this I have represented four specimens of *T. hastata* found near Cork (Pl. XLIX, figs. 13 to 16) which show what extraordinary modifications the same species may assume under similar circumstances.

It is not in my power at present to say what may have been the parent form from whence all these modifications of *T. hastata* have been derived. We shall probably trace it hereafter in the Devonian or even Silurian periods; but I am also quite aware that the generality of Palæontologists are not yet sufficiently imbued with the absolute necessity of enlarging the circle of variation to be permitted to a species, and will naturally say that "they must totally dissent from my putting such a lot of shells into a single species." I think, therefore, that should Palæontologists hereafter consider it desirable to merge the whole British Carboniferous *Terebratulæ* into *T. hastata* that the varietal designations of *Gillingensis, Sacculus,* and *Vesicularis,* might perhaps be retained to denote certain modifications in its shape.

ATHYRIS, or SPIRIGERA.

In external shape the species of this genus approach more to *Terebratula* than to any other genus, and therefore in a good or natural arrangement it should precede *Spirifera*. Nine species of *Athyris* have been provisionally retained from among the many synonyms, while the value of *A. globularis* and *A. squamigera* may still require confirmation; for of both these shells the material at my command has been very scanty: and it is even uncertain whether the identification with *A. squamigera* (de Koninck) be correct.

ATHYRIS AMBIGUA (p. 77). Plate XV, figs. 16—22; and Plate XVII, figs. 11—14.

The muscular impressions of *Athyris* have been represented in *Athyris undata* (a Devonian species), but not in quite so precise a manner as could be desired, as seen on some silicified internal casts of *A. ambigua* from the carboniferous limestone of Bakewell (Derbyshire), in the Museum of Practical Geology, and of which two enlarged illustrations will be found in Pl. XVII of the present Monograph. I have also ascertained (since the publication of my description of this species) that the spiral processes, and their intermediate connecting lamellæ are similarly disposed in *Athyris ambigua* to those of *A. pectinifera*, of which we have also given a representation in the same plate.

ATHYRIS SUBTILITA.[1] Plate I, figs. 21, 22; and Plate XVII, figs. 8, 9, 10.

In p. 18 I have described the shell as a *Terebratula*, but in p. 86 located it with *Athyris*, to which genus it belongs.

ATHYRIS LAMELLOSA (p. 79). Plate XVI, fig. 1; Plate XVII, figs. 6, 7; and Plate LI, fig. 14.

When describing this species in p. 79 I had not seen any specimens with its concentric lamelliform expansions completely preserved, and it was only subsequently that Mr. J. Wright discovered several fine examples at Little Island, near Cork (Pl. LI, fig. 14), showing that these expansions were prolonged in some specimens nearly an inch from the surface of the shell, and that they differed from those which adorned the valves of *Athyris plano-sulcata* by being somewhat irregularly plaited, or frill-like, as seen in Pl. LI, fig. 14.

ATHYRIS PLANO-SULCATA (p. 80). Plate XVI, figs. 2—15; and Plate LI, figs. 1—13.

This species appears to have varied considerably, and I have already shown that the original *A. virgoides*, M'Coy, is nothing more than an elongated shape of the present species, distorted from pressure and cleavage.

ATHYRIS CARRINGTONIANA, *Dav.* Plate LII, figs. 18—20.

Sp. Char. Shell transverse, sub-rhomboidal, with rounded extremities; wider than long; valves moderately convex, and about equally deep; hinge line, forming an obtuse angle, the greatest breadth of the shell being along the middle; beak of ventral valve small, slightly incurved; foramen circular, and contiguous to the umbone of the opposite valve; beak ridges sharply defined, leaving between them and the hinge line a narrow flattened space. A shallow longitudinal sinus or furrow extends from the extremity of the beak to the front. In the dorsal valve there exists a moderately elevated mesial fold, longitudinally divided by a shallow furrow. Externally both valves are regularly traversed by continuous concentric small ridges or striæ. Dimensions variable: two specimens have measured—
Length 8, width 14, depth 6 lines.
  „   4, „   7, „ 2 „

[1] The reference is not quite correctly given at p. 86, it should be Hall, in Howard Stansbury's 'Exploration of the Valley of the Great Salt Lake of Utah,' p. 409, pl. 2, figs. 1-2: 1852.

Obs. This interesting species has much resemblance to Phillips' Devonian *Spirifera phalæna,*[1] as well as to De Verneuil's *Terebratula Hispanica* from the Devonian rocks of Spain,[2] and for some time I felt disposed to consider the form under description as a variety of Phillips' species, but after comparing three examples of our carboniferous shell with the figures of the Devonian species, it appeared to me that our shell was more sub-rhomboidal in shape, and had not the straight hinge line of the Devonian species; the sinus in *S. phalæna* and *T. Hispanica* is also much wider and deeper than what we find in our shell, and the mesial fold is also much more deeply divided. I have therefore named the carboniferous species after its discoverer, and as an appreciation of the valuable assistance I have received from him during my examination of the many Staffordshire species he had so zealously collected. Mr. Carrington has obtained eight or nine specimens from the carboniferous limestone of Wetton, in Staffordshire.

RETZIA.

Two species have been already described, and a third has recently been found.

RETZIA RADIALIS (p. 87). Plate XVII, figs. 19—21, and Plate LI, figs. 4—9.

This shell appears to have varied considerably in shape, as well as in the number and size of its ribs, so much so that many of its variations when viewed individually, might lead us to doubt their being simple modifications of Phillips' type. After having assembled a great many specimens from the same as well as from distinct localities, I soon perceived that extreme forms, with twenty-three small ribs, and those with eleven large angular ones upon each valve, could be easily connected by intermediate links; that, for instance, some examples possessed thirteen ribs, others seventeen, nineteen, and twenty-one. In some localities, likewise, owing no doubt to peculiar circumstances, the shells were all small, while in other places they have attained half an inch in length. In Pl. LI, as well as in Pl. XVII, will be found illustrations of all the most marked variations in form hitherto observed.

RETZIA ULOTRIX (p. 88). Plate XVIII, figs. 14, 15.

This appears to be a rare species; a very perfect example has, however, been recently discovered by Mr. Carrington in the carboniferous limestone of Allstonefield, in Staffordshire. (Pl. LIV, fig. 45.)

---

[1] Figures and descriptions of the Palæozoic fossils of Cornwall, Devon, and West Somerset, p. 71, pl. xxviii, fig. 123, found at Hope, near Torquay. *S. phalæna* belongs to the genus *Athyris,* not Spirifera.

[2] 'Bulletin de la Soc. Geol. de France,' 2d serie, Tom. 2d, p. 463, pl. xiv, fig. 7. Mr. de Verneuil's *T. Hispanica,* belongs to the genus *Athyris,* and is perhaps a synonym of *A. phalæna.*

RETZIA ? CARBONARIA, *Dav.* Plate LI, fig. 3.

Spec. Char. Shell ovate, longer than wide; valves moderately and equally convex, without fold or sinus, margin of valves nearly straight; beak moderately produced, incurved and truncated by a circular foramen, slightly separated from the hinge-line by a small deltidium. Surface marked with about twenty-four small rounded ribs; shell structure minutely punctured. Interior unknown.

Length 9½, width 7½, depth 5½ lines.

Obs. A single example of this interesting species only has been hitherto discovered, and it was not until after much research and hesitation that I venture to apply to it a new specific denomination. Prof. De Koninck, to whom I submitted drawings, pronounced it quite distinct from *Retzia serpentina,* in which the striæ are much more numerous. At one time I thought it might perhaps be referred to *Terebratula Marcyi,* of Shumard ('Palæontology of the Red River of Louisiana,' Pl. I, figs. 4 and 6), as the description given of that shell by the American author, so nearly agreed with that of our British fossil, but having forwarded drawings of our specimen to Prof. Hall, that distinguished palæontologist seemed inclined to consider the English shell as belonging to a more robust species with smaller beak and fewer ribs, and that it differs likewise from *Retzia vera,* in several particulars.

Some uncertainty as to the genus to which it belongs must naturally prevail, since we are unacquainted with its interior dispositions. I am also undecided whether it belongs to *Retzia* or to Prof. Hall's sub-genus *Rhynchospira,*[1] proposed for several shells which bear a close resemblance both in general form and in the interior spires to *Retzia;* but of which the dorsal valve never presents the straight extended hinge-line. nor the ventral valve the short area common to all true species of that genus.

I am indebted to my friend Mr. Salter for the first knowledge of this new British fossil, which was obtained from the lower Carboniferous Shales of Skrinkle, Pembrokeshire, and which is now preserved in the Museum of Practical Geology.

SPIRIFERA.—Since describing the *Spirifers,* many more specimens and observations have been gradually assembled, which will necessitate the introduction of some alterations and additions to what has been written upon the subject. Thirty-seven species are described in pages 19 to 76 of our Monograph, but it must be remembered that some of these were at the time doubtfully and provisionally retained from want of sufficient grounds for rejection or adoption, and that it was only during the interval that uncertainty has been dispelled in certain cases, while a few species that had been supposed distinct subsequently proved to be varieties of some of the others. In the following list a point of

[1] 'Twelfth Annual Report of the Regents of the University of the State of New York,' No. 185, p. 29, 1859.

interrogation is placed before those names, of which the specific claims are still uncertain, and a few synonyms have been appended.

1. SPIRIFERA STRIATA, *Martin.* Sp. = *T. spirifera*, Val. apud Lamarck = *S. attenuata*, Sow. = *S. Princeps*, M'Coy.

? 2.   —   MOSQUENSIS, *Fischer.* = *C. Sowerbyi* and *C. Kleinii*, Fischer = *S. choristites*, V. Buch.

3.   —   DUPLICICOSTA, *Phillips.* = ? var. *S. humerosa*, Phil.

4.   —   TRIGONALIS, *Martin*, Sp. = *Sp. bisulcata*, Sow.; = *S. semicircularis*, Phillips; *S. calcarata*, M'Coy (not Sow.); = *S. grandicostata*, *S. planicosta*, and *S. transiens*, M'Coy; = *S. crassa*, De Koninck; = ? *S. clatharata*, M'Coy.

5.   —   CONVOLUTA, *Phillips.*

6.   —   TRIANGULARIS, *Martin*, Sp.

? 7.   —   FUSIFORMIS, *Phillips.* Very doubtful species.

8.   —   RHOMBOIDEA, *Phillips.*

9.   —   ACUTA, *Martin*, Sp.

10.   —   PLANATA, *Phillips.*

11.   —   CUSPIDATA, *Martin*, Sp.

? 12.   —   SUBCONICA, *Martin.*

13.   —   DISTANS, *Sow.* = *Sp. bicarinata*, M'Coy.

? 14.   —   MESOGONIA, M'Coy.

15.   —   PINGUIS, *Sow.* = *S. rotundatus*, Sow. (not Martin); = *S. sub-rotundatus*, M'Coy.

16.   —   OVALIS, *Phil.* = *S. exarata*, Fleming; = *S. hemispherica*, M'Coy.

17.   —   INTEGRICOSTA, *Phil.* = ? *A. rotundatus*, Martin (not Sow.) = ? *Sp. paucicosta*, M'Coy.

18.   —   TRIRADIALIS, *Phil.* = *Sp. trisulcosa* and *Sp. sex-radialis*, Phillips.

? 19.   —   REEDII, *Dav.* A very doubtful so-termed species, which will probably have to be suppressed.

20   —   GLABRA, *Martin.* Sp. = *S. oblatus* and *S. obtusus*, Sow.; = *S. linguifera*, *S. symmetrica*, and *S. decora*, Phillips; = *S. rhomboidalis*, M'Coy.

21.   —   CARLUKIENSIS, *Dav.*

22.   —   URII, *Fleming.*

23.   —   LINEATA, *Martin*, Sp. = *S. mesoloba*, Phillips; = *S. imbricata*, Sow., *S. reticulata*, and = *S. strigocephaloides*, M'Coy; = *S. Martini*, Fleming.

? 24.   —   ELLIPTICA, *Phillips.*

25. SPIRIFERINA LAMINOSA, *M'Coy*, Sp. = *S. tricornis*, De Kon.

26.   —   CRISTATA, *Schloth.* Var. *octoplicata*, Sow. = *S. partita*, Portlock, and var. *biplicata*, Dav.

? 27.   —   MINIMA, *Sow.* Doubtful species.

28.   —   INSCULPTA, *Phil.* = *S. quinqueloba*, M'Coy.

29. CYRTINA SEPTOSA, *Phillips*, Sp.

? 30.   —   DORSATA, *M'Coy.*

31.   —   CARBONARIA, *M'Coy.*

From this list it will be seen that not more than about twenty-three or four species of *Spirifera* have been satisfactorily determined, *S. elliptica*, *S. mesogonia*, *Sp. fusiformis*, *Sp. Reedii*, *Sp. minima*, and *S. subconica*, *S. mosquensis*, *C. dorsata* being probably to some extent synonyms or varieties, and not sufficiently studied from want of sufficient material. I will now add a few remarks with reference to some of the species.

SPIRIFERA STRIATA, *Martin* (p. 19). Plate II, figs. 12—21, and figs. 9—11. Referred to *S. duplicicosta*, and probably the young of *S. striata*; Plate III, figs. 2—6, and Plate LII, figs. 1, 2.

This is a very variable species, the shell is generally transverse, but sometimes it is longer than wide; and I am therefore not quite certain whether the specimens referred to *Sp. Mosquensis* do in reality belong to the Russian type. For instance, the specimen, Pl. LI, fig. 1, is certainly *Sp. striata*, and it will remain a question for future determination whether the specimens, Pl. IV, figs. 13, 14 ; and Pl. XIII, fig. 16, do really belong to *Sp. Mosquensis*. Mr. J. Wright is of opinion that the so-termed *Sp. clatharata*, M'Coy (p. 21, Pl. III, fig. 6) should be considered a synonym of *Sp. bisulcata* rather than of *Sp. striata* (although in 1855 so referred by Prof. M'Coy), for the description and figure in the 'Synopsis' agrees very closely with some of the finely-ribbed varieties of *Sp. bisulcata*, and that shell M'Coy describes in p. 120 of the last-named work, " with three or four ribs on the mesial fold." It appears evident, likewise, that *Sp. striata*, as described in the 'Synopsis,' includes both *Sp. striata* and the larger forms of *Sp. bisulcata*; *Sp. bisulcata* of the same author refers only to the young of that species.

SPIRIFERA DUPLICICOSTA, *Phillips* (p. 24). Plate III, figs. 3, 4, (?) 5—11. Plate V, figs. 35, 37, incorrectly referred to *S. trigonalis*, Plate LII, fig. 6.

This very variable species is sometimes with difficulty, and even uncertainty, distinguishable from certain shapes of *Sp. striata*. Mr. Burrow is of opinion that *Sp. humerosa*, Phillips, (p. 23 Pl. IV, figs. 15, 16), should be considered a thickened ponderous local variation of *S. duplicicosta*; and although I was at one time disposed to view it as distinct, am now more inclined to follow Mr. Burrow by placing it among the varieties of the last-named species. The mesial fold in *S. duplicicosta* is sometimes much prolonged beyond the level of the lateral portions of the valves, as seen in Pl. V, fig. 35, and Pl. LII, fig. 6 ; and it is even sometimes difficult to distinguish certain examples of *S. duplicicosta* from *S. striata*.

SPIRIFERA TRIGONALIS, *Martin.*

At one time I erroneously believed, with the generality of Palæontologists, that *Sp. trigonalis*, Martin, p. 29 ; *Sp. bisulcata*, Sow., p. 31; *Sp. crassa*, de Kon., p. 25 ; *Sp. grandicostata*, M'Coy, p. 23 ; and *Sp. transiens*, M'Coy, p. 33, were sufficiently distinct to be retained as separate species ; but a subsequent examination of a more extensive series of specimens has led me to infer that they are all modifications of a single very variable species, for which the term *trigonalis*, or *bisulcata*, should be retained ; and I am glad to say that this opinion has been already accepted by several experienced observers. No species is more variable in its general aspect, or in the number of its ribs, still every intermediate form may be found in our carboniferous limestone districts. To attempt, therefore, to describe all these variations would be endless ; but the following figures will convey an idea of its more prevalent shapes. We would therefore refer to *Sp. trigonalis*, Pl. IV, figs. 1, 2 ; Pl. V, figs. 1, 23, 24, and 38, 39; Pl. VI., figs. 1—22; Pl. VII, figs. 1—4, 7—16 ; and Pl. L, figs. 3—8 ; but I am at the same time ready to allow that if, for example, we take the winged more simple form, Pl. L, fig. 7, and then compare it with the transversely oval, rounded, thickened var. *crassa* (Pl. VI, fig. 20), the notion of both being modications of a single species will, to the generality of observers, appear absurd ; still if we find every variations connecting these extremes, are we to refuse the evidence of our eyes and senses, and to create as many species as we possess specimens ? In many examples of undoubted *Sp. trigonalis* and *S. bisulcata*, the cardinal angles are rounded so that the hinge-line is shorter than the breadth of the shell, but in the larger number of individuals these angles became more or less prolonged, and in some specimens they form long attenuated wings (Pl. L, figs. 3, 5, 6, 7, 8, and 9). In his list of the carboniferous Brachiopoda of Belgium, Prof. de Koninck admits that his *Sp. crassus*, and M'Coy's *Sp. grandicostata* are in all probability varieties of *Sp. bisculata*.[1] In its most simple shape, the sinus of *S. trigonalis* presents three longitudinal ribs, of which the central one is usually the largest (Pl. L, figs. 3 and 7) ; in other specimens there exists five, or two smaller ones, one on either side of the central rib (fig. 4), while in larger individuals we often find seven ribs, or three on either side of the central one (fig. 9), but in some specimens the ribs are more numerous and less regular in their respective widths. The mesial fold is often composed of three bifurcated ribs, or is divided by two sulci ; but here again, although this is certainly the prevalent feature, in some specimens these three ribs are more divided and the fold is sometimes not so sharply bisulcated. The shell of *Sp. trigonalis* is but rarely perfectly preserved, but when so the whole surface or ribs are finely striated and closely imbricated or decussated by numerous transverse fine spinulose or serrated ridges, as many

---

[1] 'Mémoire sur les genres et les sous-genres des Brachiopodes munis d'appendices spiraux,' par M. Davidson, traduit et augmenté de Notes par le Dr. L. A. Koninck. 'Mémoirs de la Societé Royale de Liege,' 1859.

as from twenty to thirty occupying the width of three lines, and of which an enlarged representation will be found in Pl. L, fig. 9ᵃ.

SPIRIFERA CONVOLUTA, *Phillips* (p. 35). Plate V, figs. 9—15 (2 to 8 excluded); and Plate L, figs. 1, 2.

This appears to be a rare species, of which three or four very remarkable specimens have been found by Mr. J. Rofe, at Thorneley, near Chipping (ten miles N.E. of Preston). One specimen in particular (Pl. L, fig. 1) was perfect from end to end, and measured eleven lines in length by fifty-seven in breadth, and twelve in depth. The shell occurs also in the carboniferous limestone of the neighbourhood of Wetton, in Staffordshire. I fear having too hastily coincided in the view taken by Prof. de Koninck, that *Sp. rhomboidea* (Phillips) was a synonym of *S. convoluta*. Mr. E. Dupont states he has found *S. convoluta* abundantly at Celles, near Dinant, in Belgium.

SPIRIFERA RHOMBOIDEA, *Phillips*. Plate V, figs. 2—8.

Prof. Phillips's description of this species will be found in the foot note of p. 36. A great difference is observable in the mesial fold and sinus of all the specimens I have seen of this and the preceding species. In *Sp. convoluta* it resembles much that of *Sp. trigonalis*, its three principal ribs being strongly marked, while in *Sp. rhomboidea* they are more numerous, and hardly defined.

SPIRIFERA TRIANGULARIS, *Martin* (p. 27). Plate V, figs. 16—24; and Plate L, figs. 10—17.

This is a very elegant and well-characterised species, easily distinguishable, but extremely difficult to extract from the hard limestone matrix in which it is usually imbedded: nevertheless, after much labour, Mr. Burrow has been able to procure a numerous series, among which were several examples that retained their elongated tapering wings quite perfect; but it is remarkable that when one wing became very much prolonged, the other was somewhat suddenly abbreviated, and this is very clearly discernible in the two or three of the most perfect specimens hitherto discovered (Pl. L, figs. 13, 15, 16). It is also necessary to remark, that in young shells the front was not very much produced, while in the adults the acutely angular cuneiform rib in the fold and sinus projects considerably above and beyond the regular surface of the valve (Pl. V, fig. 21).

SPIRIFERA ACUTA, *Martin.* Plate LII, figs. 6, 7.

CONCHYLIOLITHUS ANOMITES ACUTUS, *Martin.* Petrificata Derbiensia, pl. 49, figs. 15, 16, 1809.

For a long time I felt much puzzled with reference to this shell, and supposed it the young of some other species. My uncertainty has been, however, recently removed by Mr. Burrow's fortunate discovery of a number of specimens of all ages, and exactly agreeing with Martin's description and figure : " Valves convex semicircular, marked with deep, longitudinal equal striæ ; hinge patulous, straight, but not extending the whole breadth of the shell ; foramen triangular, large ; beak of the perforated valve prominent pointed, incurved ; the other short and obtuse ; margin acutely crenate, and furnished with a large angular sinus, causing a somewhat strong plicature on the surface of the valves; not frequent ; small ; limestone ; Winster and Croom Hill." None of Mr. Burrow's specimens exceeded the proportions of Martin's figure,—viz., six lines in length by about nine in breadth, each valve in adult individuals possessing from seventeen to nineteen angular ribs, the central one being at the same time the largest and most elevated of the valves. Mr. Burrow obtained his specimens from the carboniferous limestone in the neighbourhood of Settle, where the shell is not very abundant.

SPIRIFERA DISTANS *Sow.* (p. 46). Pl. VIII, figs. 1—17.

To this species I would unite *Sp. bicarinatus* (p. 47, Pl. VIII, fig. 18, and Pl. LII, fig. 4), which Prof. M'Coy established on a single very imperfect specimen from Cork, in Ireland. This specimen, now in the possession of Mr. J. Wright, of Cork, has quite the appearance assumed by certain examples of *Sp. distans*, of which it is, in all probability, a synonym. Prof. de Koninck, in his list of the Carboniferous Brachiopoda of Belgium, published in 1859, places M'Coy's *Sp. bicarinatus* among the synonyms of his *Sp. Roemerianus*, but this, I fear, is a mistake, for the original specimens of both would not lead me to a similar conclusion.

SPIRIFERA CUSPIDATA, *Martin* (p. 44). Plate VIII, figs. 19—24 ; Plate IX, figs. 1, 2.

Prof. de Koninck, myself, and others, have been led to suppose Martin's *Anomites subconicus* (p. 48, Pl. IX, fig. 3) to be a synonym of *Sp. cuspidatus.* I am, however, doubtful as to this being a correct opinion, from the fact that on perfectly preserved examples of Martin's *A. subconica*, of which Pl. LII, fig. 4, is an illustration, the entire surface of the ribs and shell (area excepted) are regularly traversed by continuous equidistant, sharp projecting laminæ, exactly similar to those which cover the surface of *Sp. laminosa* ; indeed, some examples of this last shell so closely resemble several Derbyshire specimen of *Sp. subconica* as to have puzzled me extremely, and to have almost made me consider both it and M'Coy's species as synonymous. Again, other examples of *Sp. laminosa* have so narrow an area as to differ much from Martin's shell. The ribs also in *Sp. subconica* appear relatively stronger and less numerous than in the generality of specimens of *Sp. cuspidata*. I therefore for the present, and until the question can be definitely settled by the discovery of a larger number of Derbyshire specimens of *S. subconica*, prefer to retain this last (as I have done in p. 48 of the present Monograph) as a distinct species. It is unfortunately very rare to find specimens of *Sp. cuspidatus* with the shells preserved, so that we are still uncertain whether its external sculpture was similar to that of *Sp. subconica ?*

SPIRIFERA GLABRA, *Martin* (p. 59). Plate XI, figs. 1—9; and Plate XII, figs. 1—12.

This is an excellent but most variable species, or a type round which are clustered many modifications not sufficiently marked to constitute separate species; for although the typical form of *Sp. glabra* possesses smooth valves, it is not uncommon to find in other and exceptional examples faint indications of lateral plication, obscurely flattened, or slightly rounded ribs ; the fold and sinus remaining always smooth. These modifications lead us gradually to such forms as *Sp. rhomboidalis*, M'Coy, (p. 57, Pl. XII, figs. 6, 7), which are likewise in all probability mere modifications of *Sp. glabra.*

SPIRIFERA LINEATA, *Martin* (p. 62), Plate XIII, figs. 4—13; and Plate LI, fig. 15.

When describing *Sp. lineata*, at p. 62, I had not seen any examples in which the shell was perfectly preserved, but the subsequent discovery of several excellent specimens in Scotland, as well as in India and America, has shown that externally the surface was covered with numerous concentric ridges, rarely in any place more than a line apart, but usually very much closer, and from each of which projected numerous, closely-packed spines, which thus formed a series of spiny fringes overlaying each other all over the shell. When the spines were absent, which is the general condition in which the shell is found, the surface appears marked by numerous and regularly imbricated lines, the radiating ones being produced by the small elevations from which each spine took its rise, as I have attempted to show in the enlarged figure in Pl. LI, which is very different from the irregular manner in which the spines are scattered over the surface of *Sp. Urii*, of which Pl. LI, fig. 16, is an illustration.

SPIRIFERA ELLIPITICA, *Phillips* (p. 63). Figs. 1—3.

I am not quite satisfied of having been correct while placing this species? among the varieties of *S. lineata,* its general transverse form being very constant, as far as I have hitherto seen.

SPIRIFERINA CRISTATA, *Schlotheim,* var. *octoplicata Sow.* (p. 38). Plate VII, figs. 37—47, and 60, 61; and Plate LII, figs. 9, 10, and 13.

This is certainly a very variable species. In the generality of specimens the mesial fold is composed of a single rib, which is much larger than those situated on the lateral portions of the shell; its crest is sometimes evenly rounded in all its length, but, as I have already described, becomes in many cases flattened, and even slightly longitudinally depressed as it approaches the frontal margin. Subsequently to my description of this interesting species, I became completely satisfied that *S. partita,* Portlock (p. 41, Pl. VII, figs. 60, 61) would require to be located among the synonyms of the species under description, and another remarkable modification has turned up, which I have distinguished by the varietal designation of *biplicata* (Pl. LII, figs. 11, 12). This variety, of which many examples have been found by Mr. Burrow in the lower scar limestone of Settle, in Yorkshire, and by Prof. Harkness in that of Little Island, near Cork, in Ireland, has the usual shape and character of *Sp. cristata,* or of its large carboniferous variety *octoplicata;* it presents also, according to the specimens, the same variable number of lateral ribs, viz., four, five, and eight, on the lateral portions of the shell, but the fold is no longer simply rounded or flattened, as in the typical shapes of *Sp. cristata,* but divided into two distinct ribs by a sulcus of variable depth; a well-marked rib extending likewise along the middle of the sinus, as seen in the illustrations above mentioned. That this is nothing more than a modification of the more general shape of Schlotheim's species is clearly proved by the many intermediate gradations in form which connect the specimens with rounded sinus to those with biplicated ones. In Pl. LII, fig. 9 represents a specimen of *Sp. cristata* with a more than usual angular fold; fig. 10 shows the fold slightly flattened along its crest, and divided by a slight groove, while in figs. 11 and 12 it is so much deepened as to divide the fold into two ribs. Very rarely indeed, but still as an exception, the fold has become triplicated towards the front, a fact which was not overlooked by Sowerby, since a specimen so conditioned is figured by that author along with his type-shapes of *Sp. octoplicata,* Pl. LII, fig. 13. The largest example of the var. *biplicata* that has come under my notice measured—

Length 12, width 15, depth 9 lines.

CYRTINA SEPTOSA, *Phillips* (p. 68). Plate XIV, figs. 1—10; Plate XV, figs. 1, 2; and Plate LI, fig. 17.

When publishing my description and figures of this interesting species I expressed a regret that all my efforts had proved ineffectual in making out the interior characters of the dorsal valve. Since then, thanks to the continued and zealous exertions of my indefatigable friend Mr. Burrow, the internal cast of the dorsal valve was discovered (Pl. LI, fig. 17), showing that the muscular impressions (anterior and posterior divisions of the adductor or occlusor muscle) were similarly arranged to those of *Spirifera*, and that there does not exist in that valve any septa, as in *Pentamerus*, a fact I also mentioned while describing M'Coy's so-termed *Pentamerus carbonarius*.

RHYNCHONELLA (pp. 89—112).

Eight or nine species have been provisionally retained; but the claims of *Rh. cordiformis* have not been satisfactorily established, and of *Rh.* (?) *gregaria*, but two imperfect valves have come under my examination. *Rhynchonella trilatera* is a rare shell, but several specimens have been recently found by Mr. Burrow near Settle, in Yorkshire, and some others by Mr. Carrington in the Limestone of Wetton, in Staffordshire. *Rh.* (?) *nana*, and *R. semisulcata* are at present by far too doubtful to deserve more than a passing notice. When describing and illustrating *Rh. pleurodon*, I was much puzzled with a large Rhynchonella (Pl. XXIII, fig. 22), which appeared to me to differ from the last-named shell in several particulars, and which I then doubtfully and provisionally left under *R. pleurodon*.

RHYNCHONELLA (?) CARRINGTONIANA, *Dav.* Plate XXIII, fig. 22; and Plate LIII, figs. 1, 2.

Sp. Char. Shell transversely oval, valves almost equally deep. Dorsal valve convex, with a broad mesial fold apparent only on the anterior half of the shell, where it is rarely very much elevated. About thirty-two or thirty-four radiating, rounded, simple ribs ornament this valve, of which from six to seven occupy the fold. Ventral valve longitudinally divided by a broad sinus, and marked with about the same number of ribs as in the opposite valve. Beak small and incurved; dimensions variable.

Length 15, breadth 19, depth 11 lines.

Obs. While illustrating *Rh. pleurodon* in Pl. XXIII of the present Monograph, I received from Mr. Parker, of Manchester, the loan of a large shell, fig. 22 of the same plate, which he had obtained from the Carboniferous Limestone of Twiston, in Lancashire. This shell appeared to differ so much from the many specimens of *R. pleurodon*, in my possession, that I did not venture to refer it positively to any known species. Since that period, several more similar specimens having been discovered by Mr. Carrington in the Carboniferous Limestone of Wetton, in Staffordshire, I have refigured the species, and have distinguished it by a separate denomination. When placing adult specimens of the shell under description in the same tray along with full-grown examples of *Rh. pleurodon*, both appeared clearly distinguishable, but the distinction between it and certain specimens of the Permian *Camarophoria Humbletonensis*, Howse, was not quite so apparent.[1] *Rh. (?) Carringtoniana* is more regularly transversely oval than *Rh. pleurodon*; the valves are more evenly convex; the ribs on the fold are not deflected so as to meet the corresponding margin of the opposite valve, as is the case with the last-named species. The ribs of *Rh. Carringtoniana* are also more rounded, and those on the lateral portions of the ventral valve regularly arched, and not straight, with their extremities bent upwards as in *Rh. pleurodon*. None of the ribs of *Rh. Carringtoniana* are longitudinally grooved near their extremities, as is the case with Phillips's species. When quite young, and up to a certain age, *Rh. Carringtoniana* is very slightly convex, and without any defined mesial fold.

CAMAROPHORIA (pp 113—118).

Four species have been recorded, but more abundant and better materials, with reference to *C. isorhyncha* and *C. lateralis*, must be obtained before these so-termed species can be definitely admitted. Of the first, I am acquainted with but a single imperfect Irish example, now in the possession of Sir R. Griffiths; of the second, with those only in the Cambridge Museum. *C. crumena* is a well made out species, and certainly the same as that from the Permian rocks, known under the designation of *C. Schlotheimi*. Nothing definite can be said with reference to *Camarophoria (?) proava*, which is, probably, a variation of *C. crumena*.

STROPHOMENA ANALOGA (pp. 119—123). Pl. XXVIII, figs. 1—13.

With small modifications, in detail, this species appears to have been recurrent from the Silurian and Devonian periods; the term *analoga* may be retained for the Carboniferous variety.

STREPTORHYNCHUS CRENISTRIA (p. 124).

I can add but little to what I have already said with reference to this species and its varieties; but some very interesting interiors of the dorsal valve, found by Mr. Burrow

---

[1] It is at times difficult to determine, from exteriors only, whether a shell belongs to *Rhynchonella* or to *Camarophoria*, and this is the case with the three or four examples of *R. Carringtoniana* I have been able to examine.

near Settle, in Yorkshire, are deserving of notice. In these specimens (Pl. LIII, fig. 3), the adductor or *anterior occlusor* scars of Hancock (A) are longitudinally striated, while the *posterior occlusors* (A') are clearly defined and dendritic in their markings.

ORTHIS (pp. 129—135).

I can add nothing to what has been already said with reference to the species composing this genus, except that *Orthis* (?) *antiquata* must still remain among the doubtful so-termed species, for only one or two specimens of it have been hitherto found.

PRODUCTUS (pp. 135—179).

Twenty-eight species? and a few named varieties have been here described; but, subsequently, five more new to England were discovered by Mr. Burrow in the Carboniferous Limestone of Settle, which we will at once proceed to describe.

PRODUCTUS MARGINALIS, *De Koninck* (Pl. LIII, fig. 3).

> PRODUCTUS MARGINALIS, *De Koninck*. Monographie du genre Productus, pl. xiv, fig. 7, 1847.

Sp. Char. Shell thin, circular, or sub-trapezoidal; slightly wider than long, somewhat geniculated and gibbous, with a narrow, projecting, curved margin; hinge-line rather less than the greatest width of the shell, with a small rudimentary area; beak small; ears flattened; surface wrinkled over the visceral portion of the shell, and irregularly interrupted, here and there, by prominent tubercules, which give rise to slender spines; while on the anterior portion of the curved margin the wrinkles of the visceral portion are replaced by small contiguous ribs; dorsal valve almost flat, concave only, or suddenly bent close to the margin; surface slightly marked by concentric wrinkles, which are replaced by small ribs near the margin; little pits are likewise dispersed over its surface. Interior muscular and other markings agreeing with those of the generality of Producta. Dimensions variable.

Length 8, width 9 lines.

Obs. The discovery of this Productus, as a British species, is due to Mr. Burrow; and in order to be quite certain as to its identity, the specimen here figured was fowarded to Prof. de Koninck, who confirmed the identification. In England it appears to be a rare species. It was obtained from the Lower Scar Limestone of Settle, in Yorkshire. At Visé, in Belgium, it is not rare in the state of internal casts, four or five specimens only have been found by Prof. de Koninck with the shell completely preserved.

PRODUCTUS KONINCKIANUS, *De Verneuil.* Plate LIII, fig. 7.

> PRODUCTUS CANCRINI, *De Koninck.* Descript. des Anim. foss. du Terr. Carb. de Belgique, p. 179, pl. ix, fig. 3, 1843 (not of De Verneuil and De Keyserling).
> — KONINCKIANUS, *De Verneuil.* Russia and the Ural Mount., vol. ii, pp. 253 and 274, 1845.
> — — *De Keyserling.* Reise in das Petschora Land, p. 203, pl. iv, fig. 4, 1846.
> — SPINULOSUS, *De Koninck.* Mon. du genre Productus, pl. xi, fig. 2, 1847 (not of Sowerby).

Sp. Char. Shell longer than wide, posteriorly rounded, broadest anteriorly; ventral valve evenly convex, without sinus; beak gibbous, much elevated and incurved, but not overlying the hinge line, which is less than the greatest width of the shell; ears very small, with two or three wrinkles; surface covered with numerous undulating regular, thread-like striæ, which increase from the interpolation of smaller striæ at variable distances from the beak, and which are interrupted at short intervals by small tubercules, disposed pretty regularly in quincunx, and giving rise to slender spines. Ventral valve concave, following the curves of the opposite one. Dimensions variable.

Length 9, width 8 lines.

Obs. This pretty little species has been taken for Sowerby's *P. spinulosus* by Prof. de Koninck, but from which it is completely distinct, Sowerby's shell not being longitudinally striated, as is the case with the form under description. It differs also in shape from *P. spinulosus*, which is usually transverse, while *P. Koninckianus* is generally slightly longer than broad; and, lastly, it differs again by the profile curve presented by the larger or ventral valve. It is also distinct from *Productus Cancrini*. Prof. de Koninck observes that apart from the arrangement of the spines it is a complete miniature of *P. Cora*, and that it is distinguishable from *P. arcuarius* by the absence of the transverse concentric furrow which divides the ventral valve of the last-named shell into two distinct portions.

The discovery of this species in the Lower Scar Limestone of Settle, in Yorkshire, is due to Mr. Burrow, and I am not acquainted with any other British locality. In Belgium it is very rare in the limestone of Visé, and has been found by Count Keyserling in Carboniferous Limestone on the banks of the Soiwa in the Petschora, Russia.

PRODUCTUS UNDIFERUS, *De Koninck.* Plate LIII, figs. 5, 6.

> PRODUCTUS SPINULOSUS, *De Koninck.* Descript. des Anim. foss. du Terrain Carb. de Belgique, p. 183 (partim), pl. x, fig. 4, 1843 (not *P. spinulosus*, Sow.).
> UNDIFERUS, *De Koninck.* Monograghie du genre Productus, pl. v, fig. 4, and pl. xi, fig. 5, 1847.

Sp. Char. Shell small, almost circular, about as wide as long; ventral valve gibbous; beak vaulted, but not projecting over the hinge-line, which is about as wide as the greatest breadth of the shell; ears small; dorsal valve concave, following the curves of the opposite one; valves externally marked with regular longitudinal, undulating thread-like striæ, and small more or less defined irregular concentric wrinkles. Delicate spines rise likewise here and there from the surface of the valves, and are more numerous on the ears close to the cardinal edge.

Length 6, width 6 lines.

Obs. Having sent British specimens of this shell to Prof. de Koninck, they were declared to be identical with his *Prod. undiferus.* This small species does not appear to attain the proportions of *P. undatus,* which it most resembles; its ribs appear to be proportionately smaller, and the wrinkles, where these exist, are never so large or regular as in *P. undatus.* Its margin appears also to have been broad and regularly curved in perfectly preserved specimens, as represented in Pl. LIII, fig. 6. This little shell was discovered by Mr. Burrow in the Lower Scar Limestone of Settle, the only British locality at present known. On the continent, it was found by Prof. de Koninck in the Carboniferous Limestone of Visé, as well as in the shales of Tournay, in Belgium.

PRODUCTUS NYSTIANUS, *De Koninck.* Plate LIII, fig. 9.

> PRODUCTUS NYSTIANUS, *De Koninck.* Descrip. des Animaux foss. du Ter. Carb. de Belgique, p. 202, pl. vii[bis] fig. 3; pl. ix, fig. 7, and pl. x, fig. 9. 1843. Also Monographie du genre Productus, pl. vi, fig. 4, and pl. xiv, fig. 5.

Sp. Char. Shell rather small; hinge-line straight, and as wide as the greatest width of the shell; ventral valve geniculated, semicircular, and much flattened on the posterior or visceral portion, abruptly bent towards the margin; beak very small, and hardly produced. The visceral portion is marked by numerous more or less regular undulating concentric wrinkles, interrupted here and there by projecting tubercules, while the anterior or bent portion of the valve is ornamented with small longitudinal ribs. A row of curved spines rise from and project over the cardinal edge. Dorsal valve almost flat on the visceral portion, bent near the margin, and ornamented as in the opposite one. Interior unknown.

Length 6, breadth 8 lines.

Obs. Three or four examples of this interesting species, completely agreeing with those represented by Prof. de Koninck in Plate XIV, fig. 5, of his 'Monographie,' were discovered for the first time in England by Mr. Burrow, but none of them assumed the tubuliform prolongations represented by my Belgian friend in Pl. VI, fig. 4, of the above-named work. Prof. de Koninck informs us that when adult and fully developed the shell assumes an entirely different aspect, viz., that a portion of the prolongation of the larger or ventral valve becomes dilated at the sides prior to becoming elongated and transformed into a cylindrical tube; that while this singular modification was being effected the inferior edge became elongated in a perpendicular direction to that of the anterior tube, but in a very irregular manner, and thus producing a second tube whose sides are at the same time irregular, strongly undulated, and ribbed; that these small ribs and undulations are particularly observable on the portion of the tube formed by the prolongation of the cardinal edge. I have never, however, had the advantage of seeing any of these singularly modified specimens, and which must be of very rare occurrence; for the generality of Belgian examples that have come under my notice exactly agree as to shape with those figured by Prof. de Koninck in Plate xiv of his work, as well as to those represented in the present Monograph.

Mr. Burrow found this shell in the Lower Scar Limestone of Settle, in Yorkshire, while Prof. de Koninck's specimens were obtained in the equivalent Limestone of Visé, in Belgium.

PRODUCTUS DESHAYESIANUS, *De Koninck*. Plate LIII, figs. 11, 12.

> PRODUCTUS DESHAYESIANUS, *De Koninck*. Descript. des Animaux foss. du Terr. Carb. de Belgique, p. 193, pl. x, fig. 7, 1843; and Mon. du genre Productus, pl. xiv, fig. 4.

Sp. Char. Shell small, semicircular about as wide as long; hinge-line nearly straight, and as wide as the greatest breadth of the shell. Ventral valve regularly arched, and evenly convex; surface marked by minute concentric ridges, from which rise closely-set spiny tubercules, but which become gradually less numerous towards the margin in adult individuals. Dorsal valve concave, following the curves of the ventral one.

Length 3, width $3\frac{1}{2}$ lines.

Obs. Two imperfect specimens only of this small species (?) have been hitherto discovered in England, which, having been forwarded to Prof. de Koninck for identification and comparison, were declared by him to be referable to his Belgian type. Mr. Burrow found his specimens in the Lower Scar Limestone of Settle, in Yorkshire, where the shell appears to be exceedingly rare. Prof. de Koninck procured his specimens from the equivalent Limestone of Visé, in Belgium.

PRODUCTUS STRIATUS (p. 139). Plate XXXIV, figs. 1—5; and Plate LIII, fig. 4.

Since describing this species, a specimen measuring nearly five inches in length by rather more than the same in breadth, has been discovered by Mr. Carrington in the Carboniferous Limestone of Wetton, in Staffordshire. It has also been found in the counties of Dublin, Kerry, and Leitrim, in Ireland.

PRODUCTUS SINUATUS (p. 157). Plate XXXIII, figs. 8—11.

Specimens of this interesting species have been recently found by Mr. Carrington at Wetton, in Staffordshire.

PRODUCTUS ERMINEUS, *De Koninck* (p. 164). Plate XXXIII, fig. 5.

This rare species (?) has been discovered by Mr. Carrington in the Carboniferous Lime-stone of Wetton, in Staffordshire.

PRODUCTUS ACULEATUS, *Martin* (p. 166).   Plate XXXIII, figs. 16—18 (19?), and
PRODUCTUS YOUNGIANUS, *Dav.* (p. 167).   Plate XXXIII, figs. 21—23.

Mr. Burrow is disposed to consider these two species as synonymous, and believes that there exists every possible gradation from the almost perfect smoothness or transverse lines of *aculeatus* to the strong ribs of *Youngianus;* while Messrs. Young, Armstrong, and some others are of a contrary opinion.   Mr. Young observes that if there is such a thing as value to be attached to species of the same genus, that there are good distinguishing characters between each of these species; that he has collected a great many specimens of *P. Youngianus* out of the Shale and out of the hard Limestone, where the whole of the finer external markings have been stripped off, and that he has never been able to identify it in this condition with any form of *P. aculeatus* he has ever collected or seen figured; that in all conditions perfect examples show distinctly-marked ribs, and also that these ribs are not due, as has been sometimes supposed, to the prolongation of the basis of the spine, but that they exist independently of the spines; that he has in his collection specimens that have more than their usual complement of spines, which are not so distinctly ribbed as some that have many less spines, showing, as in many other species, that spines had nothing to do with the formation of the ribs, the spines having no regular order of posi-tion in the valves, while the ribs, in all the examples that have come under his notice, show a regular order of formation; that, in addition to this peculiar structure, the general form of the shell would be a good guide in distinguishing the two species, *P. aculeatus* being a much rounder shell, with broader ear expansions than *P. Youngianus*, which in all Scotch specimens is much elongated, and with small ear expansions ; that on all his best specimens of Martin's shell there is not a trace of ribs, although the lines of growth are preserved, and he can count from eight to twelve distinct scattered spines on each specimen, while on an average-sized specimen of *P. Youngianus* there are as many as seventy to eighty spines. Such are the results of Mr. Young's careful examination of many Scottish specimens; and I must admit that when we look at a tray full of shells agreeing with Martin's type and figure of *P. aculeatus* (Pl. XXXIII, figs. 16, 17, 18, and

20 of our work), and then cast a glance at another *one* of *P. Youngianus* (figs. 21—23), a very great difference is observable between them, fig. 19 belonging, according to Mr. Young, to another species. I am quite disposed to concur with Mr. Young, that in the case of *P. Youngianus* the spines have nothing to do with the formation of the ribs, but I am still uncertain whether in the fully grown or adult condition *P. aculeatus* was not liable to become more or less ribbed towards its margin, for even in *P. Youngianus*, when quite young, no ribs are discernible, but they become apparent with the growth of the shell.

PRODUCTUS SUB-LŒVIS (p. 177). Plate XXXI, figs. 1, 2; Plate XXXII, fig. 1; and Plate LI, figs. 1, 2.

This interesting species has been found by Mr. Wardle in the Carboniferous Limestone of Caldon Low, Staffordshire. One specimen shows a row of spines projecting from a median longitudinal ridge, which extends along the ventral valve, as represented in Pl. LI, fig. 2. The visceral portion of the shell is also sometimes very irregularly wrinkled, although at other times it is comparatively smooth.

CHONETES.

I can add nothing to what has been stated in pages 179—191 of this Monograph. There are, however, a few points which will require further consideration when more ample material shall have been obtained. *C. papilionacea* has sometimes exceeded the dimensions I have given; for there exists in Sharp's collection at the Geological Society a specimen from Kendal which measures six and a half inches in breadth by three in length.

CRANIA AND DISCINA (pp. 192—198).

Nothing new has been found since my descriptions have been written.

LINGULA (pp. 199—210).

ERRATA.

At page 203, line 23, write—1. MUSCLES PRÉADDUCTEURS (*anterior* adductor), &c.

This Table shows the Shires in which Carboniferous Strata exist, as well as the amount of work done in

ENGLAND.	WALES.	SCOTLAND.	IRELAND.	ABROAD.	GENERA AND SPECIES. The point of ? denotes that the so-termed species requires further research prior to its being definitely admitted.	Yorkshire.
×	×	×	×	×	*Terebratula hastata*, Sow. Min. Con., tab. 446, figs. 1, 2, 3. Dav. Mon., pp. 11 and 213, pl. i, figs. 1—17 ; and pl. xlix, figs. 11—17 = *T. ficus*, M'Coy	×
×	...	...	...	×	— var. *Gillingensis*, Dav. Mon., pp. 17 and 213, pl. i, figs. 18—20 ; pl. xlix, figs. 19, 20...	×
×	×	×	×	×	— *sacculus*, Martin. Pet. Derb., tab. xlvi, figs. 1, 2. Dav. Mon., pp. 14 and 213, pl. i, figs. 23—30 ; pl. xlix, figs. 27—29 = *T. sufflata*, Schlotheim ...	×
×	...	×	...	×	— *vesicularis*, De Kon. Anim. Foss. Belg., p. 666, pl. lvi, fig. 10. Dav. Mon., p. 15, pl. i, figs. 25, 26, 28, 31, 32 ; and p. 213, pl. xlix, figs. 26—30	×
×	×	×	×	×	*Athyris Royssii*, L'Eveillé. Mém. S. G. France, vol. ii, pl. ii, figs. 18—20. Dav. Mon., p. 84, pl. xviii, figs. 1—11 = *S. glabristria* and *S. fimbriata*, Phil. = *A. depressa*, M'Coy ? = *A. pectinifera*, J. de C. Sow.	×
×	...	...	...	...	— *expansa*, Phillips. Geol. York., vol. ii, p. 220, pl. x, fig. 18. Dav. Mon., p. 82, pl. xvi, figs. 14, 16—18 ; pl. xvii, figs. 1—5 = *A. fimbriata*, J. de C. Sow. (not of Phillips)	×
×	...	...	...	...	— *Carringtoniana*, Dav. Mon., p. 217, pl. lii, figs. 18—20	...
? — ×	...	...	×	×	— *squamigera*, De Kon. Anim. Foss. Belg., p. 667, pl. lvi, fig. 9. Dav. Mon., p. 83, pl. xviii, figs. 12, 13	...
×	...	...	×	×	— *lamellosa*, L'Eveillé. Mém. S. G. Fr., vol. ii, pl. ii, figs. 21—23. Dav. Mon., pp. 79 and 217, pl. xvi, fig. 1 ; pl. xvii, fig. 6 ; and pl. li, fig. 14 = *S. squamosa*, Phil....	×
×	...	×	×	×	— *plano-sulcata*, Phillips. Geol. York., vol. ii, p. 220, pl. x, fig. 15. Dav. Mon., pp. 80 and 217, pl. xvi, figs. 2—13, 15 ; and pl. li, figs. 11—13 = *A. de Royssii*, De Vern. (not Phil.) = *A. oblonga*, J. de C. Sow. = *A. obtusa* and *A. paradoxa*, M'Coy = *A. virgoides*, M'Coy	×
×	×	×	×	×	— *ambigua*, Sow. Min. Con., vol. iv, p. 105, pl. 376. Dav. Mon., p. 77, pl. xv, figs. 16—22 ; pl. xvii, figs. 11—14 = *T. pentaedra*, Phil. = *A. sublobata*, Portl.	×
×	...	...	...	×	— *globularis*, Phil. Geol. York., vol. ii, p. 220, pl. x, fig. 22. Dav. Mon., p. 86, figs. 15—18 ...	×
×	...	...	...	×	— *subtilita*, Hall, in Howard Stansbury's Exploration of the Valley of the Great Salt Lake of Utah, p. 409, pl. ii, figs. 1, 2, 1852. Dav. Mon., pp. 18 and 86, pl. i, figs. 21, 22 ; pl. xvii, figs. 8—10	×
×	...	×	×	×	*Retzia radialis*, Phil. Geol. York., vol. ii, pl. xii, figs. 40, 41. Dav. Mon., pp. 87 and 218, pl. xvii, figs. 19—21 ; and pl. li, figs. 4—9 = *T. mantiæ*, De Kon. (not Sow.)	×

me in the way of collecting. Those Counties from which few Species are recorded will, no doubt, upon furthe

	ENGLAND.																	WALES.									SCOTLAND.															
	Yorkshire.	Derbyshire.	Lancashire.	Westmoreland.	Cumberland.	Durham.	Northumberland.	Isle of Man.	Herefordshire.	Staffordshire.	Shropshire.	Leicestershire.	Worcestershire.	Devonshire.	Somersetshire.	Gloucestershire.	Monmouthshire.	Pembrokeshire.	Anglesea.	Denbighshire.	Flintshire.	Montgomeryshire.	Brecknockshire.	Glamorganshire.	Carmarthenshire.	Carnarvonshire.	Lanarkshire.	Renfrewshire.	Ayrshire.	Buteshire.	Dumbartonshire.	Stirlingshire.	Dumfriesshire.	Peeblesshire.	Edinburghshire.	Linlithgowshire.	Haddingtonshire.	Fifeshire.	Berwickshire.	Kircudbrightshire.	Armagh.
i,	×	×	×	×		×	×			×	×				×	×	×		×								×	×	×		×				×	×		×			×
...	×				×																																				
i,	×	×				×			×						×				×								×		×		×				×						
5, 30	×	×				×																					×	×	×												
iii, oy	×	×	×			×			×						×				×								×	×	×	×					×	×		×			×
vi, w.	×	×	×	×							×				×																										
...											×																														
iii,		×																																							
17,	×	×	×	×																																					
17, sii, and	×	×	×	×		×	×		×						×	×											×	×	×		×				×	×		×			
22;	×	×	×	×		×	×	×	×	×	×					×			×		×		×				×	×	×	×	×	×			×	×	×	×			×
...	×	×	×				×		×																																
of 22;	×	×	×	×																																					
vii,	×	×	×						×								×										×	×	×												

one in the way of collecting. Those Counties from which few Species are recorded will, no doubt, upon furthe

	ENGLAND.																	WALES.									SCOTLAND.															
	Yorkshire.	Derbyshire.	Lancashire.	Westmoreland.	Cumberland.	Durham.	Northumberland.	Isle of Man.	Herefordshire.	Staffordshire.	Shropshire.	Leicestershire.	Worcestershire.	Devonshire.	Somersetshire.	Gloucestershire.	Monmouthshire.	Pembrokeshire.	Anglesea.	Denbighshire.	Flintshire.	Montgomeryshire.	Brecknockshire.	Glamorganshire.	Carmarthenshire.	Carnarvonshire.	Lanarkshire.	Renfrewshire.	Ayrshire.	Buteshire.	Dumbartonshire.	Stirlingshire.	Dumfriesshire.	Peeblesshire.	Edinburghshire.	Linlithgowshire.	Haddingtonshire.	Fifeshire.	Berwickshire.	Kircudbrightshire.	Armagh.
i,	×	×	×	×		×	×	×		×	×				×	×	×			×							×	×	×			×			×	×		×			×
...		×				×																																			
i,	×	×	×	×		×	×	×		×					×	×				×																					
5, 30	×	×				×																						×	×							×					
iii, oy	×	×	×					×		×	×									×	×						×	×	×			×			×	×	×	×	×		×
vi, ow.	×	×	×	×						×					×																										
.....											×																														
iii,			×																																						
17,		×	×	×	×					×																															
17, sii, and ..ne	×	×	×	×		×	×			×						×	×										×	×	×			×			×	×		×			
22 ;	×	×	×	×		×	×	×	×	×	×									×	×		×				×	×	×	×	×	×			×	×	×	×			
.....		×	×	×						×																															
of 22 ;	×	×	×	×	×																																				
vii,		×	×	×						×																	×	×	×												

rther exploration, be found to contain some more of the common Species.

	IRELAND.																													
Kircudbrightshire.	Armagh.	Cork.	Carlow.	Clare.	Cavan.	Dublin.	Donegal.	Down.	Fermanagh.	Galway.	Kerry.	Kildare.	King's County.	Limerick.	Louth.	Longford.	Leitrim.	Meath.	Mayo.	Monaghan.	Queen's County.	Roscommon.	Sligo.	Tipperary.	Tyrone.	Waterford.	Westmeath.	Wexford.	Kilkenny.	Antrim.
	×	×				×	×			×		×	×	×	×		×	×	×		×	×		×	×	×	×		×	
		×	×			×	×					×				×		×						×	×					
	×	×				×	×		×			×	×		×		×	×			×	×	×	×	×			×		
												×															×			
	×				×	×			×			×		×			×				×							×		
	×			×	×	×	×		×			×		×	×	×		×					×	×	×					
	×				×	×				×			×		×								×	×						
		×							×																					

England.	Wales.	Scotland.	Ireland.	Abroad.	GENERA AND SPECIES.
×				×	*Retzia ulotrix*, De Kon. Anim. Foss. Belg., p. 292, pl. xix, fig. 5. Dav. Mon., p. 88, pl. xviii, figs. 14, 15; and pl. liv, fig. 45
	×				?— *carbonaria*, Dav. Mon., p. 219, pl. li, fig. 3
×	×		×	×	*Spirifera striata*, Martin. Pet. Derb., tab. xxxiii. Dav. Mon., pp. 19 and 221, pl. ii, figs. 12—21; pl. iii, figs. 2—6; and pl. lii, figs. 1, 2 = *T. spirifera*, Val., apud Lamarck = *S. attenuatus*, Sow. = *princeps*, M'Coy
×			×	×	?— *Mosquensis*, Fischer de Waldheim. Prog. sur les Choristite, p. 8, No. 1, 1837. Dav. Mon., pp. 22 and 221, pl. iv, figs. 13, 14 = *C. Sowerbyi* and *Kleinii*, Fischer = *S. choristites*, V. Buch
×	×	×	×	×	— *duplicicosta*, Phillips. Geol. York., vol. ii, p. 218, pl. x, fig. 1. Dav. Mon., p. 24, pl. iii, figs. 3, 4?, 5—11; pl. v, figs. 35—37; pl. lii, fig. 6
×					— var. *humerosa*, Phillips. Geol. York., vol. ii, p. 218, pl. ix, fig. 8. Dav. Mon., pp. 23 and 221, pl. iv, figs. 15, 16
×	×	×	×	×	— *trigonalis*, Martin. Pet. Derb., tab. xxxvi, fig. 1. Dav. Mon., pp. 29 and 222, pl. iv, figs. 1, 2; pl. v, figs. 1, 23, 24, and 38, 39; pl. vi, figs. 1—22; pl. vii, figs. 1 and 4, 7—16; and pl. l, figs. 3—9 = *S. bisulcata*, Sow.; Dav. p. 31 = *Sp. crassa*, De Kon., p. 25 =? *Sp. grandicostata*, M'Coy, p. 33 = *S. transiens*, M'Coy, p. 33
×			×	×	— *convoluta*, Phillips. Geol. York., vol. ii, p. 217, pl. ix, fig. 7. Dav. Mon., pp. 35 and 223, pl. v, figs. 9—15 (2—8 excluded); pl. l, figs. 1, 2
×	×		×	×	— *triangularis*, Martin. Pet. Derb., pl. xxxvi, fig. 2. Dav. Mon., pp. 27 and 223, pl. v, figs. 16—24; and pl. l, figs. 10—17 = *S. ornithorhynca*, M'Coy
×					?— *fusiformis*, Phillips. Geol. York., vol. ii, p. 217, pl. ix, figs. 10, 11. Dav. Mon., p. 56, pl. xiii, fig. 15 (a very doubtful species)
×		×			— *rhomboidea*, Phillips. Geol. of York., vol. ii, p. 217, pl. ix, figs. 8, 9. Dav. Mon., p. 36 (footnote), and p. 223, pl. v, figs. 2—8
×					— *acuta*, Martin. Pet. Derb., pl. xlix, figs. 15, 16. Dav. Mon., p. 224, pl. lii, figs. 6, 7
×				×	— *planata*, Phillips. Geol. York., vol. ii, p. 219, pl. x, fig. 3. Dav. Mon., pl. vii, figs. 25—36
×	×		×	×	— *cuspidata*, Martin. Trans. Lin. Soc., vol. iv, p. 44, figs. 1—6. Dav. Mon., p. 44, pl. viii, figs. 19—24; pl. ix, figs. 1, 2; pl. lii, fig. 3 = *Cyrtia simplex*, M'Coy (not Phillips)
×					?— *subconica*, Martin. Pet. Derb., tab. xlvii, figs. 6—8. Dav. Mon., pp. 48 and 224, pl. ix, fig. 3; pl. lii, fig. 4
×			×	×	— *distans*, Sow. Min. Con., tab. 494, fig. 3. Dav. Mon., pp. 46 and 224, pl. viii, figs. 1—17; pl. lii, fig. 5 = *S. bicarinata*, M'Coy (p. 47, pl. viii, fig. 18)
		×			— *mesogonia*, M'Coy. Synopsis Carb. Foss. Ireland, p. 137, pl. xxii, fig. 13. Dav. Mon., p. 48, pl. vii, fig. 24

	ENGLAND.																	WALES.									SCOTLAND.													
	Yorkshire.	Derbyshire.	Lancashire.	Westmoreland.	Cumberland.	Durham.	Northumberland.	Isle of Man.	Herefordshire.	Staffordshire.	Shropshire.	Leicestershire.	Worcestershire.	Devonshire.	Somersetshire.	Gloucestershire.	Monmouthshire.	Pembrokeshire.	Anglesea.	Denbighshire.	Flintshire.	Montgomeryshire.	Brecknockshire.	Glamorganshire.	Carmarthenshire.	Carnarvonshire.	Lanarkshire.	Renfrewshire.	Ayrshire.	Buteshire.	Dumbartonshire.	Stirlingshire.	Dumfriesshire.	Peeblesshire.	Edinburghshire.	Linlithgowshire.	Haddingtonshire.	Fifeshire.	Berwickshire.	Kircudbrightshire.
	×	×								×																														
																		×																						
	×	×						×								×	×	×																						
	×	×																																						
	×	×	×	×				×				×			×		×	×		×							×		×		×				×	×		×		×
	×																																							
	×	×	×	×				×		×					×	×	×	×		×	×			×			×	×	×			×	×		×	×		×	×	×
	×	×	×							×																														
	×	×	×					×	×		×											×																		
	×																																							
	×	×						×								×																								
	×	×																																						
	×	×	×					×		×		×																												
	×	×	×	×				×		×		×				×	×			×				×																
		×						×			×																													
	×	×	×					?																																

	ENGLAND.																	WALES.									SCOTLAND.													
	Yorkshire.	Derbyshire.	Lancashire.	Westmoreland.	Cumberland.	Durham.	Northumberland.	Isle of Man.	Herefordshire.	Staffordshire.	Shropshire.	Leicestershire.	Worcestershire.	Devonshire.	Somersetshire.	Gloucestershire.	Monmouthshire.	Pembrokeshire.	Anglesea.	Denbighshire.	Flintshire.	Montgomeryshire.	Brecknockshire.	Glamorganshire.	Carmarthenshire.	Carnarvonshire.	Lanarkshire.	Renfrewshire.	Ayrshire.	Buteshire.	Dumbartonshire.	Stirlingshire.	Dumfriesshire.	Peeblesshire.	Edinburghshire.	Linlithgowshire.	Haddingtonshire.	Fifeshire.	Berwickshire.	Kircudbrightshire.
	×	×								×																														
																	×																						
	×	×					×	×				×				×		×																						
	×	×														×																								
	×	×	×	×			×	×		×		×				×	×	×		×	×	×		×			×	×	×		×	×			×	×	×	×		×
	×	×																																						
	×	×	×	×		×	×	×		×	×				×	×	×	×		×	×	×		×			×	×	×		×	×			×	×	×	×		×
	×	×	×				×	×		×								×																						
	×	×	×				×	×		×								×																						
	×																																							
	×	×	×				×			×						×																								
	×	×																																						
	×	×	×				×			×																														
	×	×	×	×						×	×				×			×						×																
		×								×		×																												
	×	×	×							?																														

IRELAND.

Armagh.	Cork.	Carlow.	Clare.	Cavan.	Dublin.	Donegal.	Down.	Fermanagh.	Galway.	Kerry.	Kildare.	King's County.	Limerick.	Louth.	Longford.	Leitrim.	Meath.	Mayo.	Monaghan.	Queen's County.	Roscommon.	Sligo.	Tipperary.	Tyrone.	Waterford.	Westmeath.	Wexford.	Kilkenny.	Antrim.
	×				×			×		×	×		×			×	×			×	×	×		×		×		×	
	×												×																
×	×				×	×					×					×	×						×						
×	×	×	×		×		×			×		×	×	×	×	×	×			×	×	×	×		×	×	×		
					×	×																							
	×									×	×																		
	×				×	×					×						×			×			×		×		×		
	×				×	×				×	×		×										×	×	×		×		
	×				×	×				×	×								×				×	×	×		×		
												×															×		

England.	Wales.	Scotland.	Ireland.	Abroad.	GENERA AND SPECIES.
×	...	×	×	×	*Spirifera pinguis*, Sow. Min. Con., vol. iii, p. 125, tab. 271. **Dav.** Mon., p. 50, pl. x, figs. 1, 2 = *S. rotundata*, Sow. = *S. subrotundata*, M'Coy
×	×	×	×	×	— *ovalis*, Phillips. Geol. York., vol. ii, p. 219, pl. x, fig. 5. Dav. Mon., p. 53, pl. ix, figs. 20—26; pl. lii, fig. 8 = *S. exarata*, Fleming = *B. hemisphærica*, M'Coy
×	...	...	×	×	— *integricosta*, Phillips. Geol. York., vol. ii, p. 219, pl. x, fig. 2. Dav. Mon., p. 55, pl. ix, figs. 13—19 = ? *A. rotundatus*, Martin = ? *S. paucicosta*, M'Coy
×	...	...	×	×	— *triradialis*, Phillips. Geol. York., vol. ii, p. 219, pl. x, fig. 7. Dav. Mon., p. 49, pl. ix, figs. 4—12 = *S. trisulcosa* and *S. sexradialis*, Phillips
×	...	...	...	...	? — *Reedii*, Dav., p. 43, pl. v, figs. 40—47 (doubtful species)
×	×	×	×	×	— *glabra*, Martin. Pet. Derb., pl. xlviii, figs. 9, 10. Dav. Mon., pp. 59 and 225, pl. ix, figs. 1—9; and pl. xii, figs. 1—5, 11, 12 = *Sp. obtusus* and *S. oblatus*, Sow.; *S. linguifera*, *S. symmetrica* and *S. decora*, Phillips = *S. rhomboidalis*, M'Coy, pl. xii, figs. 6, 7
...	...	×	...	...	— *Carlukensis*, Dav. Mon., p. 59, pl. xiii, fig. 14
×	...	×	×	×	— *Urii*, Flem. Brit. Animals, p. 376. Dav. Mon., p. 58, pl. xii, figs. 13, 14 = *S. Clannyana*, King
×	×	×	×	×	— *lineata*, Martin. Pet. Derb., tab. xxxvi, fig. 3. Dav. Mon., pp. 62 and 225, pl. xiii, figs. 4—13 (pl. li, fig. 15) = *S. mesoloba*, Phillips = *S. reticulata*, M'Coy = *S. imbricata*, Sow. = *S. Martini*, Flem. = *S. Stringocephaloides*, M'Coy
×	...	...	×	×	— *elliptica*, Phillips. Geol. York., vol. ii, p. 219, pl. x, fig. 16. Dav. Mon., p. 63, pl. xiii, figs. 1—3
×	×	×	×	×	*Spiriferina laminosa*, M'Coy. Synopsis Carb. Foss. Ireland, p. 137, pl. xxi, fig. 4. Dav. Mon., p. 36, pl. vii, figs. 17—22 = *Sp. hystericus*, De Kon. (not of Schlotheim) = *S. speciosa*, M'Coy (not Schlot.) = *S. tricornis*, De Kon.
×	×	×	×	×	— *cristata*, Schloth.; var. *Octoplicata*, Sow. Min. Con., p. 120, pl. 562, figs. 2—4; Dav. Mon., pp. 38 and 226, figs. 37—47, pl. lii, figs. 9, 10, 13 = *S. partita*, Port., p. 41, pl. vii, figs. 60, 61
×	...	...	×	...	— var. *biplicata*, Dav., p. 226, pl. lii, figs. 11, 12
×	...	...	...	...	? — *minima*, Sow. Min. Con., p. 105, tab. 377, fig. 1. Dav. Mon., p. 40, pl. vii, figs. 56—59 (a doubtful species)
×	...	×	×	×	— *insculpta*, Phil. Geol. York., vol. ii, p. 216, pl. ix, figs. 2, 3. Dav. Mon., p. 42, pl. vii, figs. 48—55; pl. lii, figs. 14, 15 = *S. crispus* and *heteroclytus*, De Kon. (not Linnæus nor Defrance) = *S. quinqueloba*, M'Coy = *S. Koninckiana*, D'Orbigny
×	...	...	...	×	*Cyrtina septosa*, Phillips. Geol. York., vol. ii, p. 216, pl. ix, fig. 7. Dav. Mon., p. 68, pl. xiv, figs. 1—10; pl. xv, figs. 1, 2; pl. l, fig. 19; pl. li, figs. 17, 18
...	...	...	×	...	? — *dorsata*, M'Coy. Syn. of the Carb. Foss. of Ireland, p. 136, pl. xxii, fig. 14

	ENGLAND.																	WALES.									SCOTLAND.													
	Yorkshire.	Derbyshire.	Lancashire.	Westmoreland.	Cumberland.	Durham.	Northumberland.	Isle of Man.	Herefordshire.	Staffordshire.	Shropshire.	Leicestershire.	Worcestershire.	Devonshire.	Somersetshire.	Gloucestershire.	Monmouthshire.	Pembrokeshire.	Anglesea.	Denbighshire.	Flintshire.	Montgomeryshire.	Brecknockshire.	Glamorganshire.	Carmarthenshire.	Carnarvonshire.	Lanarkshire.	Renfrewshire.	Ayrshire.	Buteshire.	Dumbartonshire.	Stirlingshire.	Dumfriesshire.	Peeblesshire.	Edinburghshire.	Linlithgowshire.	Haddingtonshire.	Fifeshire.	Berwickshire.	Kircudbrightshire.
	×	×	×					×			×						×										×													
	×	×	×					×		×	×						×					×							×			×					×	×		
	×	×	×					×			×						×																							
	×	×	×					×									×																							
	×																																							
	×	×	×	×		×		×			×				×	×	×	×					×				×	×	×			×			×					×
																		×									×	×	×											
	×	×	×					×			×				×		×	×									×	×	×						×			×		
	×	×	×	×		×		×			×				×	×	×	×					×				×	×	×	×	×	×			×		×	×		×
	×	×	×	×		×		×			×																													
						×		×			×				×	×		×									×		×									×	×	
	×	×						×			×				×			×									×	×	×				×		×		×	×		
	×							×			×																													
		×																																						
	×	×	×					×																			×		×											
	×	×	×		×			×			×				×	×	×										×		×											

(continued.)

	ENGLAND.																	WALES.									SCOTLAND.													
	Yorkshire.	Derbyshire.	Lancashire.	Westmoreland.	Cumberland.	Durham.	Northumberland.	Isle of Man.	Herefordshire.	Staffordshire.	Shropshire.	Leicestershire.	Worcestershire.	Devonshire.	Somersetshire.	Gloucestershire.	Monmouthshire.	Pembrokeshire.	Anglesea.	Denbighshire.	Flintshire.	Montgomeryshire.	Brecknockshire.	Glamorganshire.	Carmarthenshire.	Carnarvonshire.	Lanarkshire.	Renfrewshire.	Ayrshire.	Buteshire.	Dumbartonshire.	Stirlingshire.	Dumfriesshire.	Peeblesshire.	Edinburghshire.	Linlithgowshire.	Haddingtonshire.	Fifeshire.	Berwickshire.	Kircudbrightshire.
	×	×	×					×		×							×										×													
	×	×	×					×		×												×					×		×				×				×	×		
	×	×	×				×			×																														
	×	×	×				×			×																														
	×																																							
	×	×	×	×			×	×		×						×	×		×	×	×			×			×	×	×								×			×
																											×	×	×											
	×	×					×									×	×										×	×	×			×			×		×			
	×	×	×				×			×					×	×	×	×						×			×	×	×			×						×	×	×
	×	×	×	×																																				
							×	×		×												×					×	×											×	×
	×	×					×			×	×				×						×						×	×	×	×		×			×	×	×			
	×									×																														
		×																																						
	×	×	×				×			×																	×		×											
	×	×	×		×					×		×			×	×																								

IRELAND.

	Armagh.	Cork.	Carlow.	Clare.	Cavan.	Dublin.	Donegal.	Down.	Fermanagh.	Galway.	Kerry.	Kildare.	King's County.	Limerick.	Louth.	Longford.	Leitrim.	Meath.	Mayo.	Monaghan.	Queen's County.	Roscommon.	Sligo.	Tipperary.	Tyrone.	Waterford.	Westmeath.	Wexford.	Kilkenny.	Antrim.
	×	×		×		×	×		×		×	×		×		×	×	×				×	×	×	×	×		×		
	×		×													×	×							×						
	×	×							×															×	×					
				×		×																								
	×	×		×	×	×			×		×	×		×		×	×	×		×		×		×	×					
						×	×		×			×	×											×						
	×	×	×	×		×			×		×	×		×		×	×	×	×			×		×		×				
	×					×	×				×																			
		×				×	×				×					×						×		×		×		×		
						×	×										×					×		×		×				
		×																					×		×					
	×	×		×	×	×			×		×			×	×	×							×							
		×				×	×																					×		

ENGLAND.	WALES.	SCOTLAND.	IRELAND.	ABROAD.	GENERA AND SPECIES.
×				×	*Cyrtina carbonaria*, M'Coy. Brit. Pal. Foss., p. 442, pl. iii[D], figs. 12—18. Dav. Mon., p. 71, pl., xv, figs. 5—14
×			×	×	*Rhynchonella reniformis*, Sow. Min. Con., pl. 496, figs. 1—4. Dav. Mon., p. 90, pl. xix, figs. 1—7
×			×	×	? — *cordiformis*, Sow. Min. Con., pl. 495, fig. 2. Dav. Mon., p. 92, pl. xix, figs. 8, 9, 10 (species still doubtful)
×			×	×	— *acuminata*, Martin. Pet. Derb., pl. xxxii, figs. 7, 8. Dav. Mon., p. 93, pl. xx, figs. 1—13; pl. xxi, figs. 1—20 = *T. platyloba*, Sow. = *T. mesogonia*, Phillips
×		×	×	×	— *pugnus*, Martin. Pet. Derb., tab., xxii, figs. 4, 5. Dav. Mon., p. 97, pl. xxii, figs. 1—15 = ? *T. sulcirostris*, Phillips = ? *A. laticliva*, M'Coy
×	×	×	×	×	— *pleurodon*, Phillips. Geol. York., vol. ii, p. 222, pl. 12, figs. 25—30 (but not 16). Dav. Mon., p. 101, pl. xxiii, figs. 1—15, 16—22 ? = *T. mantiæ*, Sow. = *T. ventilabrum*, Phil. = *T. pentatoma*, De Kon. (not Fischer) = *A. triplex*, M'Coy = *T. Davreuxiana*, De Kon.
×			×		— *flexistria*, Phillips. Geol. York., vol. ii, p. 222, pl. xii, figs. 33, 34. Dav. Mon., p. 105, pl. xxiv, figs. 1—8 = *T. tumida*, Phil. = *H. heteroplycha*, M'Coy
×			×	×	— *angulata*, Linnæus. Systema Naturæ, i, pars 2, p. 1154. Dav. Mon., p. 107, pl. xix, figs. 11—16 = *T. excavata*, Phil.
×				×	— *trilatera*, De Kon. Anim. Foss. de la Belgique, p. 292, pl. xix, fig. 7. Dav. Mon., p. 109, pl. xxiv, figs. 23—26
×			×		? — ? *gregaria*, M'Coy. Synopsis Carb. Foss. of Ireland, p. 153, pl. xxii, fig. 18. Dav. Mon., p. 112, pl. xv, figs. 27, 28
×					— *Carringtoniana*, Dav. Mon., p. 227, pl. xxiii, fig. 22; and pl. liii, figs. 1, 2

N.B. *Rhynchonella* ? *nana*, M'Coy, *Rh.* ? *semisulcata*, M'Coy, and *Rh.* or *Cam. proava*, Phil., are still so doubtful that they need not be here recorded.

ENGLAND.	WALES.	SCOTLAND.	IRELAND.	ABROAD.	
×	×	×		×	*Camarophoria crumena*, Martin. Pet. Derb., pl. xxxvi, fig. 4. Dav. Mon., p. 113, pl. xxv, figs. 3—9 = *T. Schlotheimi*, V. Buch
×			×		— *globulina*, Phil. Ency. Met. Geol., vol. iv, pl. iii, fig. 3. Dav. Mon., p. 115, pl. xxiv, figs. 9—22 = *T. rhomboidea*, Phil. = *T. seminula*, Phil. ? = *H. longa*, M'Coy
×					? — *laticliva*, M'Coy. Br. Pal. Foss., p. 444, pl. iii[D], figs. 20, 21 (not *A. laticliva*, M'Coy, of the Synopsis). Dav. Mon., p. 116, pl. xxv, figs. 11, 12
		×			? — *isorhyncha*, M'Coy. Synopsis, p. 154, pl. xviii, fig. 8. Dav. Mon., p. 117, pl. xxv. figs. 1, 2
×	×	×	×	×	*Strophomena analoga*, Phillips. Geol. York., vol. ii, p. 215, pl. vii, fig. 10. Dav. Mon., p. 119, pl. xxviii, figs. 1—13 = *Lept. distorta*, Sow. = *L. multirugata*, M'Coy

(continued.)

	ENGLAND.																	WALES.									SCOTLAND.													
	Yorkshire.	Derbyshire.	Lancashire.	Westmoreland.	Cumberland.	Durham.	Northumberland.	Isle of Man.	Herefordshire.	Staffordshire.	Shropshire.	Leicestershire.	Worcestershire.	Devonshire.	Somersetshire.	Gloucestershire.	Monmouthshire.	Pembrokeshire.	Anglesea.	Denbighshire.	Flintshire.	Montgomeryshire.	Brecknockshire.	Glamorganshire.	Carmarthenshire.	Carnarvonshire.	Lanarkshire.	Renfrewshire.	Ayrshire.	Buteshire.	Dumbartonshire.	Stirlingshire.	Dumfriesshire.	Peeblesshire.	Edinburghshire.	Linlithgowshire.	Haddingtonshire.	Fifeshire.	Berwickshire.	Kircudbrightshire.
				×																																				
	×	×	×					×		×																														
	×		×							×																														
	×	×	×					×		×						×																								
	×	×	×							×						×											×	×	×							×	×		×	×
	×	×	×	×	×	×	×	×		×								×	×								×	×	×	×					×	×	×	×	×	×
	×	×	×			×				×						×																								
	×	×	×							×																														
	×	×	×							×																														
												×				×																								
			×									×				×																								
	×	×	×				×		×	×						×						×													×					
	×	×	×							×						×																								
		×						×																																
	×	×	×				×	×		×						×		×	×					×			×	×	×				×					×		

(continued.)

	ENGLAND.																	WALES.									SCOTLAND.														
	Yorkshire.	Derbyshire.	Lancashire.	Westmoreland.	Cumberland.	Durham.	Northumberland.	Isle of Man.	Herefordshire.	Staffordshire.	Shropshire.	Leicestershire.	Worcestershire.	Devonshire.	Somersetshire.	Gloucestershire.	Monmouthshire.	Pembrokeshire.	Anglesea.	Denbighshire.	Flintshire.	Montgomeryshire.	Brecknockshire.	Glamorganshire.	Carmarthenshire.	Carnarvonshire.	Lanarkshire.	Renfrewshire.	Ayrshire.	Buteshire.	Dumbartonshire.	Stirlingshire.	Dumfriesshire.	Peeblesshire.	Edinburghshire.	Linlithgowshire.	Haddingtonshire.	Fifeshire.	Berwickshire.	Kircudbrightshire.
				×																																					
	×	×	×					×		×																															
	×		×					×		×																															
	×	×	×					×		×						×																									
	×	×	×	×				×		×																	×	×	×			×	×		×		×		×	×	
	×	×	×	×	×	×	×	×		×	×					×	×	×	×								×	×	×		×	×	×		×		×	×	×	×
	×	×	×					×		×																															
	×	×						×		×																															
	×	×	×					×		×																															
										×																															
			×							×																															
	×	×	×				×	×		×											×																×				
	×	×	×							×						×																									
		×						×																																	
	×	×	×				×	×								×	×			×	×			×		×	×	×	×			×						×		×

Armagh.	Cork.	Carlow.	Clare.	Cavan.	Dublin.	Donegal.	Down.	Fermanagh.	Galway.	Kerry.	Kildare.	King's County.	Limerick.	Louth.	Longford.	Leitrim.	Meath.	Mayo.	Monaghan.	Queen's County.	Roscommon.	Sligo.	Tipperary.	Tyrone.	Waterford.	Westmeath.	Wexford.	Kilkenny.	Antrim.
	×					×					×							×					×				×		
	×										×																		
	×		×								×		×				×										×		
	×		×	×		×		×	×		×		×		×		×						×	×	×				
	×		×		×			×			×		×		×		×						×	×					
			×			×		×			×		×			×		×					×						
			×								×							×											
																	×	×					×						
	×																												
																							×						
×	×			×	×	×		×	×		×	×		×		×	×	×	×		×	×		×	×	×	×	×	×

	ENGLAND.																	WALES.									SCOTLAND.													
	Yorkshire.	Derbyshire.	Lancashire.	Westmoreland.	Cumberland.	Durham.	Northumberland.	Isle of Man.	Herefordshire.	Staffordshire.	Shropshire.	Leicestershire.	Worcestershire.	Devonshire.	Somersetshire.	Gloucestershire.	Monmouthshire.	Pembrokeshire.	Anglesea.	Denbighshire.	Flintshire.	Montgomeryshire.	Brecknockshire.	Glamorganshire.	Carmarthenshire.	Carnarvonshire.	Lanarkshire.	Renfrewshire.	Ayrshire.	Buteshire.	Dumbartonshire.	Stirlingshire.	Dumfriesshire.	Peeblesshire.	Edinburghshire.	Linlithgowshire.	Haddingtonshire.	Fifeshire.	Berwickshire.	Kircudbrightshire.
	×	×	×	×			×	×	×						×	×	×	×	×	×	×			×			×	×			×		×		×	×	×	×		
																×																								
	×	×	×	×			×			×					×	×	×			×	×				×		×	×	×	×			×							
	×	×	×	×		×	×	×	×						×	×	×			×	×				×		×	×	×	×	×			×	×	×	×	×	×	×
	×	×					×			×						×	×										×													
	×	×					×			×					×	×	×			×	×			×	×		×	×	×	×			×	×	×	×	×	×	×
	×																																							
	×	×	×							×		×			×		×	×																						
	×	×	×	×	×	×	×	×							×	×	×	×	×	×	×	×	×	×	×	×	×	×	×	×	×	×	×	×	×	×	×	×	×	×
	×	×		×						×			×		×	×		×		×	×				×	×														
	×	×			×	×	×												×								×	×	×	×	×	×								
									×		×		×																											
	×	×	×	×			×	×		×					×	×		×			×			×		×	×	×		×	×					×				
	×																																							

	ENGLAND.																	WALES.									SCOTLAND.													
	Yorkshire.	Derbyshire.	Lancashire.	Westmoreland.	Cumberland.	Durham.	Northumberland.	Isle of Man.	Herefordshire.	Staffordshire.	Shropshire.	Leicestershire.	Worcestershire.	Devonshire.	Somersetshire.	Gloucestershire.	Monmouthshire.	Pembrokeshire.	Anglesea.	Denbighshire.	Flintshire.	Montgomeryshire.	Brecknockshire.	Glamorganshire.	Carmarthenshire.	Carnarvonshire.	Lanarkshire.	Renfrewshire.	Ayrshire.	Buteshire.	Dumbartonshire.	Stirlingshire.	Dumfriesshire.	Peeblesshire.	Edinburghshire.	Linlithgowshire.	Haddingtonshire.	Fifeshire.	Berwickshire.	Kircudbrightshire.
	×	×	×	×			×	×	×	×	×					×	×	×	×		×						×	×	×			×	×		×			×	×	
																×																								
	×	×	×													×	×					×	×			×	×	×	×			×	×		×					
	×	×	×			×		×	×		×					×	×	×		×	×		×			×	×	×	×	×	×	×	×	×	×	×	×	×	×	×
	×										×																													
	×	×	×			×		×			×					×	×	×		×	×		×			×	×	×	×	×	×	×	×	×	×		×	×	×	×
	×																																							
	×	×	×					×		×					×	×				×							×	×												
	×	×	×	×	×	×	×	×		×	×				×	×	×	×		×	×	×		×		×	×	×	×		×	×	×	×	×	×		×	×	
	×		×	×							×			×	×		×		×	×			×								×	×								
	×	×	×		×	×				×	×				×	×		×		×	×			×			×	×	×	×	×	×								
								×		×		×				×	×																							
	×	×	×	×		×	×	×		×					×	×		×		×			×			×	×	×		×	×						×			
	×																																							

IRELAND.

Armagh.	Cork.	Carlow.	Clare.	Cavan.	Dublin.	Donegal.	Down.	Fermanagh.	Galway.	Kerry.	Kildare.	King's County.	Limerick.	Louth.	Longford.	Leitrim.	Meath.	Mayo.	Monaghan.	Queen's County.	Roscommon.	Sligo.	Tipperary.	Tyrone.	Waterford.	Westmeath.	Wexford.	Kilkenny.	Antrim.
×	×		×	×	×	×	×	×		×		×	×	×		×		×			×			×	×	×		×	
																								×					
							×																						
	×																×				×								
	×		×	×	×				×			×			×		×	×			×		×	×			×	×	
	×	×			×	×						×						×			×		×	×	×		×		
						×						×						×											
×	×	×			×	×	×	×		×		×						×	×			×		×					
		×				×	×	×									×	×			×			×	×				
	×			×			×			×			×			×	×	×		×	×							×	

GENERA AND SPECIES.

England.	Wales.	Scotland.	Ireland.	Abroad.	Genera and Species.
×	×	×	×	×	*Productus semireticulatus*, Martin. Pet. Derb., pl. xxxii, figs. 1, 2; and pl. xxxiii, fig. 4. Dav. Mon., p. 149, pl. xliii, figs. 1—11; pl. xliv, figs. 1—4 = *A. productus*, Martin = *P. Scoticus*, *P. Martini*, *P. antiquatus*, and *P. concinnus*, Sow. = *P. pugilus*, Phil. = *A. tubulifera*, Fischer = *P. inca*, and *P. Peruvianus*, D'Orb. = *P. flexistria*, M'Coy
×	×	×	×	×	— *costatus*, J. Sow. Min. Con., pl. 560, fig. 1. Dav. Mon., p. 152, pl. xxxii, figs. 2—9 = *P. sulcatus*, Sow. = *P. costellatus*, M'Coy
×		×			— *muricatus*, Phillips. Geol. York., pl. viii, fig. 3. Dav. Mon., p. 153, figs. 10—14
×	×	×	×	×	— *longispinus*, Sow. Min. Con., pl. lxviii, fig. 1. Dav. Mon., p. 154, pl. xxxv, figs. 5—17 = *An. echinatæ*, pars, Ure. = *P. Flemingii*, *P. spinosus*, and *P. lobatus*, Sow. = *P. elegans*, Davreux = *P. setosa*, Phil. = *P. capacii*, D'Orb. = *T. tubarius*, De Key. = *P. Wabashensis* and *P. splendens*, Norwood and Pratten
×		×		×	— *sinuatus*, De Kon. Anim. Foss. de Belgique, p. 654, pl. lvi, fig. 2. Dav. Mon., p. 157, pl. xxxiii, figs. 8—11
×			×	×	— *margaritaceus*, Phillips. Geol. York., vol. ii, p. 215, pl. viii, fig. 8. Dav. Mon., p. 159, pl. xliv, figs. 5—8 = *P. pectinoides*, Phillips
×			×	?—	*arcuarius*, De Kon. Anim. Foss. Belg., p. 171, pl. xii, fig. 10. Dav. Mon., p. 160, pl. xxxiv, fig. 17
	?		×	?—	*carbonarius*, De Kon. Anim. Foss. Belg., pl. xii *bis*, fig. 1. Dav. Mon., p. 160, pl. xxxiv, fig. 6
×		×	×	×	— *undatus*, Defrance. Dic. Sc. Nat., vol. xliii, p. 354. Dav. Mon., p. 161, pl. xxxiv, figs. 1—13 = *P. Tortilis*, M'Coy
		×			— *Wrightii*, Dav. Mon., p. 162, pl. xxxiii, figs. 6, 7
×				×	— *proboscideus*, De Vern. Bull. Soc. G. Fr., vol. xi, p. 259, pl. iii, fig. 3. Dav. Mon., p. 163, pl. xxxiii, figs. 1—4 = *C. prisca*, Goldfuss
×				×	— *ermineus*, De Kon. Anim. Foss. Belgique, p. 181, pl. x, fig. 5. Dav. Mon., p. 164, pl. xxxiii, fig. 5
×			×	×	— *tessellatus*, De Kon. Anim. Foss. Belgique, p. 192, pl. ix, fig. 2; and pl. xiii *bis*, fig. 5 (not of Phillips). Dav. Mon., p. 165, pl. xxxiii, figs. 24, 25; pl. xxxiv, fig. 14
×				×	— *marginalis*, De Kon. Mon. du Genre Productus, pl. xiv, fig. 7. Dav. Mon., p. 229, pl. liii, fig. 3
×				×	— *Nystianus*, De Kon. Desc. Anim. Foss. Belg., p. 202, pl. vii *bis*, fig. 3; pl. ix, fig. 7; pl. x, fig. 9. Dav., Mon., p. 231, pl. liii, fig. 9
×	×	×	×	×	— *aculeatus*, Martin. Pet. Derb., pl. xxxvii, figs. 9, 10. Dav. Mon., pp. 166 and 233, pl. lxiii, figs. 16—18, 20 = *P. laxispina*, Phillips

	ENGLAND.																	WALES.									SCOTLAND.													
	Yorkshire.	Derbyshire.	Lancashire.	Westmoreland.	Cumberland.	Durham.	Northumberland.	Isle of Man.	Herefordshire.	Staffordshire.	Shropshire.	Leicestershire.	Worcestershire.	Devonshire.	Somersetshire.	Gloucestershire.	Monmouthshire.	Pembrokeshire.	Anglesea.	Denbighshire.	Flintshire.	Montgomeryshire.	Brecknockshire.	Glamorganshire.	Carmarthenshire.	Carnarvonshire.	Lanarkshire.	Renfrewshire.	Ayrshire.	Buteshire.	Dumbartonshire.	Stirlingshire.	Dumfriesshire.	Peeblesshire.	Edinburghshire.	Linlithgowshire.	Haddingtonshire.	Fifeshire.	Berwickshire.	Kircudbrightshire.
	×	×	×	×		×	×	×		×	×			?	×	×	×	×		×		×	×		×		×	×	×	×	×	×	×	×	×	×	×	×	×	×
	×	×	×	×		×	×	×		×										×				×			×		×		×									
	×																												×		×									
	×	×	×			×	×		×					×		×				×	×			×		×	×	×	×	×	×	×						×	×	×
	×	×								×										×	×						×	×	×											
	×	×	×	×						×																														
	×																																							
																											?													
	×	×	×			×	×		×																		×				×	×								
	×								×																															
	×									×																														
	×		×					×		×																														
	×																																							
	×																																							
	×	×		×		×		×													×						×	×	×		×							×	×	

	ENGLAND																	WALES									SCOTLAND													
	Yorkshire.	Derbyshire.	Lancashire.	Westmoreland.	Cumberland.	Durham.	Northumberland.	Isle of Man.	Herefordshire.	Staffordshire.	Shropshire.	Leicestershire.	Worcestershire.	Devonshire.	Somersetshire.	Gloucestershire.	Monmouthshire.	Pembrokeshire.	Anglesea.	Denbighshire.	Flintshire.	Montgomeryshire.	Brecknockshire.	Glamorganshire.	Carmarthenshire.	Carnarvonshire.	Lanarkshire.	Renfrewshire.	Ayrshire.	Buteshire.	Dumbartonshire.	Stirlingshire.	Dumfriesshire.	Peeblesshire.	Edinburghshire.	Linlithgowshire.	Haddingtonshire.	Fifeshire.	Berwickshire.	Kircudbrightshire.
	×	×	×	×		×	×	×		×	×			?	×	×	×	×		×	×	×	×	×	×	×	×	×	×	×	×	×	×	×	×	×	×	×	×	
	×	×	×	×		×	×	×		×	×									×				×			×	×	×	×	×	×								
	×																												×			×								
	×	×	×			×	×		×						×	×		×	×				×		×	×	×	×	×	×	×					×	×	×		
	×	×								×																×	×	×												
	×	×	×	×						×																														
	×																																							
																											?													
	×	×	×			×	×			×																	×			×	×									
	×								×																															
	×									×																														
	×		×					×		×																														
	×	×																																						
	×																																							
	×	×		×			×			×										×							×	×	×			×						×	×	

Armagh.	Cork.	Carlow.	Clare.	Cavan.	Dublin.	Donegal.	Down.	Fermanagh.	Galway.	Kerry.	Kildare.	King's County.	Limerick.	Louth.	Longford.	Leitrim.	Meath.	Mayo.	Monaghan.	Queen's County.	Roscommon.	Sligo.	Tipperary.	Tyrone.	Waterford.	Westmeath.	Wexford.	Kilkenny.	Antrim.

IRELAND.

England.	Wales.	Scotland.	Ireland.	Abroad.	GENERA AND SPECIES.
×	...	×	...	...	*Productus Youngianus*, Dav. Mon., pp. 167 and 233, pl. xxxiii, figs. 21—23 ....
×	...	...	...	×	— *Koninckianus*, De Verneuil. Russia and Ural, vol. ii, p. 253. Dav. Mon., p. 230, pl. liii, fig. 7 = *P. spinulosus*, De Kon. (not Sow.) ....
×	×	×	×	×	— *pustulosus*, Phillips. Geol. York., vol. ii, pl. vii, fig. 15. Dav. Mon., p. 168, pl. xli, figs. 1—6; pl. lxii, figs. 1—4 = *P. rugata* and *P. ovalis*, Phil. = *P. pyxidiformis*, De Kon. ....
×	×	×	×	×	— *scabriculus*, Martin. Pet. Derb., p. 8, pl. xxxvi, fig. 5. Dav. Mon., p. 169, pl. xlii, figs. 5—8 = *P. quincuncialis*, Phil. = *P. corbis*, Potiez ....
×	×	×	×	×	— *fimbriatus*, Sow. Min. Con., pl. 459, fig. 1. Dav. Mon., p. 171, pl. xxxiii, figs. 12—15; pl. xliv, fig. 15 = *S. marsupit*, Davreux, ? = *P. laciniata*, M'Coy ....
×	×	×	×	×	— *punctatus*, Martin. Pet. Derb., pl. xxxvii, fig. 6 (7, 8, *exclusa*). Dav. Mon., p. 172, pl. xliv, figs. 9—16 = *T. rugosa*, Parkinson = *A. thecarius*, Schloth. = *Lept. sulcata*, Fischer = *P. concentricus*, Potiez = *S. tubulospina*, M'Chesney = var. ? *P. elegans*, M'Coy ....
×	...	...	...	×	— *Keyserlingianus*, De Koninck. Mon. du Genre Productus, pl. xiv, fig. 6. Dav. Mon., p. 174, pl. xxxiv, figs. 15, 16 ....
×	×	×	×	×	— *spinulosus*, Sow. Min. Con., pl. lxviii, fig. 3. Dav. Mon., p. 175, pl. xxxiv, figs. 18—21 = *P. granulosus*, Phillips = *P. papillatus*, De Kon. = *P. cancrini*, De Kon. (not of De Verneuil) ....
×	...	...	...	×	? — *Deshaysianus*, De Kon. Anim. Foss. Belg., p. 193, pl. x, fig. 7. Dav. Mon., p. 232, pl. liii, figs. 11, 12 ....
×	...	...	...	×	— *plicatilis*, Sow. Min. Con., pl. 459, fig. 2. Dav. Mon., p. 176, pl. xxxi, figs. 3—5 = *Lept. polymorpha*, Muenster ....
×	×	...	...	×	— *sub-lævis*, De Kon. Ann. Foss. Belgique, pl. x, fig. 1. Dav. Mon., pp. 177 and 234, pl. xxxi, figs. 1, 2; pl. xxxii, fig. 1; pl. li, figs. 1, 2 ? = *Stroph. antiquata*, Potiez = *P. Christiani*, De Kon. ....
×	...	×	×	×	— *mesolobus*, Phillips. Geol. York., vol. ii, pl. vii, figs. 12, 13. Dav. Mon., p. 178, pl. xxxi, figs. 6—9 ....
×	×	...	×	×	*Chonetes comoides*, Sow. Min. Con., pl. 329. Dav. Mon., p. 180, pl. xlv, fig. 7 (? 1—6); pl. xlvi, fig. 1 ....
×	×	...	×	×	— *papilionacea*, Phillips. Geol. York., vol. ii, pl. ii, fig. 6. Dav. Mon., p. 182, pl. xlvi, figs. 3—6 = *P. flabelliformis*, Lister = *P. multidentata*, M'Coy = ? *P. papyracea*, M'Coy ....
×	...	...	...	×	? — *Dalmaniana*, De Kon. Anim. Foss. Belg., pl. xiii, fig. 3; pl. xiii *bis*, fig. 2. Dav. Mon., p. 183, pl. xlvi, fig. 7 ....
×	...	×	×	×	— *Buchiana*, De Kon. Anim. Foss. Belg., pl. xiii, fig. 1. Dav. Mon., p. 184, pl. xlvii, figs. 1—7 and 28; pl. lii, fig. 21 = *L. crassistria*, M'Coy ....

	ENGLAND.																	WALES.									SCOTLAND.													
	Yorkshire.	Derbyshire.	Lancashire.	Westmoreland.	Cumberland.	Durham.	Northumberland.	Isle of Man.	Herefordshire.	Staffordshire.	Shropshire.	Leicestershire.	Worcestershire.	Devonshire.	Somersetshire.	Gloucestershire.	Monmouthshire.	Pembrokeshire.	Anglesea.	Denbighshire.	Flintshire.	Montgomeryshire.	Brecknockshire.	Glamorganshire.	Carmarthenshire.	Carnarvonshire.	Lanarkshire.	Renfrewshire.	Ayrshire.	Buteshire.	Dumbartonshire.	Stirlingshire.	Dumfriesshire.	Peeblesshire.	Edinburghshire.	Linlithgowshire.	Haddingtonshire.	Fifeshire.	Berwickshire.	Kircudbrightshire.
	×									×																	×	×	×			×				×		×		
	×																																							
	×		×	×				×		×					×	×				×							×	×	×		×				×	×	×	×		
	×	×	×		×		×			×	×			×	×	×	×			×							×	×	×		×				×	×	×	×		
	×	×	×	×			×	×		×					×					×	×			×			×	×	×		×				×	×	×			
	×	×	×		×		×	×		×					×					×	×			×			×	×	×		×				×	×	×			
	×									×																														
	×	×	×				×			×			×							×				×			×	×	×						×	×				
	×																																							
	×	×	×							×																														
				×																	×																			
	×	×	×	×						×																	×	×	×		×							×		×
															×	×	×			×	×																			
	×	×	×	×						×			×		×	×		×		×	×																			
	×							×		×																														
	×			×						×																	×													

	ENGLAND.																	WALES.									SCOTLAND.														
	Yorkshire.	Derbyshire.	Lancashire.	Westmoreland.	Cumberland.	Durham.	Northumberland.	Isle of Man.	Herefordshire.	Staffordshire.	Shropshire.	Leicestershire.	Worcestershire.	Devonshire.	Somersetshire.	Gloucestershire.	Monmouthshire.	Pembrokeshire.	Anglesea.	Denbighshire.	Flintshire.	Montgomeryshire.	Brecknockshire.	Glamorganshire.	Carmarthenshire.	Carnarvonshire.	Lanarkshire.	Renfrewshire.	Ayrshire.	Buteshire.	Dumbartonshire.	Stirlingshire.	Dumfriesshire.	Peeblesshire.	Edinburghshire.	Linlithgowshire.	Haddingtonshire.	Fifeshire.	Berwickshire.	Kircudbrightshire.
	×	×								×																	×	×	×			×				×		×		
	×																																							
	×	×	×	×						×						×											×	×	×			×				×	×	×		
	×	×	×	×			×	×			×									×	×						×	×	×			×				×		×		
	×	×	×	×		×				×				×						×	×			×			×	×	×			×				×		×		
	×									×																														
	×	×	×							×						×				×	×															×		×		
	×																																							
	×	×	×							×																														
				×						×									×																					
	×	×	×							×						×								×			×	×	×			×								×
															×	×			×	×																				
	×	×	×						×	×		×			×	×		×																						
	×						×			×																														
	×			×						×																	×													

IRELAND.

Armagh.	Cork.	Carlow.	Clare.	Cavan.	Dublin.	Donegal.	Down.	Fermanagh.	Galway.	Kerry.	Kildare.	King's County.	Limerick.	Louth.	Longford.	Leitrim.	Meath.	Mayo.	Monaghan.	Queen's County.	Roscommon.	Sligo.	Tipperary.	Tyrone.	Waterford.	Westmeath.	Wexford.	Kilkenny.	Antrim.

GENERA AND SPECIES.

England.	Wales.	Scotland.	Ireland.	Abroad.	
×	×	×	×	×	*Chonetes Hardrensis*, Phillips. Fig. and Des. of the Pal. Foss. of Cornwall, p. 138, pl. xl, fig. 104. Dav. Mon., p. 186, pl. xlvii, figs. 12—18 ....
×	...	×	×	...	— *polita*, M'Coy. Brit. Pal. Foss., p. 456, pl. iii°, fig. 30. Dav. Mon., p. 190, pl. xlvii, figs. 8—11 ....
					N.B. Several very doubtful species of Chonetes are not here tabulated (see p. 188).
×	...	×	×	×	*Crania quadrata*, M'Coy. Synopsis, pl. xx, fig. 1. Dav. Mon., p. 194, pl. xlviii, figs. 1—13 ....
×	...	...	×	×	? — *Ryckholtiana*, De Kon. Anim. Foss. Belgique, pl. xxiii, fig. 5. Dav. Mon., p. 195, pl. xlviii, figs. 15, 16 (17?) = *C. vesicularis*, M'Coy ....
...	...	...	×	...	? — *trigonalis*, M'Coy. Synopsis, pl. xx, fig. 2. Dav. Mon., p. 196, pl. xlviii, fig. 14 ....
×	×	×	×	×	*Discina nitida*, Phillips. Geol. York., vol. ii, p. 221, pl. ix, figs. 10—13. Dav. Mon., p. 197, pl. xlviii, figs. 18—25 = *D. cincta*, Portlock = *D. bulla*, M'Coy = *D. Koninckii*, Geinitz
...	...	...	×	...	? — *Davreuxiana*, De Kon. Anim. Foss. Belg., p. 306, pl. xxi, fig. 4. Dav. Mon., p. 198, pl. xlviii, fig. 26 ....
×	...	×	×	×	*Lingula squamiformis*, Phillips. Geol. York., vol. ii, pl. ix, fig. 14. Dav. Mon., p. 205, pl. xlix, figs. 1—10
...	...	×	...	...	— *Scotica*, Dav. Mon., p. 207, pl. xlviii, figs. 27, 28 ....
×	×	×	×	×	— *mytiloides*, Sow. Min. Con., tab. xix, figs. 1, 2. Dav. Mon., p. 207, pl. xlviii, figs. 29—36 = *L. elliptica*, *L. marginata*, and *L. parallela*, Phillips ....
×	...	...	...	×	? — *Credneri*. Gunij Verst. der Zech., pl. iv, figs. 23—29. Dav. Mon., p. 209, pl. xlviii, figs. 38—40
					N.B. This is probably a synonym of *L. mytiloides*.
×	...	...	...	...	? — *latior*, M'Coy. Br. Pal. Foss., pl. iii°, fig. 23. Dav. Mon., p. 210, pl. xlviii, fig. 37 ....

(continued).

	ENGLAND.																	WALES.									SCOTLAND.													
	Yorkshire.	Derbyshire.	Lancashire.	Westmoreland.	Cumberland.	Durham.	Northumberland.	Isle of Man.	Herefordshire.	Staffordshire.	Shropshire.	Leicestershire.	Worcestershire.	Devonshire.	Somersetshire.	Gloucestershire.	Monmouthshire.	Pembrokeshire.	Anglesea.	Denbighshire.	Flintshire.	Montgomeryshire.	Brecknockshire.	Glamorganshire.	Carmarthenshire.	Carnarvonshire.	Lanarkshire.	Renfrewshire.	Ayrshire.	Buteshire.	Dumbartonshire.	Stirlingshire.	Dumfriesshire.	Peeblesshire.	Edinburghshire.	Linlithgowshire.	Haddingtonshire.	Fifeshire.	Berwickshire.	Kircudbrightshire.
	×	...	...	×	...	...	×	×	...	...	×	...	...	...	×	×	×	×	...	×	×	...	×	×	...	×	×	×	×	×	×	×	...	...	...	×	×	×	...	...
	...	...	...	...	...	...	×	...	...	...	...	...	...	...	...	...	...	...	...	...	...	...	...	...	...	...	...	...	...	...	...	×	...	...	...	...	...	...	...	...
	...	...	...	...	...	...	...	...	...	...	...	...	...	...	×	...	...	...	...	...	...	...	...	...	...	...	×	×	×	...	...	×	...	...	...	...	...	...	...	×
	...	×	...	...	...	...	...	...	...	...	...	...	...	...	...	...	...	...	...	...	...	...	...	...	...	...	...	...	...	...	...	...	...	...	...	...	...	...	...	...
	...	...	...	...	...	...	...	...	...	...	...	...	...	...	...	...	...	...	...	...	...	...	...	...	...	...	...	...	...	...	...	...	...	...	...	...	...	...	...	...
	×	×	×	...	...	...	×	...	...	×	×	...	×	...	×	×	×	...	...	...	...	...	×	...	...	...	×	×	×	...	×	...	...	×	...	×	×	×	×	...
	?	...	×	...	...	...	×	...	...	...	...	...	...	...	...	...	...	...	...	...	...	...	...	...	...	...	×	×	...	...	×	×	...	×	...	...	×	...	×	...
	...	...	...	...	...	...	...	...	...	...	...	...	...	...	...	...	...	...	...	...	...	...	...	...	...	...	×	...	...	...	...	...	...	...	...	...	...	...	...	...
	×	×	×	...	...	...	×	...	...	...	×	...	...	...	×	×	×	...	...	...	...	...	...	×	...	...	×	×	×	...	...	×	...	...	...	...	...	...	×	×
	...	...	...	×	...	×	...	...	...	...	...	...	...	...	...	...	...	...	...	...	...	...	...	...	...	...	...	...	...	...	...	...	...	...	...	...	...	...	...	...
	...	...	×	...	...	×	...	...	...	...	...	...	...	...	...	...	...	...	...	...	...	...	...	...	...	...	...	...	...	...	...	...	...	...	...	...	...	...	...	...

	ENGLAND																	WALES									SCOTLAND													
	Yorkshire.	Derbyshire.	Lancashire.	Westmoreland.	Cumberland.	Durham.	Northumberland.	Isle of Man.	Herefordshire.	Staffordshire.	Shropshire.	Leicestershire.	Worcestershire.	Devonshire.	Somersetshire.	Gloucestershire.	Monmouthshire.	Pembrokeshire.	Anglesea.	Denbighshire.	Flintshire.	Montgomeryshire.	Brecknockshire.	Glamorganshire.	Carmarthenshire.	Carnarvonshire.	Lanarkshire.	Renfrewshire.	Ayrshire.	Buteshire.	Dumbartonshire.	Stirlingshire.	Dumfriesshire.	Peeblesshire.	Edinburghshire.	Linlithgowshire.	Haddingtonshire.	Fifeshire.	Berwickshire.	Kircudbrightshire.
	×			×			×	×					×		×	×	×	×		×	×		×	×		×	×	×	×	×	×	×					×	×	×	
							×																									×								
															×												×	×	×			×								×
		×																																						
	×	×	×				×			×	×		×		×	×	×						×				×	×	×			×			×		×	×	×	
	?		×				×																				×	×				×	×			×		×	×	
																											×													
	×	×	×				×						×			×	×								×		×	×	×			×			×			×	×	
			×			×																																		
			×			×																																		

Armagh.	Cork.	Carlow.	Clare.	Cavan.	Dublin.	Donegal.	Down.	Fermanagh.	Galway.	Kerry.	Kildare.	King's County.	Limerick.	Louth.	Longford.	Leitrim.	Meath.	Mayo.	Monaghan.	Queen's County.	Roscommon.	Sligo.	Tipperary.	Tyrone.	Waterford.	Westmeath.	Wexford.	Kilkenny.	Antrim.
	×	×			×	×										×	×	×			×	×		×	×		×		
																				×									
						×																							
											×																		
						×																							
×				×	×	×																×	×						
×																													
								×																					×
					×																	×							

EXPLANATION OF THE TABLES AND LIST OF LOCALITIES.

The Carboniferous system occupies so large an area in Great Britain, that it appeared desirable to tabulate the amount of work done in collecting its Brachiopoda, and to correctly define our present knowledge with reference to their distribution. The labour required to effect this object has been very great; and although the results are no doubt far from complete, or entirely satisfactory, my tables will, I trust, serve as a groundwork to which may be hereafter added the fruit of further search in the various counties there inscribed.

When my tables had been almost completed, Mr. Salter suggested, that instead of counties, it might be preferable to divide Great Britain into Carboniferous districts, and to give the range of species in them, including in these districts the Carboniferous Limestone, Millstone-grit, and Coal, somewhat as follows:

1, The Scotch Basin; 2, Northumberland, Durham, and north of the Tees (the line of the Eden and the Tees forming a good boundary to separate from No. 2); the 3rd, or Yorkshire and North Lancaster, as far south as Wharfedale; 3a, Cumberland or Whitehaven; 4, Derbyshire, with what are called the Yorkshire and Lancashire Coal-fields on each flank; 5, North Wales and Anglesea; 6, Shropshire and the Forest of Wyre, Staffordshire and Leicestershire patches; 7, South Wales Basin; 8, Forest of Dean, Bristol, and the Mendips; 9, Devonshire; 10, Isle of Man; 11, Ireland; the last also being similarly divided.

I should have preferred arranging my tables into such natural boundaries; but besides certain difficulties, it would have necessitated another kind of research, as for many months previous I had been arranging the species in their respective counties, and which for practical purposes may not be devoid of utility.

In preparing these tables, and the following lists of localities, no trouble has been spared, for in addition to my own personal researches, which have extended over five years, I have availed myself of all the assistance that could be obtained.[1]

---

[1] Mr. Salter and Mr. Etheridge placed the Geological Survey manuscript lists and specimens before me, and kindly assisted with their personal observations; Mr. Waterhouse and Mr. S. P. Woodward, with their usual urbanity and desire to turn the national collection to public use, afforded me every facility to examine at leisure all the Carboniferous species in the British Museum, where Gilbertson's and Sowerby's original collections are now carefully preserved. The Geological Society's stores were also examined, as well as the collection made by the late Mr. D. Sharpe, and I have had the loan of thousands

The work in connection with British Carboniferous Brachiopoda must not, however, be supposed exhausted, for there is much still to be learnt and achieved by future observers; for out of the one hundred and twenty species enumerated in my tables, from fifteen to eighteen have not been sufficiently studied, from want of satisfactory or sufficient material, and these may hereafter prove to be partly or entirely synonyms, varieties, or variations in shape of some of the others; so that I do not consider that many more than about one hundred good species have been proved to exist in Great Britain.

Seventy-one British counties have, up to the present time, afforded Carboniferous Brachiopoda, and the following numerical statements must be considered to represent the present state of our knowledge, as some counties have been much more searched than others, so that with time these numbers will no doubt be notably modified. It will however be interesting, I think, to record the state of our information up to the early portion of 1862.

Number of Species hitherto recorded from each County.

ENGLAND.		Herefordshire	2
Yorkshire	90	Staffordshire	78
Derbyshire	76	Shropshire	19
Lancashire	69	Worcestershire	3
Westmoreland	31	Cheshire	2
Cumberland	6	Somersetshire	32
Durham	33	Monmouthshire	13
Northumberland	42	Gloucestershire	40
Isle of Man	50	Leicestershire	7

of specimens belonging to many private collections; the following gentlemen having assisted to the utmost of their power in the working out of the lists, &c.:

IN ENGLAND—Mr. E. Wood, of Richmond; Mr. Burrow, of Settle; Mr. Reed and Mr. Dallas, of York, Yorkshire; Mr. Carrington, of Wetton; Mr. Wardle, of Leek, Staffordshire; Mr. Tate, of Alnwick, Northumberland; Mr. Hutchinson, of Durham; Mr. Binney, Mr. Ormerod, Mr. Parker, and Dr. Fleming, of Manchester; Mr. Rofe, of Preston, and Mr. Froggatt, of Stockport, Lancashire; the Rev. W. Coleman, of Ashby, Leicestershire; Prof. Sedgwick, Cambridge; Prof. Phillips, Oxford; Dr. Bowerbank, Prof. Tennant, and Mr. Rodwell, London; Mr. Walton, and Mr. Moore, Bath; Mr. W. Sanders, and Mr. W. W. Stoddart, Clifton; the Rev. J. G. Cumming, of the Isle of Man; Dr. Bevan, of Beaufort, Monmouthshire; Mr. Mushen, Birmingham, &c.

In SCOTLAND—The late Rev. J. Fleming, and H. Miller, Mr. Page, and Mr. Geikie, of Edinburgh; Mr. Young, Mr. J. Armstrong, Mr. J. Thomson, Mr. W. Johnston, Mr. Fraser, Mr. Crosskey, and Mr. Bennie, of Glasgow; Drs. Rankin and Slimon, of Carluke and Lesmahago; Mr. W. Grossart, of Salsburgh, Lanarkshire; Prof. Nicol, of Aberdeen.

In IRELAND—Mr. Kelly and Mr. Baily of the Geological Survey of Ireland; Mr. Carte, of the Royal Dublin Museum; Sir R. Griffith, and Mr. Byron, of Dublin; Mr. J. Wright and Prof. Harkness, Cork; and Prof. de Koninck, has also communicated to me the results of his examination of our British species.

To these gentlemen I beg to return my warmest thanks and acknowledgments for all the kind and generous assistance they have so liberally bestowed, and to whose help many of the results here recorded are mainly due.

WALES.

Pembrokeshire	9
Anglesea	5
Carnarvonshire	8
Montgomeryshire	3
Denbighshire	24
Flintshire	23
Brecknockshire	7
Glamorganshire	15
Carmarthenshire	4

SCOTLAND.

Lanarkshire	46
Renfrewshire	38
Ayrshire	42
Buteshire	11
Dumbartonshire	18
Stirlingshire	41
Dumfriesshire	3
Peeblesshire	4
Edinburghshire	18
Linlithgowshire	26
Haddingtonshire	20
Fifeshire	26
Berwickshire	14
Kirkudbrightshire	7

IRELAND.

Armagh	19
Cork	57
Carlow	13
Clare	12
Cavan	13
Dublin	52
Donegal	46
Down	7
Fermanagh	31
Galway	2
Kerry	23
Kildare	44
King's County	4
Limerick	30
Louth	7
Longford	24
Leitrim	34
Meath	39
Mayo	22
Monaghan	4
Queen's County	5
Roscommon	26
Sligo	13
Tipperary	24
Tyrone	35
Waterford	26
Westmeath	8
Wexford	25
Kilkenny	2
Antrim	1

The species which have been found in the larger number of counties, or which have had the greatest range, are:

Terebratula hastata, found in	39 counties.	Orthis, resupinata, in	45 counties.
Athyris Royssii	35 ,,	— Michelini	37 ,,
— planosulcata	29 ,,	Productus giganteus	43 ,,
— ambigua	31 ,,	— semireticulatus	57 ,,
Spirifera striata	25 ,,	— longispinus	40 ,,
— trigonalis=bisulcata	48 ,,	— pustulosus	32 ,,
— glabra	37 ,,	— scabriculus	40 ,,
— lineata	41 ,,	— fimbriatus	32 ,,
Rhynchonella pleurodon	35 ,,	— punctatus	38 ,,
Strophomena analoga	38 ,,	Chonetes Hardrensis	36 ,,
Streptorhynchus crenistria	50 ,,	Discina nitida	25 ,,

All the other species are more sparingly distributed, as may be seen by a glance at the tables.

We will now mention some of the localities where Carboniferous species have been found, and class them by counties for convenience, commencing by those of

ENGLAND.

In England eighteen counties have afforded about 112 species, and is therefore richer in this respect than Wales, Scotland, or Ireland. By far the larger number of species are derived from the Carboniferous Limestone (Lower Scar Limestone, and its accompanying shales); the Yoredale rocks, Millstone-grit and Coal Formation having offered a much smaller proportion. Carboniferous Brachiopoda have been noticed in England as early as 1685-1692; for several recognisable species will be found in Listers' 'Historia Sive Synopsis Methodica Conchyliorum et Tubularum Marinarum,' vol. iv, in fol., cum tab œneis; but it was not until 1809 that they were seriously collected or studied, and in p. 7 of the present monograph, as well as in the body of the work, references to those authors who have published upon the subject will be found.

YORKSHIRE.—Settle; Clattering Dykes, and Malham Moor, Otterburn, all at about six miles distant from Settle, and all in the Craven district; Craco, Bunsall, Grassington, Greenhow Hill, localities on or near the river Wharfe; Whitewell, eight miles west of Clitheroe; Sykes, five miles from Whitewell; Slaidburn, Newton—these four last are in Bolland proper; Withgill, two miles from Clitheroe, and the same distance from Mitton; Richmond; Gilling, three miles north of Richmond; Marslie, five miles west of Richmond; Washton, three miles north-west of Richmond; Downholm, five miles west of Richmond; Barton, six miles north of Richmond; Askrigg; Thornton, three miles south of Askrigg; Aysgarth, six miles south-east of Askrigg; West Witton, four miles west of Leyburn; Kettlewell; Cray, one mile north of Kettlewell; Linton, six miles south of Kettlewell; Thorp, seven miles south of Kettlewell. For this list of localities I am indebted to Mr. E. Wood, of Richmond; to Mr. J. Parker, of Manchester; and to Mr. Burrow, of Settle.

LANCASHIRE.—Clitheroe; Chatburn, two miles, Downham, three miles, and Twiston, five miles north-east of Clitheroe; Worston; Harbour in the township of Thornley, near Longridge, six miles, and Harbour eight miles north-east of Preston; Thornley, near Chipping, about ten miles north-east of Preston; Ulverston; Scales, near Ulverston Conishead; Kirby Lonsdale, on the borders of Westmoreland. These localities have been made out by Mr. J. Parker and Mr. Rofe.

WESTMORELAND.—Kendal.

CUMBERLAND.—Chawk, near Rose Castle; Buxton Fall. Poltross Burn; Bird Oswuld; Combe Crag; Bank Head, Harehill—these localities have been furnished me by Mr. Tate, and are within one to five and a half miles of Lanercost.

DURHAM.—Stanhope limestone quarry; Silvertongue Mine, near Muggleswick; Hysehope Burn; Muggleswick; Waskerley Mine; Muggleswick; Thimbleby Hill, near Stanhope; Rookhopedale; East Gate, near Stanhope; Bishopley quarries; Buffside, parish of Edmondbyers; (these localities have been mentioned to me by Mr. G. Tate, and Mr. T. Hutchinson); Wolsingham (Sow.)

STAFFORDSHIRE.—Wetton parish generally; Wetton Hill, about ten miles from Leek; Narrowdale, Gateham, Allstonefield; Beeston Tor; Butterton, eight miles east, and Mixon, five miles north-east of Leek; Ilam, near Dovedale, twelve miles from Leek; Waterhouses, about five miles on Ashbourne road; Bed of the Manyfold, near Wetton; Ecton, near Warslow, about ten miles from Leek; Grindon parish, Caldon Low quarries, at about seven miles from Leek, in the direction of Ashbourne. Longnor is in Staffordshire, but on Millstone-grit; the limestone is on the Derbyshire side of the Dove, at this side of its boundary. These localities have been well searched by Messrs. S. Carrington, T. Wardle, and J. Parker. A very fine series of these fossils, collected during several years by Mr. Carrington, are preserved in the museum of the late T. Bateman, Esq., of Lomberdale House, Youlgrave, Staffordshire.

NORTHUMBERLAND.   This list has been communicated by Mr. G. Tate, of Alnwick.

Sea Coast north of the Coquet.

Spittal	near Tweedmouth.				Newton-by-the-Sea	from Chathill	4 miles	S.E.
Scremerstone	from Berwick	2 miles	S. by E.		Embleton	from Christon		
Cheswick	,,	,,	4	,,	,,		Bank	2 ,,	E.
Fenham	,,	Beal	2	,,	,,	Dunstan	,, ,,	3 ,,	S.E.
Budle	,,	Belford	3	,,	E.	Craster	,, Alnwick	6 ,,	N.E.
North Sunderland	Chathill	3½	,,	N.E.	Howick	,, ,,	5¼ ,,	E.N.E.
Beadnell	North Sunder-				Alnmouth	,, ,,	4½ ,,	E. by S.
	land	2	,,	S.				

Islands—Lindisfarne and Farne.

Inland.

Hetton	from Lowick	3 miles	S.		Christon Bank	from Alnwick	6 ,,	N.N.E.
Ford	,,	,,	4½ ,,	W.	Rock	,, ,,	4½ ,,	N. by E.
Chatton	,,	Belford	5 ,,	S.W.	Rennington	,, ,,	3½ ,,	N. by E.
Belford					Little Mill	,, Alnwick	4 ,,	N.E.
Spindlestone	,,	Belford	3 ,,	E. by S.	Denwick Lane	,, Alnwick	2½ ,	N.E.
Lucker	,,	,,	4 ,,	S.E.	Denwick Mill	,, ,,	1½ ,,	E.

Ratcheugh	from Alnwick	2½	miles	E. by N.	Kyloe	from Belford	5½ ,,	N.W.
Calishes	,; ,,	2	,,	S.E.	Eglingham	,, Alnwick	6 ,,	N.W.
Shilbottle	,, ,,	3	,,	S.S.E.	Hohberlaw	,, ,,	1 ,,	S. by W.
Whittle	,, ,,	4½	,,	S. by E.	Alnwick Moor	,, ,,	1 ,,	S.S.W.
Newton-on-the-Moor	,,	5	,,	S.	Rugley	,, ,,	2 ,,	S.
Framlington	,, Felton	4	,,	W. by N.	Lemmington	,, ,,	3½ ,,	S.S.W.

South of the Coquet.

Wards-Hill	from Framlington	4	miles	S.W.	Tone	from Bellingham	4½ miles	S.E.
Whitton	,, Rothbury	½	,,	S.W.	Buteland	,, ,,	3 ,,	S.E.
Tosson	,, ,,	2	,,	W.	Four Laws	,, ,,	4 ,,	E.S.E.
Grasslees	,, ,,	5½	,,	W. by S.	Risingham	,,	3	N.E.
Horsley Birks	,, Framlington	3	,,	S.	Keilder	,, ,,	15 ,,	N.N.W.
Hartington	,, Kirkwhelp- ington	3	,,	N. by E.	Lewis Burn	,, ,,	13 ,,	N. by N.
					Plashetts	,, ,,	11 ,,	N.N.W.
Sandhoe	,, Hexham	2½	,,	N.E.	Otterstone Lee	,, ,,	11 ,,	W. by W.
Belsay	,, Stamfordham	4	,,	N. by E.	Billing Burn	,, ,,	10 ,,	W.N.W.
Stamfordhand	,, Newcastle	12	,,	N.N.W.	Whickhope	,, ,,	11 ,,	W.
Sweethope	,, Kirkwhelp- ington	3	,,	S.W.	Falstone	,, ,,	8 ,,	W.N.W.
Carter Fell	Sources of the Reed, near the Borders				Bellingham	On North Tyne		
Shittleheugh	from Otterburn	2	,,	N.W.	Harlow Hill	from Newcastle	11 ,,	W. by N.
Otterburn	On Reed Water				Brunton	,, Hexham	3½ ,,	N.W.
					Fallow Field	,, ,,	3 ,,	N.W.
Redesdale	,, Bellingham	4	,,	E.	Chesterholm	,, ,,	10 ,,	W. by N.
					Haltwhistle	On South Tyne.		

CHESHIRE.—Carboniferous rocks occur sparingly in this county, and it was only quite recently that Mr. J. Parker, of Manchester, found Brachiopoda, viz., *Strophomena analoga*, and some spines of *Producta* in Carboniferous Limestone, immediately under the Millstone grit at Newbold, Aspburg, near Congleton, in Cheshire. Mr. Binney is inclined to believe that some Lingulæ may perhaps occur among the Goniatites and Pectens of Dunkenfield?

DERBYSHIRE.—Parkhill, in the parish of Earl Sterndale, near Jericho Church, Sterndale; Helter Hill, Crowdecote, Pilsbury Hartington—the above close to the river Dove (above Dovedale proper); Birch Quarry, Ashford; Bowdale-house, Bakewell, Matlock-bridge, Royston, Corwen, Alport, Grindlow near Tidswell, Kniveton (four miles north-north-east of Ashburn), Parwich, near Ballidon; Tickenhall; Blue John Cavern, Cave Dale, close to Castleton. These localities are all well known, and have afforded rich stores of valuable fossils.

SHROPSHIRE.—Oswestry, Ilanymynack Hill, near Oswestry (this last locality is in

Shropshire, just on the border, in fact; the boundary line is south, and Mont-gomeryshire runs through the hill of Ilanfyllin).   Steeraway, Coal-brook-dale, Lilles Hall, Wellington; Treflack Oswestry; Old Statch Wrekin, Clee Hill. Several of these localities have been studied by the Geological Survey and other geologists.

LEICESTERSHIRE.—Ashby-de-la-Zouch, Breedon Hill and Breedon Cloud Wood; Barrow Hill, Asgathorpe, Gracedieu. These localities (in limestone, all more or less dolo-mized, and forming five distinct outlines in the Red Marl) have been studied by the Rev. W. Coleman; see also Hull's 'Report on the Geology of the Leicestershire Coal Field.'

WORCESTERSHIRE.—In coal measures near Dudley.

HEREFORDSHIRE.—Houl Hill, near Ross.   The other side of the river is Chepstow, ii. Monmouthshire. (From unpublished lists of the 'Geological Survey.')

MONMOUTHSHIRE.—Rhymney, Beaufort, Glan Rhymney, or Rhymney Gate; Cwm-Bryn-ddu, Dowlais, Clydach, Pontypool; Ebbwvale, Chepstow.   All these localities have been well studied by Dr. Bevan.

GLOUCESTERSHIRE.—River Avon, Cook's Folly Wood, Clifton rocks, &c.   Westbury; Olveston; Alveston, Tytherington; Cromhall; Tortworth; Wickwar; Chipping Sodbury; Mitcheldean; Coleford, Briavels, Granham Rocks; Under Lansdown, Bath; Wick, near Bath.

DEVONSHIRE.—Westleigh, Brushford, Pilton, Coddon Hill.   The exact position and palæontological contents of the Carboniferous group in Devonshire does not appear to have been as yet completely worked out.   It is believed by some geologists that the upper part of the Pilton group may perhaps belong to the Carboniferous series.

SOMERSETSHIRE.—Leigh Woods, opposite Cock's Folly and Clifton; Broadfield Down (Clevedon) near Bristol; Weston-super-Mare; Portishead; Sims Hill, Broadfield Down, Bristol; Wrington; Burrington Coombe; Axbridge; South side of the Mendips; Nunney, near Frome; Binegar; Charter House, Mendip Hills; Banwell; Sheep Mayswood, Broadfield Down, Bristol; Broadfield Farm, north-east of Wrington; Blagdon; Burrington Ham, south-west of Blagdon; Lower Farm, south of Blagdon; Stoke Farm, three miles north of Wells; West Horrington, north-east of Wells; Penhill House, north-east of Wells; Whatley House, near Frome; Vallis, near Frome; Whatcomb Farm, near Frome; Cannington, near Bridgewater;

Uphill; Cheddar. For the knowledge of many of these localities, as well as for some in Gloucestershire, I am indebted to the Geological Survey, to Messrs. Salter, Etheridge, Moore, Walton, and W. Stoddart, &c.

WALES.

In Wales, nine counties have hitherto afforded about forty species; Denbighshire, Flintshire, and Glamorganshire, having produced the largest number.

The following are the principal localities with which I am at present acquainted, and are to some extent taken from the unpublished lists of the Geological Survey of that portion of Great Britain:

PEMBROKESHIRE.—Skrinkle (lower black Carboniferous shales), Tenby; Caldy, Giltar Point. Pembroke dockyard.

ANGLESEA.—Llynback, six miles east of Llanerchymedd, Pencaint, Llangefri.

CARNARVONSHIRE.—Great Ormes Head.

DENBIGHSHIRE.—Langollen Crags; Tyfyn-uchaf, near Ruabon; Chirk.

FLINTSHIRE.—Mold; Bryn-davin-mold; Halken Mountain; Holywell.

BRECKNOCKSHIRE.—Not far from Rymney Gate.

GLAMORGANSHIRE.—Cowbridge; Castle Mumbles; Newton.

CARMARTHENSHIRE.—Cromanmon, north of Curnammon.

MONTGOMERYSHIRE.—Lanfyllin. This locality is close to the boundary line of Shropshire, and near Oswestry.

SCOTLAND.

In Scotland fourteen counties have afforded fifty species;[1] and it has been calculated by Prof. Nicol that the Carboniferous strata cover nearly a seventeenth of the entire surface of the country; but it is very difficult to form a correct estimate, on account of the numerous breaks from intrusive igneous rocks, rendering mapping very complex. It is, however, in the central portion of Scotland that the rocks which we are now describing

[1] All the species and their localities have been described in my monograph of the Carboniferous Brachiopoda of Scotland, published in the 'Geologist' for 1860.

occupy the greatest surface; they form there a wide sub-parallel band of nearly one hundred miles in length by some fifty in breadth, extending from the northern portion of the Frith of Forth to the Clyde, and as far as the extremity of Cantyre. No portion of the system appears to have been discovered in the north: but in the south there exists a narrow band, or separate patches, which extend along the frontiers of Scotland and England, from Berwick to near Kircudbright, on the Solway Frith.

Scottish Carboniferous deposits differ, however, from strata of a similar age, existing both in England and Ireland, in the manner in which the various beds of encrinal and coralline limestones are intercalated with coal-beds and bituminous schists in the lower parts of the system. In no single locality do we find a section in which all the beds occur in regular and uninterrupted succession; the absence of some or the thinning-out of others constitute local differences which may always be expected and duly considered. Thus in Lanarkshire generally, as well as in other parts of the Clydesdale coal-field, the Carboniferous strata have been divided into four principal groups, viz.—1. The Upper Coal series. 2. The Upper Limestone series. 3. The Lower Coal measures. 4. The Lower Limestone series. In all but the Upper Coal series Brachiopoda have been found; they appear, however, more numerous in the second and fourth divisions.

At p. 6, we alluded to David Ure's valuable work published in 1793, in which twelve species of Carboniferous Brachiopoda have been described and figured; and it would appear from an extract taken from George Crawford's 'History of Renfrewshire,' that in the beginning of last century there was a collector of fossils (the Rev. Robert Wodrow, who died in 1757) in Renfrewshire, and that though Ure was the first that figured and described Scottish fossils, he was not the first upon record that collected them, and indeed from their great abundance one cannot feel surprised that they should have attracted some notice, although they could not be understood at a period prior to the introduction of the science of Palæontology.

List of Localities in Scotland where Carboniferous Brachiopoda have been found.

LANARKSHIRE.

	Distance and direction from Carluke Church.	Stratigraphic position below the Ell coal.	Nature of strata.
Belston Place Burn	1¼ miles N.	160 fathoms.	Slaty ironstone.
Belston Place Burn	1¼ „ N.	173 „	Ironstone shales.
Gare Limestone	2 „ N.E.	239 „	
Westerhouse	3 „ E.N.E.		Old shale heaps.
Bashaw	1½ „ N.E.		
Whiteshaw	¼ „ W.		
Belston Burn Limestone	½ „ N.E.	265 „	Limestone and shales.
Maggy Limestone	.	300 „	
Brocks Hole	1 „ E.		
Below Whiteshawbridge	1 „ W.		Ironstone and shales.
Near Chapel	2 „ S.		

32

	Distance and direction from Carluke Church.	Stratigraphic position below the Ell coal.	Nature of strata.
Lingula Ironstone	.	317 fathoms.	
Braidwood Gill	2 miles S.		Ironstone and shale.
Lingula Limestone	2 ,, S.	337 ,,	
Hallcraig Bridge	1½ ,, W.		
Raes Gill	2 ,, W.		Limestone and shales.
Langshaw Burn	1 ,, S.E.		
Braidwood Burn	2 ,, S.		
1st Kingshaw Limestone	.	338 ,,	
Hallcraig Bridge	1½ ,, W.		Limestone and shales.
Kingshaw	1 ,, N.E.		
2nd Kingshaw Limestone	.	341 ,,	
Hallcraig Bridge	1½ ,, W.		
Langshaw	1 ,, S.E.		Limestone and shales.
Kingshaw	1 ,, N.E.		
1st Calmy Limestone	.	343 ,,	
Raes Gill	2 ,, W.		
Braidwood	1½ ,, S.		
Langshaw	1 ,, S.E.		Limestone and shales.
Waygateshaw	1 ,, S.		
Headsmuir	1¼ ,, S.E.		
Raes Gill Limestone	.	354 ,,	
Raes Gill	2 ,, W.		
Waygateshaw	1 ,, S.		
Braidwood	1½ ,, S.		Alternate beds of ironstone and shales.
Langshaw	1 ,, S.E.		
Kilcadzow	3 ,, E.		
Hill Head	1 ,, E.		
Hosie's Limestone	.	356 ,,	
Hillhead	1 ,, E.		
Raes Gill	2 ,, W.		
Waygateshaw	1 ,, S.		Limestone and shale.
Braidwood	1½ ,, S.		
Mosside	1 ,, N.E.		
2nd Calmy Limestone	.	371 ,,	
Braidwood	1 ,, S.		
Mosside	1 ,, N.E.		Limestone and shales.
Kilcadzow	3 ,, E.		
Main Limestone	.	375 ,,	
Braidwood	1½ ,, S.		
Langshaw	1 ,, S.E.		
Mosside	1 ,, N.E.		Limestone and shales.
Bashaw	1½ ,, N.E.		
Kilcadzow	3 ,, E.		
Shelly Limestone	.	391 ,,	
Braidwood Gill	2 ,, S.		Limestone.
Nellfield Burn	2 ,, S.E.		

	Distance and direction from Carluke Church.	Stratigraphic position below the Ell coal.	Nature of strata.
Productus Limestone	.	397 fathoms.	
Braidwood Gill	2 miles S.		
Nellfield Burn	2 ,, S.E.		Limestone and shales.
Near Yuildshields	2 ,, E.		
Ironstone Beds	.	410 ,,	
Nellfield Burn	2 ,, S.E.		Ironstone and shales.

The foregoing list embraces strata in descending order where Brachiopoda and other fossils have been found in Carluke parish, and for which I am indebted to a local inquirer, whose knowledge of the district and its localities has extended over thirty years.

Brockley	6 miles S. of Lesmahago	Limestone and shale.
Coalburn	4 ,, S. ,,	,,
Brown Hill	2 ,, S. ,,	,,
Middleholm	2 ,, S.W. ,,	,,
Moat	3 ,, E. ,,	,,
Hall Hill	3½ ,, N.E. ,,	,,
Auchenbeg	3 ,, S. ,,	,,
Kersegill	1½ ,, N. ,,	,,
Birkwood	2 ,, N. ,,	,,
Dykehead	3 ,, N.W. ,,	,,
Auchenheath	3 ,, N. ,,	Ironstone and shale.
Den	3 ,, N. ,,	Limestone and shale.
Dalgow	3 ,, W. ,,	,,
Flat	5 ,, N.E. ,,	,,
Crossford	2 ,, S. of Braidwood	,,
Gallowhill	¾ ,, E. of Strathavon	,,
Limekiln Burn	} 3½ ,, S.W. of Hamilton	Limestone.
Boghead		
Auchentibber	1½ ,, S.W. of High Blantyre	Limestone and shale.
Calderside Mines	2 ,, S.W. ,,	
Brankamhall, Calderwood	1¼ ,, S. of East Kilbride	
Capelrig, Calderwood	1 ,, S. ,,	
Limekilns, near East Kilbride		
Lickprivick	2 ,, S.W. ,,	Limestone and shale.
Hermyres	2 ,, W. ,,	
Thorntonhall	2½ ,, W. ,,	
Parliamentary Road, corner of North Frederick Street, Glasgow (exposed during building operation in 1857)		Calcareous sandstone and shale.
Robroyston	2 ,, N.E. of Glasgow	Old shale heaps.

Bedlay	.	.	.		
Chryston	.	.	.	} 6 to 7 miles N.E. of Glasgow	Limestone and shale.
Garnkirk	.	.	.		
Moodiesburn	.	.	.		

Shales above limestone at Bishopbriggs, 3 miles N. of Glasgow.

The Lanarkshire localities have been carefully explored by Mr. J. Armstrong, Mr. J. Thomson, Mr. R. Slimon, Mr. Young, Mr. Bennie, &c., and comprise likewise those quoted by David Ure in his 'History of East Kelbride,' &c.

STIRLINGSHIRE.

Calmy limestone and shales, Balquarhage, 2 miles S.S.E. of Lennoxtown.

Corrieburn, on Campsie Hill, 4 miles N.E. of Kirkintilloch—limestone, ironstone, and shales.

Dark gray limestone and shale, 22 fathoms above Campsie main limestone, South Hill pits, and Barraston, near Lennoxtown.

Shales above Campsie main limestone, Schiliengow, near Lennoxtown.

Campsie main limestone, Schiliengow, Ferrot's and Gloratt quarries, North Hill, and Alum Work mines and Craigend Muir, South Hill, all near Lennoxtown.

Shelly limestone, ironstone, and shale, Balgrochan Burn, ½ mile N. of Lennoxtown.

Limestone and shale, Mill Burn, near Lennoxtown.

Ironstone and shale, Balglass Burn, near Lennoxtown.

Limestone, ironstone, and shales, Craigenglen and Glenwine, 2 miles S.W. of Lennoxtown.

In the foregoing list are enumerated all the localities from which Brachiopoda have been obtained in the Campsie district.

Banton .	.	.	. 2 miles E. of Kilsyth	Limestone.
Murray's Hall	.	.	. S.W. of Stirling.	

All the Stirlingshire localities have been minutely examined by Mr. Young.

DUMBARTONSHIRE.

Castlecary	.	.	. } Near Cumbernauld	Limestone and shale.
Netherwood	.	.	.	
Duntocher	.	.	. 9¼ miles N.W. of Glasgow	Bed of limestone and shale, near sandstone quarry.

RENFREWSHIRE.

Howood .	.	.	. 4 miles W. of Paisley	Limestone and shale.
Wauk Mill Glen	.	.	. Barrhead	„
Hurlet .	.	.	. 7¼ miles S.W. of Glasgow	„
Orchard .	.	.	. 1 „ E. of Thornliebank	„

Davieland quarry . . .	Near Thornliebank	Limestone and-shale.
Arden quarry . . .	,,	,,

These as well as the Ayrshire localities have been carefully searched by Mr. J. Thomson, Mr. J. Armstrong, and others.

AYRSHIRE.

West Broadstone . . .	1 mile S. of Beith	Limestone and shale.
Roughwood . . .	Near Beith	,,
Auchenskeigh . . .	2 miles S. of Dalry	,,
Highfield Quarry . . .	1 ,, N.E. ,,	,,
Linn Spout . . .	Near Dalry	,,
Bowertrapping . . .	,,	,,
Gateside . . .	Near Beith	,,
Goldcraig . . .	1 mile E. of Kilwinning	,,
Monkredding . . .	1½ ,, ,,	,,
Hallerhirst . . .	Near Stevenston	,,
Craigie . . .	Near Kilmarnock	,,
Cessnock . . .	1 mile S.E. of Galston	,,
Alton . . .	2 miles N. ,,	
Moscow . . .	3 ,, N. ,,	
Nethernewton . . .	3 ,, N.E. ,,	
Hyndberry Bank . . .	2 ,, N.E. ,,	
Meadowfoot . . .	5 ,, E. of Darvel near Drumclog	
Gainford . . .	2½ ,, E. of Stewarton	
Bruntland . . .	1 ,, E. of Fenwick	
Mulloch Hill . . .	New Dailly	

EDINBURGHSHIRE.

Gilmerton, near Edinburgh.
Wardie, ,,
Dryden, 6 miles S. of Edinburgh.
Carlops, 14 miles S. ,,
Joppa, near Portobello.
Roman Camp, near Dalkeith.
Cousland.
Magazine, 6 miles S.E. of Dalkeith.
Esperston, 2 miles S.E. of Temple.

Crichton Dean, Crichton Castle.
Penicuick.
Cornton, near Penicuick.
Mount Lothian, 3 miles S.E. of Penicuick.
Leven Seat,
Addiewell,
Scola Burn, } S.W. of West Calder.
Baad's Mill,

PEEBLESSHIRE.

Bents.
Lamancha.

Whim.
Whitfield.

HADDINGTONSHIRE.

Prestonpans.

Aberlady.

Longniddry.

Jerusalem.

Salton.

Kidlaw.

The Vaults, E. of Dunbar.

Skateraw   „

Cat Craig   „

East Barns   „

Saughton, 4 miles W. of Haddington.

LINLITHGOWSHIRE.

Kinniel,

Dykeneuk, } W. of Borrowstownness.

Craigenbuck,

Tod's Mill, River Avon.

Caribber, S.W. of Linlithgow.

Bowden Hill, S.W.   „

Bathgate Hills.

Balbardie, near Bathgate.

Blackburn, S.E.   „

Breichwater, above Breichdyke.

Hillhouse, 1 mile S. of Linlithgow.

Tartraven, 3 miles S.E.   „

FIFESHIRE.

Ladedda,

Wilkieston, } 3 miles S.W. of St. Andrew's.

Winthank,

Craig Hartle, near St. Andrew's.

Craighall,

Cult's Hill, } 3 miles S.W. of Cupar.

Forthar,

St. Monance, 3 miles W. of Anstruther.

Strathkenny, St. Andrew's.

Dumbarnie, near Largo.

Chapel,

Bogie, } Near Kirkcaldy.

Inverteil,

Seafield, } Near Kirkcaldy.

Sunnybank, N. of Inverkeithing.

Parkend, N.E.   „

Brucefield, S.E. of Dunfermline.

Rescobie, N.   „

Duloch, E.   „

Charlestown.

Rosyth, W. of the Castle.

Crombie Point.

Bucklyre, N. of Aberdour.

BERWICKSHIRE.

Cove at Cockburnspath.

Marshall Meadows, 3 miles N. of Berwick.

Coast between Lammerton and Berwick.

DUMFRIESSHIRE.

Closeburn.

Hollows, 4 miles S. of Langholm.

KIRCUDBRIGHTSHIRE.

Coast of Arbigland, Parish of Kirkbean.

This locality has been explored by Mr. John Steven, of Glasgow.

BUTESHIRE.

Corrie, Arran.

Salt Pans.

For much information relative to the localities of the last nine Counties, I am indebted to Mr. Geikie, Mr. Page, Mr. Tate, Prof. Ramsay, Mr. Fraser, the late Dr. Fleming, and

H. Miller; as well as to Sir R. Murchison, and Mr. Salter, who have kindly allowed me access to the lists and specimens assembled during the survey of part of Scotland.

IRELAND.

In Ireland thirty counties have afforded about seventy-nine species, and is, next to England, the portion of Great Britain which has hitherto produced the largest number of species. It is possible that a few of these seventy-nine will turn out, when better known, to be synonyms, and that a few others may occur; but all my researches and efforts, as well as those of several friends in Ireland, have not hitherto succeeded in detecting a larger number, and I have already given my reasons why so many of those recorded in the 'Synopsis' must be rejected.

The portion of my table devoted to Irish species is founded on a personal examination of the specimens collected during many years by Mr. Kelly and others for Sir R. Griffith, and from which Prof. M'Coy's 'Synopsis of the Characters of the Carboniferous Fossils of Ireland' (1844) originated. I have also seen General Portlock's specimens, now forming part of the Museum of Practical Geology in London, and have examined many other specimens from various Irish private collections, as well as from the Geological Survey of Ireland, in addition to a small series in my own possession. It is, however, to Mr. Kelly, and to Mr. Joseph Wright, of Cork, that the distribution of the species in the larger number of the Irish counties is mainly due, as it is to them that I am indebted for most of the information and specimens I possess. As it is the case with England, some of the Irish counties have been more carefully searched than others; thus, for example, those of Dublin, Kildare, and Cork, have hitherto afforded the largest number of species. I have also availed myself of much information contained in Mr. Kelly's valuable paper 'On Localities of Fossils of the Carboniferous Limestone of Ireland,' 1855, and in the explanations of the Geological Survey of Ireland, No. 102, 122, 197, 198, and 153. Mr. Kelly divides the Carboniferous system of Ireland into—1, Old Red Sandstone; 2, Calciferous Slate; 3, Limestone; 4, Coal. I have, however, elsewhere objected to the term *Old* Red Sandstone being made use of for a division of the Carboniferous system, as it is evident that the term *Old* Red Sandstone cannot be retained or made use of to designate at the same time a Silurian, Devonian, and Carboniferous rock; the term *Old Red Sandstone* being now retained for a Devonian rock older than the Irish Red and Yellow Sandstone, which constitute the first or lowest division of the system. These Irish Sandstones, at Kildress and elsewhere, are full of Carboniferous, and not Devonian fossils; the same species, occurring in the Calciferous Slates, Carboniferous Limestone, and Shales. I have, therefore, suggested that geologists should drop the term "old" in their subdivisions of the Carboniferous group, and distinguish their lowest or first division by the designation of—1, *Lower*

Carboniferous Red and Yellow Sandstone. Mr. Kelly has, moreover, informed me, that in Ireland this red rock is not that which predominates, that it averages about one thousand feet in thickness, and is not much exposed, being usually covered with Limestone, except at the outcrop; that the 2nd, or Calciferous Slate, is not considerable in thickness, and that, in the best developed places (Clonea and Dungarvan), half of it is made up of bands of Limestone, the other half Calcareous Slate. The fossils in both he states to be inseparable, so that the Calciferous Slate and Mountain Limestone might be considered as one division, but that it is, perhaps, more correct, as a lithological distinction, to separate them into two. The Carboniferous, or Hibernian Limestone, is fifty feet thick at Drumquin, in Tyrone, and about 1500 feet thick at Black Head, in Clare; it occupies about 20,000 square miles in Ireland; while the coal measures are 2000 feet or more. Such are Mr. Kelly's views relative to the subdivisions of the Carboniferous system in Ireland. The great bulk of the specific forms among the Brachiopoda are found in the Calciferous Slates and Mountain Limestone, but few species occurring in the Red and Yellow Sandstones, or in the Coal measures. I include, also, in the Carboniferous Limestone, those bands of Limestone south of the Blackwater River, such as those of Cork, which have a strong cleavage (the fossils they contain being usually much contorted). I do so because the fifty-eight species of Brachiopoda, discovered in them by Mr. J. Wright and other geologists, are all the same as those common and characteristic to the Carboniferous Limestone of other parts of Ireland, as well as of England, Scotland, and the Continent generally, and which will be found enumerated, after careful identification, in the column of the table devoted to the county of Cork. With these preliminary observations I will now give the list of localities drawn up for this Monograph by Mr. Kelly; Mr. Joseph Wright having added those of Cork and from some other counties with which he was acquainted.

County of ARMAGH.—Annahugh (Limestone), six miles north-east of Armagh; Armagh, about the town; Ballygasey, four miles north of Armagh; Benburb, six miles north-west of Armagh; Calragh, five miles north-west of Armagh; Down, a quarter of a mile south-west of the town; Drummanmore, one mile north-east of Armagh; Kilmore, six miles north-east of Armagh; Tullyard, one mile north of Armagh. (Carboniferous Limestone in all these localities.)

CORK.—Little Island, four miles east of Cork; Windmill quarry is situated at the southern extremity of Cork; Midleton, thirteen miles east of Cork; Blackrock, two miles east of Cork; Carrigtwohill, eight miles east of Cork; Rafeen, five miles south-east of Cork; Mallow, on the River Blackwater; Glounthane, four miles east of Cork; Carrigaline, six miles south-east of Cork; Ballywalter, two miles north of Castletownroche; Castletownroche, eight miles north-east of Mallow; Streamhill, three miles north of Doneraile; Ringaskiddy, eight miles south-east of Cork; Fort William, one mile and a half south-west of Doneraile; Annagh, four miles south-west of Charleville; Araglin Bridge, two miles north-east of Fermoy (the rock is yellowish sandstone); Banteer, three miles south of Kanturk; Castlecreagh, one mile east of Doneraile; Doneraile town stands on Fossiliferous limestone; Tankardstown, six miles north-east of Doneraile. These localites are almost all in limestone.

CARLOW.—Bannaghagole, two miles west of Leighlinbridge (fossils abundant in limestone, covered by the coal rocks of Castlecomer); Old Leighlin, two miles west of Leighlinbridge (the limestone occurs here, and at its junction with the overlying Millstone grits of the Castlecomer district fossils are numerous, and some beautiful casts are found in old excavations); Raheendoran, four miles south-west of Carlow (limestone).

CLARE.—Clifden, two miles west of Corofin (Millstone grit); Cloonlara, three miles north-east of Limerick; Kilmacduagh, five miles south-west of Gort (limestone); Moymore, seven miles east of Ennis; Scariff, eight miles north-west of Kilaloe.

CAVAN.—Aghbay, one mile and a quarter south-west of Swanlinbar Village; Alteen, one mile north-west of Swanlinbar (limestone and shale); Ballyconnel, four miles west of Belturbet; Clonkeiffy, five miles south-west of Virgina; Countenan, one mile north-west of Stradone (arenaceous limestone); Kilcar, two miles south-west of Belturbet (Killeshandra limestone); Laragh, one mile north of Stradone; Swellan, one mile east of Cavan.

DUBLIN.—Ballintree, one mile north of Rush; Ballykea, two miles south of Skerries; Curkeen, two miles south of Skerries; Drumslattery, two miles south of Skerries; Howth, nine miles east of Dublin; Lane, two miles south-east of Skerries Malahide, shore very fossiliferous; Milverton, one mile south-west of Skerries; Oldtown, two miles north-west of Swords; Poulscadden joins the village of Howth; Salmon, three miles south of Balbriggan; St. Douloughs, five miles north-east of Dublin.

DONEGAL.—Abbeylands, one mile north-west of Ballyshannon (*Orthis Michelini* occurs here, along with *Strept. crenistria*, four inches in diameter); Ardloughill, two miles south-east of Ballyshannon; Ballybodonnel, ten miles west of Donegal; Bruckless, one mile north of Dunkineely; Bundoran, three miles south-west of Ballyshannon; Doorin, seven miles west of Donegal; Dunkineely, a village ten miles west of Donegal; Finner, three miles south-west of Ballyshannon; Greaghs, three miles south-east of the town of Donegal; Killoghtee, one mile south of Dunkineely; Lisnapaste, five miles south of Donegal; Rahans Bay, one

33

mile south-west of Dunkineely; Spierstown, two miles east of Donegal; St. John's Point, fourteen miles north-west of Donegal; Tinnycahill, two miles east of Donegal.

DOWN.—Castle Espie, two miles south-east of Comber; Cultra, five miles north-east of Belfast.

FERMANAGH.—Ardatrave, two miles south-west of Kesh; Agharainy, one mile south of Kesh; Belmore Mountain, six miles south-west of Enniskillen; Boa Island, in north end of Lough Erne; Bohevny, one mile north-east of Church Hill; Bunaninver, three miles south of Kesh; Carn, three miles south-east of Kesh; Carrickoughter, two miles north-west of Kesh; Carrowntreemall, ten miles south-west of Enniskillen; Clareview, two miles south-west of Kesh; Cleenishgarve, an island in North Lough Erne; Corlave, three miles north-west of Kesh; Cornagrade, half a mile east of Enniskillen; Crevenish, one mile south-east of Kesh; Deerpark, two miles south-east of Kesh; Derrygonelly, eleven miles north-west from Enniskillen; Derrynacapple, four miles north-east of Kesh; Ederney, two miles east of Kesh; Kesh, twelve miles north-east of Enniskillen; Killycloghy, two miles south-west of Lisbellaw; Knockninny, ten miles south-east of Enniskillen; Leam, two miles east of Tempo; Ring, two miles north-east of Enniskillen; Shean, a mile north of Church Hill.

GALWAY.—Athenry, ten miles east of Galway; Ballinfoyle, one mile north-east of Galway; Ballyhanry, five miles west of Portumna; Caheratrim, three miles south-west of Loughrea; Cappaghmoyle, four miles north-east of Athenry; Carrowntobber, two miles north-east of Athenry (Cong limestone all round the town); Cregganore, six miles south-west of Loughrea.

KERRY.—Ballymacelligot, three miles east of Tralee; Castle Island, quarries round the town; Currens, six miles east of Tralee; near Farmer's Bridge; Castlemaine, nine miles south of Tralee.

KILDARE.—Ardclough, eight miles west of Dublin, near the Grand Canal; Boston, six miles north of Kildare; Millicent, four miles north of Naas.

KING'S COUNTY.—Banagher, quarries about the town.

LIMERICK.—Kilmallock, quarries about the town; Croagh, three miles north-east of Rathkeale; Kyletaun, one mile north of Rathkeale; Blossomhill, one and a half miles north-east of Rathkeale; Ballingarrane, two miles north of Rathkeale,

Doohylebeg, two miles north-east of Rathkeale; Stoneville, one mile north-west of Rathkeale; Fanningstown, three miles north-west of Croom; Caherass, one mile north-west of Croom; Rathkeale, all about the town; Adare, ten miles south-west of Limerick; Castleconnel, seven miles north-east of Limerick.

LOUTH.—Carlingford, quarries about the town; Kilcurry, four miles north-west of Dundalk; Knockagh, three miles north-west of Dundalk.

LONGFORD.—Ballymahon, quarries about the town; Carrickboy, five miles north-east of Ballymahon; Carrickduff, two miles north of Granard; Granard has fossils in the rocks about it; Kilcommock, three miles north-west of Ballymahon; Mullawornia, two miles north-west of Ballymahon; Rathcline, two miles south of Lanesborough; Shrule, four miles north of Ballymahon; Tirlicken, three miles north-west of Ballymahon.

LEITRIM.—Aghamore, five miles south-west of Ballyshannon; Blacklion, twelve miles west of Enniskillen; Manorhamilton; Mohill, a small town on black slate fossiliferous; Ussaun, half a mile west of Mohill.

MEATH.—Altmush, two miles north of Nobber; Ardagh, five miles south of Carrickmacross; Ballyhoe, five miles south of Carrickmacross; Balsitric, three miles east of Nobber; Castletown, four miles south of Trim; Cregg, two miles north of Nobber; Cruicetown, two miles west of Nobber; Cusackstown, six miles south-east of Navan; Flemingstown, six miles west of Balbriggan; Horath, three miles north of Kells; Laracor, two miles south of Trim; Mullaghfin, two miles west of Duleek; Rathgillen, two miles north of Nobber; Crusserath, three miles south of Drogheda.

MAYO.—Ballina, quarries about the town; Ballinglen, six miles north-west of Killala; Bunatrahir, one mile north-west of Ballycastle; Crosspatrick, one mile south-east of Killala; Cuilmore, three miles east of Claremorris; Kilbride, two miles north-east of Bally Castle; Kilcummin, five miles north-west of Killala; Killogunra, two miles south-west of Killala; Killybrone, one mile north-west of Killala; Larganmore, fourteen miles west of Crossmolina.

MONAGHAN.—Clonturk, three miles south-east of Carrickmacross; Dundonagh, six miles north of Monaghan; Killyrean Upper, two miles north-east of Emyvale; Leck, two miles north of Glasslough; Monaghan Town, quarries about it; Mullaghboy, two miles east of Emyvale; Mullaliss, two miles east of Emyvale.

QUEEN'S COUNTY.—Aghafin, one mile south-west of Castletown; Burris, two miles north-east of Maryborough; Ringstown, one mile north-east of Mountrath; Roundwood, three miles north-west of Mountrath; Tinnekill, three miles north-east of Mountmelick.

ROSCOMMON.—Carrownanalt, two miles north-east of Keadue; Cartronaglogh, half a mile north of Keadue; Cleen, four miles north-east of Boyle; Drumdoe, four miles north of Boyle; Grangemore, three miles south-west of Boyle; Killukin, one mile south-west of Carrickonshannon; Lacken, three miles south-east of Athleague; Lisardrea, two miles south-west of Boyle; Moore, three miles east of Ballinasloe; Rathmoyle, five miles south of French Park; Termon joins Boyle.

SLIGO.—Ballinafad, five miles north of Boyle; Ballymeeny, two miles south-east of Easky; Bunowna, in the river and quarries about Easky; Carrowmably, four miles south-east of Easky; Carrowmacrory, six miles east of Easky; Carrowmore, four miles south-west of Coolaney; Cashelboy, twelve miles west of Sligo; Culleenamore, five miles west of Sligo; Kilglass, eight miles north of Ballina; Magheranore, two miles east of Tobercorry; Streedagh, ten miles north-west of Sligo.

TIPPERARY.—Carrigahorig, three miles south-east of Portumna; Nenagh, in limestone quarries about the town; Clonmel, limestone in the vicinity of the town.

TYRONE.—Aghintain, two miles west of Clogher; Aghnaglogh, two miles north-west of Clogher; Annaghilla, three miles south-west of Ballygawley; Ballymacan, two miles south-west of Clogher; Cavansallagh, two miles north-west of Drumquin; Claraghmore, one mile north-east of Drumquin; Clare, half a mile east of Cookstown; Cookstown, the town stands on limestone; Derryloran joins Cookstown; Donaghrisk, two miles south-east of Cookstown; Drumowen, two miles west of Drumquin; Drumscraw, one mile south-west of Drumquin; Edenasop, five miles south-west of Castlederg; Kildress, two miles west of Cookstown; Killymeal, half a mile east of Dungannon; Knockonny, half a mile north of Ballygawley; Lackagh, one mile west of Drumquin; Lismore, half a mile south-west of Clogher; Magherenny, one mile south-east of Drumquin; Mullaghtinny, a mile east of Clogher; Rahoran, two miles north of Fivemiletown.

WATERFORD.—Ardoe, five miles east of Youghal; Ballinacourty, three miles east of Dungarvan; Ballyduff, two miles west of Dungarvan; Clonea, three miles north-east of Dungarvan; Curragh, one mile north of Ardmore; Killinamack, three miles south-west of Clonmel; Whiting Bay, two miles east of Youghal.

WESTMEATH.—Cornadowagh, seven miles west of Ballymahon; Athlone, two miles south-east of the town.

WEXFORD.—Hook Head, on the east side of Waterford Harbour; Drinagh, one mile south of Wexford.

KILLENNY.—Dunkit, three miles north of Waterford; Kilkenny; marble quarries.

ANTRIM.—Tornaroan, one mile and a half east of Ballycastle, on the shore at highwater mark.

CARBONIFEROUS BRACHIOPODA BEYOND THE LIMITS OF GREAT BRITAIN.

Ninety-three of the species mentioned in my tables have been found in various countries beyond the limits of Great Britain, and perhaps a larger number will be obtained after a more extended search, for the species of many distant regions, and of even European countries (Belgium excepted), are but imperfectly known. It is to Prof. de Koninck's admirable researches that we are mainly indebted for our knowledge of the palæontological richness of the Carboniferous rocks of Belgium, which he has ably elaborated during many years. The number of Carboniferous Brachiopoda discovered by our friend in his country but slightly exceed (?) those hitherto found in Great Britain. Mr. Edward Dupont has also assembled from the neighbourhood of Dinant nearly ninety species of Brachiopoda,[1] and which, with very few exceptions, are specifically the same as those which occur in Great Britain, so that while Belgium contains some forms hitherto unknown to our strata, Great Britain numbers likewise a few species not hitherto discovered in Belgium.

The French Carboniferous species have not yet been studied in a satisfactory manner, but there, as in Russia and in other parts of Europe, many of our British species occur, along with a few forms special to the district.

Having recently examined and described the Carboniferous Brachiopoda of the Punjab (India),[2] I found that out of twenty-eight species, at least thirteen were common to

---

[1] "Notice sur les Gites de fossiles du Calcaire des bandes Carbonifères de Florennes et de Dinant," 'Bulletins de l'Académie Royale de Belgique,' 2d ser., vol. xii, No. 12, 1861.

Great credit is due to this young naturalist, who by dint of labour and perseverance has, within a comparatively short period, assembled upwards of ten thousand specimens representing some five hundred species from the Carboniferous rocks of his district.

[2] 'Quarterly Journal of the Geol. Soc.,' vol. xviii, p. 25, 1862.

European rocks of the same period, although several of these have in India attained larger proportions; and among which we may mention, *Athyris Royssii, A. subtilita, Retzia radialis,* var. *Grandicosta, Spirifera striata, Sp. lineata, S. octoplicata, Rh. pleurodon, Orthis resupinata, Strept. crenistria, Prod. striatus, P. semireticulatus, P. longispinus;* and a further search in these distant regions will, no doubt, bring to light a larger number of the common species.

The Australian and Tasmanian Carboniferous rocks have also afforded their quota of common species, for, although the forms from these continents have not yet been sufficiently examined, still, from a glance I have given to collections sent home from Bundaba and Port Stephen in Australia, as well as from Van Diemen's Land, I have already been able to recognise *T. hastata, Sp. striata, Sp. glabra, S. lineata, Rh. pleurodon, Strept. crenistria, Orthis Michelini, Prod. Cora,* &c. If again, and by a rapid stride, we should find ourselves cast on some of the Spitzbergian frozen coasts, we would there pick up several of our common species, such as *Sp. Octoplicatus, St. crenistria, Prod. semireticulatus, P. costatus,* &c., along with other forms not known in Britain, for we cannot expect to find all the same species repeated and assembled everywhere; some forms were more localized than others. Even in Great Britain, we find certain species in England that do not appear to have existed either in Scotland or Ireland, and *vice versá;* it is not, therefore, surprising that in other countries the same order of things should prevail. If we cast a glance at the prodigiously extended Carboniferous regions of America, we shall there also find a vast per-centage of species identical with our own, but which, in many cases, have received new names from our American cousins. Possessing an extensive series of American Carboniferous species, for which I am indebted to the kindness of Mr. Worthen, as well as to that of some other American geologists, and having compared these with our British specimens, I may mention, from among others not yet sufficiently studied, the following few as being identical with our own—*T. sacculus, A. ambigua, A. subtilita, A. plano-sulcata, A. lamellosa, A. Royssii, Retzia radialis, Sp. striata, S. trigonalis, Sp. lineata, Sp. Urii, S. octoplicata, Rh. pleurodon, Orthis Michelini, Strept. crenistria, Prod. Cora, P. punctatus, P. longispinus, P. semireticulatus, P. scabriculus, P. costatus, Crania quadrata, Discina nitida, Lingula mytiloides,* &c., and I entirely concur with the observation made by Sir R. Murchison, at page 324 of his celebrated "Siluria," that "The specific identity of so many of the Brachiopoda of the marine or Lower Carboniferous rocks situated at enormous distances in latitude from one another (*e. g.* from the Arctic circle to within a few degrees of the Equator), is an additional and striking proof of the general uniformity of temperature and condition during this epoch."

I must now conclude this lengthened inquiry into the history of British Carboniferous species, but sincerely hope that local observers will continue the study I have so far sketched out, for much remains still to be achieved, which time and search alone can accomplish.

APPENDIX

CARBONIFEROUS AND PERMIAN MONOGRAPHS.

Plate LIV.

It is to be regretted that the series of Monographs on British Fossil Brachiopoda, as well as those relating to other classes, published by the Palæontographical Society, had not commenced with the species of the Silurian system, and progressed regularly upwards to the Tertiary period. By such a mode of proceeding important advantages would have been obtained, from the possibility of tracing with more certainty and regularity the recurrence of certain species, and thus have obviated the unfortunate necessity of occasional alterations and repetitions to the parts already printed. The recurrence of certain species in one or more systems is a subject of much importance, and is now an acknowledged fact, but while there appears in some minds a manifest disinclination to admit such a thing as possible, others have exaggerated the occurrence, and thus done more harm than good. It would certainly be very agreeable and advantageous were there no recurrent species, and that all the forms of each system or zone were characteristic to it; but such not being the case, we must unavoidably abandon preconceived ideas, and endeavour to carefully trace the recurrence. I will not at present allude to those species which are common to the Devonian and Carboniferous systems, as I have not yet critically examined the Devonian forms, and shall therefore reserve what I might have to say upon the subject until a future period, but it will be desirable to briefly refer once more to those that appear to be common to the Carboniferous and Permian epochs.

Several observers, among whom we may mention M. de Verneuil, Profs. King and Morris, Messrs. Howse, Kirkby, and myself, have more than once alluded to the strong resemblance—nay, identity—of certain Carboniferous and Permian species,[1] but

---

[1] My observations upon the recurrent forms in question will be found in various parts of my Carboniferous Monograph, as well as in some numbers of 'The Geologist.' Mr. Kirkby's views are incorporated in his excellent paper, "On the recurrence of *Lingula Credneri*," &c., 'Journal of the Geological Society,' for March, 1858.

as Prof. King appears to question the correctness of some of our statements[1] it will be desirable to represent, side by side (in Pl. LIV) those forms that appear to be common to the Carboniferous and Permian deposits of our island.

Carboniferous Names.   *Permian Names.*

TEREBRATULA SACCULUS, *Martin*, sp. 1809.   TEREBRATULA SUFFLATA,. *Schlotheim*, sp.
Dav., pl. liv, fig. 5.   1816. *Dav.*, pl. liv, fig. 6.

The identity of the Carboniferous and Permian shell is so complete that Schlotheim's denomination will require to be located among the synonyms of *S. sacculus*. Prof. King admits this identity.

TEREBRATULA HASTATA, *Sow.*, sp. 1824.   ? TEREBRATULA ELONGATA, *Schlotheim*, sp.
Dav., pl. liv, figs. 1 and 3.   1816. *Dav.*, pl. liv, figs. 2 and 4.

It is an unquestionable fact that many specimens of the Carboniferous and Permian shells are undistinguishable, as will be at once perceived by a glance at the specimens or figures selected from among others for illustration; but I must confess that more difference is shown between the greater number of typical *T. hastata* and *T. elongata* than is here represented, the strong resemblance being the exception, and more especially observable between that variety of *T. hastata* we have termed *Gillingensis* and *T. elongata* proper. It must also be observed, that it is often impossible to distinguish certain examples of *T. sacculus* and *T. hastata*, which appear to merge the one into the other, and the same may be said with reference to *T. sufflata* and *T. elongata*, and this proves how intimately connected are all the British forms of Carboniferous and Permian Terebratula. The subject relating to the identity of *T. hastata* and *T. elongata* may therefore, for the present, remain an open question.

ATHYRIS ROYSSII, *L'Eveillé*, sp. 1835. *Dav.*,   ATHYRIS PECTINIFERA, *J. de C. Sow.*, 1840.
pl. liv, fig. 8.   *Dav.*, pl. liv, fig. 9.

M. De Verneuil and Prof. King have both alluded to the resemblance which appears to exist between these shells; and although I thought at one time that sufficient differences might perhaps be established between them, I am now very much afraid that they will have to be merged into a single species, and, if so, Sowerby's denomination would require to give way to that of L'Eveillé. *A. pectinifera* does not appear to have attained the large proportions of a full grown *A. Royssii*, but size alone cannot be made use of as a distinguishing character between species which otherwise resemble each other, and in the

---

[1] King, "On certain species of Permian Shells said to be found in Carboniferous Rocks," 'Edinburgh Philosophical Journal,' vol. xiv, new series, p. 37, &c., 1861, and vol. xv, April, 1862.

present case more especially, for Permian shells as a rule are of smaller size than Carboniferous ones.

Carboniferous Names.	Permian Names.
SPIRIFERA URII, *Fleming*, sp. 1828. *Dav.*, pl. liv, fig. 14.	SPIRIFERA CLANNYANA, *King*, sp. 1848. *Dav.*, pl. liv, fig. 15.

Prof. King denies the identity of these two shells, although allowing them to be closely related; but after having compared many examples of the Carboniferous and Permian species? neither Mr. Kirkby nor myself could perceive any valid specific difference between them.[1]

SPIRIFERINA OCTOPLICATA, *Sow.*, sp. 1827. *Dav.*, pl. liv, figs. 10 and 12.	SPIRIFERINA CRISTATA, *Schloth.*, sp. 1816. *Dav.*, pl. liv, figs. 11 and 13.

It appears to me, as well as to my friend Mr. Kirkby, that there cannot exist a doubt as to these shells belonging to a single species, for which Schlotheim's name should be preferred. None of the Permian examples that have come under our observation have attained the dimensions of certain full-grown Carboniferous specimens, but the general character in both is specifically the same.

CAMAROPHORIA CRUMENA, *Martin*, sp. 1809. *Dav.*, pl. liv, figs. 16, 17, 18.	CAMAROPHORIA SCHLOTHEIMI, *v. Buch*, sp. 1834. *Dav.*, pl. liv, fig. 19.

I feel satisfied that my identification of the Permian *C. Schlotheimi* with Martin's *Anomites crumena* is correct. Prof. King has admitted that the last-named shell is a *Camarophoria*, but believes it specifically different to *C. Schlotheimi*. I have, however, found Permian specimens agreeing very closely with Martin's imperfect figure, and am not acquainted with any other Carboniferous Camarophoria to which the Permian shell could be assimilated than the one we recognise as *crumena*. In *C. Schlotheimi* the mesial fold varies in width and elevation according to the number of ribs which ornament its surface, these varying usually from two to seven, and from one to six in the sinus, while in the Carboniferous shell the same differences are observable, as may be seen in figs. 3—9 of Pl. XXV. It is, therefore, evident that one of the two above-mentioned names will have to be erased, and as Martin's species possesses claims to priority, its name will have to be retained for the Permian as well as the Carboniferous specimens.

[1] The term *recurrent* has been applied by palæontologists to such species as occur in more than one formation or system of strata. In this sense I use it here. For instance, as I consider *Spirifera unguiculus*, Sow. of the Devonian, *Sp. Urii* of the Carboniferous, and *Sp. Clannyana*, king of the Permian system, to be same species, it is said to be recurrent, because it reappears or recurs in two distinct groups of strata after its first appearance in the Devonian system. The term also necessarily implies that the species *existed during the whole of the period* that is included between its first and last appearance.

CAMAROPHORIA RHOMBOIDEA, *Phillips*, sp. 1836. *Dav.*, pl. liv, figs. 20—22.

CAMAROPHORIA GLOBULINA, *Phillips*, sp. 1834. *Dav.*, pl. liv, figs. 23—25.

After a very attentive comparison of many specimens of these shells from the Carboniferous and Permian deposits, I cannot perceive the smallest distinguishing features; specimens of equal size resemble each other, as may be seen from my illustrations. The Carboniferous form, when full grown, has attained somewhat larger proportions than any I have observed from the Permian rocks. The term *globulina*, from having been first proposed, should be employed for both, and that of *rhomboidea* be placed among the synonyms.

Carboniferous Names.
DISCINA NITIDA, *Phillips*, sp. 1836. *Dav.*, pl. liv, fig. 26.

Permian Names.
DISCINA KONINCKII, *Geinitz*, 1848. *Dav.*, pl. liv, fig. 27.

Having examined and compared with much attention a number of Carboniferous and Permian specimens of these two so-termed species, I have come to the conclusion that there exists no specific difference between them, and that the term *nitida* will have to be made use of for the Permian shell. Many large and well-preserved specimens of the Permian *Discina* have been collected by Messrs. R. J. Manson, T. Parker, and E. Gower, in the compact Limestone of East Thickley, near Darlington, which are identically similar to others from the Carboniferous shales of Capel Rig in Lanarkshire; in all we could perceive the same contour, the same degrees of convexity and external sculpture.

LINGULA MYTILOIDES, *Sow.*, 1812. *Dav.*, pl. liv, figs. 28—31.

LINGULA CREDNERI, *Geinitz*, 1848. *Dav.*, pl. liv, figs. 32—34.

In his excellent paper (already quoted) Mr. Kirkby has satisfactorily proved that the Permian *Lingula Credneri* is found in the Carboniferous Rocks of England, a fact I can likewise attest from personal observation; but I must go a step further, by suggesting that the Permian Lingula is doubtfully distinguishable from Sowerby's *L. mytiloides*, which was likewise described from specimens obtained in the Carboniferous Shales of Wolsingham in the County of Durham. Mr. Tate is of opinion that *L. Credneri* is nothing more than a small variety or race of Sowerby's species.

It is not my present desire to enter into further details with reference to certain other species which may possibly be recurrent, but I cannot help mentioning that it is not quite certain that *Crania Kirkbyi* is distinct from *C. quadrata*, nor that *Spiriferina multiplicata* and *Sp. cristata* should be separated.

No new British Permian species of Brachiopoda has been discovered since the publication of my monograph, notwithstanding the most zealous researches of several naturalists. The compact Limestone of East Thickley, near Darlington, has been well searched by Messrs. R. T. Manson, T. Parker, and E. Gower, who have succeeded in procuring some excellent specimens of several species, such as *T. elongata, Cam. crumena, Sp. alata, Strept. pelargonatus, Stroph. Goldfussii, St. Morrisiana, Prod. horridus, Discina nitida* (D. Koninckii), and *Lingula mytiloides* or *Credneri*, so that the total number of Permian species of Brachiopoda occurring in Britain would, according to my views, not exceed sixteen or seventeen species, of which about half would be common to the Carboniferous and Permian periods.

As far, therefore, as the Brachiopoda are concerned, there exists a very intimate relationship between the Carboniferous and Permian species, and, as Messrs. Kirkby and R. Jones have already shown (and are ready to show still further), that species from other classes partake of the same identity. The Permian strata are, therefore, the natural continuation of the Great Carboniferous period, although it may be desirable to preserve the term Permian as originally given by the celebrated authors of 'Russia in Europe and the Ural Mountains Geographically Illustrated' to those strata which succeed the highest portion of the Carboniferous series. The term "Dyas" recently proposed as a substitute for that of "Permian" appears to me to be a very unfortunate idea, for, besides being incorrect in its meaning, it is, in reality, only a synonym, with which science is, alas, becoming so heavily burdened.

STROPHALOSIA.

I am still of opinion that all our known British Permian *Strophalosias* should be referred to two species only.

1. *Strophalosia Goldfussii.* — Notwithstanding Prof. Geinitz's contrary opinion,[1] I believe that the shells referred to Münster's species in p. 39 of my Permian Monograph have been correctly identified, and am also of opinion that what Geinitz terms *St. excavata* from the Zechstein dolomite of Pössneck, and the typical *St. Goldfussii* from the Untere Zechstein of Trebnitz, near Gera, belong to the same species; but with this difference, that at the last-named place, the shell is found perfect with all its spines, while at Pössneck the specimens are either casts, or in a bad state of preservation, and it is in this decorticated condition that the shell is usually found in England.

2. The second species has been referred to *Strophalosia lamellosa* of Geinitz? but I am not so certain that this identification is strictly correct. Prof. Geinitz objects to the idea of uniting King's *St. Morrisiana* with his *St. lamellosa*, and considers the two as

[1] 'Dyas oder die Zechsteinformation und das Rothliegende,' part i, p. 96, 1861.

distinct species; while Messrs. Howse and Kirkby are still disposed to maintain the conclusions arrived at in p. 44 to 49 of my Permian Monograph. All our testiferous specimens of *St. Morrisiana* from Tunstall Hill, and casts from Humbleton, nowhere appear to have possessed that immense number of closely packed spines which are observable on the ventral valves of Geinitz's typical examples of *St. lamellosa* from Trebnitz; and indeed when I compare typical specimens of the same valve of *St. Goldfusii* and *St. lamellosa* from Trebnitz sent to me by Dr. Geinitz, I am at a loss to detect any difference between them; in both the spines appear equally numerous, and to be similarly implanted, so much so that it would not be possible to distinguish the two, had we not the smaller valve which in *St. Goldfusii* is covered with spines, while that of *St. lamellosa* is traversed by concentric laminæ of growth, usually individualised and ornamented with fine radiating striæ. In none of the British examples of *St. Morrisiana* that have come under my notice do we perceive that immense abundance of spines which are visible on the ventral valve of *St. lamellosa* from Trebnitz; on the contrary they are comparatively few, adpressed, and allowing one to perceive between them fine radiating striæ, while in the smaller valve we do not observe those prominent laminæ of growth visible in Geinitz's typical specimens of *St. lamellosa*. It is, therefore, uncertain whether we are quite justified while placing King's *St. Morrisiana* among the synonyms of *St. lamellosa*. The subject had better therefore, for the present be left an open question.

CRANIA KIRKBYI, *Dav.* Permian Mon., page 49; and Carb. Mon., Pl. liv, fig. 35—38.

When describing this species? the exterior of the unattached or upper valve had alone been discovered, and although most of the specimens had their external surface roughly granulated, I have since been led to surmise that this appearance is perhaps due to the decomposition of parts of the shell, for in two or three more solid and better preserved specimens, the surface was almost smooth, or marked only by a few concentric lines of growth. The apex is more or less sub-marginal, while in some specimens it is almost central. It varies also much in shape, as do all *Crania* which live attached to submarine bodies. The interior of the upper valve has also been found by Mr. Kirkby, and of which illustrations will be found in Plate LIV of the present volume. So great is the resemblance of some specimens of this Permian Crania to others of *C. quadrata* from the Carboniferous Rocks of Scotland and Ireland, that I am somewhat undecided whether I should still retain the denomination applied to the Permian shell.

INDEX TO VOL. II.

(Names in italic are synonyms, a ? before indicates probable synonyms or doubtful species.)

FINIS.

PRINTED BY J. E. ADLARD, BARTHOLOMEW CLOSE.

APPENDIX A.

SINCE the concluding pages as well as the index to this Monograph have been printed, some additional information has been obtained, which, I think, had better be recorded, and I must again apologise for being obliged to add a second appendix to this Monograph.

SCOTLAND.

During his summer excursions Mr. James Thomson, of Glasgow, was induced to examine some portions of Cantire or Kintyre, and in Tirfergus Glen, four and a half miles south-south-west from Campbeltown, he discovered Carboniferous Limestone, much altered by heat, and tilted up by an extensive outburst of trap on one side and of porphyry on the other.[1] In this Carboniferous rock thirty-five species of Mollusca have been discovered by Mr. Thomson, among which are the following Brachiopoda:—*Athyris ambigua, Sp. Urii, Rh. pleurodon, Strep. crenistria* and var. *radialis, Prod. latissimus, P. semireticulatus, P. costatus* and var. *muricatus, P. scabriculus, P. aculeatus, P. longispinus,* and *Chonetes Hardrensis*—thirteen species.

ENGLAND AND WALES.

Mr. D. C. Davies, of Oswestry, has obtained the following additional species in the north-west of Shropshire and in Denbighshire :[2]

---

[1] These beds, until the last few years, were thought to belong to the Liassic formation, but they have been tinted as Carboniferous in Prof. Nichols's new geological map of Scotland, as well as in that recently published by Sir R. Murchison and Mr. Geikie.

[2] Much information relative to the geology of these two counties will be found in Mr. D. C. Davies's interesting paper published in the 'Oswestry Advertiser and Montgomeryshire Mercury' for June 12th, 1861.

SHROPSHIRE.—*Ter. sacculus, Athyris expansa, A. globularis, A. plano-sulcata, Spirifera duplicicosta, S. glabra, S. ovalis, Sp. pinguis, Cyrtina carbonaria, Rh. acuminata, Camarophoria globulina, Strophomena analoga, Orthis resupinata, Prod. Cora, P. Keyserlingiana, P. punctatus, P. sulcatus, P. spinulosus, P. sinuatus, P. undatus, C. papilionacea, P. Llangollensis,* and an undetermined *Lingula;* so that instead of nineteen species (recorded at page 244), as many as forty species will have been discovered up to the present time in Shropshire.

DENBIGHSHIRE.—*Athyris globularis, A. expansa, A. plano-sulcata, Sp. octoplicata, Sp. pinguis, Rh. pugnus, Cam. globulina, Prod. aculeatus, P. fimbriatus, P. Keyserlingiana, P. latissimus, P. margaritaceus, P. scabriculus, P. sinuatus, P. Youngianus,* and *Ch. papilionacea,* have to be added to the twenty-four species recorded in our table, so that here also we have about the same number of species as we find recorded for Shropshire.

Fossiliferous localities in the North Wales belt of Carboniferous Limestone, counties Shropshire and Denbighshire, beginning at the southern termination, and proceeding north and west.—The Limestone consists of three principal divisions, viz., "Lower beds," a series of pale-coloured beds, separated by thin shales, and interstratified with several layers of red marl; "Middle beds," of gray crystalline Limestone; and "Upper beds," alternations of layers sometimes similar to the middle bed, and presenting a less splintering structure, and suitable for various architectural purposes. This list has been communicated to me by Mr. D. C. Davies, of Oswestry.

Localities, north-west Shropshire.	Characteristic Fossils and General Observations.[1]
Llanymynech, 6 miles S. of Oswestry..................	Lower beds: *Prod. Llangollensis, Ter. hastata, Prod. Cora.*
Porthywaen, 4½ miles S.W. of Oswestry...............	Lower beds: *P. Llangollensis, T. hastata, T. sacculus, Cyrtina carbonaria,* &c. Upper beds: *P. giganteus, P. latissimus, Spirifers,* &c.
Treflach, 3 miles W. of Oswestry.......................	*Orthis Micheleni, Sp. octoplicata,* and nearly all the fossils peculiar to the upper beds; but the quarries, which are not much worked, have been diligently searched.
Pentregaer, 4 miles N.W. of Oswestry..................	*Chonetes papilionacea, C. Hardrensis, Spiriferinas,* and the smaller *Producta.*
Lawnt and Craig-y-rhiw, 3½ miles N.W. of Oswestry	*Chonetes Hardrensis, Prod. spinulosus,* and *P. longispinus.*
Selattyn Hills, W. of Lawnt, and N.N.W. of Oswestry	*Strept. crenistria, Prod. fimbriatus, P. latissimus, P. giganteus, Orthis Michelini, Athyris plano-sulcata,* &c.
Bronygarth, 2 miles W. of Chirk	Ordinary fossils.

[1] The foregoing remarks are confined to remains of Brachiopoda. Other fossils are found at most of the places mentioned.

Localities, Denbighshire.	Characteristic Fossils and General Observations.
Llangollen..	In dark "upper beds," near Trevor, *Sp. trigonalis, S. duplicicosta,* &c. In the "middle beds," various *Producta* and *Spirifera,* but difficult to extract from the matrix. In the "lower beds," at the same place, small *Terebratulæ, Prod. Cora, P. semireticulatus,* var. *Martini.* In the "lowest beds," continued on towards Minera, *Prod. Llangollensis.* In the "topmost beds," extending towards the same place, *Athyris Royssii,* with abundance of various *Producta, P. fimbriatus, P. Martini, Sp. glabra, Rhynchonella,* &c.
Fron, between Chirk and Llangollen	
Trevor, near Llangollen	
Eglwseg Ridge, near Llangollen	
Head of the Vale of Clwyd, ¼ mile S. of Llomfour Chapel	*Prod. latissimus, Rh. pleurodon,* and the usual upper fossils.
Caergwrle Hills, between Wrexham and Molef	The common fossils of the upper beds.
Abergele, North Wales...................................	*Prod. Cora, P. semireticulatus, C. papilionacea.* Best quarries are at Llysfaen.
Great Ormes Head, near Llandudno	Upper beds' usual fossils, in shale and débris of mine shafts, on surface of the cropping beds.
Slope S. and escarpment W. of telegraph	
Hafod, 2 miles W. of Corwen, N. Wales..............	An outlier. Beds correspond to those of the main belt; upper portion worked. *P. giganteus, P. latissimus, P. scabriculus, Spiriferidæ, Rhynchonella,* &c.

DERBYSHIRE.—Add to the localities given at page 248, Low Fields, near Middleton, by Yolgrave.

STAFFORDSHIRE.—Page 244, instead of seventy-eight species write eighty-five, and to the tables add *Athyris subtilita, Rhyn. Wettonensis,* Dav., *Streptorhynchus Kellii, Prod. Carringtoniana,* Dav., *Prod. Koninckiana,* De Verneuil.

GLOUCESTERSHIRE.—Add to the tables, *Chonetes Buchiana,* found by W. W. Stoddart near Bristol.

RHYNCHONELLA WETTONENSIS, *Dav.*   Plate LV, figs. 1—3.

Sp. Char.   Shell transversely oval, wider than long; ventral valve more convex than the
dorsal one, with a mesial fold of greater or lesser elevation, commencing at about the
middle of the valve, and extending to the front; beak small, angular, but slightly produced,
and incurved with a minute circular foramen, placed under its extremity.   Dorsal valve
moderately convex, with a sinus of greater or lesser depth, commencing close to the
extremity of the umbone, and extending to the front, where it attains its greatest breadth
and depth.   Each valve is ornamented with small, radiating ribs or striæ, while numerous
concentric lines of growth occur at irregular intervals on the surface of the valves.   Interior
unknown.

Average dimensions: length 8, width 11, depth 5 lines.

Obs.   This Rhynchonella is at once distinguishable from all its congeners in the
Carboniferous period on account of its peculiar shape and character, occasioned by the fold
existing in the ventral valve and the sinus in the dorsal one.   This arrangement is, however, known to exist in some Devonian, Jurassic, and Cretaceous Rhynchonellæ, but is not
of common occurrence.   In very young specimens the fold and sinus are but slightly marked,
but they become very apparent and developed with age; the striæ also in some aged
examples are hardly visible, but are well marked in the larger number of specimens.

This remarkable shell occurs by myriads in the Carboniferous Limestones at Narrowdale, in the parish of Allstonefield, not two miles from Dove, in Staffordshire, and is one
of Mr. Carrington's most interesting discoveries.[1]

PRODUCTUS CARRINGTONIANA, *Dav.*   Plate LV, fig. 5.

Sp. Char.   Shell somewhat sub-orbicular or transversely semicircular, wider than long;
hinge-line straight, and about as long as the width of the shell; ventral valve regularly
and evenly convex; beak small, hardly produced beyond the hinge-line; ears small;

---

[1] In a letter dated 11th September, 1862, Mr. Carrington writes, "I had a fortunate find yesterday,
having been quarrying for several days on a hill side to the depth of three feet, I arrived at a fault which
originally had been an open cleft in the ocean bed.   This bed had been filled with materials differing from
the sides which bounded the cleft.   The fossils were peculiar, consisting of thousands of closely packed
specimens of *Lingula mytiloides, Discina nitida, Rhyn. Wettonensis* (Dav., n. sp.), *Productus Carringtoniana* (Dav., n. sp.), and of a large variety of *Spirifera Carlukiensis* (Dav.); and it would appear to me
that the cleft had been perpendicular at the time of their existence, as it is now (?), and that the
Brachiopoda must have been attached to its sides, as the great abundance of the specimens are found in a
conglomerate of one inch or more all the way down.   The small Spirifer occurs by myriads, and there are
also some fine Pectens.   In the limestone bounding the cleft we find *Prod. giganteus, P. striatus,
Bellerophon, Nautilus,* &c."

surface entirely covered with tolerably regular, but more or less interrupted, sub-parallel, concentric ridges or wrinkles. Slender spines rise here and there from the surface of the ventral valve, and are more numerous on the ears. Dorsal valve moderately concave, and ornamented as in the opposite one.

Average dimensions: length 11, breadth 13 lines.

Obs. This *Productus* appears to be easily distinguishable from the other British Carboniferous species of the genus by its shape and sculpture, and I am supported in this view by Prof. de Koninck, to whom I submitted several specimens. None of the examples hitherto discovered by Mr. Carrington in the Narrowdale Cleft appears to have exceeded the dimensions given above, nor is it near so common as are *Rh. Wettonensis*, *Discina nitida*, and *Spirifer Carlukiensis*.

PRODUCTUS OR CHONETES COMOIDES, *Sow.* (p. 180).

While describing Sowerby's *Productus* (or *Chonetes*) *comoides*, I did not omit to refer to the several difficulties in the way of a satisfactory determination of this important species. I also mentioned that it had been questioned by certain palæontologists whether the shells (Pl. XLV, figs. 1—6) really belonged to Sowerby's species, and I gave the reasons why I had united them to *C. comoides*. Feeling, however, uncertain as to the correctness of this view, I requested Mr. D. C. Davies, of Oswestry, to kindly endeavour to search for more specimens of the Llangollen form (Pl. XLV, figs. 1—6; Pl. LV, figs. 9, 10), and especially for examples which would show the interior of the dorsal or smaller valve. Mr. Davies's success was complete in this respect, for at the close of the present summer he forwarded to me seven examples, of which one was a bivalve shell, and of which the valves could be separated (Pl. LV, figs. 9, 10). I felt much interested with this important discovery, for the interior of both valves proved in the most satisfactory manner that, notwithstanding the double area and strong articulating hinge-teeth, all the interior dispositions were those of *Productus*, and this has proved once more that *Chonetes, Aulosteges*, and *Strophalonia*, cannot be considered in any other light than sub-genera or section of *Productus*, and cannot claim generic value. Profiting by Prof. de Koninck's passage through London, I requested him to accompany me to the British Museum, in order that we might have a complete and minute examination of my Llangollen specimens with Sowerby's *Prod. comoides* and *P. hemisphæricus*.

The result of this examination was that we agreed completely that Sowerby's *P. hemis-phæricus* was nothing more than a modification of age and specimen of Martin's *Productus giganteus*, as I had already stated it to be at p. 144 of this Monograph, and that it is specifically distinct from *P. comoides* as well as from the Llangollen form. Prof. de Koninck was, moreover, of opinion that the Llangollen form and *P. comoides* should be considered as distinct species, and adduced the much larger area, but slightly produced beak beyond and above the cardinal edge, in *C. comoides*, as good specific distinctions whereby to separate it from the Llangollen shell, which possesses a very large, rounded beak and narrow area. I am not, however, quite so certain as to the absolute value of these characters in the forms under discussion; but as my distinguished friend so strongly advocates their separation, I will provisionally adhere to his view, and retain the term *comoides* for Sowerby's type, and that of *Llangollensis* for those represented in Pl. XLV, figs. 1—6; and Pl. LV, figs. 9, 10. They will be provisionally characterised in the following manner.

PRODUCTUS COMOIDES, *Sow.* Pl. XLV, fig. 7; Pl. XLVI, fig. 1; and Pl. LV, figs. 6, 7, 8.

Sp. Char. Shell large, transversely semicircular, concavo-convex; hinge-line straight, nearly as long as the width of the shell; ventral valve convex, very thick and wide; beak not protruding, or but very slightly so, beyond the level of the cardinal edge; fissure triangular and wide, partly arched over by a small pseudo-deltidium. Dorsal valve moderately concave, much thinner than the opposite one; area narrower than that of the ventral valve. Externally, both valves are covered with exceedingly fine and contiguous longitudinal striæ (four or five occupying the breadth of a single line). In the interior of the ventral valve (the only one known) and under the extremity of the beak there exists a pyriform muscular depression, which extends to about half the length of the valve (Pl. LV, fig. 8). This cavity is longitudinally divided into three almost equal portions; the central division contains, on a wide, flattened elevation, two pair of muscular impressions, situated one above the other. Those nearest to the extremity of the beak are due to the adductor (A), while the smaller, circular or oval pair (c) are supposed to have afforded another point of attachment to the same muscle. On either side of this central elevation are situated larger and deeper scars, which are due to the cardinal or divaricator muscle.

Dimensions variable; the largest specimen I have seen measured $3\frac{1}{2}$ inches in length by rather more than 6 inches in breadth, while Sowerby's typical specimen, when entire, must have measured 3 inches in length by somewhat less than $4\frac{1}{2}$ inches in width.

The grounds for locating this species in the sub-genus *Chonetes* have not been clearly made out. I will therefore leave it provisionally with *Productus*, as was originally done by Sowerby.

Sowerby states that his specimens (from which my description was taken) were from the wayboards between the limestone under the coal at Llangaveni, in Anglesea.

PRODUCTUS LLANGOLLENSIS, n. sp.? Plate XLV, figs. 1—6; Plate LV, figs. 9—10.

Sp. Char. Shell large, almost circular or transversely semicircular, concavo-convex; hinge-lines straight, usually shorter than the width of the shell. Ventral valve enormously thick and ponderous, very convex; beak rounded and much developed; area narrow, rarely exceeding a line in width; fissure triangular. Dorsal valve moderately concave; area very narrow; surface of both valves finely striated, three of these occupying the width of a line.

The ventral valve in some examples has attained nearly one inch in thickness. In the interior, on each side of the fissure, are two strong, projecting teeth, and under these commences a large, pyriform, muscular cavity, which extends to beyond half the length of the valve, and occupies about a third of its inner surface, the greatest breadth being situated towards the centre of the shell. In this depression three distinct pair of muscular impressions are visible, which are separated to a greater or lesser extent by three longitudinal ridges, the first central pair, or those situated nearest to the extremity of the beak, occupy the two sides of the central ridge, and which ridge is far more prominent or elevated than the other two; these impressions, which also extend sometimes a little beyond the limits of the central ridge, are due to the adductor (A); and still lower down, towards the centre of the valve, are two smaller, sharply defined, obliquely oval-shaped scars (c) (c in *Productus*, Pl. XXXVII, fig. 1), and which are supposed to be due to another attachment of the same muscle (?); while outside of these there exists a large, elongated impression, referable to the divaricator or cardinal muscle. It is somewhat singular that the small accessory adductor impression (c) had not been hitherto observed, either by M. de Koninck or by myself, but in the specimens recently found they are unmistakably defined.

If we now compare the interior of this valve with that of several species of *Productus*, we shall find that, excepting the dental processes, the muscular impressions would agree in all essential conditions with those of *Productus*. The relative position of the adductor and divaricator impressions varies somewhat in different species of the genus, as we have already described; thus, in *P. giganteus*, *P. longispinus*, &c., the divaricators are situated immediately *under* and outside of the adductor, while in *P. pustulosus*, *P. humerosus*, &c., the adductor is located between the two divaricator impressions, as in the case of *P. comoides*.

In the interior of the *dorsal valve*, under the large V-shaped cardinal process, there exists a longitudinal ridge, which extends to two thirds the length of the valve, being widest, rounded, and grooved near its origin under the cardinal process, but becoming narrower and more elevated towards its extremity. On either side of the cardinal process there exists depressions for the reception of the teeth of the opposite valve, while on either side of the upper part of the central ridge are situated large, wide, ramified impressions, due to the adductor muscle, while outside and in front of these are the two so-termed reniform impressions,[1] and a little under the adductor scars above mentioned may be noticed a small, conical eminence (z), which will also be seen in the interior of the same valve of *P. giganteus*.

This is the first notice that has been given of the interior of the dorsal valve of this remarkable species, and which corresponds so exactly with what we find in similar valves of other *Producta* that no further comments upon its generic claims appear to be necessary.   While in ignorance of the interior character of the dorsal valve, I had, along with other palæontologists, been led to place this shell in the sub-genus *Chonetes*, having attributed undue importance to the area and teeth, which had until then been considered peculiar to *Chonetes*, and not to *Productus ;* but since the discovery of a well-defined area in *P. sinuatus*, as well as occasional hinge-areas in *P semireticulatus*, *P. punctatus*, &c., we cannot claim the area as a permanent distinguishing feature between *Productus* and *Chonetes*.   Mr. D. C. Davies states in his paper already referred to that *P. Llangollensis* exists in great plenty in the lower beds of the Carboniferous series at the base of the cliffs near Llangollen, and that it is also found, though not so abundantly, at about the same level at Llanymynech and Porthywaen; that in the Eglwseg cliffs it lies in a shale bed, from which beautiful specimens of the interior of the ventral valve may be obtained.   It is, however, important to notice that the generality of specimens are so imperfect round the margin as to lead one to imagine that they had been drifted from some distance to the spot where they are at present found.

CHONETES CONCENTRICUS, *De Kon.*   Plate LV, fig. 13.

        CHONETES CONCENTRICA, *De Koninck.*   Monographie du genre Chonetes, p. 186, pl. xx,
                            fig. 19, 1847.

Shell marginally semicircular, about twice as wide as long, flat in the casts; hinge-line straight, angle rounded.   External surface covered with numerous sub-regular, concentric ridges (thirty-seven being present on a specimen one inch and a quarter in length). The largest specimen at present known measures $1\frac{1}{2}$ inch in length by 3 inches in width.

Obs.   In the Carboniferous Limestone at Clatteringwell Quarry, Bishop's Hill, Kenness Wood, Kinross, in Scotland, were recently found a number of external and imperfect internal impressions of a shell which neither Mr. Salter nor myself were at the time able to determine.   Upon showing the specimen here described, and belonging to the Geolo-

---

[1] Prof. Suess advocates the opinion that the reniform impressions are equivalent to the sunken oral processes of Thecidium (?).

gical Survey, to Prof. L. de Koninck, he at once appeared disposed to refer it to a large variety of *Chonetes concentrica;* but as none of the Belgian specimens from the Lower Carboniferous Limestone of Visé have attained more than about one third of the dimensions of our Scottish examples, and as the concentric ridges are so much more numerous in these last, I have adopted our friend's identification with some reserve, and especially so as our Scottish material is very imperfect, and hardly sufficient to justify a positive conclusion. *Sp. ovalis, Strep. crenistria, Prod. scabriculus,* &c., occur in the same locality.

Prof. de Koninck describes his *C. concentrica* as small, transversely semicircular; ventral valve very slightly convex, dorsal one very feebly concave, surface of each valve ornamented with from twelve to fifteen concentric ridges.

In the explanation of Pl. XLV, with figs. 1—6, write *"Productus Llangollensis"* instead of " *Chonetes comoides.*"

Before concluding this Appendix I must briefly allude to several casts and impressions of a *Chonetes* recently discovered by Mr. J. Kirkby, in a Lower *Permian* Limestone, at Hartley quarry, Sunderland, in Durham, and which will thus add another species to our list of British Permian Brachiopoda.

These casts exactly agree in shape and dimensions with Phillips's *Chonetes Hardrensis,* and to which species I should have at once referred them had I been able to satisfy myself that the external markings, striation, &c., were those of the Carboniferous shell.

These casts are smooth, and show no evidence of striation, and some examples, without considering dimensions, bear some resemblance to Schauroth's Permian *Chonetes Davidsoni.*[1]

All I can, therefore, at present say with regard to these interesting specimens is that the shell is marginally semicircular, wider than long, plano-convex; hinge-line straight, and either a little shorter or as long as the width of the shell; that each valve is provided with a sub-parallel area, but which is widest in the ventral one, and divided in the middle by a small fissure, partly arched over by a pseudo-deltidium. The ventral valve is mode-

---

[1] " Ein neuer Beitrag zur Paläontologie des deutschen Zechsteingebirges " ('Abdruck. a. d. Zeitschr. d. deutschen Geologischen Gesellschaft,' Jahrg, 1856, pl. xi, fig. 1).

Baron Schauroth's *C. Davidsoni* appears to be a small shell, not exceeding three lines in length by about four in breadth, is semicircular, but comparatively and proportionately longer than wide than is *C. Hardrensis;* concavo-convex, with six slanting spines on the cardinal edge. The surface is marked with numerous concentric lines of growth, which are crossed by some radiating striæ. It is now an undoubted fact that *Chonetes* has continued to exist since the Lower Silurian epoch up to the Permian one inclusive.

rately convex, the dorsal one gently concave, and about eight slanting spines rise from the cardinal edge.   Surface markings unknown.   Length 6, breadth 8 lines.   In Pl. LV, fig. 16, will be found a correct representation of these casts, which Mr. Kirkby found to be associated with a large variety of *Sp. Urii*, and with some Permian species belonging to other classes.   Mr. Kirkby assures me that the bed containing these fossils is truly Permian.   The Rothleigende, Marlstone, and about fifty feet of the Lower Limestone, are all to be seen in the same quarry below it.

INDEX TO THE APPENDIX.

PLATE I.

(Carboniferous Species.)

Fig.

1. Terebratula hastata, *Sowerby.* From the original example figured in the 'Mineral Conchology.' *Loc.* Limerick. Collection of J. de C. Sowerby.

2. — — The largest specimen I have seen from Millecent. Collection of Mr. Griffith.

3. — — A very elongated example from Derbyshire, also figured by Professor Phillips. British Museum.

4. — — Ventral valve. *Loc.* Kildare.

5. — — From Settle, Yorkshire. Collection of Mr. Reed.

6. — — A beautiful example from Park Hill, Longnor, Derbyshire, in which bands or stripes due to colour are still preserved. Museum of the Geological Survey.

7. — — A short, very pentagonal form. In the collection of the late Mr. Daniel Sharpe.

8. — — A young ovate example from Longnor, Derbyshire. Museum of Practical Geology.

9. — — A young specimen (enlarged), showing remains of colour; kindly presented to me by Mr. E. Wood, of Richmond.

10. — — Interior of the dorsal valve, showing that the loop was short.

11. — — A short and unusual shape from the Valley of the Maine, County Kerry. Museum of the Geological Society.

12. — — *T. virgoides,* M'Coy. From pl. iii D, fig. 23, of M'Coy's 'British Palæozoic Fossils.'

13. — — var. *ficus,* M'Coy? A very elongated variety, convex, and without mesial depression on either valve. From Park Hill, Longnor, Derbyshire. Museum of Practical Geology.

14. — — Another similar example in the collection of Professor Phillips.

15. — — *T. ficus,* M'Coy. From pl. iii D, fig. 22, of the 'British Palæozoic Fossils.' Derbyshire.

16. — — With remains of colour; from Park Hill, Longnor. Museum of Practical Geology.

17. — ? Derbyshire. Cambridge Museum.

18, 19, 20. — Gillingensis, *Dav.* From Gilling, in Yorkshire. Collection of Mr. E. Wood, of Richmond.

21, 22. ? — subtilita, *Hall.* From Mayen Wais. Collection of Professor Phillips.

23. — sacculus, *Martin.* Pet. Derb. pl. lxiv, figs. 1, 2.

24. — — A large example from Bolland. Collection of Mr. Read.

25, 26, 28, 31, 32. Terebratula vesicularis, *De Koninck.* Yorkshire. Collection of Mr. E. Wood.

27, 29, 30. — sacculus, *Martin.* Park Hill, Longnor, Derbyshire.

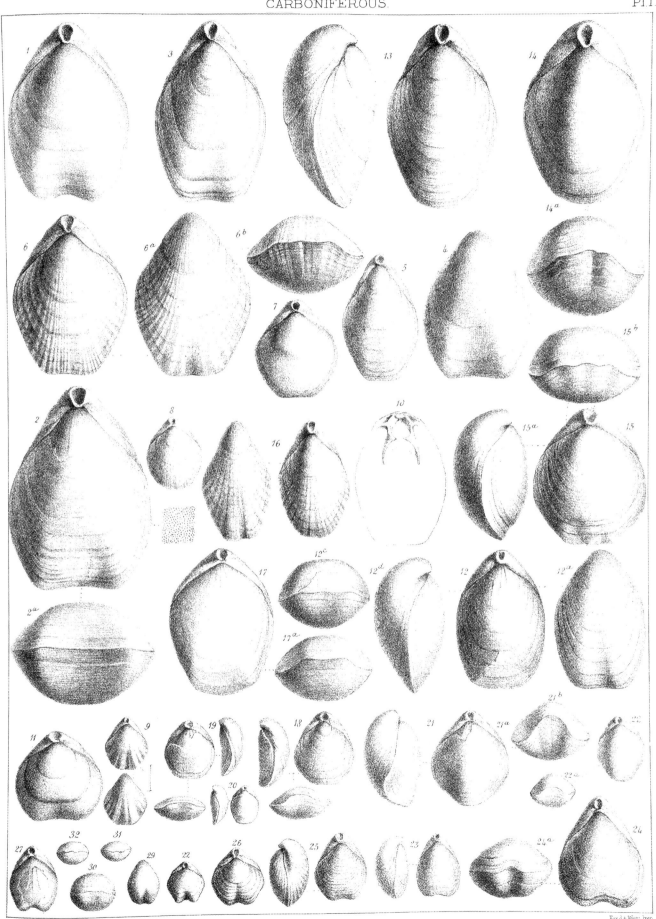

PLATE II.

(Carboniferous Species.)

Fig.

1—8. Terebratula vesicularis, *De Koninck*. Different varieties of shape. Figs. 1, 2, 3, 4, 6 from the Yoredale Rocks, Wensleydale, Yorkshire, and collection of Mr. E. Wood. Figs. 5, 8 from Pilsbury Castle, Longnor, Derbyshire. Museum of the Geological Survey. Fig. 7 enlarged from a specimen in the Cambridge Museum, erroneously identified with *Ter. seminula* of Phillips by Professor M'Coy.

9, 10, 11. Spirifera duplicicosta, *Phillips*. Young individuals? according to Professor Phillips and De Koninck, from Thorpe, Cloud Hill, Dovedale, and Park Hill, Longnor; but these very transverse specimens lead me to doubt the identification, and perhaps they should be considered the young of *Sp. striata*.

12, 13. — striata, *Martin*, var. *attenuata*, Sowerby. These two examples are drawn from the original specimens figured in the 'Mineral Conchology' as *Sp. attenuata*. Kildare, Ireland.

14. — — Millecent. Collection of the late Mr. Daniel Sharpe.

15, 16. — — Var. with very small ribs. From Dovedale, Derbyshire. Museum of Practical Geology.

17. — — From the Craven district, Yorkshire, and collection of Mr. E. Wood.

18. — — A remarkable variety, with very deep sinus, and produced mesial fold. Bolland, in the collection of Mr. J. de C. Sowerby. Another similar example may be seen in the British Museum.

19. — — Interior of the dorsal valve, with perfect spiral appendages.

20. — — This illustration is taken from a beautifully perfect specimen in the Cambridge Museum, in which the matrix enveloping the spirals had been carefully removed by the means of acid; the shell and its spiral lamellæ being silicified. The spirals are here represented, as seen by removing a portion of the dorsal valve; so that the back and front view of these appendages are completely illustrated in figs. 19 and 20.

21. — — Interior of the ventral valve, to show the muscular impressions, &c. From a specimen in the British Museum.

Pl. II.

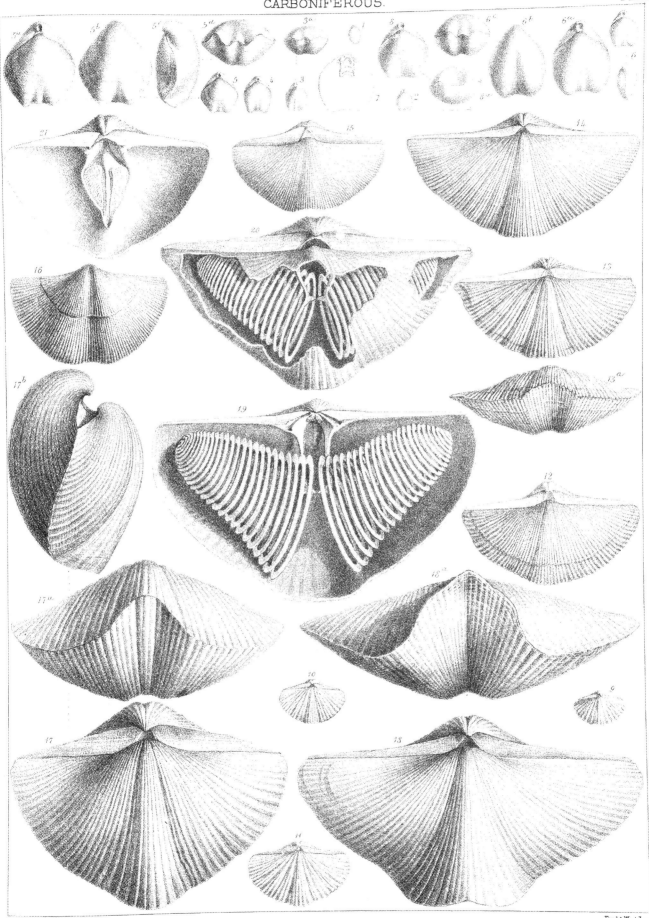

PLATE III.

(Carboniferous Species.)

Fig.

1. TEREBRATULA GILLINGENSIS, *Dav.* From Westlothian. Collection of Dr. Fleming.

2. SPIRIFERA STRIATA, *Bolland.* Gilbertson's collection in British Museum. This is the largest example of the species at present known in England, and is the *Sp. princeps* of M'Coy.

3. — — var. *attenuata.* From Millecent.

4. — — From a specimen in grit, near Richmond, Yorkshire. Collection of Mr. E. Wood.

5. — — From Ireland, Museum of Practical Geology. Mr. Kelly informs me that similar specimens are not uncommon at Cornacarron, near Inniskillen.

6. — — This figure is taken from the 'Synopsis of Carb. Fossils of Ireland,' pl. xix, fig. 9, and represents M'Coy's *Sp. clatharata.*

7. — DUPLICICOSTA, *Phillips.* From the Wensleydale district, Yorkshire, and collection of Mr. E. Wood.

8. — — This figure is taken from Phillips's 'Geology of Yorkshire,' vol. ii, pl. x, fig. 1, and is the author's type.

9, 10. — — Two typical shapes; fig. 9 from Park Hill, Longnor, Derbyshire; fig. 10 from Poolvash, Isle of Man: Museum of Practical Geology.

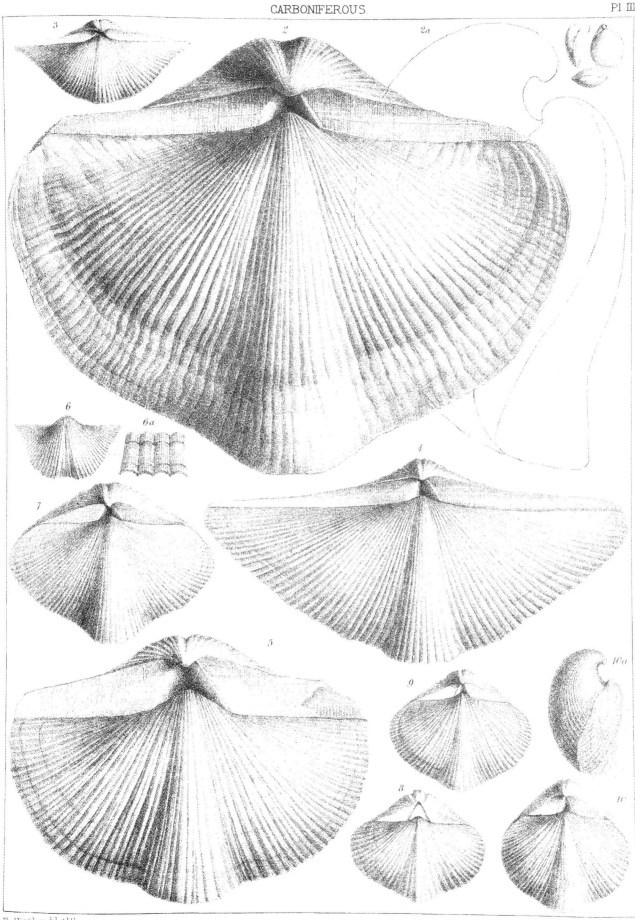

Tho.ˢ Davidson, del. et lith.

Ford & West, Imp.

PLATE IV.

(Carboniferous Species.)

Fig.

1. Spirifera bisulcata, var., *Sowerby*. Bolland. Collection of M. De Koninck.

2. — transiens, *M'Coy*. Taken from the 'Synopsis of Carboniferous Fossils of Ireland,' tab. xix, fig. 14.

3. — duplicicosta, *Phillips*. This figure is taken from the 'Synopsis of Carboniferous Fossils of Ireland,' pl. xxii, fig. 12, where it is described under the name of *Sp. furcata*, M'Coy, but is evidently the same as *duplicicosta*.

4. — — A var. with wide ribs, from Park Hill, Longnor; in the Museum of Practical Geology.

5—11. — — Varieties of age, as admitted by Professor Phillips and M. De Koninck. Fig. 5 is from Bolland, collection of Professor Phillips; 6, 7, 9, 10 from Park Hill, Longnor, Derbyshire, Museum of Practical Geology; fig. 8, a very elongated shape from Bolland; fig. 11 is what Professor M'Coy has described and illustrated as *Sp. fasciculata* in tab. iii D, fig. 25, of the 'British Palæozoic Fossils.'

12. — penguis, *Sowerby?* *Sp. paucicostata*, M'Coy. These figures are taken from the 'British Palæozoic Fossils,' pl. iii D, fig. 26. Derbyshire; in the Museum of Cambridge. It is probably a young shell of Sowerby's species.

13. — mosquensis, *Fischer*. Derbyshire. Cambridge Museum.

14. — — From Little Ireland (Ireland). Collection of the Royal Dublin Society. (See also pl. ix, fig. 4.)

15. — humerosa, *Phillips*. This illustration is drawn from the original specimen on which Professor Phillips founded his species. 'Geology of Yorkshire,' vol. ii, pl. xi, fig. 8; and is from Greenhow Hill. Collection of Professor Phillips.

16. — — A perfect example, in the collection of Mr. E. Wood, from the Wensleydale district, Yorkshire. (See also pl. ix, fig. 5.)

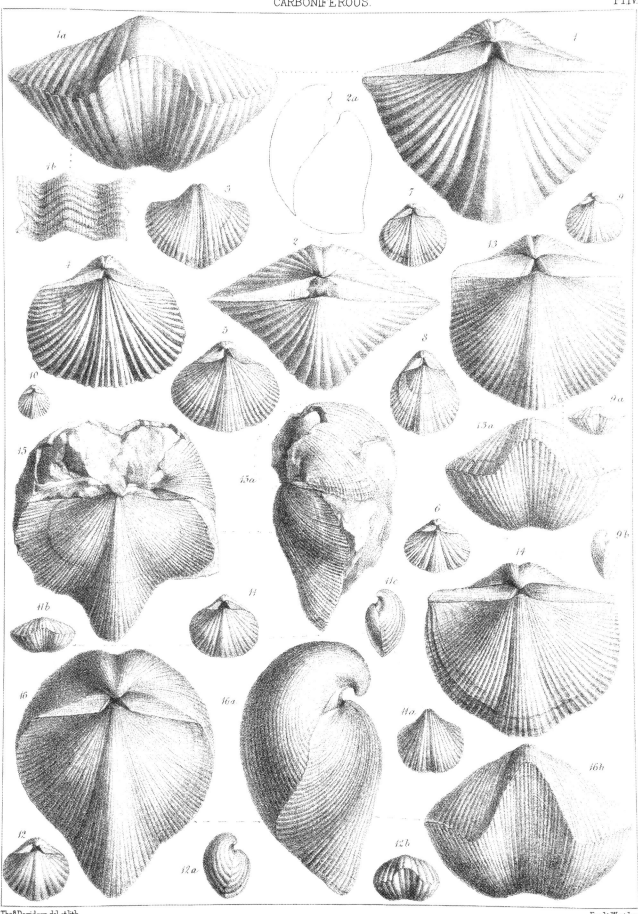

Thos Davidson,del.et lith. Ford&West,Imp.

PLATE. V.

1. Spirifera bisulcata, *Sowerby*. Yorkshire. Collection of Mr. Reed.

2, 3.   —   convoluta, var. *rhomboidea*, Phillips. Drawn from the original specimen figured in the 'Geology of Yorkshire,' vol. ii, tab. ix, figs. 8 and 9.

4—6. —   —   From Lancashire. Collection of Mr. Parker, of Manchester.

7, 8. —   —   From Bolland. Collection of Professor Phillips.

9—11. —   —   *Phillips*. From the original example; 'Geology of Yorkshire,' vol. ii, pl. ix, fig. 7. Bolland. British Museum.

12, 13. —   —   From Clitheroe quarry, Lancashire. Collection of Mr. Parker, of Manchester.

14, 15. —   —   From Kildare. Museum of the Royal Dublin Society.

16, 17. Spirifera triangularis, *Martin*. The original example figured in the 'Pet. Derb.,' pl. xxxvi, fig. 2; also 'Min. Con.,' tab. 562, figs. 5 and 6. From Derbyshire. Collection of Mr. J. de C. Sowerby.

18—21. —   —   A perfect individual. From Yorkshire. Collection of Mr. Reed, of York.

22.   —   —   An imperfect example. From the Wensleydale district, Yorkshire.

23. Spirifera triangularis = sp. *ornithorhyncha*, M'Coy. This is one of the original examples upon which Professor M'Coy founded his species. *Loc.* Millecent (Clare). Collection of Mr. Griffith.

24. —   —   From Bolland; in the collection of Mr. Reed.

25. —   trigonalis, *Martin*. After the original figure, 'Pet. Derb.,' tab. xxxvi, fig. 1. Derbyshire.

26. —   —   Cast of the interior of the dorsal valve, with beak of the ventral one; from Bakewell, Derbyshire, presented to me by Mr. Binny. A. Adductor impressions.

27. —   —   Cast of the interior of the ventral valve. A. Adductor. R. Cardinal muscular impressions. o. Ovarian spaces.

28. —   —   From Buxton.

29—32. —   From a beautiful example in the collection of Mr. Tate. Denwick, Northumberland.

33. —   —   From Courland, near Dalkeith.

34. —   —   Interior, showing the spirals; from the original example figured in the ' Min. Con.,' tab. 265, fig. 1, under the erroneous denomination of *Sp. bisulcatus.* Derbyshire.

35—37.   ?   A remarkable and unusual form; from Campsie, near Glasgow; in the collection of Mr. A. Cowan.

38, 39. Spirifera grandicostata, *M'Coy*. After the author's illustrations in the 'British Palæozoic Fossils, pl. iiiᴅ, fig. 29.

40—42. Spirifera Reedii, *Davidson*. British Museum.

43—48. —   —   Two young examples; from Settle, Yorkshire. Collection of Mr. Reed.

Thos Davidson del. et lith W.West Imp.

PLATE VI.

(Carboniferous Species.)

Fig.

1. Spirifera bisulcata, *Sowerby*, var. *semicircularis*, Phillips. From the original illustration, 'Geol. of Yorkshire,' vol. ii, tab. ix, fig. 15. Bolland.

2. — — var. *semicircularis*. Bolland. Collection of Mr. Muschen.

3—4. — — In the collection of Professor Phillips, labelled *Sp. semicircularis*. Bolland.

5. — — From Twiston, Lancashire. Collection of Mr. Parker.

6—9. — — From the original type specimen, 'Min. Con.,' tab. 494, fig. 1. Collection of Mr. J. de C. Sowerby.

10. — — A very large example from Dovedale, Derbyshire; in the Museum of Practical Geology.

11. — — A specimen with very large ribs, from the Wensleydale district, Yorkshire. Collection of Mr. E. Wood.

12. — — From Derbyshire; several of the ribs are bifurcated.

13, 14. — — From the specimen, 'Min. Con.,' tab. 265, fig. 3, erroneously described by Sowerby as *Sp. trigonalis*.

15. — — From Craigenglen (Campsie), Scotland. Collection of Mr. Young.

16. — — Ventral valve, from Barrhead, Scotland. Collection of Mr. A. Cowan. The fine imbricated lines which cover the costæ are beautifully seen in this example.

17. — — From a specimen in the collection of Mr. J. de C. Sowerby. Derbyshire.

18. — — A very transverse young specimen, from the Wensleydale district, Yorkshire. Collection of Mr. E. Wood, of Richmond.

19. — — A very transverse specimen, in the collection of Dr. Reed, of York. Yorkshire.

20—22. Spirifera crassa, *De Koninck* = *Brachythyris planicosta*, M'Coy. This is the original specimen, imperfectly illustrated in the 'Synopsis of the Carboniferous Fossils of Ireland,' tab. xxi, fig. 5. *Loc.* Mullaghfin, Duluk, Ireland.

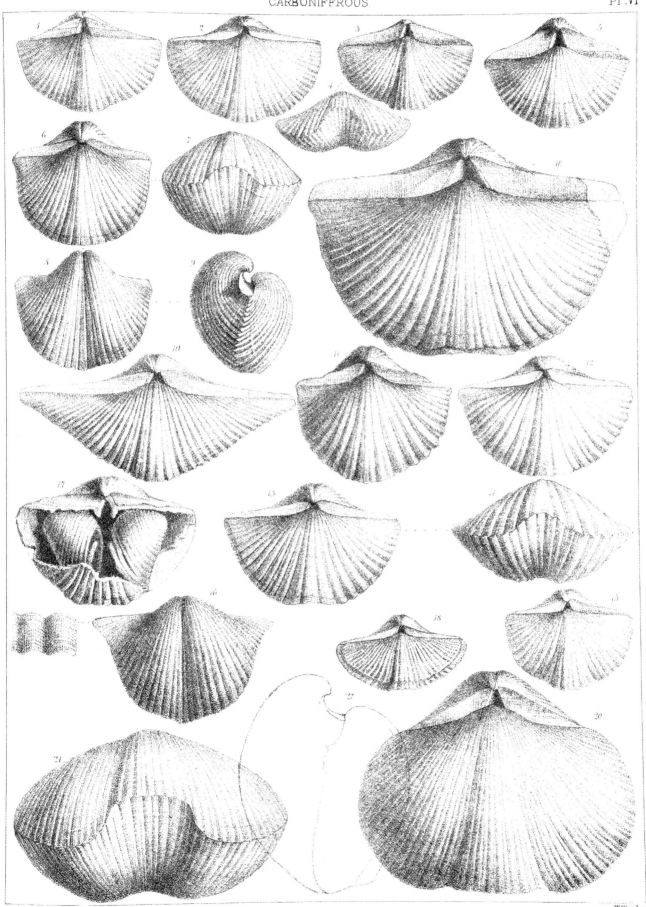

PLATE VII.

(Carboniferous Species.)

Fig.

1, 2. Spirifera crassa ? *De Koninck.* Millecent, Ireland.

3. — — In this specimen no mesial fold is observable. From the lower limestone of Milverton, Skerrie, Ireland. Collection of Mr. Griffith.

4. — bisulcata. A malformation. This illustration is taken from the specimen named *Sp. calcarata* by M'Coy. 'Synopsis of the Carboniferous Fossils of Ireland,' pl. xxi, fig. 3.

5, 6. Spirifera acuta, *Martin.* From the 'Pet. Derb.,' tab. xlix, figs. 15, 16. *Loc.* Winster, Derbyshire.

7. Spirifera grandicostata, *M'Coy.* Bolland. British Museum.

8—11. — — A beautiful example from Bolland. Collection of Mr. Muschen.

12. — — Front view of a specimen from Park Hill, Longnor, Derbyshire. Museum of Practical Geology.

13. — — A very large ribbed specimen, from Bolland. Collection of Mr. E. Wood.

14, 15, 16. — A young shell, from Park Hill, Longnor. Museum of Practical Geology. Young shells. 15. Collection of Mr. E. Wood.

17. Spirifera laminosa, *M'Coy.* Drawn from two examples in the collection of Mr. Griffith. Hook, Ireland.

18, 19. — — From Hook. Museum of the Geological Survey of Ireland.

20. — — A young shell from Hook.

21, 22. — — A var. from Derbyshire. Cambridge Museum.

23. Spirifera decemcostata, *M'Coy.* From the original specimen. 'Synopsis of the Carboniferous Fossils of Ireland,' pl. xxii, fig. 9. Millecent; Kildare. Collection of Mr. Griffith.

24. — mesogonia, *M'Coy.* 'Synopsis of the Carboniferous Fossils of Ireland,' tab. xxii, fig. 13. Ireland.

25—28. Spirifera planata, *Phillips.* Bolland. Collection of Mr. Muschen.

29. — — Professor Phillips's type. British Museum. Bolland.

30—32. — — Collection of Professor Phillips. Bolland.

33, 34. — — A circular specimen in the collection of Mr. Muschen.

35. — — Collection of Professor Phillips. Bolland.

36. — — Young shell in the collection of Mr. Muschen.

37—41. Spiriferina cristata, var. *octoplicata*, Sowerby. From the original examples. 'Min. Con.,' tab. 562, figs. 2 and 3. Derbyshire, and collection of Mr. J. de C. Sowerby.

42—44. — — From the shores of Lough Hill, County Sligo. Collection of the Geological Society. Fig. 44 enlarged, to show the cardinal area, cardinal process, and central septum of the ventral valve.

45—46. — — From Westlothian, Scotland. Collection of Dr. Fleming.

47. — — From the shores of Lough Hill, County of Sligo. Collection of the Geological Society.

48, 49, 50. — insculpta, *Phillips.* Young shells from Longnor, Derbyshire. Museum of Practical Geology.

51. — — From the specimen named *Sp. quinquelobas*, M'Coy. 'Synopsis,' tab. xxii, fig. 7. From Ardagh, Drumcondra, Ireland. Collection of Mr. Griffith.

52—55. — — Adult. Yorkshire. Museum of Practical Geology.

56—59. — minima, *Sowerby.* From the two figured types. 'Min. Con.,' tab. 377. Bakewell. Collection of Mr. J. de C. Sowerby.

60—61. — partita, *Portlock.* Copies from the 'Rep. of the Geol. of Londonderry, &c.,' tab. xxxviii, fig. 3. Ireland.

PLATE VIII.

(Carboniferous Species.)

[1] At p. 45, I stated that no specimen of *Sp. cuspidata* I had hitherto been able to examine possessed its deltidium, and that I considered it was in all probability not perforated by a circular foramen, as in true types of the sub-genus *Cyrtia*.   Subsequently, however, Mr. J. P. Woodward showed me the internal cast of the ventral valve of a specimen in the British Museum, thought to have belonged to *Sp. cuspidata*, and derived from the dolomitic Carboniferous limestone of Breedon hill (pl. ix, figs. 1 and 1 *a*) in which there is evidence that the deltidium was in reality perforated by a circular foramen, as in *Cyrtia*.

Pl VIII

Tho.⁹ Davidson del et lith.

W West Imp

PLATE IX.

(Carboniferous Species.)

1. Spirifera cuspidata (?), *Martin*. Internal cast of a ventral valve. This specimen denotes the existence of a tubular perforation (*f*) in the deltidium of the larger valve; from the Magnesian Limestone of the Carboniferous period of Breedon Hill. British Museum. Fig. 1*a* is from a gutta percha impression from the same, to show the position of the dental or rostral plates.

2. — — Bolland. Collection of Professor Phillips.

3. — subconica, *Martin*, sp. From the 'Petrif. Derb.,' tab. xlvii, figs. 6, 7, 8. Middleton, Derbyshire.

4, 5, 6. — triradialis, var. *sexradialis*, Phillips. Bolland. British Museum.

7. — — A large example. From the same locality and collection.

8. — — Ibid. Collection of Professor Phillips.

9. — — Ibid. Museum of Practical Geology.

10. — — Ibid. Collection of Professor Phillips.

11. — — Ibid. Collection of Mr. E. Wood.

12. — — var. *trisulcosa*, Phillips. 'Geol. of York.,' vol. ii, pl. x, fig. 6.

13—18. — integricosta, *Phillips*. Craven district. Collection of Mr. E. Wood.

19. — — A young shell, closely resembling Martin's representation of *Anomites rotundatus*.

20. — ovalis, *Phillips*. A very large and fine example from the Wensleydale district. Collection of Mr. E. Wood. Another almost similar is to be seen in the British Museum.

21. — — From Derbyshire. Cambridge Museum.

22. — — An imperfect example. From Corieburn, Campsie, Sterlingshire.

23. — — A young shell. From Bolland.

24. — — ? *exarata*, Fleming. From the original fragment in the collection of the late Dr. Fleming. West Lothian, Scotland.

25, 26. — — = *hemisphærica*, M'Coy. Craven district. Fig. 25 from the collection of Mr. Reed, of York.

27. — integricosta ? *Anomites rotundatus*, Martin. From the original figure in the 'Petrif. Derb.,' tab. 48, fig. 11, 12. Middleton.

Tho.ˢ Davidson del et lith.   W. West Imp.

PLATE X.

1—7. Spirifera pinguis, *Sowerby*. Different specimens and ages. From Millecent, Ireland.

8—12. — — var. *rotundata*, Sow. (not Martin) = *subrotundata*, M'Coy. From Millecent, Ireland, fig. 9. From Yorkshire.

Pl. X

Tho.ˢ Davidson del et lith.

W. West Imp.

PLATE XI.

(Carboniferous Species.)

1—2. Spirifera glabra, *Martin*. Very large specimen. From the Wensleydale district, in the collection of Mr. E. Wood.

3. — — From Millecent, Ireland.

4, 5. — — Young shells. From Bolland.

6. — — var. *symmetrica*, Phillips.

7. — — From Hill-head, Carluke parish, Lanarkshire, Scotland.

8. — — = var. *oblata*, Sow. From the original specimen in the collection of Mr. J. de C. Sowerby.

9. — — Fragment, showing the dimensions of the spiral appendages. From the original specimen represented under another aspect in the 'Min. Con.,' tab. 268, fig. 1.

10. — lineata, *Martin* = Sp. *mesoloba*, Phillips. This illustration is copied from pl. x, fig. 14, in the 'Geol. of Yorkshire.'

Tho.ˢ Davidson del et lith.

W. West Imp.

PLATE XII.

(Carboniferous Species.)

1, 2. Spirifera glabra, *Martin.*   Unusual varieties from Bolland.   In the collections of Messrs. J. de C. Sowerby and Muschen.

3.   —   —   Var. with obscurely marked ribs on the lateral portions of the shell.   Bolland.

4, 5.   —   —   Var. *linguifera,* Phillips.   Fig. 4, from the original specimen represented in the 'Geology of Yorkshire,' and now in the British Museum; 5, from the cabinet of Professor Phillips.

6, 7.   —   rhomboidea, *M'Coy.*   Fig. 6, from the original specimen figured in the 'Synopsis,' coll. of Mr. Griffith; 7, from Millecent, Ireland.

8, 9.   —   glabra, ? Var. from Lowick.   Fig. 9, in the collection of Mr. Tate, of Alnwick.

10.   —   ?   From Yorkshire, in the cabinet of Mr. E. Wood.   Figs. 8, 9, 10, are very unusual shapes, and it is with considerable hesitation that they are referred to.   *Sp. Glabra,* of Martin.

11, 12. —   —   Var. *decora,* Phillips.   Fig. 11 is taken from the representation published in the 'Geology of Yorkshire;' 12, from a very fine specimen from Bolland, in the collection of Mr. Muschen, of Birmingham, and every variation in shape may be found connecting this specimen to the normal shape of *Sp. glabra.*

13, 14. —   urii, *Fleming.*   From the carboniferous shales of Carluke, Lanarkshire.   14 *a, b, c, d, e,* are enlarged representations.

15, 16. —   stringocephaloides, *M'Coy.*   Fig. 15, from the representation published in the 'Synopsis;' 16, from Leighlin, Ireland, in the Museum of the Royal Dublin Society.   It is probable that these specimens belong to exceptional varieties of *Sp. lineata,* Martin.

Pl XII.

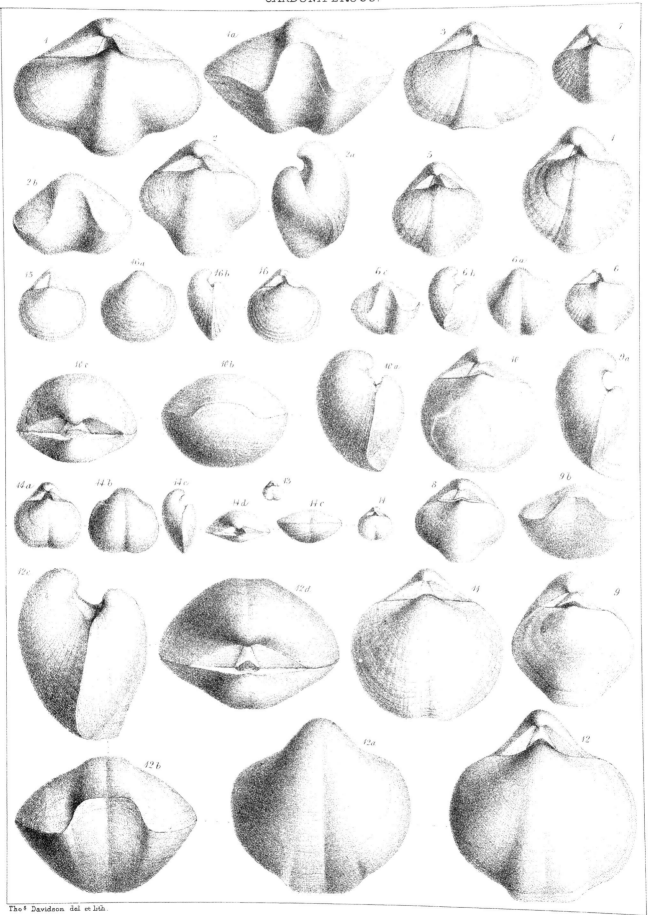

PLATE XIII.

(Carboniferous Species.)

1, 2, 3. Spirifera elliptica, *Phillips*. Ireland. Fig. 1, from the original specimen in the British Museum; 2, from the collection of Mr. Rose, of Edinburgh; 3, a young shell from Millecent.

4. — lineata, *Martin*. From the original representation in the 'Petrificata Derbiensia.'

5. — — From Dryden, near Edinburgh.

6—10. — — Different ages. Fig. 6, from Millecent; 7, Lowick; 8, a specimen very finely reticulated, Craven district; 9, a very large example from Dovedale, Derbyshire, Museum of Practical Geology; 10, profile view of an aged individual from Bolland, collection of Mr. Muschen.

11—12. — — var. *reticulata*, Sow. Fig. 11, from Millecent, Ireland; 12, Dovedale, Derbyshire, Museum of Practical Geology.

13. — — var. *reticulata*, M'Coy. From the original figures in the 'Synopsis.'

14. — carlukiensis, *Dav.* From the Carboniferous shales of Carluke, Lanarkshire. 14, natural size; 14 *a*, *b*, *c*, enlarged.

15. — fusiformis, *Phillips*. From the original specimen in the British Museum. Bolland.

16. — mosquensis, *Fischer*. A fine specimen. From Ragreagh, Limerick. in the Museum of Geological Survey of Ireland.

Pl. XIII.

Tho.ˢ Davidson del et lith.

W. West Imp.

PLATE XIV.

(Carboniferous Species.)

1—5. Cyrtina septosa, sp. *Phillips.* From Park Hill, Longnor, Derbyshire. In the Museum of Practical Geology.

6. — — Interior of the ventral valve.

7. — — Specimen from which a portion of the shell has been removed so as to expose the combined mesial septa (s) and dental plates (r) seen in profile.

8. — — Fragments of the ventral valve of a specimen from Ribble Head. In the collection of Professor Phillips. To show the position of the conjoined septa (s), and dental plates (r).

9. — — Another fragment of the same valve in the cabinet of Mr. Reed, of York, (s) conjoined septa.

10. — — An internal cast of the ventral valve in red dolomitic limestone. From Ashby de la Zouch. Museum of Practical Geology.

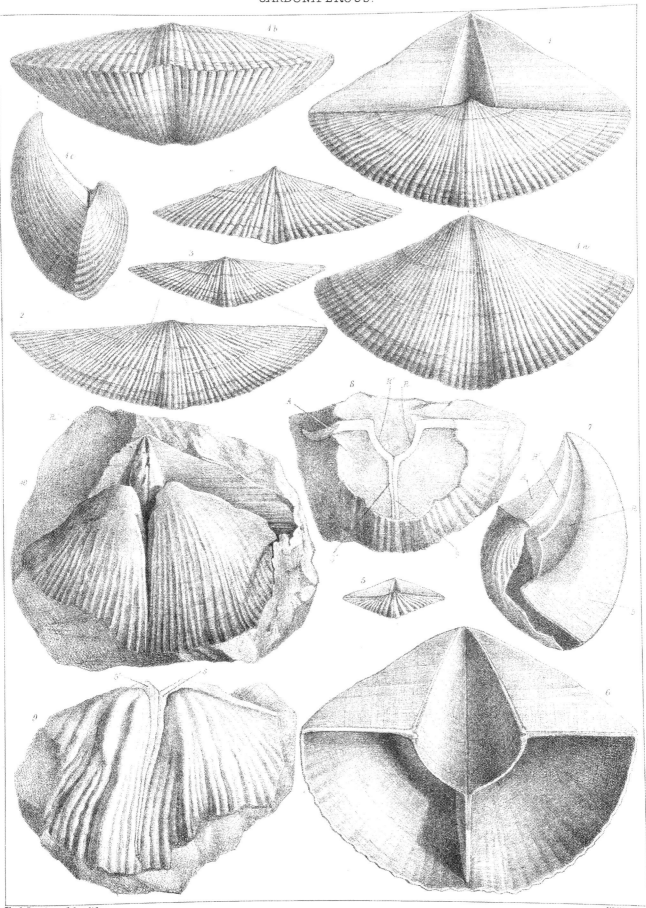

Tho.ˢ Davidson del et lith.   W. West Imp.

PLATE XV.

(Carboniferous Species.)

1. Cyrtina septosa, *Phillips*.   From the original representation in the 'Geology of Yorkshire.'

2. — — A fragment of ventral valve from Visé, Belgium, in the collection of Professor L. de Koninck, to show the inequality in width of the ribs.

3. — Dorsata, *M'Coy*.   From the original figure in the 'Synopsis.'

4. — — These figures are partly restored from a fragmentary specimen from Cork, Ireland, in the collection of Mr. Griffith.

5. ? Carbonarius, *M'Coy*.   From the figures in the work on 'British Palæozoic Fossils.'

6. — — A specimen from Kendal, in the Museum of Practical Geology. A portion of the shell of the dorsal valve being removed, a single median slit is observable in the cast, and not two, as would be the case in a true *Pentamerus*.

7, 10. — — Different specimens from Kendal.

11, 12. — — Interior of the ventral valve; 12, section of the same. Kendal.

13, 14. — — Two sections, one horizontal the other transversal, published by Professor M'Coy, in his work on 'British Palæozoic Fossils.'

15—26. Athyris ambigua, *Sow*.   Fig. 15 from Bolland, British Museum; 16, 17, in the collection of Professor Phillips; 18—24, from the Carboniferous shales of Carluke, Lanarkshire; 25 = *Ter. pentaëdra*, Phillips; 26 = *A. trilobata*, M'Coy, from the figure in the 'Synopsis.'

27, 28. ? Gregaria, *M'Coy*. · Fig. 27, from the figure in the 'Synopsis.' Ireland; 28, another specimen in the cabinet of Mr. Griffith.

Pl. XV

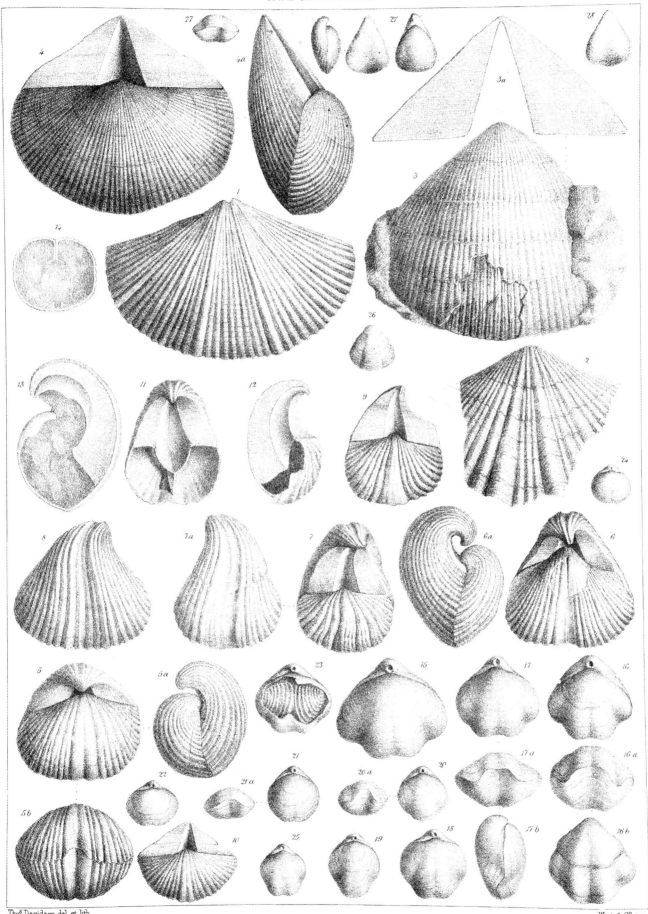

Thos Davidson del. et lith.

West & Co. imp.

PLATE XVI.

(Carboniferous Species.)

1. Athyris lamellosa, *L'Eveillé = squamosa, Phillips,* from Dovedale, Derbyshire. Museum of Practical Geology.

2. — planosulcata, *Phillips,* from the original illustration in the 'Geology of Yorkshire.'

3. — — Ventral valve with marginal expansions, entire. From Preston. British Museum.

4, 5, 6.— — Specimens from Bolland without their expansions, fig. 4. Collection of Mr. E. Wood. Fig. 5, from Longnor, Derbyshire. Museum of Practical Geology.

7, 8, 9.— — Specimens with portions of their expansions, 7, showing the spiral appendages and marginal lamellæ, Derbyshire. Fig. 9, from Longnor. Museum of Practical Geology.

10, 11. — — From Ireland. Fig. 11, is one of Professor M'Coy's representations of his *Antinoconchus paradoxus,* (fig. 6, of the 'Synopsis.')

12. — — var. *obtusa,* M'Coy. From the figure in the 'Synopsis.'

13. — — var. *oblonga,* Sow. From the figure in the 'Mineral Conchology.'

15. — — From Lanarkshire, Scotland.

14, 16, 18. — expansa, *Phillips.* From Kendal. Figs. 14, 16, 17, from the collection of the late Mr. D. Sharpe; now forming part of the Museum of the Geological Society. Fig. 18, in the cabinet of Mr. Reed, of York.

Thos Davidson del et lith.

W.West imp.

PLATE XVII.

1, 2. ATHYRIS EXPANSA, *Phillips*. Two remarkable malformations from the Carboniferous limestone of Hittor Hill and Longnor, in Derbyshire. Collection of the School of Mines.

3, 4. — — A regular and typical specimen from Settle, in Yorkshire, collection of Mr. Burrow.

5. — — A malformation from the same locality, from the same collection as 1 and 2.

6, 7. — LAMELLOSA, *L'Eveillé*. From the Carboniferous limestone of Settle. In the collection of Mr. Burrow. These specimens are remarkable, as they show that in the dorsal valve there existed first a sinus, which soon became converted into a mesial fold.

8. — SUBTILITA, *Hall*. From near Bolland. British Museum.

9, 10. — — From the Carboniferous limestone of Kendal, in Westmoreland.

11. — AMBIGUA, *Sowerby*. A very large example from Ireland.

12. — — Interior of the dorsal valve (enlarged), from Bakewell. This figure has been completed from several specimens in the Museum of the School of Mines.

13. — — Ventral valve (enlarged) from silicified internal casts, from Bakewell in Derbyshire, and Museum of the School of Mines. A, ADDUCTOR or OCCLUSOR, R, DIVARICATOR, muscular impressions.

14. — — Dorsal valve of the same specimen as 13. A A, quadruple impression of the adductor or occlusor muscle.

15. — GLOBULARIS, *Phillips*. From the Carboniferous limestone of Bolland, and Phillips's original specimen in the British Museum.

16, 17, 18. — — From Settle in Yorkshire, collection of Mr. Burrow.

19. RETZIA RADIALIS, *Phillips*. From the original specimen in the British Museum. Bolland.

20, 21. — — From Settle, Yorkshire. Collection of Mr. Burrow.

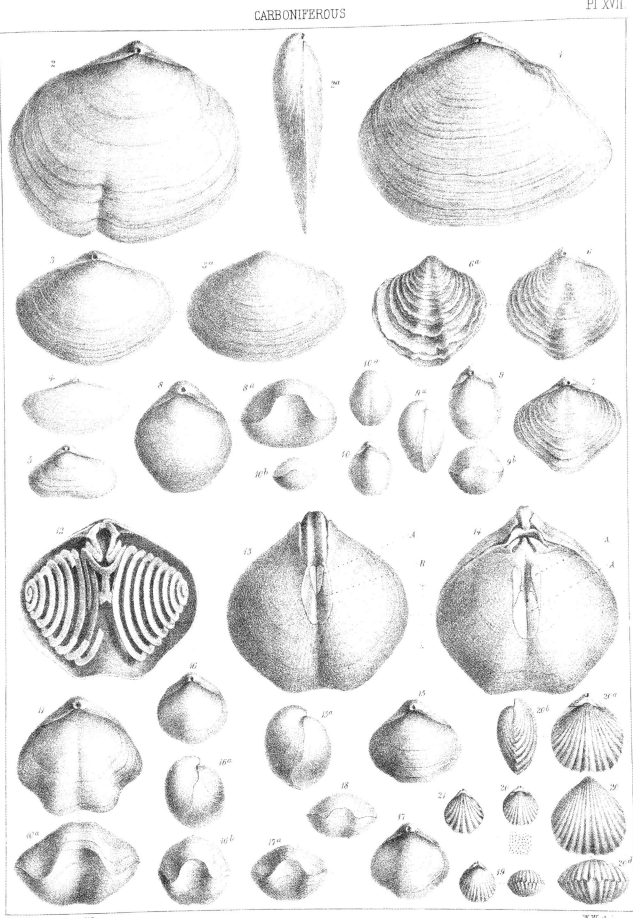

PLATE XVIII.

1. ATHYRIS ROYSSII, *L'Eveillé.* This figure represents the original example of Phillips's *Spirifera glabristria*, from the Carboniferous limestone of Bolland. British Museum. The outer layer of the shell, with its spiny investment, is absent.

3—7. — — From Millecent, Ireland. The outer layer of the shell is absent.

8—10. — — Three specimens showing a portion of the spiny investment, from Carboniferous shales. Ulverstone, fig. 11, being an enlarged representation, to better explain the character of the spines.

11. — — This is drawn from Phillips's original example of *Sp. fimbriata*, and was communicated by the author.

12. — SQUAMIGERA, *De Koninck.* From the Carboniferous limestone of Millecent in Ireland. I have seen many examples of this shell from both Ireland and England; and Prof. de Koninck, to whom I communicated specimens, believes that it might perhaps be referable to his species. The outer or reticulated surface of the shell is, however, absent.

13. — — From the Carboniferous limestone of Ballina, County Dungarvan, Ireland. In the collection of Sir R. Griffith. This is the specimen which Prof. M'Coy described (in his 'Synopsis'), as the *Martinia phalæna* of Phillips, but beside not being a spirifer, it does not belong to Phillips's Devonian shell. 13 shows a small portion of the imbricated surface.

14. RETZIA ULSTRIX, *De Koninck.* From the Carboniferous limestone of Bolland. British Museum.

15. — — From the Carboniferous limestone of Wetton, in Derbyshire, and collection of the School of Mines. This shell has been identified by Prof. de Koninck as belonging to his species.

Pl. XVIII.

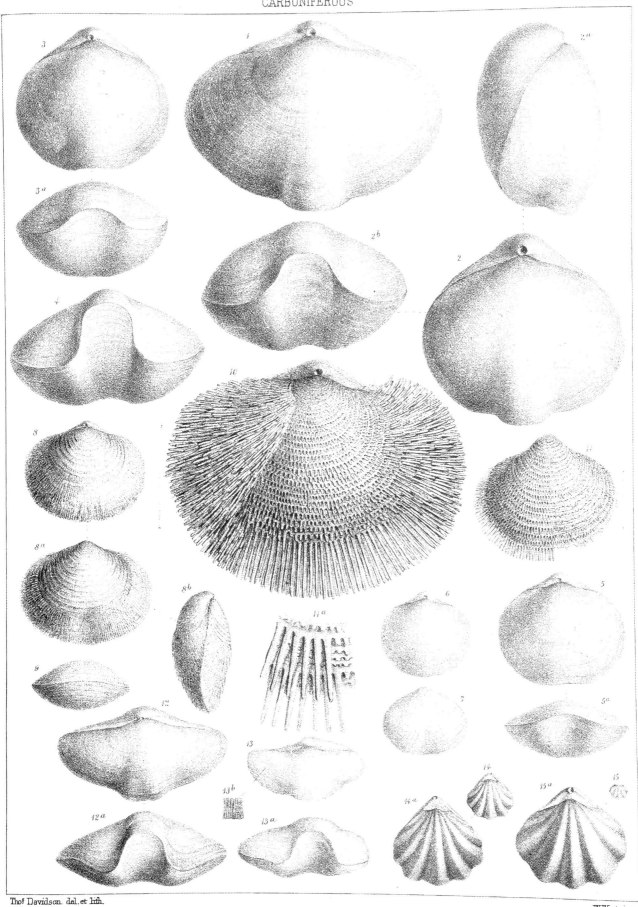

PLATE XIX.

1. RHYNCHONELLA RENIFORMIS, *Sowerby*. A large example from the Carboniferous limestone of Kildare in Ireland, with three ribs on the mesial fold.

2. — — In the collection of Mr. E. Wood, with four ribs on the mesial fold.

3, 7. — — Different specimens from Millecent, in Ireland; fig. 4 alone being from Castleton, in Derbyshire, and collection of Dr. Bowerbank.

8. — CORDIFORMIS, *Sowerby*. From the original example in the collection of Mr. J. de C. Sowerby.

9, 10. — ? From the Carboniferous limestone of Kildare, in Ireland.

11, 13, 14, 15. } — ANGULATA, *Linnæus*, sp. From the Carboniferous limestone of the Isle of Man, in the collection of the Rev. J. G. Cumming.

12. — — From Yorkshire, in the collection of the School of Mines.

16. — — From Settle, in Yorkshire, in the collection of Mr. Burrow.

N.B. This series illustrates the variations in number of the ribs.

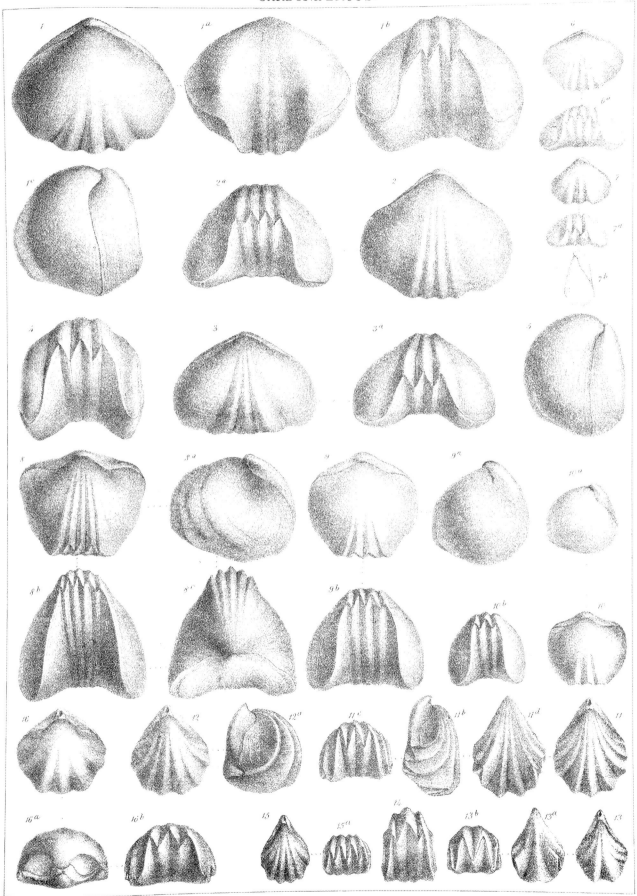

Thos Davidson del & lith.

W West imp

PLATE XX.

1, 2. Rhynchonella acuminata, *Martin*, sp. Typical shape. Carboniferous limestone, Clitheroe, Lancashire. Gilbertsonian Collection. British Museum.

3. — — From the Isle of Man, and collection of the Rev. J. G. Cumming.

4. — — From Park Hill, Longnor, Derbyshire, and collection of the School of Mines.

5, 7, 8. — — From Settle, in Yorkshire. Collection of Mr. Burrow.

6. — — Interior of the dorsal valve.

9, 10. — — Isle of Man. Collection of the Rev. J. G. Cumming.

11. — — From Yorkshire. Collection of Mr. E. Wood.

13. — — From Kildare, in the collection of Professor Phillips.

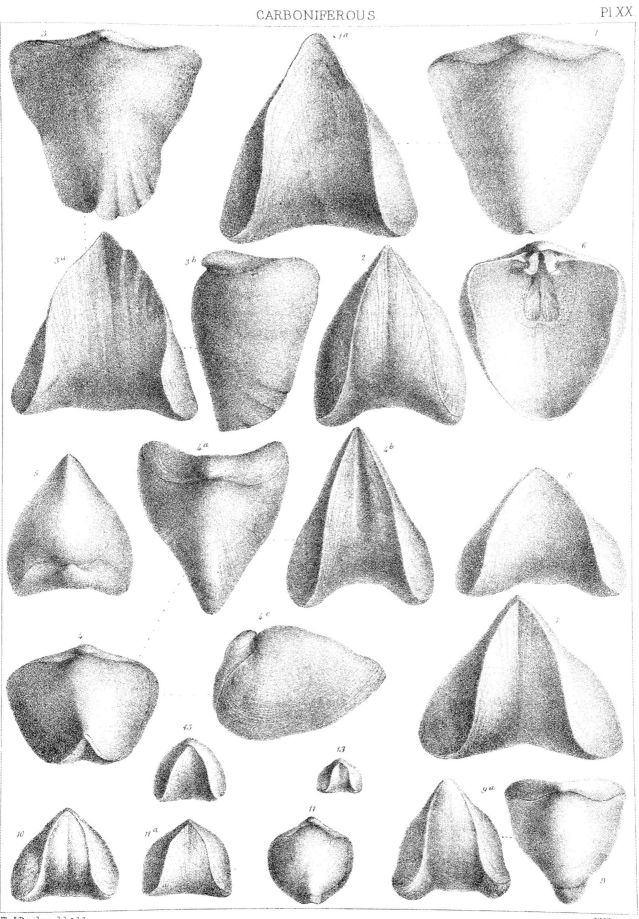

Tho.ˢ Davidson del. & lith.

W West imp.

PLATE XXI.

1. RHYNCHONELLA ACUMINATA, var. MESOGONIA, *Phillips.* From the Carboniferous limestone of Clitheroe.

2. — — From the same locality. Gilbertsonian collection. British Museum.

3. — — var. with two ribs on the fold; same collection and locality.

4 — 13. — — var. *plicata.* A number of variations in shape more or less plicated, from Clitheroe. Gilbertsonian collection. British Museum.

14 — 20. — — var. *platiloba*, Sowerby. From Clitheroe and Kildare.

Pl. XXI.

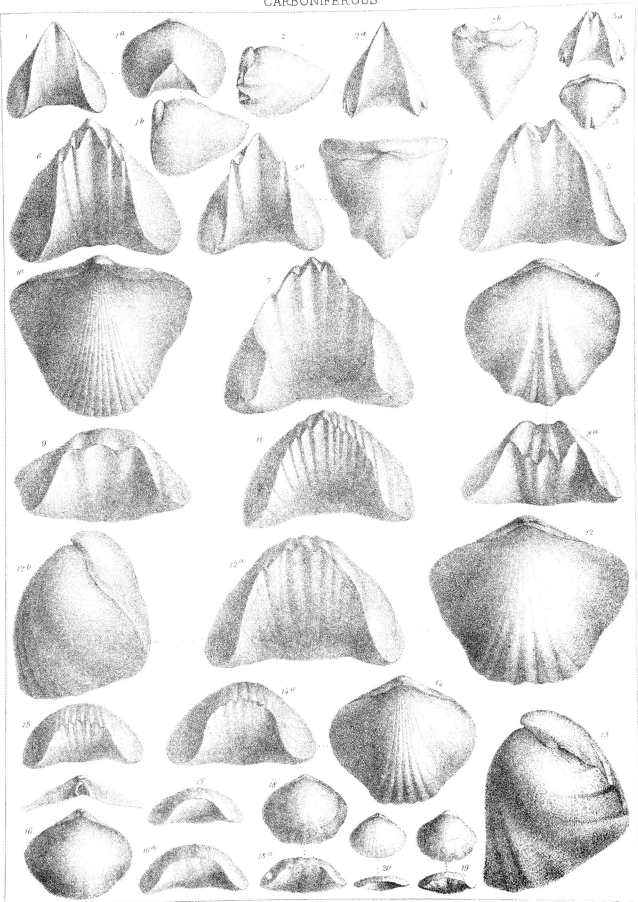

Thos Davidson del. & lith.

W. West imp.

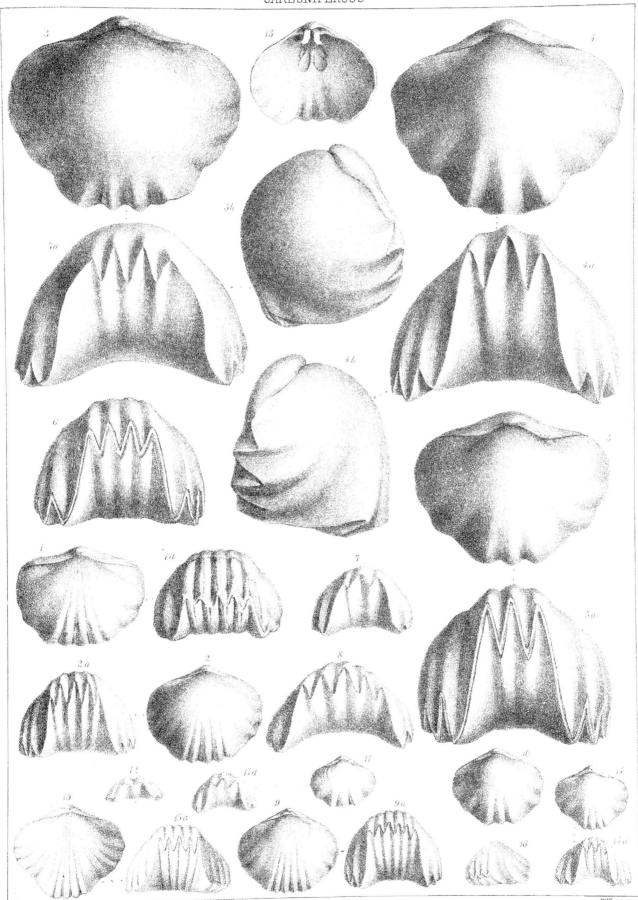

Tho.' Davidson del et lith

W. West imp.

PLATE XXIII.

1, 2. RHYNCHONELLA PLEURODON, *Phillips.* From the Carboniferous limestone of Bolland.

3—5. — — From Bolland and Settle.

6. — — A specimen with nine ribs on the mesial fold. Bolland. British Museum.

7. — — From Kerry, Ireland.

8—11. — — Young shells from Settle, in Yorkshire.

12. — — A globose variety, in which the ribs do not attain the extremity of the beaks.

13, 14. — — var. *T. ventilabrum*, Phillips. After the figure in the 'Geology of Yorkshire.'

15. — — = *Terebratula Mantiæ*, Sow. From the original specimen in the collection of Mr. J. de C. Sowerby. It is a malformation of *Rh. pleurodon*.

16. — — var. *triplex*, M'Coy. From the Carboniferous shale, near Carluke, Lanarkshire, Scotland.

17. — — From Professor M'Coy's type of *A. triplex*.

18. — — A malformation, the sinus being twisted to one side. Carluke.

19—21. — — var. *Davreuxiana*, De Koninck. Carboniferous limestone, Gilling, Yorkshire, in the collection of Mr. E. Wood.

22. — ? From the Carboniferous limestone of Twiston, Lancashire, and collection of Mr. Parker. I am uncertain as to the species to which this remarkable specimen should be referred, a single example only having been hitherto discovered.

Pl. XXIII.

PLATE XXIV.

1. RHYNCHONELLA FLEXISTIRIA, *Phillips.* = *T. tumida*, Phillips.   From the original figure, 'Geology of Yorkshire,' tab. 12, fig. 35.

2. — — A very fine specimen from the Carboniferous limestone of Clitheroe, and collection of Mr. M. Parker.

3. — — From Phillips's original specimen of *T. tumida*, British Museum.   Bolland.

4. — — From Millecent, Ireland.

5. — — = *Rh. heteroplycha*, M'Coy.   'British Palæozoic Fossils,' tab. 3 D, fig. 19.

6. — — From Professor Phillips's original specimen, *T. flexistria.* Bolland.   British Museum.

7, — — From Professor Phillips's figure of *T. flexistria*, 'Geology of Yorkshire,' tab. 12, figs. 33, 34.   This was not very correctly drawn in the work last mentioned : fig. 6 of our plate is a more correct representation.

8. — — From Bolland.   A curious variety.   A careful examination of the specimens, figs. 1 to 8, leads me to believe that they are all referable to a single species.

9, 10. CAMAROPHORIA GLOBULINA, *Phillips*, var. *rhomboida*, Phillips.   From tab. 12, figs. 18, 19, 20, of the 'Geology of Yorkshire,' and Carboniferous limestone of Bolland.

11, 12. — — Two specimens from the Bolland district and British Museum.   I believe the example, fig. 12, to be the original specimen on which Phillips's *T. rhomboida* was founded.

13, 14. — — *T. seminula*, Phillips's 'Geology of Yorkshire,' tab. 12, figs. 21—23.

15, 16. — — Were drawn from Phillips's original specimen of *T. seminula*.   Carboniferous limestone.   Bolland district.   British Museum.   It appears to me that *T. seminula* is the fry or young of *T. rhomboida*, Phillips; and this last is, to all appearance, the same as the Permian *C. globulina*.

17. — — *Rh. longa*, in M'Coy's 'British Palæozoic Fossils,' pl. 3 D, fig. 24.

18—22. — — Different specimens and varieties from the Carboniferous limestone of Longnor, in Derbyshire.

23—25. RHYNCHONELLA ? TRILATERA, *De Koninck*. From the Carboniferous limestone of Derbyshire.   British Museum.

26. — — From the Carboniferous limestone of Alstonfield. Museum of Practical Geology.

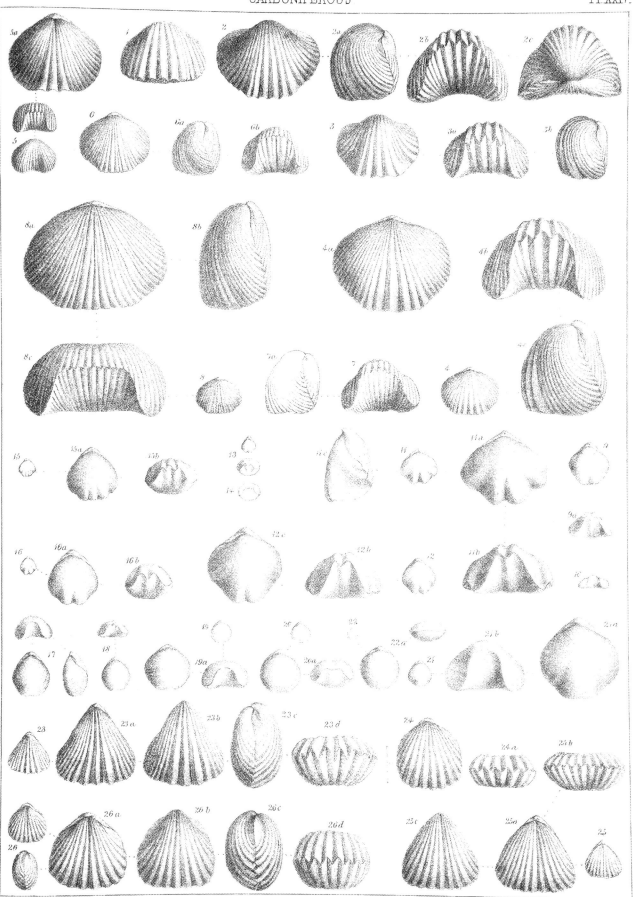

PLATE XXV.

1. CAMAROPHORIA ? ISORHYNCHA, *M'Coy*.   From the Carboniferous limestone of Cookstown, Tyrone, Ireland;   and collection of Sir R. Griffith.

2. —   —   From M'Coy's figure in the 'Synopsis of the Carboniferous Fossils of Ireland,' tab. 18, fig. 8.

3. CAMAROPHORIA CRUMENA, *Martin*.   Sp. 'Petrif. Derb.,' tab. 36, fig. 4.   Carboniferous limestone of Derbyshire.

4. —   —   From Settle, Yorkshire.

5. —   —   From West Lothian, Scotland.   Collection of the late Dr. Fleming.

6. —   —   From the Carboniferous limestone of Dovedale, Derbyshire, and Museum of Practical Geology.

7, 8. —   —   From Settle, Yorkshire.

9. —   —   From Bolland.   British Museum.   This series of specimens or figures illustrates the variation in the number of ribs on the mesial fold.

10. CAMAROPHORIA ? PROAVA, *Phillips*.   From the original specimen in the British Museum.   It is, however, uncertain whether this specimen belongs to the genus *Camarophoria*, and whether it may not be specifically the same, but an exceptional shape of *C. crumena*.

11, 12. CAMAROPHORIA ? or RHYNCHONELLA ? LATICLIVA, *M'Coy*.   'British Palæozoic Fossils,' tab. 3D, figs. 20, 21.

13. RHYNCHONELLA ? SEMISULCATA, *M'Coy*.   'Synopsis of the Carboniferous Fossils of Ireland,' tab. 22, fig. 15.   Walterstown, Skreen, Ireland.   May not this belong to the same species as *Rh. proava*, Phillips ?

14. RHYNCHONELLA LATERALIS, *Sow.*, 'Min. Con.,' tab. 83, fig. 1.

15.   —   NANA, *M'Coy*.   'Synopsis of the Carboniferous Fossils of Ireland,' tab. 23, fig. 19.   Rahoran, Ireland.

16. STREPTORHYNCHUS CRENISTRIA, Var. *Radialis*, *Phillips*.   From the Carboniferous limestone of Whatley, near Frome, Somersetshire.   British Museum.

17, 18. —   —   From Gare, in Lanarkshire, Scotland.   17A, interior of the ventral valve;   18A, interior of the dorsal valve.

19. —   —   Var. *Arachnoidea*, Phillips 'Geol. of Yorkshire,' Pl. II. fig. 4.

20. —   —   In Sandstone at Haltwhistle, from a specimen in the collection of Professor Phillips.

21. —   —   From Rutchugh, Northumberland.   Similar specimens occur in the Kildress red sandstone.

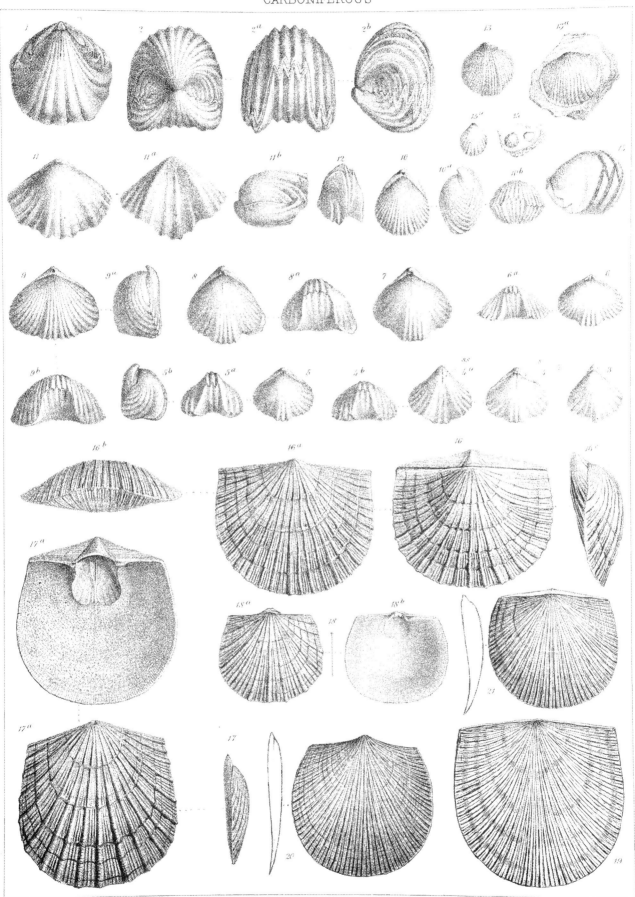

PLATE XXVI.

1. STREPTORHYNCHUS CRENISTRIA, *Phillips.* A very large example from the Carboniferous limestone of Kendal, in the collection of the Geological Society.

2. — — From Denwell, Northumberland, and collection of Mr. Tate. 2a, a fragment of the shell magnified.

3. — — From the Carboniferous limestone of Lime-kilns above Queensferry, Fifeshire, and collection of the late H. Miller.

4. — — = *O. caduca,* M'Coy. 'Synopsis,' tab. 22, fig. 6. I have examined the original example, which is nothing more than a flattened valve of *A. crenistria.* From Rahoran, Fivemiletown, Ireland, and collection of Sir R. Griffith.

5. — — Interior of the ventral valve from the Carboniferous shell of Hook Point, county of Wexford, in the collection of the Geological Society.

6. — — Interior of the dorsal valve, from Denwell, Northumberland, collection of Mr. G. Tate.

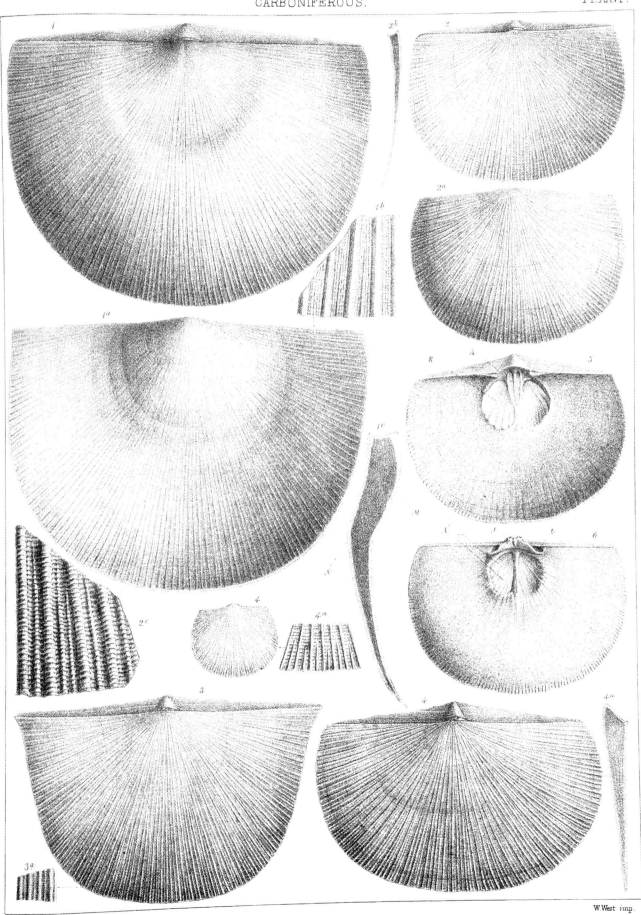

Thos Davidson. del. et lith.

W.West imp.

PLATE XXVII.

CARBONIFEROUS SPECIES.

Fig.

1. Streptorhynchus crenistria, *Phillips*. Carboniferous limestone, Kildare, Ireland.

2, 3, 4.   „   „   var. senilis, *Phillips*. Fig. 2. The original specimen, C. limestone, Bolland. British Museum.

5.   „   „   From Park Hill, Longnor, Derbyshire. Museum of Practical Geology.

6, 7.   „   „   A fragment of dorsal valve, showing the cardinal process and dental sockets.

8.   „   „   var. Kellii, *M'Coy*. The original example, C. limestone, Monaghan, Ireland.

9.   „   „   var. cylindrica, *M'Coy*. The original specimen, from Arenaceous limestone of Castle Espie, Comber, Ireland.

10.   „   „   var. quadrata, *M'Coy*. The original specimen, Calp of Ballintrillick, Bundoran, Ireland. 8, 9, 10. From Sir R. Griffith's collection.

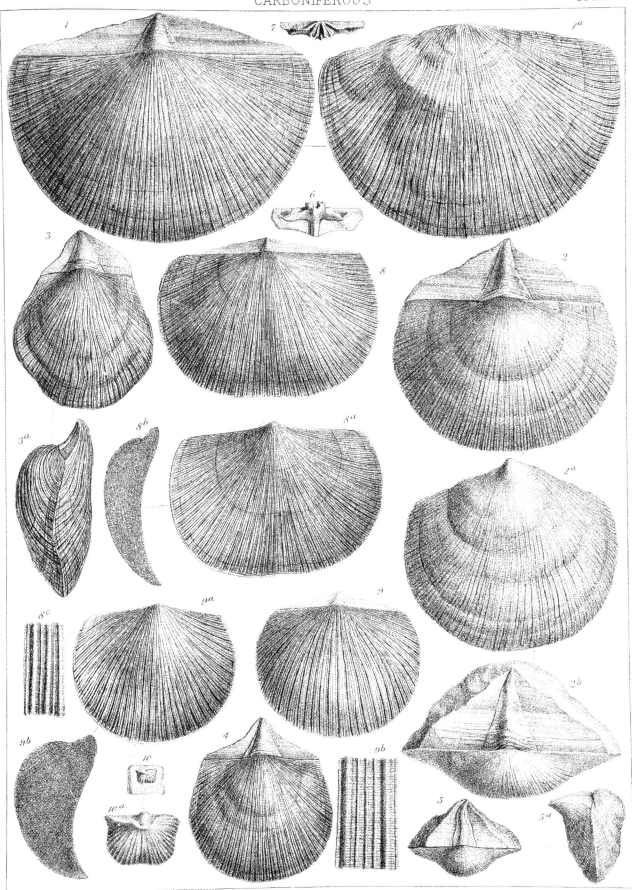

Tho.ˢ Davidson del. & lith.

W.West imp

PLATE XXVIII.

Fig.

1 to 6.   Strophomena analoga, *Phillips.*   Different specimens from the Carboniferous limestone of Kildare, in Ireland; Longnor, in Derbyshire; and Corrieburn, Stirlingshire.

7.   „   „   var. distorta, *Sow.*   Lanarkshire, Scotland.

8.   „   „   „   A very young shell of the same.

9.   „   „   Interior of the ventral valve, in which the foraminal aperture is cicatrized.

10.   „   „   Interior of the dorsal valve.

11.   „   „   Internal cast, seen from the front, showing the vascular impressions (enlarged).

12.   „   „   A fragment of the ventral valve, showing the position of the foraminal aperture.

13.   „   „   The same, seen from the exterior, showing the foramen.

14.   Orthis Keyserlingiana, *De Koninck.*   From the Carboniferous limestone of Settle, in Yorkshire.

15.   Orthis? antiquata, *Phillips.* From the figures in the 'Geology of Yorkshire.' $15^{a, b, c, d}$. Enlarged representations, drawn from the original specimens, Bolland. British Museum.

Pl XXVIII.

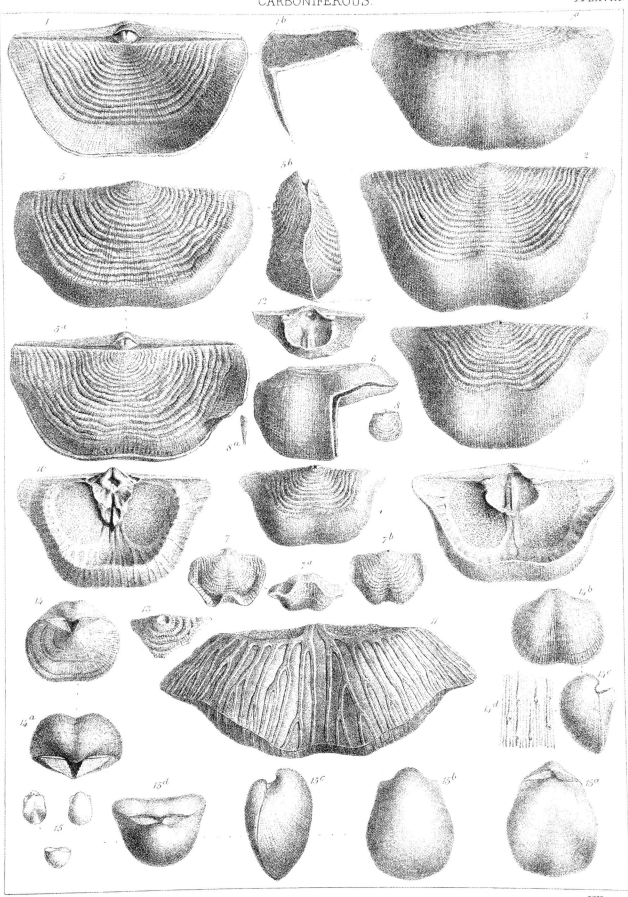

Tho.ˢ Davidson del. & lith.

W. West imp

PLATE XXIX.

Fig.

1. Orthis resupinata, *Martin*. From the Carboniferous limestone of Bolland. British Museum.

2. „ „ A very large example from Yorkshire.

3. „ „ A globose variety from Withgill, in Yorkshire, and collection of Mr. Parker, of Manchester.

4. „ „ Another example, seen from the front.

5. „ „ var. *gibbera, Portlock*. From the lower limestone of Cornacarrow, Enniskillen, and collection of Sir R. Griffith.

6, 7. „ „ var. *connivens, Phillips*. Carboniferous limestone of Little Island, near Cork, Ireland, and collection of Sir R. Griffith.

Pl. XXIX

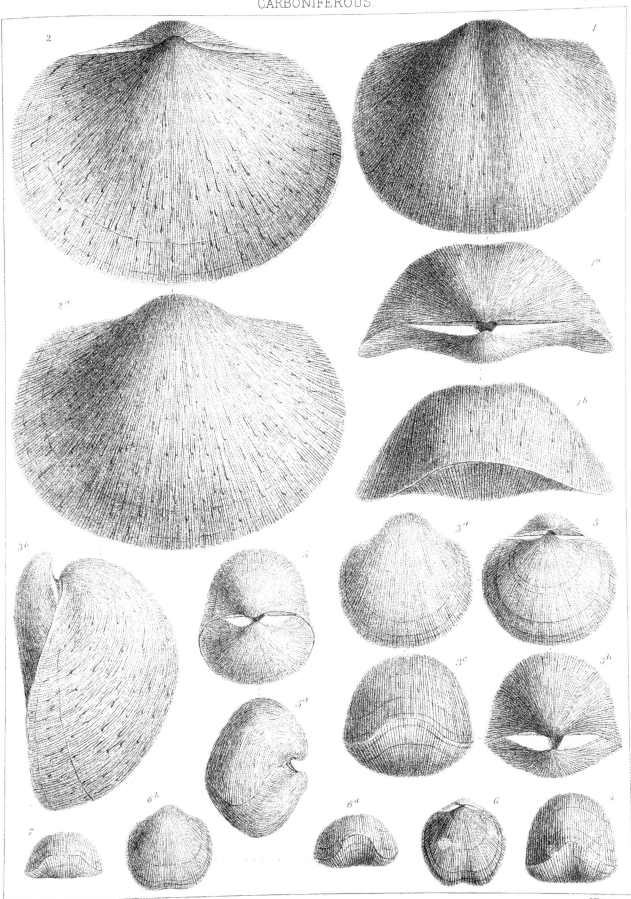

PLATE XXX.

CARBONIFEROUS SPECIES.

Fig.

1. Orthis resupinata, *Martin.* Carboniferous limestone, Millecent, Ireland. 1^d. Portion of the surface, enlarged.

2.   ,,   ,, Interior of the dorsal valve, Derbyshire.

3.   ,,   ,, Interior of the ventral valve. A, Occlusor; R, divaricator, muscular impressions. Both 2 and 3 belong to the same specimen.

4.   ,,   ,, Interior of the dorsal valve. A, Adductor or occlusor muscular impressions; *j*, cardinal process. Carboniferous shales, Ulverston, Lincolnshire.

5.   ,,   ,, Interior or ventral valve, shales near Settle.

6, 7, 8, 9. Orthis Michelini, *L'Eveillé.* 6. From the Carboniferous shales of Clattering Dykes, Malham Moor, Yorkshire. 7. From shales, Gateside, near Beith, Scotland. 8, 9. From Carboniferous limestone, Millecent, Ireland.

10.   ,,   ,, Interior of the dorsal valve (enlarged). *j*, Cardinal process; A, adductor or occlusor muscular impressions. Gateside, Ayrshire.

11.   ,,   ,, Interior of the ventral valve (enlarged), same locality. A, Adductor or occlusor; R, cardinal or divaricator, muscular impressions. Mr. Hancock is of opinion that the upper portion of this impression, or that nearest the beak, may belong to the ventral adjustor, and that N may also perhaps be caused by the pedicle. O, Ovarian spaces.

12.   ,,   ,, ? Carboniferous limestone, Settle, Yorkshire. A similar shell to this is described under *O. Michelini* by Prof. de Koninck, in his work on the 'Animaux fossiles de la Belgique,' pl. xiii, fig. 8, but was afterwards believed to be, perhaps, different. Having examined many examples, I am disposed to consider it only a slightly modified condition of L'Eveillé's species.

13. Streptorhynchus crenistria, var. senilis. A very globose example, from the Carboniferous limestone of Bowertrapping, near Dalry, in Ayrshire, and collection of Mr. J. Thomson.

14.   ,,   ,, var. senilis (deprived of its outer surface or sculpture). From the Carboniferous limestone of Settle, Yorkshire.

15.   ,,   ,, This is the valve described as *Leptæna anomala,* Sow., in pl. 615, fig. 1^b (not ^{a, c} or ^d) of the 'Mineral Conchology,' and as *Strophalosia striata* by Prof. Morris, at p. 155 of his 'Catalogue;' but an attentive study of the specimen and of other similar examples has convinced me that it is none other than an imperfect, contorted valve of *S. senilis,* of which the outer layer or sculpture has been obliterated. Traces of the striæ may, however, be noticed here and there upon its surface.

Pl. XXX.

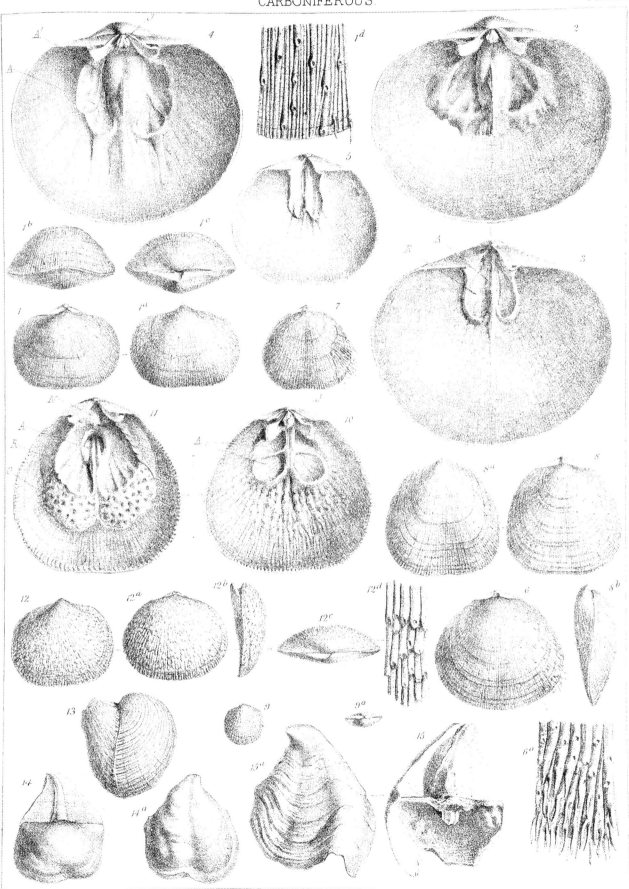

Thos Davidson del. & lith.

W.West imp.

PLATE XXXI.

CARBONIFEROUS SPECIES.

Fig.

1. Productus sub-lævis, *De Koninck*. Stated to be from the Carboniferous lime-stone near Leek, in Staffordshire (but probably from Clitheroe, in Lancashire). Coll. of Dr. Bowerbank.

2.   ,,   ,, From the Carboniferous limestone of Llangollen. Museum of Practical Geology, London.

3.   ,, plicatilis, *Sow.* From the limestone of Longnor, Derbyshire. Museum of Practical Geology.

4.   ,,   ,, From the limestone of Settle, in Yorkshire.

5.   ,,   ,, Internal cast of the ventral valve, from Longnor, in Derbyshire, showing the position of the adductor and divaricator muscular impressions.

6.   ,, mesolobus, *Phillips.* A remarkably circular example from the Carboniferous limestone of Settle, in Yorkshire.

7, 8.   ,,   ,, Two specimens of the more common shape, from Yorkshire.

9.   ,,   ,, From the Glarat limestone, Campsie, Stirlingshire, Scotland.

Pl. XXXI.

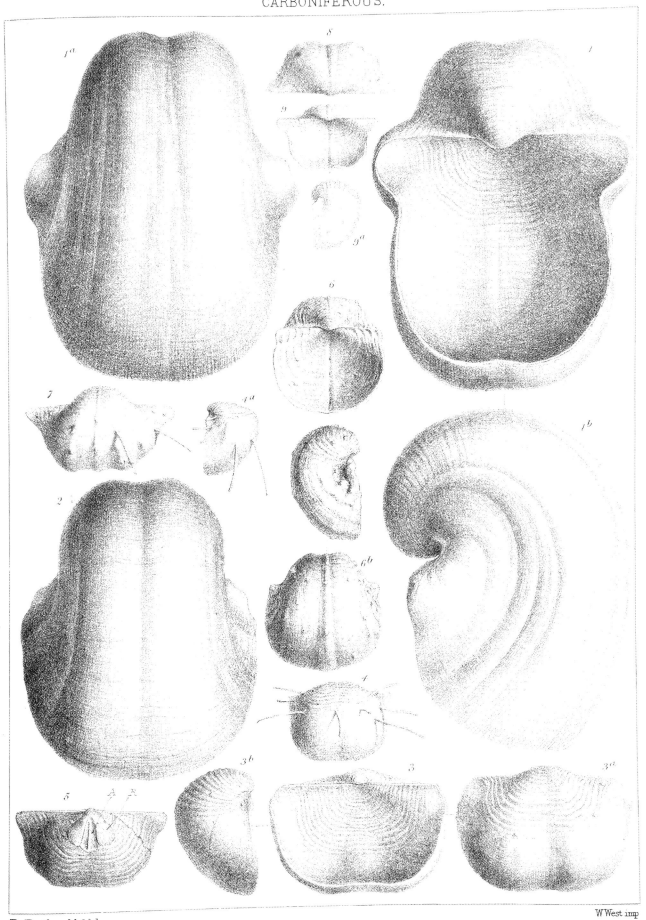

Tho⁵ Davidson del. & lith.

W. West imp.

PLATE XXXII.

CARBONIFEROUS SPECIES.

FIG.

1. Productus sub-lævis ? These are Prof. de Koninck's figures of *P. Christiani*, but which appear to me only a slightly modified condition of *P. sub-lævis* ? Carboniferous limestone, England (the exact locality is not given, but was, perhaps, Clitheroe, in Lancashire.)

2. ,, costatus, *Sow.* A specimen, with its spines, from Carboniferous shales near Glasgow.

3. ,, ,, From Richmond, Yorkshire. Collection of Mr. S. Wood.

4. ,, ,, From the limestone of Settle, in Yorkshire.

5, 6, 7. ,, ,, Three different specimens from Richmond, in Yorkshire, and Lesmahago, Lanarkshire, Scotland.

8. ,, ,, Interior of the dorsal valve, from Scotland, and collection of Mr. J. Thomson.

9. ,, ,, A very fine, ribbed specimen, forming a passage into the variety *P. muricatus*, from Howood, Renfrewshire, Scotland, and collection of Mr. J. Thomson.

10, 11. ,, muricatus, *Phillips.* Fig. 10 is taken from the figure in the 'Geology of Yorkshire,' 11 from the original specimen found at Harelaw, and now in the York Museum.

12. ,, ,, From the limestone of Gateside, near Beith, in Ayrshire.

13. ,, ,, From Corrieburn, Stirlingshire, and collection of Mr. J. Young.

14. ,, ,, Another example from the same locality, enlarged.

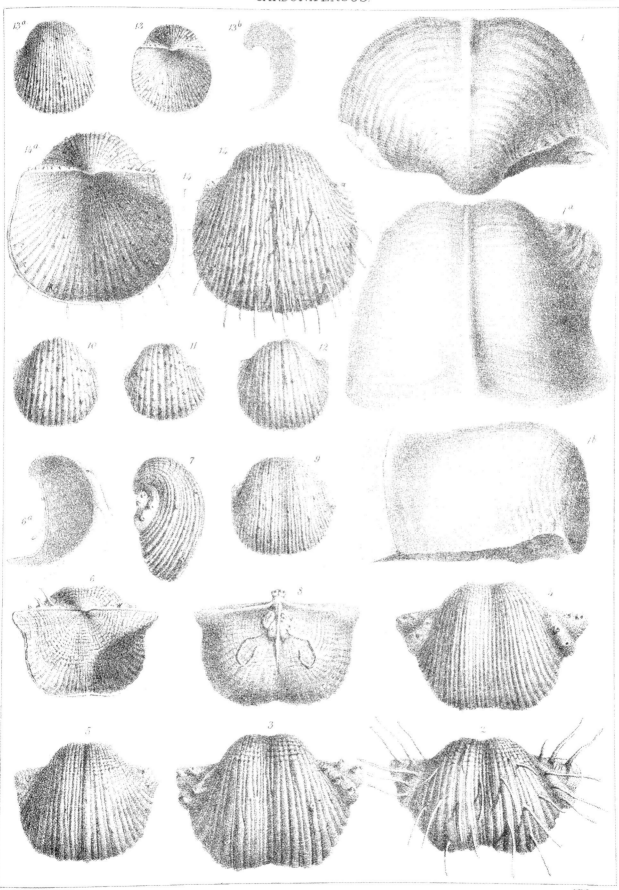

Thos Davidson del & lith   W West imp

PLATE XXXIII.

CARBONIFEROUS SPECIES.

FIG.

1 to 4. **Productus proboscideus**, *De Verneuil.* Carboniferous limestone, Settle, Yorkshire. I avail myself of the present opportunity to thank my zealous friend, Mr· Burrow, for the liberal and generous manner with which he has presented me with his best and, by me, figured specimens of *Producta*.

5.   „   ermineus, *De Koninck.* Carboniferous limestone, Settle, Yorkshire.

6, 7.   „   Wrightii, *Dav.* Carboniferous limestone, Middleton, near Cork, Ireland. Fig. 6 in the collection of Mr. J. Wright, of Cork.

8, 9.   „   sinuatus, *De Koninck.* Fig. 8, nat. size; $8^{a, b, c}$ enlarged. 9. Enlarged view of the ventral valve, as seen from the beak, to show the position of the cardinal spines, &c. Carboniferous limestone, Settle, Yorkshire.

10, 11.   „   „   Internal casts of both valves, showing the position of the adductor (A) and divaricator (R) muscular impressions, as well as the reniform impressions (z).

12.   „   fimbriatus, *Sow.* Ventral valve, with its spines. Carboniferous limestone, Settle, Yorkshire.

13, 14.   „   „   Two examples deprived of their spines, from the same locality.

15.   „   „ 15^a, interior of the dorsal valve; 15^c, the same, viewed in profile; 15^b, the same specimen, seen from the exterior, and showing how the short spines are arranged in regular rows. Carboniferous shales, near Settle, Yorkshire.

16.   „   aculeatus, *Martin.* From the original example. 16^b· A carefully enlarged illustration. Sowerby's collection, British Museum.

Tho⁵ Davidson del. & lith.

W West imp

PLATE XXXIV.

CARBONIFEROUS SPECIES.

Fig.

1. Productus striatus, *Fischer*. Carboniferous limestone, Longnor, Derbyshire. Museum of Practical Geology.

2.   ,,   ,,   Limestone, Settle, Yorkshire.

3.   ,,   ,,   ,,   Derbyshire.

4, 5.   ,,   ,,   ,,   ,,   Malformations. Museum of Practical Geology.

6.   ,,   carbonarius, *De Koniuck*. Carboniferous limestone north of Glasgow. Museum of Practical Geology.

7.   ,,   undatus, *Defrance*. Derbyshire. Collection of Dr. Fleming, of Manchester.

8.   ,,   ,,   Limestone, Poolwash, Isle of Man.

9.   ,,   ,,   ,,   Settle, Yorkshire.

10, 11, 12.   ,,   ,,   ,,   Campsie, Stirlingshire.

13.   ,,   ,,   *P. tortilis*, M'Coy. Limestone, Tullanaguiggy, Fermanagh, Ireland, and collection of Sir R. Griffith.

14.   ,,   tessellatus, *De Koninck*. Limestone, Kildare. British Museum.

15, 16.   ,,   Keyserlingianus, *De Koninck*. Limestone, Settle, Yorkshire.

17.   ,,   arcuarius, *De Koninck*. Limestone, Settle, Yorkshire.

18.   ,,   spinulosus, *Sow*. The original specimen. Linlithgowshire, Scotland, and collection of the late Dr. Fleming.

19.   ,,   ,,   From the Carboniferous limestone near Lesmahago, and collection of Dr. Slimon.

20.   ,,   *P. granulosus*, Phillips. Carboniferous limestone, Settle, Yorkshire.

21.   ,,   ,,   Phillips's figure of *P. granulosus*.

Pl. XXXIV.

PLATE XXXV.

CARBONIFEROUS SPECIES.

FIG.

1. Productus latissimus, *Sow.* Carboniferous limestone, Dalry, Ayrshire.

2. ,, ,, ,, Derbyshire. Collection of Dr. Fleming, of Manchester.

3. ,, ,, Interior of the dorsal valve, from near Carluke, in Lanarkshire.

4. ,, ,, Interior of the ventral valve, from Broadstone, near Beith, Ayrshire, and collection of Mr. J. Armstrong.

5. ,, longispinus, *Sow.* From the original example. West Lothian, and collection of the late Dr. Fleming.

6. ,, ,, Ventral valve, with its elongated spines.

7. ,, ,, From Yorkshire.

8. ,, ,, From Craigie, near Kilmarnock, and collection of Mr. J. Thomson.

9. ,, ,, $9^{a\cdot}$ Interior of the ventral valve (enlarged); $9^{b\cdot}$ interior of the dorsal valve. From the Carboniferous shales of Capelrig, Lanarkshire, Scotland.

10. ,, ,, Interior of the dorsal valve. From shales under the main limestone, Campsie, Stirlingshire, and collection of Mr. J. Young.

11. ,, ,, From East Kilbride, Lanarkshire.

12. ,, ,, From the original example of *P. Flemingii*, Sow. West Lothian, Scotland.

13. ,, ,, Specimens in Arenaceous limestone, from Rutcheugh, Northumberland.

14. ,, ,, var. *P. lobatus*, Sow. From Capelrig, Lanarkshire.

15, 16. ,, ,, var. *P. setosa*, Phillips. From the original figures in the ' Geology of Yorkshire.' These are certainly only different states of *P. longispinus*.

17. ,, ,, var. *P. spinosus*, Sow. These figures are drawn and slightly restored from the original example in the collection of the late Dr. Fleming, Scotland.

18, 19. ,, ,, Two exceptional shapes of *P. longispinus*, from Carboniferous shales near Carluke, in Lanarkshire.

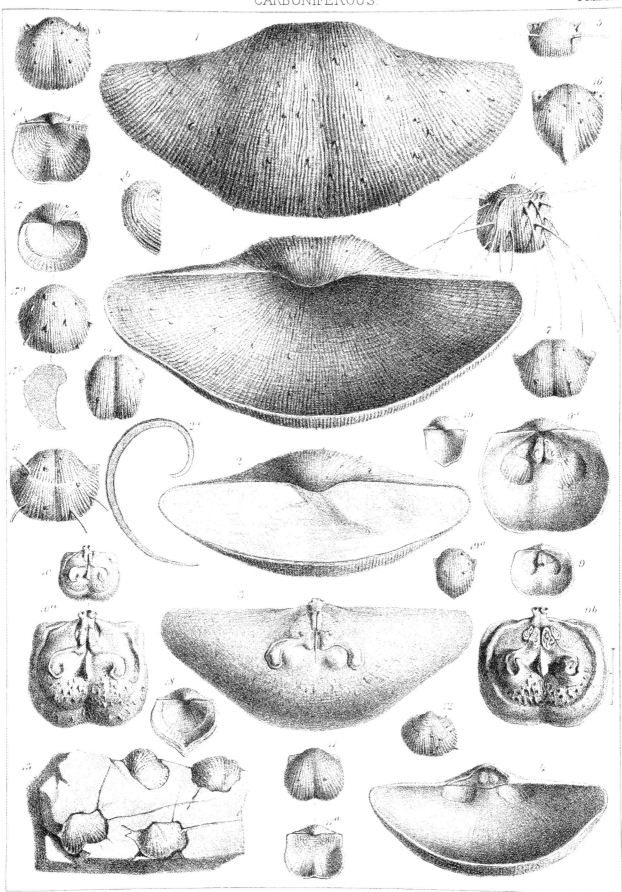

Tho⁵ Davidson del. & lith.

W.West imp

PLATE XXXVI.

CARBONIFEROUS SPECIES.

Fig.

1.   Productus humerosus, *Sow.*   1ᵃ. Internal cast of the ventral valve, from the Carboniferous magnesian limestone of Breedon, Leicestershire. Museum of the Geological Society.   1ᵇ. Interior of the ventral valve (a portion of the beak being removed to show the muscular impressions), from a gutta-percha impression.

2.   „   „   Internal cast, from the same locality. 2 shows the interior of the dorsal valve and beak of ventral one; 2ᵃ the ventral valve, viewed in profile.

3.   „   personatus, *Sow.*   From the original internal cast. The specific claims of this so-termed species cannot be decided, on account of the imperfection of the material at our command. Kendal, Westmoreland.   British Museum.

4.   „   cora, *D'Orbigny*.   Carboniferous limestone, Settle, Yorkshire.

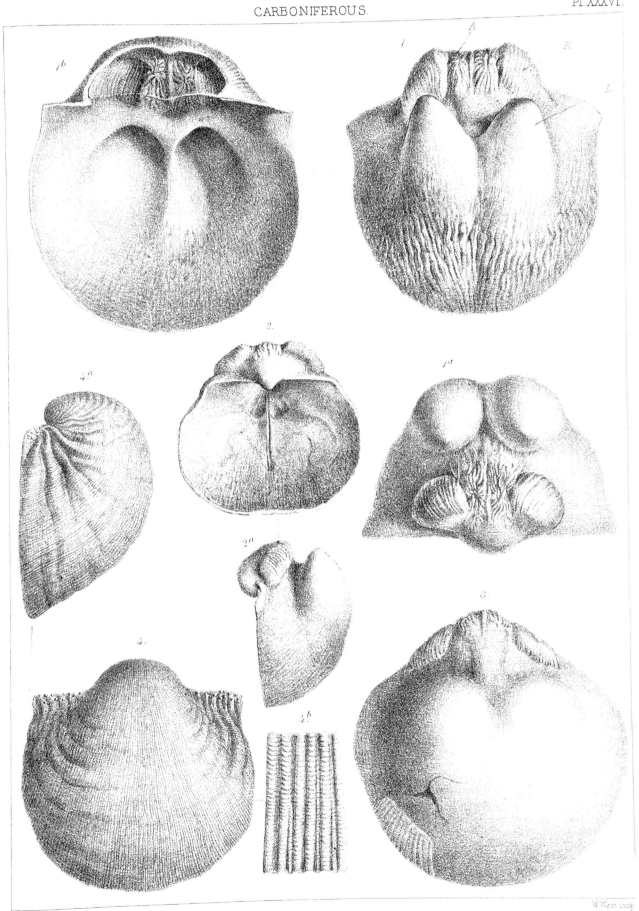

PLATE XXXVII.

Fig.

1. Productus giganteus, *Martin.* Interior of the ventral valve, from which a portion of the beak has been removed, so as to exhibit the umbonal cavity. A and c, Adductor or *occlusor;* R, cardinal or *divaricator;* muscular impressions; L, cavity occupied by the spiral arms.

2. ,, ,, Interior of the *dorsal valve.* J, Cardinal process; A, adductor or *occlusor,* muscular impressions; w, projections to which Mr. S. P. Woodward supposes the oral arms to have been attached (?); x, reniform impressions; z, eminences corresponding to the hollows (L) in the ventral valve.

3. ,, ,, Hinge-line and cardinal process of dorsal valve.

These drawings are taken from valves belonging to the same individual, which was obtained from the Carboniferous limestone Llangollen, and is in the Museum of Practical Geology. It is one of the most instructive specimens which I have hitherto examined. The cardinal process was, however, so much imbedded in the matrix that it could not be developed, so that the deficiency was completed from another example in the British Museum.

4. ,, ,, Ideal section or both valves (slightly improved) from the figure published by Mr. S. P. Woodward, at p. 233 of his 'Manual of the Mollusca.' The letters refer to the same parts in the other specimens.

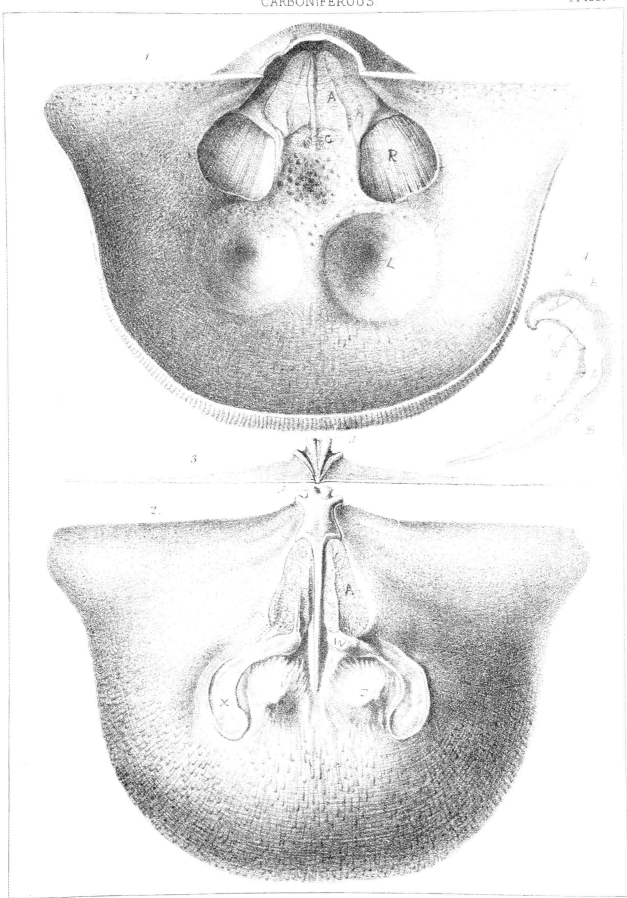

Thos Davidson del.& lith.

W.West imp.

PLATE XXXVIII.

Productus giganteus, *Martin*. A very large example, from the Carboniferous limestone of Derbyshire, in the collection of Prof. Tennant.

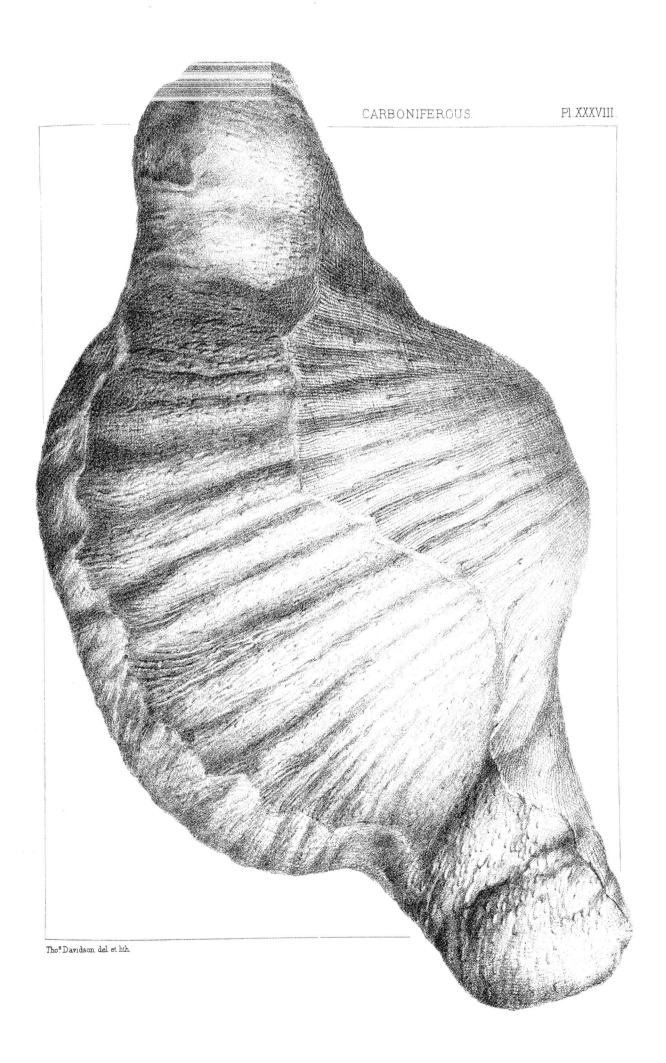

Thos Davidson. del. et lith.

PLATE XXXIX.

CARBONIFEROUS SPECIES.

Fig.

1. Productus giganteus, *Martin*. Dorsal valve and beak of ventral one. Carboniferous limestone near Richmond, Yorkshire, and collection of Mr. E. Wood.

2.   ,,   ,,   A smaller specimen. Same locality.

3.   ,,   ,,   A very circular variety.

4.   ,,   ,,   An evenly convex variety, *P. maximus*, M'Coy. Carboniferous limestone near Cork, Ireland.

5.   ,,   ,,   A young example. Yorkshire.

Pl. XXXIX.

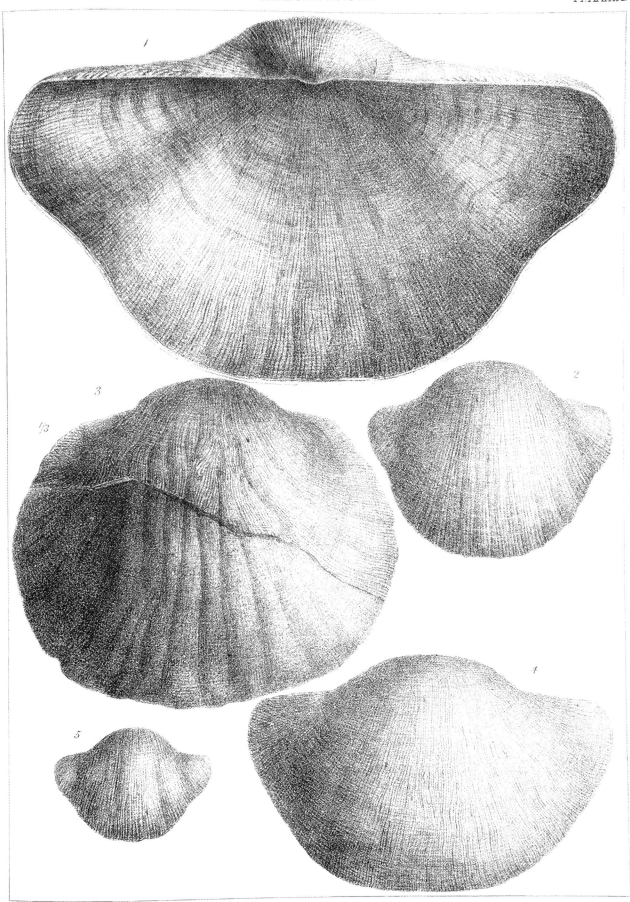

Tho.ᵍ Davidson del.et lith W.West imp.

PLATE XL.

Fig.

1. Productus giganteus, *Martin*. Internal cast of the ventral valve, from the Carboniferous limestone of Lowick, Northumberland, and from the same specimen which was represented under another aspect in Professor King's 'Permian Monograph.' The letters are the same which have been made use of in the interior, Pl. XXXVII, fig. 1.

2. „ „ var. *P. Edelburgensis*, Phillips. From a specimen in the Carboniferous limestone of Yorkshire.

3. „ „ A decorticated specimen or cast, from the Carboniferous limestone of Thornton, Wensleydale, showing a peculiar interruption and bifurcation of ribs.

4 to 8. „ „ var. ? *P. hemisphæricus*, Sow. From the Carboniferous limestone of Craven, in Yorkshire, where the shell occurs by millions.

9. „ „ From one of the original and typical specimens of *P. hemisphæricus*, Sow., from the mountain limestone of Mynidd Craig, near Kidwelly, in Carmarthenshire.

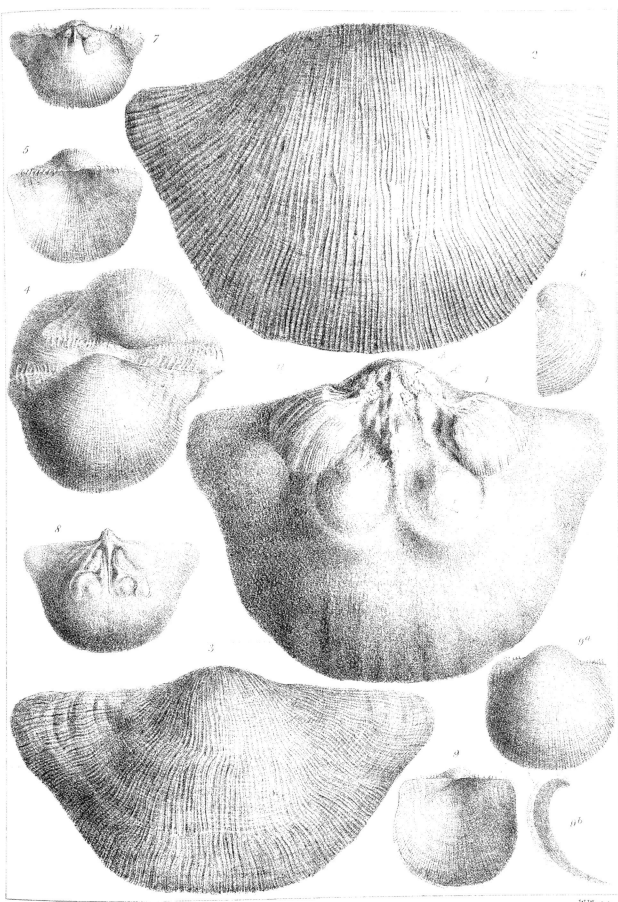

Thos Davidson del et lith

W West imp.

PLATE XLI.

CARBONIFEROUS SPECIES.

Fig.

1. Productus pustulosus, *Phillips.* The original specimen, from the mountain limestone of Bolland. Gilbertson's collection, British Museum.

2.   ,,   ,,   A very large example, from the Carboniferous limestone of Derbyshire. Museum of Practical Geology.

3.   ,,   ,,   From the Carboniferous limestone of Craven. Collection of Mr. E. Wood.

4.   ,,   ,,   Interior of the dorsal valve, from the Carboniferous limestone of Yorkshire.

5.   ,,   ,,   Internal cast of the ventral valve, seen from the umbonal portion. Shores of Lough Gill, County Sligo. Museum of the Geological Society.

6.   ,,   ,,   This is the *P. rugata* of Phillips, not well represented in the 'Geology of Yorkshire.' Bolland.

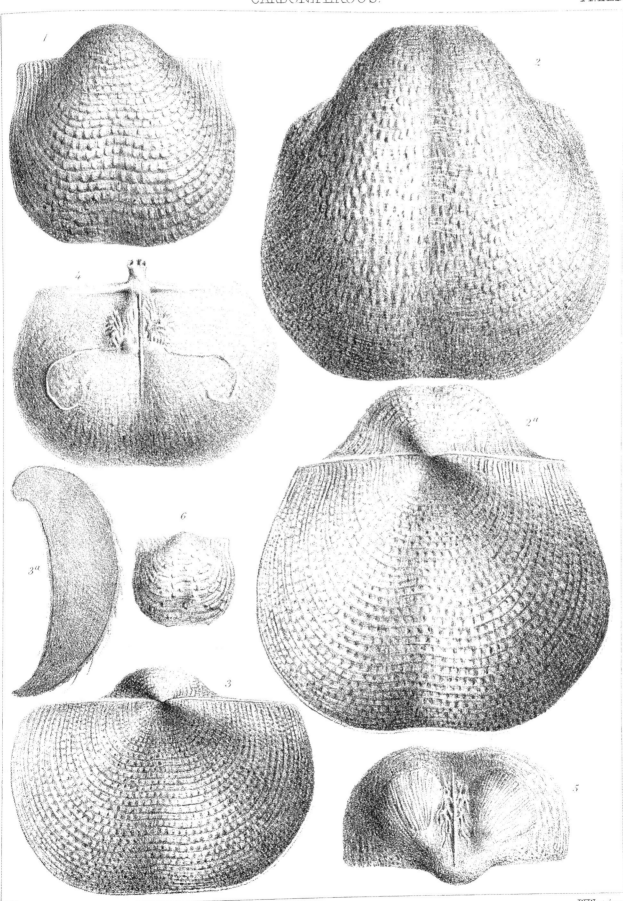

PLATE XLII.

Fig.

1, 2.   Productus pustulosus, *Phillips*.   From the Carboniferous limestone of Kildare, in Ireland.

3.   „   „   *P. ovalis*, Phillips, from the Carboniferous limestone of Bolland.

4.   „   „   *P. pyxidiformis*, De Koninck, from the Carboniferous limestone of Yorkshire.

5.   „   scabriculus, *Martin*.   From the Carboniferous limestone of Carluke, Lanarkshire, Scotland.

6.   „   „   *P. quincuncialis*, Phillips.   Carboniferous limestone of Yorkshire.

7.   „   „   A very large example from the Craven district, in Yorkshire, and collection of Mr. E. Wood.

8, 8ª.   „   „   Interior of the dorsal valve, showing the curiously divided median ridge, from the Carboniferous shales of Lanarkshire, Scotland.

9.   „   cora, *D'Orbigny*.   From a specimen showing the concentric wrinkles sometimes present in the dorsal valve. Derbyshire?

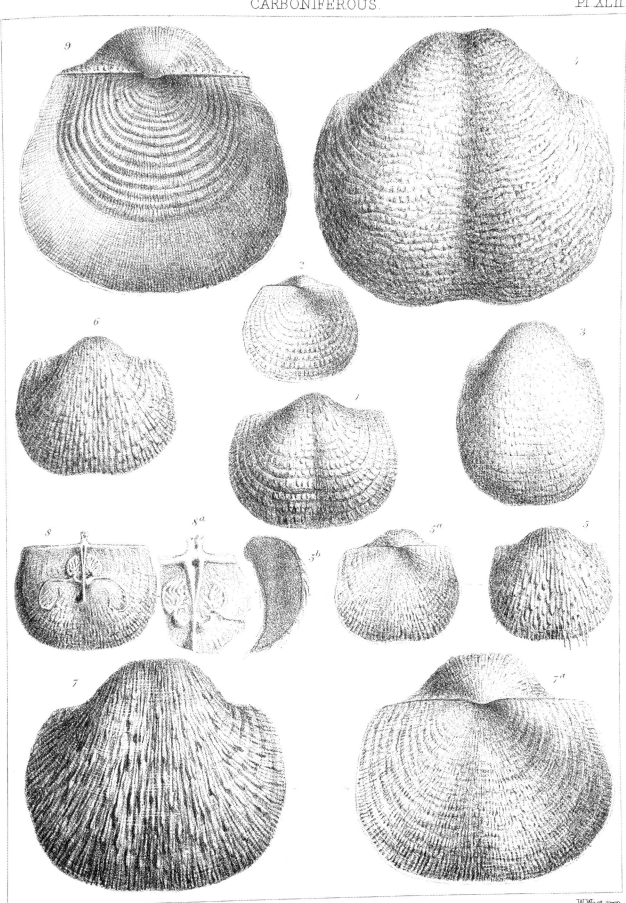

Tho⁵ Davidson del & lith

W West imp.

PLATE XLIII.

Fɪɢ.

1. Productus semireticulatus *Martin.* From the Carboniferous limestone of Kildare. 1ᵃ· A large spine.

2.   ,,   ,, A specimen agreeing with Phillips's *P. pugilis,* from the Carboniferous limestone of Stirlingshire.

2ᵃ.   ,,   ,, From the Carboniferous limestone of Nellfield, Lanarkshire.

3.   ,,   ,, Portion of a ventral valve, showing a long spine *in situ.* Limestone, Kildare, Ireland, and Museum of Practical Geology.

4.   ,,   ,, Profile of a specimen from Kildare.

5.   ,,   ,, Portion of a specimen (enlarged), in which a small, but distinct, area and a pseudo-deltidium can be perceived. Locality unknown, probably Derbyshire. British Museum.

6.   ,,   ,, var. *Martini,* Sow. Carboniferous limestone, Settle, Yorkshire.

7.   ,,   ,,   ,, From Park Hill, Longnor, Derbyshire, and Museum of Practical Geology.

8.   ,,   ,,   ,, Interior of dorsal valve, from shales near Settle, in Yorkshire.

9.   ,,   ,, var. *concinna,* Sow. From the Carboniferous limestone of Campsie, Stirlingshire.

10.   ,,   ,,   ,, A specimen of the same, to show how the shell becomes at times fractured.

Pl. XLIII.

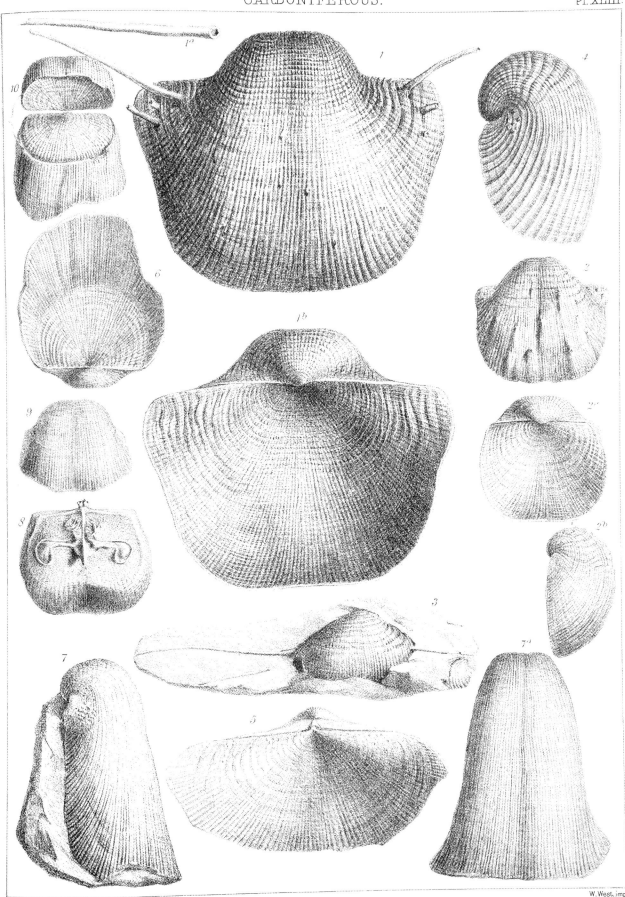

PLATE XLIV.

CARBONIFEROUS SPECIES.

Fig.

1. Productus semireticulatus, *Martin*. Interior of the dorsal valve, from the Carboniferous shales of Calderside, E. Kilbride, Scotland.

2.   „   „ Enlarged fragment of another specimen, from Redesdale, Northumberland.

3.   „   „ Fragments (enlarged) to show the cardinal process, as seen from the exterior of dorsal valve.

4.   „ Interior of the ventral valve, from Calderside, E. Kilbride, Lanarkshire, Scotland. Beautifully preserved interiors of both valves occur in a shale near Settle, in Yorkshire.

5.   „ margaritaceus, *Phillips*. From the Carboniferous limestone of Settle, in Yorkshire.

6.   „   „ The typical specimen, from Florence Court, and collection of Prof. Phillips.

7.   „   „ A young shell, from Settle.

8.   „   „ Phillips's *P. pectinoides*. Carboniferous limestone, Bolland, Yorkshire.

9.   „ punctatus, *Martin*. From the Carboniferous limestone of Wensleydale, Yorkshire.

10.   „   „ From the limestone of Ayrshire, Scotland. 10^{b} Spiny surface, enlarged.

11.   „   „ From Settle, Yorkshire.

12.   „   „ A very elongated example, from the limestone of West Broadstone, Ayrshire, Scotland.

13.   „   „ A young shell, from Bolland.

14.   „   „ An oval variation, from Wensleydale, Yorkshire.

15.   „   „ var. *elegans*, M'Coy. Carboniferous shales of Craigenglen, Stirlingshire.

16.   „   „ Interior of the dorsal valve, from shales near Settle, Yorkshire.

17.   „   „ Internal cast of the ventral valve, as seen from the beak.

18.   „ fimbricatus? var. *laciniatus*, M'Coy. From the Carboniferous limestone of Kendal, in Westmoreland.

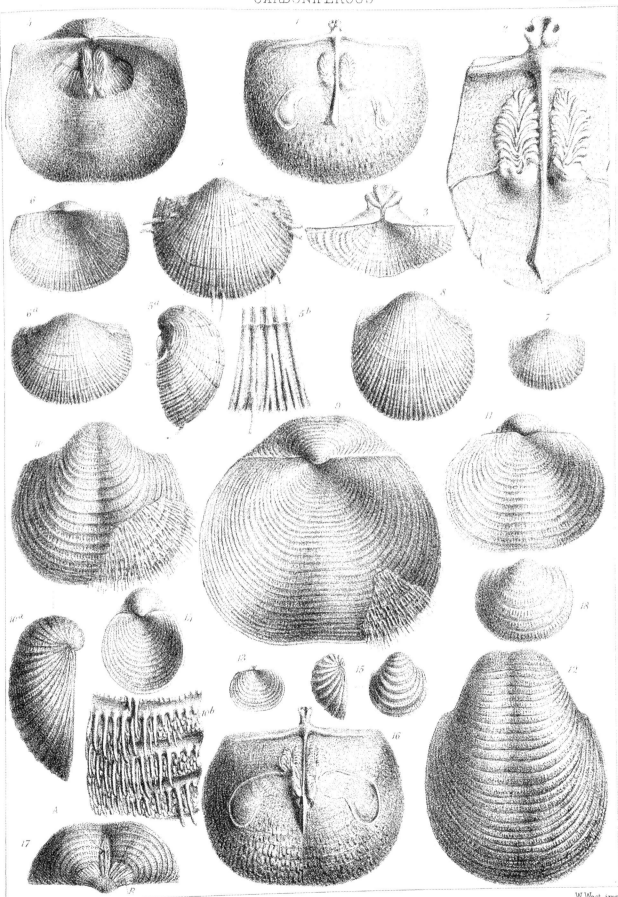

Thos Davidson del et lith.

W.West imp

PLATE XLV.

Pl. XLV.

Thos Davidson, del et lith.

W. West, imp.

PLATE XLVI.

CARBONIFEROUS SPECIES.

Fig.

1, 2. Chonetes comoides? *Sow.*   From the Carboniferous limestone of Bundoran, in Donegal, and collection of the Geological Society.

3, 4.   „   papilionacea, *Phillips.*   From Phillips's original example. Carboniferous limestone, Bolland. A portion of one wing has been restored.

5.   „   „   From Carboniferous shales near Settle, in Yorkshire. 5$^{b.}$ A portion of the shell enlarged ; o, orifices left by the broken spines ; P are pits left on the cast by the asperities which covered the inner surface of the shell.

6.   „   „   Internal cast, showing traces of the muscular impressions, from Ireland.

7.   „   Dalmaniana, *De Koninck.*   From the Carboniferous limestone of Settle, in Yorkshire.

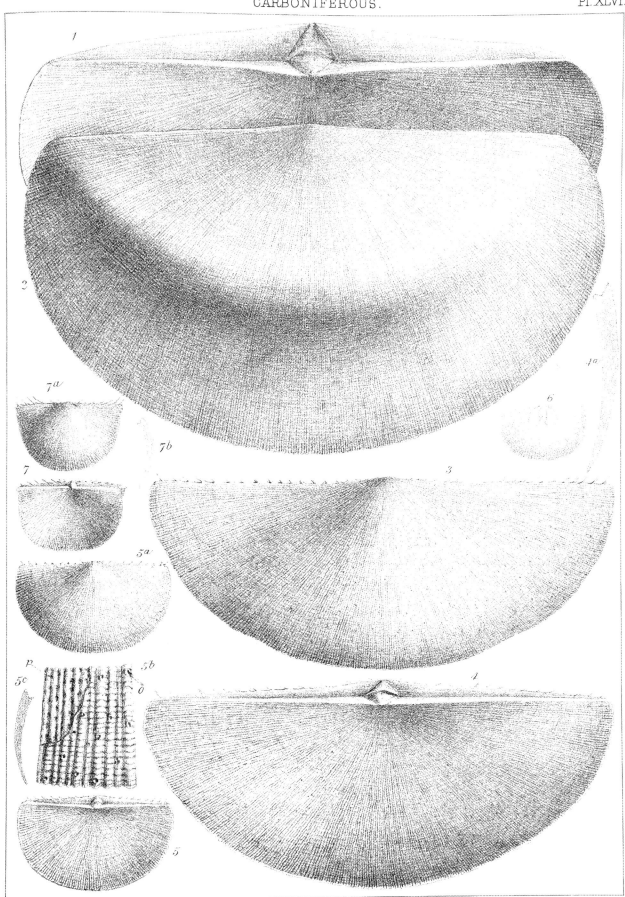

Tho⁵ Davidson, del & lith.

W.West, imp.

PLATE XLVII.

Fig.

1. Chonetes Buchiana, *De Koninck.* From the Carboniferous limestone of Settle, in Yorkshire.

2. „ „ Variety with strong ribs, from the Carboniferous shales of Malham Moor.

3. „ „ From Gare, Lanarkshire, Scotland. $3^{b,c,d}$ Enlarged.

4. „ „ A very large and transverse example. Rutcheugh, Northumberland.

5, 6. „ „ Interior of the ventral and dorsal valves, enlarged, from specimens found at Malham Moor by Mr. Burrow.

7. „ „ var. interstriata. From the Carboniferous limestone of Settle, discovered by Mr. Burrow. 7^{a} Enlarged.

8. „ polita, *M'Coy.* From the Carboniferous shales of Craigenglen, Stirlingshire, in Scotland, and collection of Mr. J. Young. $8^{b,c}$ Enlarged.

9. „ „ Another example, from the same locality.

10. „ „ Prof. M'Coy's original figure.

11. „ „ Interior of the dorsal valve, from the Craigenglen beds, Stirlingshire, Scotland.

12. „ Hardrensis, *Phillips.* A typical example, from the Carboniferous shales of East Barns, near Dunbar, in Scotland. 12^{c} Half of one of the valves, carefully enlarged.

13. „ „ A more coarsely striated variety, from Settle, in Yorkshire.

14. „ „ Interior of the dorsal valve, enlarged, from the Carboniferous shales of Calderside, Lanarkshire, Scotland, and collection of Mr. J. Thomson.

15. „ „ Interior of the ventral valve, enlarged, from Capelrig, in Lanarkshire.

16. „ „ Interior of the dorsal valve, corresponding to fig. 15. Same locality, and collection of Mr. J. Armstrong.

17. „ „ A young specimen, from Craigie, near Kilmarnock. Collection of Mr. J. Thomson. 17^{a} Enlarged.

18. „ „ A small specimen, from Settle. 18^{a} Enlarged.

19. „ „ var. laguessiana, *De Koninck?* From the Carboniferous shales of Newton-on-the-More, Northumberland. 19^{a} Enlarged.

Pl XLVII.

PLATE XLVIII.

CARBONIFEROUS SPECIES.

Fig.

1, 2.   *Crania quadrata*, M'Coy.   Fig. 1 from the original figure in the ' Synopsis ;' fig. 2 from the original specimen in the collection of Sir R. Griffith, Rahans Bay, Dunkineely, Ireland.

3—12.   „   „   Different Scottish examples, from the Carboniferous shales of Carluke, Capelrig, Calderside, &c. These figures show the exterior of the upper valve and the interior of the lower or attached one. Figs. 6, 11, and 12, show how irregular the shell sometimes becomes from being too closely clustered round some marine object.

13.   „   „   Interior of the upper valve, from Capelrig, collection of Mr. J. Armstrong.

14.   *Crania ? trigonalis*, M'Coy.   From the original specimen in the collection of Sir R. Griffith.   14ᵃ·   Magnified view, from Lisnapaste, Ballintra.

15.   „   ? *Ryckholtiana*, De Kon. = *Crania vesiculosa*, M'Coy, from the figure in the ' Synopsis of the Carboniferous Fossils of Ireland.'   Carboniferous limestone, Millecent, Ireland.

16.   „   „   Another example, in the collection of Mr. Humphreys.

17.   „   ?   „   From Castleton, in Derbyshire.

18—25.   *Discina nitida.*   Different specimens, chiefly from the Carboniferous shales of Capelrig, Auchentibber, and Carluke, in Scotland.   Fig. 18 represents the var. *D. bulla*, M'Coy, from Gare, Lanarkshire.   Fig. 20ᵇ is a portion of the shell enlarged.

26.   „   *Davreuxiana*, De Koninck?   From the Carboniferous limestone of Little Island, near Cork, Ireland, collection of Mr. J. Wright.

27, 28.   *Lingula Scotica*, Dav.   Fig. 28 from the Carboniferous shales of Hall Hill, near Lesmahago, and collection of Dr. Slimon.   Fig. 29 from Gare, collection of Mr. J. Armstrong.

29, 30.   „   *mytiloides*, Sow.   From the original figures in the ' Mineral Conchology.'   Walsingham, Durham; British Museum.

31—33.   „   „   From Craigenglen and Capelrig, Scotland.

34.   „   „   = *L. elliptica*, Phillips.   From the original figures in the ' Geol. of Yorkshire.'

35.   „   „   = *L. parallela*, Phil.   From the original figures in the ' Geol. of Yorkshire.'

36.   „   „   = *L. marginata*, Phil.   From the original figure in the ' Geol. of Yorkshire.'

37.   „   *latior*, M'Coy.   From the original figure in the ' British Palæozoic Fossils.'

38—40.   „   *Credneri.*   From drawings by Mr. Kirkby.   Carboniferous shales of Rynope Winning, near Sunderland, collection of Mr. Kirkby.

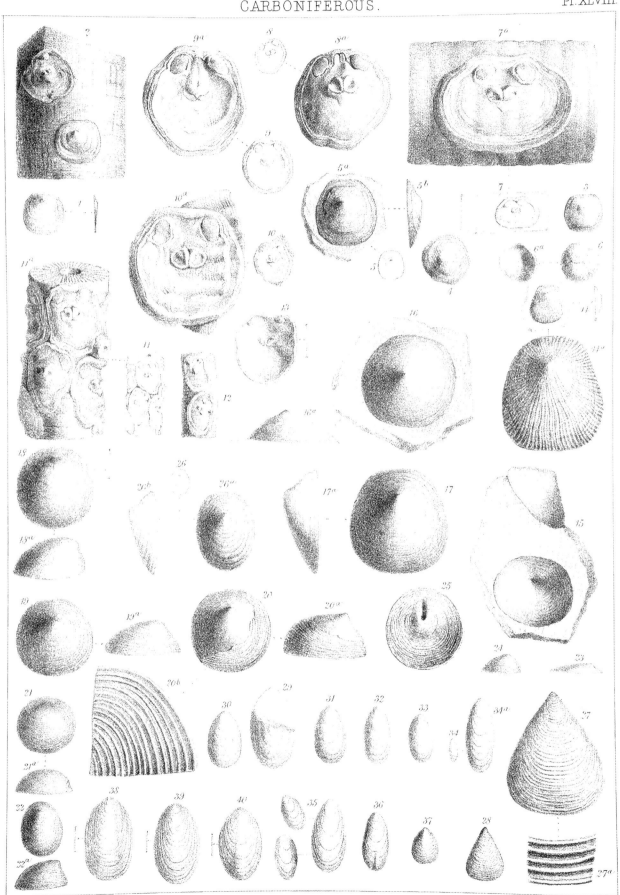

Thos Davidson del. & lith.

W. West imp.

PLATE XLIX.

CARBONIFEROUS SPECIES.

Fig.

1. *Lingula squamiformis,* Phillips. The original specimen. Bolland; Gilbertsonian collection, British Museum.

2.   ,,   ,, From Carboniferous shales, near Glasgow. Museum of Practical Geology.

3, 4, 5.   ,,   ,, From near Carluke and Lesmahago, Lanarkshire.

6.   ,,   ,, In Carboniferous shales. This fine bivalve specimen was found by Mr. Rodwell at about a mile to the east of Bally Castle, on the north coast of Antrim, in Ireland.

7.   ,,   ,, A portion of the shell enlarged, to show its sculpture.

8.   ,,   ,, Large, crushed, specimens, from an ironstone bed one mile north of Glasgow.

9.   ,,   ,, M. Interior of the ventral valve. N. An internal cast of the dorsal one, from 341 fathoms below " Ell Coal," in the parish of Carluke, Lanarkshire.

10.   ,,   ,, Interior of the ventral valve, from Lemmington, Northumberland, collection of Mr. Tate.

(Supplementary illustrations.)

11. *Terebratula hastata.* A very large, full-grown specimen. This shell is labelled " Bolland" in Dr. Bowerbank's collection.

12.   ,,   ,, With colour-markings. Settle, Yorkshire.

13—16.   ,,   ,, These figures are taken from specimens distorted by pressure or cleavage, to show what extraordinary modifications a same species may assume in fossilization under peculiar circumstances. From the Carboniferous limestone of Cork, in Ireland.

17.   ,,   ,, Interior, showing the loop. Settle, Yorkshire.

18—20.   ,,   ,, Var. *Gillingensis,* Dav.

21—26.   ,,   ,, Passage forms connecting *T. hastata* with *T. vesicularis.* From the Carboniferous limestone of Bowertrapping, three miles south of Dalry, in Scotland. Fig. 26 in the collection of Mr. R. Galloway, of Paisley.

27—30.   ,,   ,, Passage shapes between *T. sacculus* and *T. vesicularis.* From Settle and Gilling, in Yorkshire.

Pl. XLIX.

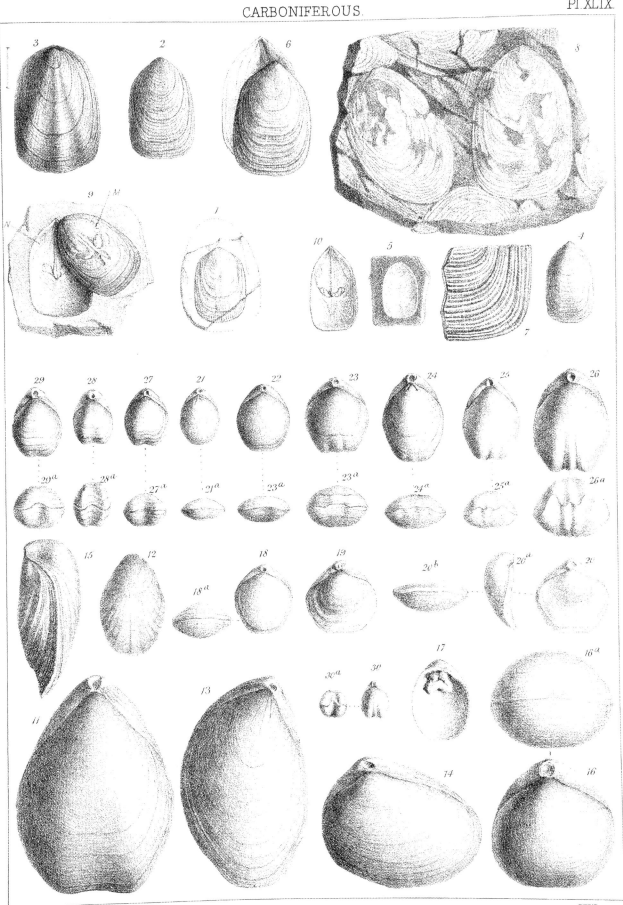

PLATE L.

(*Supplementary illustrations.*)

FIG.

1, 2.   *Spirifera convoluta*, Phillips.   Two very large and fine examples from the Carboniferous limestone of Thornely, near Chipping, Lancashire.   Fig. 1 is perfect, and from the collection of Mr. J. Rofe, of Preston.

3—9.   „   *trigonalis*.   Various specimens to illustrate certain modifications or passage shapes connecting *Sp. trigonalis* and *Sp. bisulcata*.   Fig. 3 from Brockley, near Lesmahago; fig. 4 from Cousland, near Dalkeith, Scotland; figs. 5, 6, 7, showing the gradual prolongation of the wings in certain individuals, from the Lower Scar limestone of Settle, in Yorkshire; fig. 8 from Derbyshire, in the collection of Dr. Fleming, of Manchester; fig. 9 from Barrhead, Renfrewshire, Scotland.

10—18.   „   *triangularis*.   A fine series of specimens obtained by Mr. Burrow from the Lower Scar limestone of Settle, in Yorkshire.

19.   *Cyrtina septosa*.   A large specimen.   From the Lower Scar limestone of Settle, in Yorkshire, and collection of Mr. Burrow.

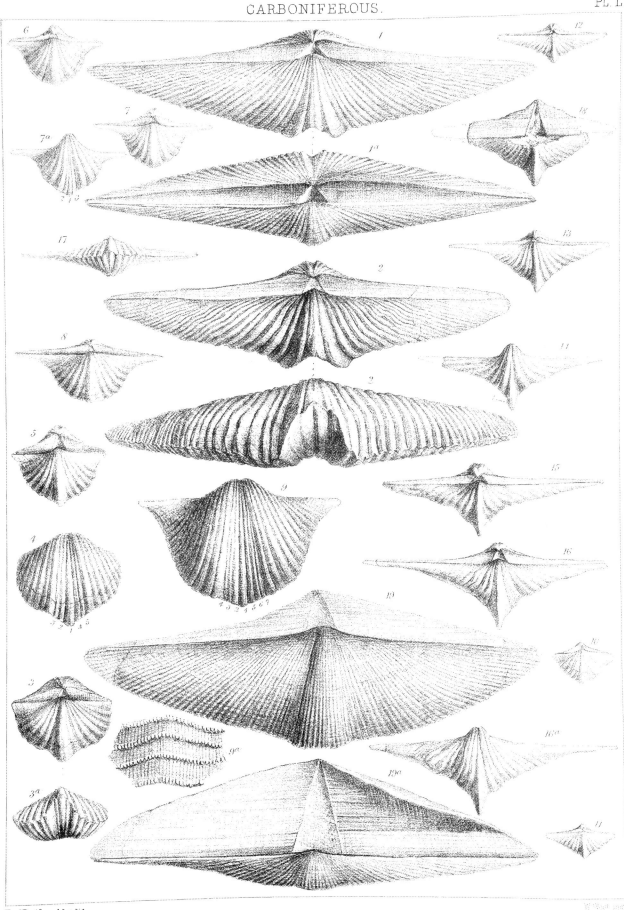

PLATE LI.

CARBONIFEROUS SPECIES.

(Supplementary illustrations.)

Fig.

1, 2.   *Productus sub-lævis.* From the Carboniferous limestone of Clitheroe, in Lancashire. Fig. 1 in the collection of Mr. J. Parker, of Manchester.

3.   *Retzia* or *Rhynchospira Carbonaria*, Dav. From the lower black Carboniferous shales of Skrinkle, Pembrokeshire, and Museum of Practical Geology.

4—9.   *Retzia radialis*, Phillips. Various modifications in shape. From Yorkshire, Derbyshire, and Scotland.

10.   „   ?   Uncertain form. From Derbyshire.

11—13.   *Athyris plano-sulcata*, Phillips. Specimens distorted by pressure and cleavage. From the Carboniferous limestone of Cork. Fig. 11 is the original specimen from which M'Coy's *A. virgoides* of the 'Synopsis' was created, now in the collection of Mr. J. Wright, of Cork. Fig. 13 shows portions of the fringe.

14.   *Athyris lamellosa*, L'Eveilé. With portions of its fringe-like expansions. From the Carboniferous limestone of Little Island, Cork, Ireland.

15.   *Spirifera lineata*, Martin. Showing the manner in which the rows of spines succeed each other over the surface of the valves. 15ᵃ portion enlarged. From the Carboniferous shales of West Broadstone, near Beith, Ayrshire.

16.   „   *Urii.* Natural size, showing its spiny investment. Fig. 16ᵃ a portion enlarged. Hill Head, Lanarkshire.

17.   *Cyrtina septosa*, Phillips. The shell being removed from the dorsal valve, shows upon the internal cast the impression of the posterior (A) and anterior (A') adductor muscles. From Settle, Yorkshire.

18.   „   „   A very transverse young specimen. From Wetton, in Staffordshire.

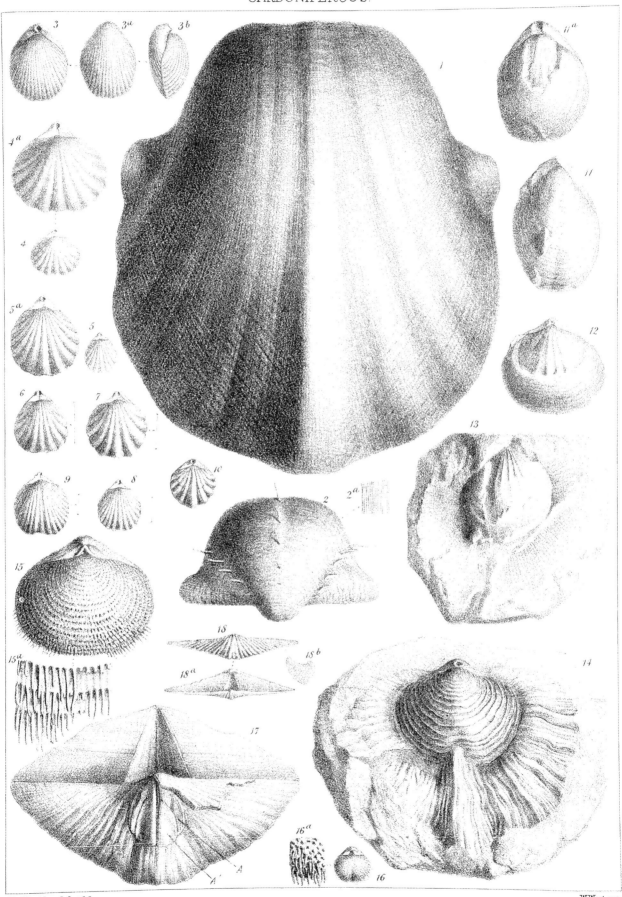

PLATE LII.

(*Supplementary illustrations.*)

1. *Spirifera striata*, Martin. An elongated specimen from the Lower Scar limestone of Settle, in Yorkshire, collection of Mr. Burrow.

2.   ,,   ,,   A very transverse example, *Sp. attenuata*, Sow. From Little Island, Cork, collection of Mr. J. Wright.

3.   ,, *cuspidata*, Martin. From the Lower Scar limestone of Settle, in Yorkshire.

4.   ,, *subconica*, Martin. A remarkable specimen, showing its beautiful laminated sculpture, from the Carboniferous limestone of Wetton, Staffordshire, collection of Mr. Carrington. 4ª, portion of the ornate surface enlarged.

5.   ,, *distans*, Sow. This specimen was labelled *Sp. bicarinata* by Professor M'Coy, and is from the Carboniferous limestone of Cork, in Ireland, collection of Mr. J. Wright.

6.   ,, *duplicicosta*, Phillips. With a prolonged mesial fold, and resembling the figure in Pl. V, fig. 35, of the present Monograph, there erroneously termed *Sp. trigonalis*. From the Carboniferous limestone of Derbyshire, collection of Dr. Fleming, of Manchester.

7.   ,, ? ,, From the Carboniferous limestone of Allstonefield, near Wetton, in Staffordshire. This is a curious form, of which I have seen several specimens, and in which the apex of the beak of the ventral valve is at a lower level than the umbone of the dorsal valve; it is, perhaps, a malformation of *Sp. trigonalis* or *bisulcata*, as some of the specimens have much the appearance of that species.

8.   ,, *ovalis*, Phillips. Interior of the dorsal valve, to show the compressed shape assumed by the spirals. I have seen several examples so disposed in this as well as in *Sp. integricosta*. From the Carboniferous limestone of Wetton, in Staffordshire.

9, 10. *Spiriferina cristata*, var. *octoplicata*, Sow. From Brockley, near Lesmahago, Lanarkshire. Fig. 10 from Wetton, in Staffordshire.

11, 12.   ,, *octoplicata*, var. *biplicata*, Dav. From the Lower Scar limestone, Settle, in Yorkshire.

13.   ,,   ,, A specimen with three plaits on the mesial fold (very uncommon), figured by Sowerby in the 'Min. Conchology.'

14, 15.   ,, *insculpta*, Phillips. Two shapes from the Lower Scar limestone of Settle.

16, 17.   ,, *acuta*, Martin. Fig. 16 the original figure from a Derbyshire specimen. Fig. 17 from Settle.

18—20. *Athyris Carringtoniana*, Dav. Carboniferous limestone of Wetton, in Staffordshire.

21. *Chonetes Buchiana*, De Kon. A large specimen from the Lower Scar limestone of Settle. The specimens from the limestone have their ribs usually more numerous and smaller than those which occur in the shales of Malham Moor, near Settle.

22. *Camarophoria?* ——— From the limestone of Wetton, in Staffordshire; it resembles much, but in large certain specimens of *Camarophoria crumena.* ?

Pl. LII.

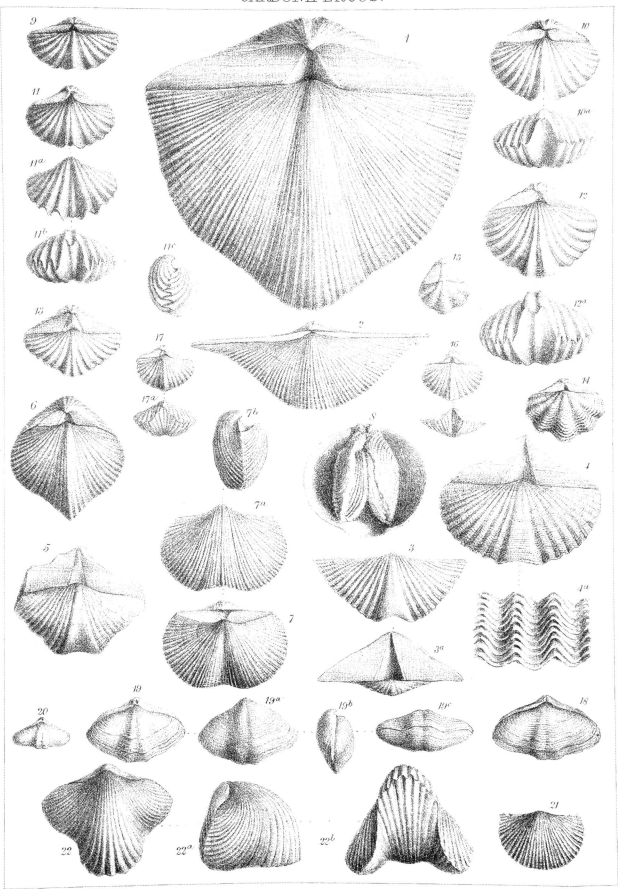

Thos Davidson. del. et lith.

W. West imp.

PLATE LIII.

CARBONIFEROUS SPECIES.

(*Supplementary illustrations.*)

FIG.

1, 2.   *Rhynchonella Carringtoniana*, Dav. Carboniferous limestone ; Wetton, Staffordshire.

3.   *Streptorhynchus crenistria*, Phillips. Interior of the dorsal valve from Malham Moor, in Yorkshire, showing in a beautiful manner the muscular impressions : A′, posterior adductor or occlusor ; A, anterior adductor or occlusor.

4.   *Productus striatus*, Fischer. A very large example from Wetton, in Staffordshire, and collection of Mr. Carrington.

5, 6.   „   *undiferus*, De Koninck. Lower Scar limestone of Settle, in Yorkshire. 5^a and 6^d enlarged.

7.   „   *Koninckianus*, De Verneuil. Settle. $7^{b,\ c}$ enlarged.

8.   „   *marginalis*, De Koninck. Settle. 8^d enlarged.

9.   „   *Nystianus*, De Koninck. Settle. 9^c and 9^d enlarged.

10.   „   *aculeatus*, Martin. With its spines preserved. From the Carboniferous shales, near Settle. Enlarged.

11, 12.   „   *Deshayesianus*, De Koninck. ? Fig. 12, Professor Koninck's original specimen. 11^a enlarged. Settle, Yorkshire.

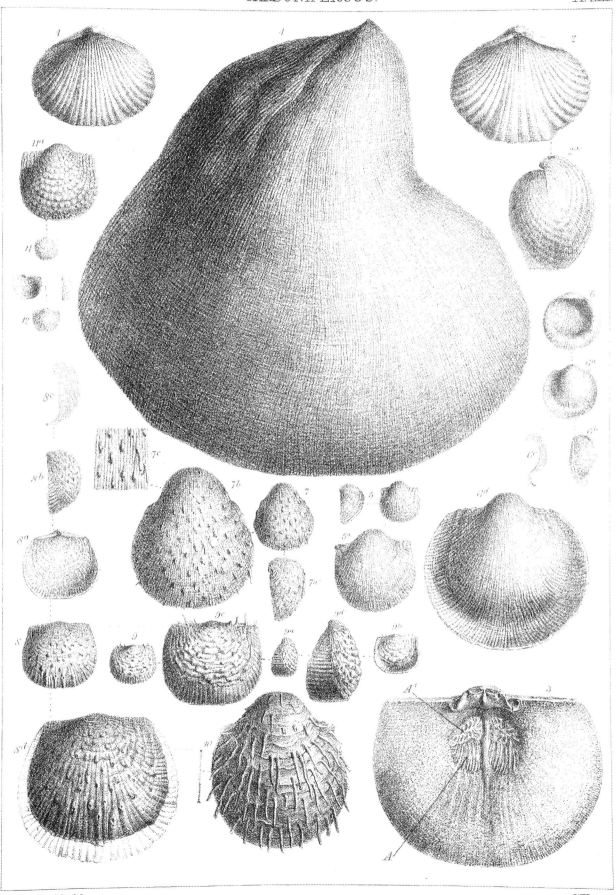

PLATE LIV.

CARBONIFEROUS AND PERMIAN RECURRENT SPECIES.

FIG.

1. TEREBRATULA HASTATA, *Sow.* Carboniferous limestone, Settle, Yorkshire.

2.   ,,   elongata, *Schlotheim.* Permian limestone, Humbleton Hill, Durham. Collection of Mr. J. Kirkby.

3.   ,,   hastata, *Sow.;* var. Gillingensis, *Dav.* Carboniferous limestone, Gilling, Yorkshire.

4.   ,,   elongata, *Schlotheim.* Permian limestone, Tunstall Hill.

5.   ,,   SACCULUS, *Martin.* Carboniferous limestone, Bolland, Yorkshire.

6.   ,,   sufflata, *Schlotheim.* Permian limestone, Tunstall Hill, Durham.

8. ATHYRIS ROYSSII, *L'Eveillé.* Carboniferous shales, Brockley, near Lesmahago, Lanarkshire, Scotland.

9.   ,,   pectinifera, *J. de C. Sow.* Permian limestone, Humbleton, Durham.

10 and 12. Spiriferina octoplicata, *Sow.* Carboniferous limestone and shale. Fig. 10, Sowerby's type ('Min. Con. Tab,' 562, fig. 2), Derbyshire. Fig. 12 from East Kilbride, Lanarkshire, Scotland.

11 and 13.   ,,   CRISTATA, *Schlotheim.* Permian limestone of Tunstall Hill, collection of Mr. J. Kirkby.

14. SPIRIFERA URII, *Fleming.* Carboniferous shales near Carluke, Lanarkshire, Scotland.

15.   ,,   Clannyana, *King.* Permian limestone, Tunstall Hill, Durham.

16, 17, 18. CAMAROPHORIA CRUMENA, *Martin,* sp. Carboniferous limestone of Derbyshire, and of Settle, in Yorkshire.

19.   ,,   Sclotheimi, *Von Buch.* Permian limestone, Tunstall Hill.

20, 21, 22.   ,,   rhomboidea, *Phillips.* Carboniferous limestone, Settle, Yorkshire.

23, 24, 25.   ,,   GLOBULINA, *Phillips.* Permian limestone, Tunstall Hill.

26. DISCINA NITIDA, *Phillips.* Carboniferous shales, Capelrig, Lanarkshire, Scotland.

27.   ,,   Koninckii, *Geinitz.* Permian compact limestone, East Thickley, Durham.

28, 29, 30, 31. LINGULA MYTILOIDES, *Sow.* Carboniferous limestone and shales. Figs. 28, 29, Sowerby's types, from Walsingham, Durham. Var. *L. Credneri,* Figs. 30, 31, from the Coal measures of Ryhope, Winning, near Sunderland, Durham, collection of Mr. J. Kirkby.

32, 33, 34.   ,,   Credneri, *Geinitz.* Permian marl slate and lower beds of the compact limestone of Ferry Hill, and East Trickley, Durham.

35, 36, 37, 38. Crania Kirkbyi, *Dav.* Permian limestone, Tunstall Hill, collection of Mr. J. Kirkby.

39, 40, 41.   ,,   QUADRATA, *M'Coy.* Carboniferous shales of Capelrig, East Kilbride, Lanarkshire.

42. Streptorhynchus pelargonatus, *Schlotheim.* Permian compact limestone of East Trichley. A very large example.

43. Strophalosia Goldfusii, *Münster.* Interior of the dorsal valve. A strongly marked example from the Permian limestone of Tunstall Hill, collection of Mr. J. Kirkby.

44. Productus horridus, *Sow.* A specimen in the Permian compact limestone of East Thickley, showing very long spines. Collection of Mr. T. Parker, of Darlington.

45. Retzia ulstrix, *De Koninck.* A very fine and perfect example, discovered by Mr. S. Carrington in the Carboniferous limestone of Wetton, in Staffordshire.

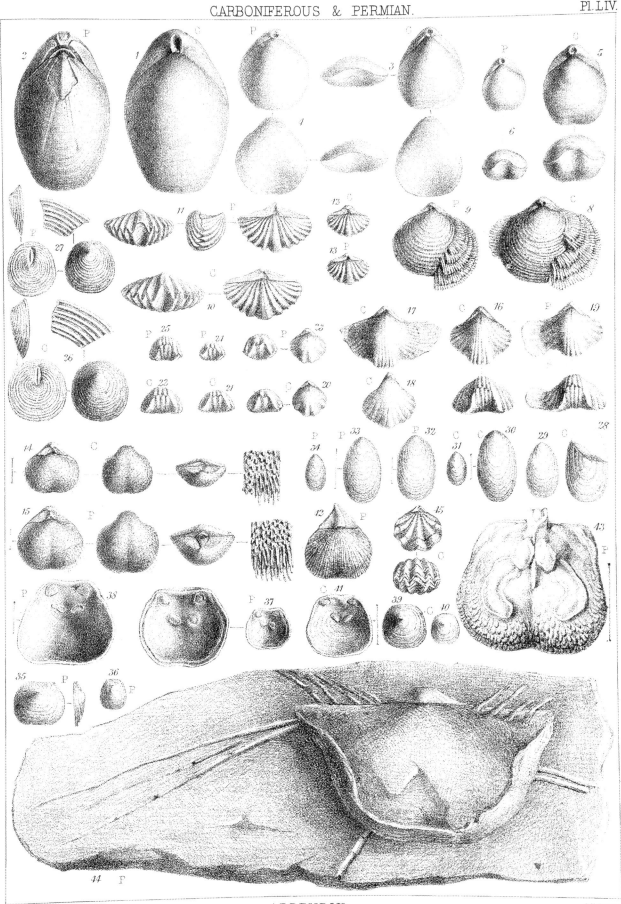

Thos Davidson delt & lith.

APPENDIX.

W West imp.

PLATE LV.

CARBONIFEROUS SPECIES.

FIG.

1, 2, 3. *Rhynchonella Wettonensis,* Dav. Three specimens from Narrowdale, near Wetton, Staffordshire.

4.   *Athyris*   ? From the Carboniferous Limestone of Wetton. This specimen is here figured, but not named or described, on account of the material at hand not being sufficient to warrant the establishing of a new species. It may, perhaps, be an abnormal form of *Athyris plano-sulcata*?

5. *Productus Carringtoniana,* Dav. Narrowdale, Staffordshire.

6, 7.   ,,   *comoides,* Sow. The original examples ('Min. Con.,' tab. 329, lower figures) from the Wayboards, between the limestone under the coal, at Llangaveni, in Anglesea. British Museum.

8.   ,,   ,,   ,, ('Min. Con.,' tab. 329, the upper figures). From the same locality and collection. Since this specimen was drawn by Sowerby in his 'Mineral Conchology' I have removed the matrix, so as to expose the interior muscular impressions. The upper attachment of the adductor partially seen (A), of the lower ones (c) are clearly exposed. R. Cardinal or divaricator impressions. British Museum.

9, 10.   ,,   *Llangollensis,* Dav. Interior of the dorsal and ventral valves, belonging to the same specimen recently found by Mr. D. C. Davies, of Oswestry, in the Carboniferous limestone of Llangollen, Denbighshire; 9 represents the ventral valve, and shows the small area, teeth, adductor (B and c), as well as the cardinal or divaricator impressions (R). Fig. 10 represents the dorsal valve, and shows the cardinal process, adductor impressions (A)' reniform markings (x), and the eminences (z). This is the only specimen hitherto discovered of the interior of the dorsal valve.

11.   ,,   *Martini,* Sow. A very remarkable specimen, showing a peculiar shape and modification of striation near the margin. From the Carboniferous limestone of Wetton, Staffordshire.

12. *Chonetes Buchiana,* De Koninck. A small variety, recently found in great abundance by Mr. W. W. Stoddart in the Carboniferous limestone near Bristol. Some of the specimens are perfectly preserved, and show that the entire surface of the valves were ornamented with minute, contiguous, concentric striæ, and we may also observe along the surface of the ribs the basis of numerous spines.

13.   ,,   *concentrica,* De Koninck? Impression of the exterior of one of the valves, from the Carboniferous limestone of Clatteringwell quarry, Bishop's Hill, Kenness Wood, Kinross, Scotland. Museum of Practical Geology.

14, 15. *Spirifera Carlukiensis,* Dav. A large variety, recently discovered by Mr. S. Carrington in the Carboniferous limestone of Narrowdale, near Wetton, in Staffordshire, in which locality it occurs by myriads.

16.   *Chonetes*   ? From a bed of lower Permian limestone at Hartley quarry, Sunderland, Durham, recently discovered by Mr. Kirkby. It is impossible, from the state in which the shell is found, to decide whether it belongs to *C. Davidsoni,* Schauroth, or to *C. Hardrensis.*

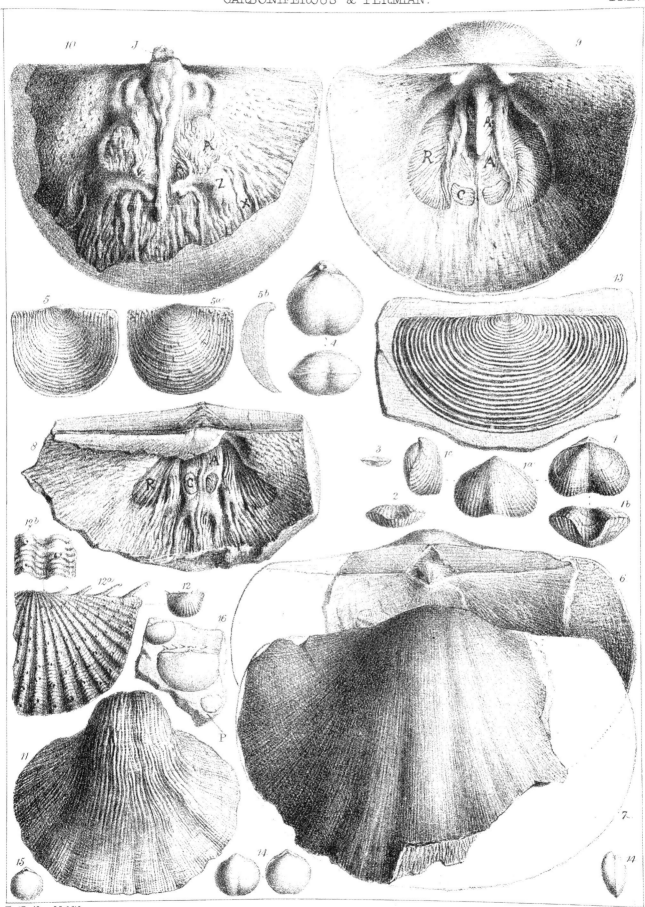

Thos Davidson del. & lith.

W.West imp.

Printed in the United States
By Bookmasters